PROBABILITY

COMPLEMENT LAW	$P(A') = 1 - P(A)$
ADDITIVE LAW	$P(A \cup B) = P(A) + P(B) - P(A \cap B)$
CONDITIONAL PROBABILITY	$P(A/B) = \dfrac{P(A \cap B)}{P(B)}$
MULTIPLICATION LAW	$P(A \cap B) = P(A/B)P(B)$ $\quad\quad\quad\quad = P(A)P(B)$ if A and B are independent

PROBABILITY DISTRIBUTIONS

FOR DISCRETE y	$E(y) = \Sigma y p(y)$ and $V(y) = \Sigma y^2 p(y) - \mu^2$ where $\mu = E(y)$
BINOMIAL DISTRIBUTION	$p(k) = \dbinom{n}{k} \pi^k (1 - \pi)^{n-k}$ $E(y) = n\pi$ $V(y) = n\pi(1 - \pi)$
NORMAL DISTRIBUTION	Probabilities are found in the standard normal probability table.
STANDARD SCORE	$Z = \dfrac{\text{score} - \text{mean}}{\text{standard deviation}}$

EXPLORING
Statistics
A MODERN INTRODUCTION
TO DATA ANALYSIS AND INFERENCE

LARRY J. KITCHENS
APPALACHIAN STATE UNIVERSITY

WEST PUBLISHING COMPANY
ST. PAUL ▲ NEW YORK ▲ LOS ANGELES ▲ SAN FRANCISCO

DESIGNER: Kathi Townes
PRODUCTION COORDINATOR: Schneider & Co.
COVER ART AND DESIGN: Kathi Townes

LIBRARY OF CONGRESS CATALOGING-IN-PUBLICATION DATA

Kitchens, Larry J.
  Exploring statistics.

  Includes index.
  1. Statistics.   I. Title.
QA276.12.K586   1987   519.5   86-24585
ISBN 0-314-28498-2
1st Reprint—1988

Output examples from SAS procedures, Version 5 and later releases, Copyright © 1985, SAS Institute Inc.

Output from Minitab printed with permission of Minitab, Inc., State College, PA. Minitab is a registered trademark of Minitab, Inc.

This book is dedicated to my Mother and Father who have impressed upon me that an education is among the most valued gifts anyone can ever receive.

Contents

5

RANDOM VARIABLES AND PROBABILITY DISTRIBUTIONS 164

6

DESCRIBING A DISTRIBUTION 207

7

SAMPLING DISTRIBUTIONS AND THE CENTRAL LIMIT THEOREM 246

8

ESTIMATION 274

9

HYPOTHESIS TESTING 323

Index

PREFACE

Exploring Statistics is intended as a one-semester or two-quarter introduction to modern statistics. A course in college algebra is the only prerequisite. The book gives a comprehensive treatment of introductory statistics, including the same general topics covered in any beginning statistics book, but with a new approach that uses the Exploratory Data Analysis (EDA) techniques of J. W. Tukey. Unlike other introductory texts, this book uses the EDA techniques consistently throughout.

Exploratory Data Analysis has gained recognition as an important practice in analyzing statistical data. With it, we are able to detect certain peculiarities about the data that the traditional techniques may overlook. The EDA tools are easy to use and give valuable insight into the data. They serve as excellent indicators of when the classical inference procedures are inadequate. These tools have been long used by skilled data analysts, but only recently have they been accepted as descriptive techniques to be presented in an introductory textbook.

Another emphasis of this text is the more careful attention given to the choice of parameter and estimator. In the past, most statistical inference procedures concerning the center of a distribution have been devoted to the population mean μ with the sample mean $\bar{y}$ being its estimate. There are other possibilities to consider. For instance, given a highly skewed distribution, it is common knowledge that the population median is a better measure of the center of the distribution than is the population mean. Therefore, techniques to estimate it should be presented in an introductory course. Even if we wish to estimate the population mean, is $\bar{y}$ always the appropriate estimator? When we deal with "real" data, the desirable characteristics of $\bar{y}$ may deteriorate under certain circumstances. In this book those circumstances are identified using the EDA tools and then alternative procedures are suggested.

IMPORTANT FEATURES

1. Each chapter begins with a preview that gives direction to the chapter and relates it to the previous chapter. Then a "Statistical Insight" application taken from a general interest news article is given. The topic selected for the "Statistical Insight" relates to the material covered in the chapter.

2. An abundance of examples relevant to the various disciplines is given so that applications in a variety of settings can be seen. Each section ends with a set of exercises to reinforce the ideas just learned. Most examples and exercises have been taken from actual surveys, reports, and experiments.

3. The text provides a blend of traditional and exploratory procedures with emphasis placed on analysis and interpretation of data.

4. In addition to the exploratory tools, special features include robust inference procedures, normal probability plots, the resistant line, and residual analysis.

5. Chapter 1 provides a study of sample surveys and experimental design seldom found in an introductory textbook.

6. Examples to illustrate the graphic capabilities of SAS/GRAPH® are included in Chapter 2.

7. Each chapter has an optional computer section in which the statistical procedures are illustrated using the Minitab® computer package. Although the students may never independently analyze their own data, they will be asked to interpret statistical reports that are produced by a computer. Thus, this book exposes the student to computer outputs and shows them how to read the reports. For hands-on experience four data sets are included in Appendix A.

8. The book gives a thorough development of the rationale for using each statistical procedure. Each procedure is presented with the assumptions that are necessary and the indications for when the procedure is appropriate.

9. The *p*-value approach to hypothesis testing is used throughout the book.

10. Most introductory texts have a separate chapter on nonparametric procedures, which is set apart from the other material, usually at the end of the book. In this book the nonparametric procedures are integrated throughout, and suggestions for using them appear when appropriate.

11. At the end of each chapter, all of the "Key Concepts" are summarized followed by a list of "Learning Goals" and a set of "Questions for Review."

12. The text is accompanied by a Student Study Guide, an Instructor's Manual, and microcomputer software. The Student Study Guide presents a unique discussion of math anxiety and how to deal with it. The Guide also gives suggestions for developing effective study skills and tells how to prepare for a test. In addition, there is a review of basic algebra, practice test with answers, and solutions to the review and supplementary exercises in the text.

ORGANIZATION

The book is divided into thirteen chapters that allow a reasonable amount of flexibility in the choice of topics to be covered. A discussion of the features of each chapter and the approximate time that should be devoted to the chapter appears in the Instructor's Manual. A possible outline of a standard one-semester course would include Chapters 1–9 as core material followed by Chapter 10, or 11, or 12. Chapter 13 could easily follow Chapter 10 if time permitted. In order to cover these topics in a reasonable amount of time, certain topics may have to be omitted. Clearly those sections marked optional could be skipped. Also, without loss of continuity, the robust inference procedures of Chapters 8, 9, 10, and 11 could be skipped.

Obviously additional materials could be covered in a two-quarter course. The complete book could easily be covered in a two-semester course.

ACKNOWLEDGEMENTS   I would like to acknowledge the following reviewers for their valued comments and encouragement during the preparation of this book:

Roy Erickson
Michigan State University

Stuart Friedman
California State Polytechnic University at Pomona

Donald Hotchkiss
Iowa State University

Terry Hughes
Arizona State University

Harry Khamis
Wright State University

Jerry Lefkowitz
formerly of *Pennsylvania State University*

Eric Lubot
Bergen Community College

David Lund
University of Wisconsin

Edward Markowski
Old Dominion University

Jeff Mock
Diablo Valley College

David Moore
Purdue University

Susan Reiland
formerly of *North Carolina State University*

John Rice
University of California at San Diego

David Robinson
St. Cloud State University

Rodolpho Serrano
California State College—Bakersfield

Howard Tucker
University of California at Irvine

David Turner
Utah State University

Dan Voss
Wright State University

Susan Reiland should be singled out for her superb job of line-by-line reviewing that caught errors that no one else knew were there. I would also like to express my appreciation to the people at West Publishing, especially my editor, Pat Fitzgerald, for his encouragement and support throughout the project, and also Deborah Schneider of Schneider & Company for her diligent work during the production stage.

I am grateful to the Literary Executor of the late Sir Ronald A. Fisher, F.R.S., to Dr. Frank Yates, F.R.S., and the Longman Group Ltd., London, for permission to reprint Tables IIi and III from their book *Statistical Tables for Biological, Agricultural and Medical Research* (6th Edition 1974).

A special thank you goes to the many students at Appalachian State University who have struggled with me through the early drafts of this text. Their input has made this a better text and, I hope, one that will give future students a better understanding of statistics. Also I would like to thank those colleagues of mine who thought enough of the book to use it in their classes in the early stages of its preparation, in particular Dr. Gary Kader, who also prepared the solutions for the Instructor's Manual.

Finally, a loving thank you to my beautiful children for being themselves and to my wife, Anita, without whom this book would never have become a reality.

1

Collecting Data

Statistical training is necessary and important for many reasons. Obviously, you cannot become a research statistician after having had only an introductory course in statistics. Yet those engaged in social sciences, physical sciences, medicine, education, and similar areas must be able to read, to interpret, and to apply the results of a statistical analysis of research data. An introductory course in statistics is necessary so that you, the student, can learn the terminology and understand the techniques that are used. In this chapter you will be introduced to some basic definitions and two general areas of statistics—sample surveys and experimental design.

Top 20 Discoveries in Science

According to the November, 1984 issue of *Science 84* magazine, U.S. scientists have judged Statistics as one of the most important scientific discoveries made since 1900. After polling the leading U.S. scientists, the magazine concluded that the discoveries that have most changed our lives are:

Antibiotics
Atomic fission
The big-bang theory
Birth control pills
Blood types
The computer
DNA
Drugs for mental illness
Einstein's theory of relativity
The IQ test
The laser
Networks (such as the phone system)
Pesticides
Plant breeding
Plastics
Statistics
The Taung skull
Television
The transistor
The vacuum tube

Statistics is a common denominator of many of the other discoveries. The IQ test is continually being analyzed with statistical tools. Data from plant breeding experiments are analyzed with statistics. The effects of penicillin and other antibiotics on humans are studied with statistics. The birth control pill and drugs for mental illness are studied by comparing the effects on a control group and an experimental group. Professionals in almost all fields of study use statistics to analyze their data. Now we will see how statistics applies to these and other fields of study.

Introduction

When we first hear the word *statistics,* we may think of sport statistics such as batting averages, free throw averages, or yards rushing. We might also think of the results of a national poll, which are published in the local newspaper, or government figures on unemployment, which are presented on television. Applications of statistics appear in almost every discipline, especially in psychology, sociology, education, political science, business, and the natural sciences. Psychologists may be interested in studying the IQ scores of first graders or the aptitude scores of entering college freshmen. Sociologists might be interested in the proportion of migrant workers in their state who have an eighth-grade education. An educator might wish to evaluate two different teaching methods. A political scientist may evaluate the percentage of voters in favor of a certain candidate. Sources of income and expenditures for the federal budget are of interest to the business statistician. A physicist may want to evaluate the reliability of the communications network for a spacecraft using statistical techniques.

Statistics is a major branch of mathematics that might be defined as follows:

Definition

Statistics is the science of collecting data, organizing or describing it, and ultimately drawing conclusions from it.

This book will address the following three areas of statistics, all of which are equally important.

1. Collecting data
2. Describing data
3. Drawing inferences from the data

Collecting data seems simple but may be the most difficult. Anyone can interview a portion of the public and get responses, but do the responses obtained truly represent the attitudes of the entire body from which the data come? The successful statistician spends many hours determining the best procedure to collect data so that it will be representative of the whole. We will see that the inference made about the whole is only as good as the collected data.

Once the data are collected they must be organized and summarized. This area of statistics has become known as *descriptive statistics.* Traditionally, this involved calculating summary measures of the data and perhaps drawing a graph or two, but with today's computer techniques, very elaborate graphs can be drawn that illustrate the data in a fashion that previously was not possible.

An inference can be described as a generalization from incomplete information. Thus, *inferential statistics* is used to draw a generalization about the whole from the data we have collected and organized. Collecting and organizing data will sometimes lead to possible inferences that can be made, but usually the inferences are formulated in the form of hypotheses and then the data are collected either to verify or deny the truth of the hypotheses.

A statistical problem involves studying some *characteristic* associated with a group of objects commonly called *units* or *subjects.* The subjects may be first graders whose IQ's we wish to study. Or the subjects may be migrant workers whose education levels are of interest. The elements of study do not have to be people. For example, they might be the components of the communications network of a spacecraft, and we are interested in the percent that are defective. The units might be hospital beds, and we wish to find the average occupancy rate. The entire group of objects of interest is called a *population*.

Definition

The **population** is the collection of all objects that are of interest to the statistician. The elements in the population may be called units or subjects.

In many situations the population is extremely large, and it would be very difficult, if not impossible, to investigate the characteristic of interest for each unit of the population. Therefore, we investigate a portion of the population called a *sample*.

Definition

A **sample** is a finite portion (subset) of the population that is used to study the characteristics of concern in the population.

A characteristic being studied that is associated with each unit in the population is called a *variable*. The collection of values it assumes in the sample is called the *data*.

Definitions

A **variable** is any characteristic that can be measured on each unit in the population.   *Question*

An **observation** is a value that the variable assumes for a single unit in the sample.   *answer*

The collection of observations is called the **data.**

Statistic is a calculation done upon an observation). A fact

In the following examples, we will identify the population, the sample, and a variable of interest.

EXAMPLE 1.1

A TV news commentator recently reported that a Gallup poll of 1,850 adults revealed that 38% of the nation approved of the President's foreign policy. The population would be all adults in the nation. The sample would be the 1,850 that were surveyed. The variable of interest would be the opinion of the subject on foreign policy.

EXAMPLE 1.2

An admissions office at a university is interested in using this year's freshman class to develop an admission formula for all new applicants. The population would consist of all students who would apply for admission. The sample would be this year's freshman class. Variables of interest might be an achievement test score or the high-school rank.

EXAMPLE 1.3

A scientist is investigating the effectiveness of a new drug to relieve the symptoms of the common cold. She administers the drug to 100 adults. The population would consist of all adults who would try the drug in the future. The sample would be the 100 chosen adults. A variable of interest might be the time required for the drug to relieve a particular cold symptom.

EXAMPLE 1.4

To determine the percent of males and females that smoke, the Surgeon General mailed questionnaires to 5,000 people across the United States. The population includes all U.S. residents. The sample consists of those of the 5,000 who returned their questionnaire. A variable of interest is the smoking habits of the subjects.

Often, the study of statistics can be classified into two general areas: *sample surveys* and *experiments*. The Gallup poll mentioned in Example 1.1 is an example of a survey. The already existing opinions of a group of people are solicited. An experiment, on the other hand, attempts to determine a cause-and-effect relationship between two or more variables. In the experiment in Example 1.3, the scientist administers a drug to the subject in an attempt to study its effect on the common cold.

The validity and accuracy of a survey or experiment are measured by the *precision* (or the lack of precision) and *bias*.

Definition

Bias is a systematic tendency of the sample to misrepresent the population.

Suppose we wish to study the attitudes of the adult population toward sexual harassment. If a sample consists of 70% females and 30% males and their attitudes toward sexual harassment differ, then we would say the sample is biased toward females. The male segment of the population is not properly represented in the sample.

Definition

Lack of precision means that in repeated sampling the values obtained tend to be widely scattered or spread out. Hence, a result obtained from one sample usually cannot be duplicated with another sample.

Suppose a radar device is used to catch speeders on an interstate highway. To check the device, an automobile is set to travel at a fixed speed of 65 miles per hour. On five attempts the radar device recorded speeds of 61.2, 69.6, 60.4, 63.8, and 67.0. Clearly, the radar device is inaccurate; we would say it has a lack of precision. The average of the above five readings is 64.4, which is reasonably close to the actual speed of 65 mph. Thus, there is little bias, but there is a definite lack of precision.

EXAMPLE 1.5

Suppose the highway department is testing four radar devices. Each device records five different measurements on a car traveling 65 miles per hour. Following are the readings of the four radar devices:

	Readings					Average
Radar A	67.3	66.8	71.2	69.5	68.2	68.6
Radar B	63.5	64.0	62.6	62.8	63.1	63.2
Radar C	63.7	68.1	63.2	65.1	66.4	65.3
Radar D	64.6	65.3	64.9	65.5	65.7	65.2

Classify each radar device according to bias and precision.

SOLUTION   Figure 1.1 is a dot plot of the measurements for the four devices on a common line graph. We can see that the readings for radar device A are all high and scattered. We would say that device A is positively biased and has low precision. Device B is negatively biased but has high precision. Device C is unbiased but low in precision. Finally, device D is the most desirable; it is unbiased and high in precision.

FIGURE 1.1

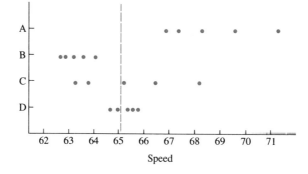

As stated, precision and bias are important considerations when studying the topics of sample survey and experimentation. In the following sections, we will take a closer look at the sample survey and the experiment.

For Exercises 1.1 through 1.6, describe

(a) the population of interest

(b) the sample

(c) a variable of interest

1.1 To investigate the attitudes of college students about abortion, a statistics class selected 500 students from their university and obtained their opinions on abortion.

1.2 In polling 1,000 residents of a neighborhood, it was found that 460 agreed with the neighborhood zoning ordinance.

1.3 Out of 1,000 accidents investigated by the highway patrol, 526 were found to be alcohol related.

1.4 To estimate the percent of students that experience an emotional tragedy during the school year, a psychology professor sampled all of his classes.

1.5 The admissions office at a university sampled 200 freshmen to determine whether the university was their first choice.

1.6 Advice columnist Ann Landers asked her readers to clip a questionnaire about sexual attitudes out of the newspaper, complete it, and return it to her.

1.7 In Exercises 1.1 through 1.6, indicate which ones you think would possibly have a biased sample.

1.8 According to the 1985 *World Almanac*, the number of deaths due to major aircraft disasters in the 10-year period from 1974 through 1983 was 5,811. This compares with 3,520 deaths in the previous 10-year span from 1964 through 1973. Does this mean that it is becoming more dangerous to travel by air? Is the population of air travelers during the first 10-year period comparable to the population of air travelers during the second 10-year period?

1.9 Is the total number of fatalities per year a reasonable variable with which to measure the danger of air travel? Can you think of a more valid measure?

1.10 Classify each of the following as either a sample survey or an experiment.

(a) A national TV network asked voters to indicate for whom they voted as they exited the polling booth.

(b) Terminally ill patients were divided into two groups, with one group receiving medication A and the other group receiving medication B. After a period of time each subject's improvement was assessed.

(c) The federal government monthly conducts its Current Population Survey of a sample of households.

1.11 A manufacturer of bathroom scales wants to test three new models. A weight of 100 pounds is placed on each of the models on five different occasions. The following measurements are recorded:

Model A	96.4	96.1	95.8	95.9	96.2
Model B	99.8	100.2	99.7	100.0	100.3
Model C	104.6	103.9	105.4	102.6	106.3

Evaluate each of the models according to bias and precision.

1.2 Sample Surveys

National surveys are conducted almost daily. In some situations different organizations will survey the same issue and get varying results. The reliability and success of a survey are determined by the accuracy with which the sample was obtained.

The 1936 Presidential Election

A historical example of a national survey that yielded unreliable results occurred in the 1936 presidential election. A national magazine of the time, called the *Literary Digest*, had successfully predicted every presidential election since 1916. For the 1936 election, they mailed out 10 million sample ballots of which over 2.3 million were returned. It was the largest number of people ever to respond to a survey. From the responses on the 2.3 million cards, the *Digest* predicted that Alf Landon would defeat Franklin D. Roosevelt by a margin of 57% to 43%. It predicted that Roosevelt would get only 161 of the 531 electoral votes. As it turned out, FDR received 523 of the 531 electoral votes for a landslide victory with 62% of the popular vote.

The *Digest* prediction (the worst ever made in a national poll) was inaccurate for several reasons. First, it was a voluntary survey; only 23% of those receiving cards returned them. Generally voluntary surveys are *biased* because only those who feel strongly about the issue will bother to respond. Second, the sample of names was taken from subscription lists of magazines, telephone directories, and automobile registrations. This was clearly biased toward those with higher incomes and better educations. This is especially important because this poll was made during the depression era. Simply stated, the sample was not *representative* of the population. It had an overrepresentation of the upper middle class and the more highly educated. There were many laborers and farm workers at that time, and it is likely that they were not properly represented in the sample. This marked the end of a 20-year period in which the *Literary Digest* had successfully predicted the presidents. The *Digest* went bankrupt shortly thereafter.

George Gallup, who received his doctorate in journalism from the University of Iowa in the early 1930s, observed the bias in the *Digest*'s poll and predicted that Roosevelt would win the election with 56% of the popular vote. This was the beginning of the nationally recognized Gallup poll. For his doctoral dissertation, Gallup devised new sampling procedures for estimating newspaper readership. Using these same techniques, he predicted the Roosevelt victory. He has been very successful in subsequent presidential polls.

The 1948 Presidential Election

One exception to Gallup's success occurred in the 1948 Truman–Dewey election. All the major pollsters—Gallup, Crossley, and Roper—predicted that Thomas Dewey would defeat Harry S. Truman. Table 1.1 compares their predictions with the actual results.

TABLE 1.1
1948 Presidential election

	Truman	Dewey	Others
Gallup poll	44.5	49.5	6.0
Crossley poll	44.8	49.9	5.3
Roper poll	37.1	52.2	10.7
Actual results	49.5	45.1	5.4

SOURCE: F. Mosteller, *The Pre-election Polls of 1948*, Report to the Committee on Analysis of Pre-election Polls and Forecasts, (Washington D.C.: Social Science Research Council, Bulletin 60, 1949): 17.

The major reason for the incorrect prediction of the 1948 election was that the pollsters failed to account for late shifts in voter preference. The last survey was made two weeks prior to the election. During those two weeks Truman gained enough support to defeat Dewey. Two weeks prior to the election approximately 15% of the voters were undecided, and it is reported that three-fourths of the undecided vote went to Truman.

From the 1948 election, pollsters have learned to keep a close watch on the "undecided voter" and to revise their predictions up to the last day preceding the election. Another important lesson learned in the Roosevelt/Landon election is that the size of the sample is not nearly as important as sound sampling techniques. The *Literary Digest* had a sample of over 2.3 million when 1,500 would have sufficed if the sample were truly representative of the population.

Lou Harris of the Harris poll has also obtained national recognition as a reliable pollster. His popularity grew as he worked for John Kennedy in his 1960 presidential campaign. Like Gallup, Harris has a syndicated newspaper column in which he publishes the results of his public opinion polls.

The A. C. Nielsen Company is a different type of polling organization that is very influential. They provide a service for the major television networks by conducting a ratings survey on all network television shows. The "ratings" help the networks decide which shows will be renewed and which will be canceled. Top-rated shows such as a Super Bowl (see Figure 1.2) command as much as $1,000,000 for 1 minute of commercial advertisement, whereas lower-rated shows sell commercial time for as little as $25,000 a minute. Thus, it is obvious why a low-rated show would be canceled within weeks, the network hoping that its replacement would become the season's new hit. The ratings indicate the percent of viewers watching the show and are obtained from a random sample of about 1,500 TV homes.

▲ Currently a rating point represents 859,000 television households. For the week ending June 28, 1986, the *Cosby Show* was ranked first with a rating of 23.4, which translates to more than 20 million households.

FIGURE 1.2

ABC nears $1 million per minute

By Rudy Martzke
USA TODAY

ABC is two minutes shy of selling out the first $1 million-a-minute Super Bowl telecast.

Larrye Barrett, sports sales manager for ABC-TV, said Thursday that two 30-second spots remain unsold at $525,000 each on the Jan. 20 game telecast, one 30-second spot in the pregame show at $135,000 and another 30-second spot in the post-game show at $300,000.

Among the major advertisers on ABC's first Super Bowl telecast, between San Francisco and Miami, are Anheuser-Busch, Nissan, Ford and IBM.

"People buy the Super Bowl because it's the highest-rated show of the year," he said. "Like *Roots*, the event guarantees at least a 40 rating. Six of the top 10 shows of all time are Super Bowls.

"Clients also like the Super Bowl because it's viewed as wholesome and favorable entertainment," he said. "There's no sex or violence or drugs."

Super Bowl advertisers

Advertisers for each portion of ABC's Jan. 20 Super Bowl XIX telecast (All times EST):

Pregame I (4–5 p.m.)
■ **Cost:** $85,000 per 30 seconds (10-minute total)
■ **Advertisers:** IBM, Miller Beer, Volkswagen, National Dairy, Johnson Controls, Michelin, Owens Corning, AJAY

Pregame II (5–5:45 p.m.)
■ **Cost:** $135,000 per 30 seconds (7½-minute total)
■ **Advertisers:** IBM, Miller Beer, Volkswagen, National Dairy, United Airlines, Federal Express, Allstate

Kick-off Special I (5:45–6 p.m.)
■ **Cost:** $250,000 per 30 seconds (2½-minute total)
■ **Advertisers:** Anheuser Busch

Kick-off Special II (6–6:15 p.m.)
■ **Cost:** $325,000 per 30 seconds (3 minutes total)
■ **Advertisers:** McDonald's, Sharp, U.S. Armed Forces, Michelin, IBM

Super Bowl game telecast (6:15–9:30 p.m.)
■ **Cost:** $525,000 per 30 seconds or $1 million a minute (25-minute total)
■ **Advertisers:** Anheuser Busch, IBM, Masterlock, McDonald's, Minolta, Nissan, Sharp, Soloflex, Stroh Brewery, U.S. Marines, National Dairy, Northwestern Mutual Life, Sony, Johnson Controls, Ford, Cullinet Computer, Computerland, ITT, National Food Processors, American Home Products, Coca-Cola

Postgame (9:30–10 p.m.)
■ **Cost:** $300,000 per 30 seconds (4-minute total)
■ **Advertisers:** IBM, Miller Beer, Volkswagen

The all-time highest rating for a Super Bowl is the 49.1 (percentage of TV homes) and audience of 110 million for Super Bowl XVI in 1982 between San Francisco and Cincinnati.

"Because of the matchup between the Dolphins and 49ers," said Barrett, "we have the potential for a record-setting Super Bowl."

SOURCE: *USA Today*, January 11, 1985. Copyright, 1985 *USA Today*. Reprinted with permission.

The source of a sample survey is extremely important. The Gallup, Harris, and Nielsen organizations are considered reliable sources of information, but there are less reliable organizations. For example, in a Senate primary race, it was reported that a polling firm offered their client two surveys: one for use with the press for publicity and fund-raising and the other a confidential survey reporting how the client actually stood in the state (Roll and Contril, 1972).

Instead of trying to classify a survey organization as a reliable source of information, it would be wiser to judge the survey itself. A properly designed survey would report the following:

1. A description of the population sampled
2. A description of the method of contact for interviews
3. The response rate
4. The exact wording of the questions
5. The timing of the interview
6. The size of the sample (or the margin of error) and the sampling design

The following discussion of the preceding topics is not intended to train people to become professional pollsters, but rather to point out possible shortcomings of a survey and to help detect poorly designed surveys. It is also intended to boost confidence in the reliability of the properly designed survey. For example, many people criticize the Nielsen ratings because the sample size appears to be small. But if the sampling is scientific, we will see that the sample size used by Nielsen is certainly adequate.

Population

It is important that the population of concern be properly identified so that it can be determined whether the collected sample tends to be representative of the population. For example, if we wish to estimate the present cost of housing in a major city, would a sample of home loans obtained from savings and loan associations be representative of the population? Possibly not, because the population of concern would consist of all homes in the city that are for sale. The sample would be representative of the population of homes that were mortgaged through a savings and loan. Clearly, we are talking about two different populations—all homes that are for sale and all homes mortgaged through a savings and loan. On the other hand, would a sample of homes listed by the realtors of the city be representative? This might be more representative of the population of concern, but not all sales are made through realtors. We could say, however, that the sample is representative of the population of homes sold by realtors in the city. Properly identifying the population might also suggest the method of contact.

Method of Contact

Possible methods of contact are mail, telephone, and personal interview. A mail survey will be time-consuming and will often have a low response rate. The sample will possibly be biased, because a larger percentage of those who feel strongly about the issue will respond. Also, negative or socially unacceptable opinions are more likely to be voiced through mailed surveys than through any other type.

Telephone surveys are easily conducted, relatively inexpensive, and they can yield timely results. A word of caution though—11% of American households do not have telephones and it is estimated that 16% of all residential phones are unlisted (see Figure 1.3).

The personal interview has a higher response rate and generally less bias than

FIGURE 1.3

USA SNAPSHOTS

A look at statistics that shape the nation

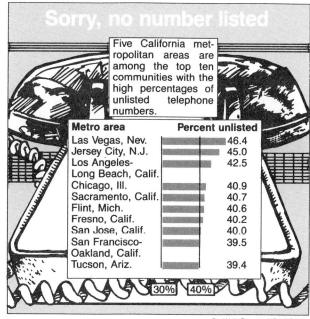

Sorry, no number listed

Five California metropolitan areas are among the top ten communities with the high percentages of unlisted telephone numbers.

Metro area	Percent unlisted
Las Vegas, Nev.	46.4
Jersey City, N.J.	45.0
Los Angeles-Long Beach, Calif.	42.5
Chicago, Ill.	40.9
Sacramento, Calif.	40.7
Flint, Mich.	40.6
Fresno, Calif.	40.2
San Jose, Calif.	40.0
San Francisco-Oakland, Calif.	39.5
Tucson, Ariz.	39.4

By Web Bryant, USA TODAY

▲ The problem of unlisted phone numbers can be avoided by random-digit dialing.

SOURCE: *USA Today*, October 12, 1984. Copyright, 1984 *USA Today*. Reprinted with permission. Data from Survey Sampling Inc.

either the mailed or telephone survey. However, personal interview surveys are expensive and time-consuming. The interviewer must be trained and paid for time and travel. Care must be taken in choosing the interviewer; he or she should be knowledgeable of the issues raised in the survey, and should not be overbearing or suggestive in presenting the questions.

Of the three methods of contact, the personal interview is the best, but the telephone survey is the most widely used because it is timely and relatively inexpensive.

Response Rate

Regardless of the method of contact, one should be aware of the response rate. A carefully designed mailed questionnaire can increase the response rate. Properly trained interviewers can increase response rate in both the telephone and the personal interview.

One must be prepared to re-call those subjects who were unavailable at the initial contact. Generally, those people who are not at home in the evening tend to have a different lifestyle from those who are and possibly would respond differently to a survey. Every effort must be made to contact those in the original sample and obtain as high a response rate as possible.

Questions Asked

The response rate can be increased with carefully designed questions. A survey should not ask too many questions, and ambiguous and confusing questions should be eliminated. A trial run is a good way to identify troublesome questions. Questions should be posed in a neutral sense rather than a negative or positive sense so as to avoid leading the subject. For example, the question "Do you agree that U.S. financial aid to

El Salvador should be reduced?" leads the subject. A better way of posing the question would be "Are you for or against reducing U.S. financial aid to El Salvador?" To get a broader response you might ask, "What is your opinion on financial aid to El Salvador?"

EXAMPLE 1.6

During the Vietnam War in 1967, President Johnson ordered the bombing of Hanoi and Haiphong. Two members of Congress conducted independent surveys of the issue in their respective districts. Member A posed the question as follows: "Do you approve of the recent decision to extend bombing raids in North Vietnam aimed at the strategic supply depots around Hanoi and Haiphong?"

Sixty-five percent favored the decision. Member B posed the question as follows: "Do you believe the U.S. should bomb Hanoi and Haiphong?" Only 14% favored the decision (Wheeler, 1976).

It is hard to compare these two figures, because they were not obtained from random samples but rather from specific voting districts. The difference is so dramatic, however, that one is led to believe that the wording of the question had a definite effect on the outcome of the survey.

Timing

The timing of a survey often affects the results. A 1971 survey by the National Football League showed that football was the nation's most popular spectator sport. The commissioner of baseball disputed the claim because the survey was conducted a week before the Super Bowl! A survey taken a week before the World Series would likely show baseball to be the most popular spectator sport (Moore, 1979). The commercial television networks are well aware of ratings week conducted by the A. C. Nielsen Company. Many viewers find that the best shows of the season are telecast opposite one another during ratings week.

EXAMPLE 1.7

In 1975, a Gallup poll showed that President Ford was ahead of Senator Hubert Humphrey in the 1976 presidential race, 51% to 39%. A Harris poll showed that Humphrey was leading Ford 52% to 41%. The difference in the two predictions can be attributed to the timing of the two surveys. The Harris survey was conducted after a cabinet shake-up and after Ronald Reagan announced he would oppose Ford for the Republican nomination and before Ford traveled to China. The Gallup poll was conducted after Ford left for Peking. Presidential popularity usually increases after foreign travel (Haack, 1979).

Sample Size

The size of the sample used in a survey is determined by the accuracy desired and the available resources. The larger the sample size the more accurate the results, but usually only a limited amount of time and money is available. The method of contact will also be a factor in the choice of the sample size. Clearly a telephone survey would be less expensive than a personal interview. We will see later that the sample size is computed from the desired *margin of error*. If we listen closely when the results of a poll are given, we will hear the newscaster say, "The margin of error was plus or minus 3 percentage points." If it were desirable to decrease the margin of error from 3 to 2 percentage points, the sample size would have to be increased substantially. At some point the gain in accuracy is not enough to justify the larger sample size, which is why most national polls have a sample size of fewer than 2,000 subjects. Gaining one more percentage point of accuracy is not worth the time and cost involved in obtaining a significantly larger sample.

Sample Design

The results of a survey will be only as good as the sample taken. If the sample is not representative of the population, then the results of the survey will be useless and misleading. If the sample is representative of the population, then the results of the survey will be accurate and useful. Thus we strive for a representative sample. But representative in what sense—IQ, height, weight, age, race? We hope the sample is representative of the population in those factors that are relevant to the study. If we are conducting a political survey, then the weight of the subject is not relevant; however, the race of the subject would be relevant. We would want the sample to be representative with respect to race. If the population consists of 30% blacks, then our sample should consist of 30% blacks.

EXAMPLE 1.8

Suppose we wish to study the television viewing habits of the general population. We would like our sample to be representative of the population with respect to what variables?

SOLUTION   Relevant variables might be education level, age, income level, IQ, and occupation of subject.

EXAMPLE 1.9

In a political poll, list relevant variables in which we would like our sample to be representative of the population.

SOLUTION   Relevant variables in a political poll might be sex, age, race, income level, education, and political party affiliation.

In the next section we will study techniques that will enhance the chances of obtaining a sample that is representative of the population.

EXERCISES 1.2

In Exercises 1.12 through 1.15, identify the most general population for which each sample given is representative.

1.12 A sample of size 10,000 was obtained from the subscription list of *Time* magazine.

1.13 A sample of size 200 was obtained from a list of students living in dorms at a small midwestern university.

1.14 A sample of size 1,000 was obtained from those who attended a rock concert in the Superdome.

1.15 A sample of size 1,000 was obtained from those who attended a football game in the Superdome.

1.16 To obtain opinions on a local bond issue, an interviewer randomly selected households throughout the city. At one of his chosen households, no one was at home. Should he have interviewed a neighbor who was at home?

1.17 A student reported that one evening he received a call from Sarasota, Florida. The caller verified the student's phone number and then identified himself as being with the A. C. Nielsen Company. Following are the questions asked:

(1) How many TV sets are in the house?

(2) How many are turned on?

(3) What network are you watching?

(4) What program are you watching?

(5) Are you hooked up to a cable system?

(a) Do the questions seem appropriate for a TV rating survey?

(b) Do you think the Nielsen ratings reflect the preferences of the viewing population? Why or why not?

(c) Can you think of other methods that the A. C. Nielsen Company might use to gather data for rating television shows?

1.18 To get the public's opinion of nuclear reactors, a utility company mailed a questionnaire to all of its customers. Along with the questionnaire was a pamphlet describing nuclear reactors and the potential source of energy they could supply. List a strong point and a weak point of such a questionnaire and comment on whether or not the results of the questionnaire will represent the true feelings of the population.

1.19 "Since over half of all fatal traffic accidents involve alcohol, don't you think the penalties for drunk driving should be increased?"

Why is this question not appropriate for a questionnaire on the drunk driving issue? Should the "over half" figure be mentioned at all? How might the question be changed to make it appropriate for the questionnaire?

1.20 A public and a private university conducted independent surveys of their respective student bodies on the issue of drug abuse. Would the results of the two surveys be similar? Why or why not? Could the results be combined to represent all college students?

1.21 In a questionnaire to constituents, a member of Congress posed the following question:

To deal with the problems in Central America, the United States should

(a) increase both economic and military aid to those countries that protect the rights of their citizens

(b) continue or increase economic assistance, but stop all military assistance

(c) stop all U.S. aid

Does this question lead the respondent in any way? Should it be reworded in order to not lead the respondent? If so, how would you reword it?

1.3   Sampling Techniques

In some situations, the population of concern may not be the same as the sampled population. For example, we may be interested in all families that are poverty stricken but may be able to sample only those who are receiving welfare benefits. The population of all poverty-level families is called the *target population,* and the population of welfare recipients is the *sampled population.*

Definitions

The **target population** is the population under study. The **sampled population** is the population from which the sample is obtained.

In many cases the target population and sampled population will be the same population. There are cases in which they differ, however. In the preceding poverty example, the two differ because we would be unable to identify all families who are poverty stricken and thus could not sample that population. However, a sample could be obtained from a list of the welfare recipients.

The target population just mentioned is much larger than the sampled population because not all poverty-stricken families receive welfare. For an example of the target population being smaller than the sampled population, consider the voter preference toward the candidates in an election. The target population would be the population of votes to be cast in the election. But the only available population might be a list of registered voters. Clearly the target population would be smaller than the sampled population because not all registered voters vote. Even though the two differ in size there will be no problems if the characteristics of the two are similar.

A listing of all the elements in the sampled population is sometimes referred to as a *sampling frame* or simply a *frame*. If there is a frame, and it agrees closely with the target population, then it would be possible to sample the whole population.

▲ It might be said that a census is an *attempt* to sample the entire population.

Definition

A **census** is a sample consisting of the entire population.

Because information is desired about the population, why not take a census? The census is appropriate in certain circumstances; however, in most cases the sample is preferred for the following reasons.

Cost.   Clearly, it will cost more to measure all subjects in the population. In most cases the cost of a census would be so high that the study could not be done. Thus, a sample would be better than doing nothing at all.

Time.   Because the population is usually very large, just the time involved in measuring all subjects would prohibit the study. Usually speed is of utmost importance. For example, a survey of voter preference that takes 6 months to collect is of little value. Opinions are likely to change over that period of time.

Destructive Testing.   In some studies, just the act of taking a measurement destroys the unit on which the observation is taken. For example, measuring the length of useful life of a battery involves operating the battery until it fails. Clearly, in this case, taking a census depletes the inventory.

Inaccessible Population Units.   Suppose one wishes to study the effectiveness of a new miracle drug to prevent the common cold. All people who may someday use this miracle drug cannot be determined and thus will not be available for observation. A sample will have to be used for the study.

Inaccuracy of a Census.   Because a census is such a difficult undertaking, systematic biases (a tendency to underestimate or overestimate) sometimes appear. Accounting for millions of people is going to lead to some inaccuracies simply because of the magnitude of the job. It would be much easier and probably more accurate to sample an inventory rather than have someone count thousands of items.

Considering the preceding reasons, a census is rarely practical and one must resort to a sample. We now consider ways to obtain a sample.

Convenience Sample   Just as the name suggests, a convenience sample is a sample that is convenient to obtain. It consists of those units in the population that are easily accessible. For a sample of college students, one might use an existing introductory psychology class. For a sample of voters, one might interview people at a shopping center. For opinions on a city government issue, one might use the responses voluntarily phoned in on a TV talk show.

Usually the results of a convenience sample are biased in that they overrepresent or underrepresent elements of the population. The letters received by a member of Congress would be considered to be a convenience sample. The opinions expressed in the letters will rarely represent the attitudes of the entire constituency, because only those who feel strongly about an issue will bother to write a letter. In general, we can say that a convenience sample will result in a situation where the sampled population will not be the same as the target population.

EXAMPLE 1.10

A local TV station conducts a weekly telephone survey in conjunction with the evening news. On one occasion the question posed was "Should the salaries of local fire fighters be comparable to that of the local police?"

Callers were able to phone in their responses (yes or no) throughout the evening, and the results were announced on the 11 o'clock news. How much credibility should be attached to the results of this survey?

SOLUTION   Little or none! First, it was a voluntary survey and only those who felt strongly (fire fighters and their families) about the issue would bother to call. Also, there was no way of monitoring the number of times a person called. Therefore, those who called did not constitute a representative sample of the residents of the city. It should be added that such polls tend to mislead the general population. The naive citizen believes that the results are reliable when, in fact, they misrepresent the true attitude of the population. The results of this survey most likely would have been biased toward the fire fighters.

The bias introduced in a sampling procedure is called *sampling bias*. To reduce the effects of sampling bias, we have the random sample.

Simple Random Sample

Definition

A **simple random sample** (SRS) of size n is a sample in which each sample of size n has the same chance of being chosen.

The aim of a simple random sample (SRS) is to obtain a sample that is representative of the population and not biased in any way. In this way, no one would have an advantage over anyone else in the population. There are several ways to obtain a SRS. A popular method is to use *physical mixing,* also referred to as *lottery sampling.* Each element in the sampling frame is given an identification tag. The tags are thoroughly mixed in a barrel, and the sample is selected from the barrel, one observation at a time. If the mixing is complete, then each element remaining in the population has the same chance of being selected at each draw. If n tags are drawn, then that group of n is just as likely as any other group of n, resulting in a SRS of size n.

1970 Draft Lottery

Although physical mixing seems simple, it is difficult to get a complete mixing. A historical example of a problem with physical mixing occurred in the 1970 draft lottery. The draft lottery was intended to select birth dates randomly. Men born on the first date chosen would be the first drafted, those born on the next date chosen would be drafted next, and so on. The 1970 lottery was for 18 year olds—those born in 1952, which was a leap year. Cylindrical capsules, each with a slip of paper inside corresponding to a day of the year, were placed in a large box. The 31 capsules corresponding to January were placed in the box first, stirred, and then pushed to one side. Next, the 29 February capsules were placed in the box and mixed in with the January capsules. The capsules for succeeding months were placed in the box and mixed with those already in the box. The box was shaken several times and the capsules were drawn one at a time. As it turned out, the dates in the later part of the year were the first drawn because their capsules were put in the box last and not mixed as thoroughly as the capsules corresponding to the early months. For example, the January capsules were mixed 11 times with the other capsules, whereas the December capsules were only mixed once. Thus, the probability of being drafted was greater for young men born the later part of the year. In 1971, the draft lottery was turned over to the statisticians at the National Bureau of Standards.* They now utilize a *Table of Random Digits*.

Table B1 in Appendix B is a portion of a table of random digits. To use the table, a sampling frame is needed so that the elements of the population can be enumerated. If the sampling frame consists of 100 elements, you will use two digits (00 to 99) of each number in the table. If there are 1,000 elements, you will use three digits (000 to 999) of each number in the table and so on. Suppose the sampling frame has 10,000 elements numbered from 0000 to 9999. Pick a random starting place in Table B1 and read the numbers systematically. Four digits of each number will correspond to a chosen element from the sampling frame. For example, suppose we choose column 8 of row 20. There we see 186614. Using the first four digits we would pick element 1866 from the sampling frame. The next number down corresponds to element 8929. We continue down the list until we have a sample of the desired size. Because the digits are random, we could have used the last four digits of each number and instead of proceeding down the table we could have gone up or across; it makes no difference. If you lose your place, pick another random starting place and continue. If a number is repeated, throw it out because it has already been selected. If the sampling frame has, for example, only 7,000 elements, you will still select four-digit numbers and throw out those that exceed 6,999. Using a table of random digits is one of the best ways to obtain a SRS.

EXAMPLE 1.11

Suppose a club has 50 members. Use Table B1 to select a random sample of size 5.

SOLUTION The club members are enumerated from 00 to 49. Enter Table 1 at a random starting place, say row 28 and column 7; there we find 259740. Because only two digits are needed we have a choice; let us use the last two digits. So we would choose the club member number 40. The next number down is 882190, which corresponds to club member number 90. But there is no 90, so we skip to the next number down, which is 437448 and choose club member 48. Continuing down we next pick club member 19. The next number is 549448, which also corresponds to club

*See Appendix A3 for the results of the 1970 and 1971 draft lotteries.

▲ To avoid skipping numbers, each
club member could have been
assigned two numbers, one from 00
to 49 and the other from 50 to 99.

member 48. Because we have already used 48 we continue to the next number, which corresponds to club member 35. Finally, our last club member chosen is number 32.

Another method of obtaining a SRS is the systematic sampling procedure.

Systematic Sample

A systematic sample is appropriate when we have a sampling frame with the elements listed in random order. Suppose we have a list of 10,000 students ordered by their social security numbers. Because these numbers are somewhat random we could start at a random place in the first 100 names on the list and then select every one-hundredth name. The result would be a random sample of size 100.

Definition

A **systematic sample** is obtained by starting at a random position and selecting every kth element from the population until the desired sample size is obtained.

It is possible to use systematic sampling without a tangible sampling frame. In conjunction with a census, to obtain a 10% sample of homes, the U.S. Bureau of the Census might start at a random place and select every tenth home for further study. (In addition to a complete census, the Census Bureau will use statistical sampling to select a random sample for more extensive study.) In this manner it is possible to sample the households in a city without knowing how many households there are in the city.

Systematic sampling should be avoided when the data in the population are of a periodic nature. For example, suppose it is desirable to sample the daily receipts of a large department store. A one-in-seven systematic sample would result in a biased sample because the sample would consist of the daily receipts for one particular day of the week. This problem could be avoided by taking a one-in-ten systematic sample. However, it is best to use some method other than systematic sampling when the population is periodic.

Stratified Sample

A simple random sample (SRS) might result in a sample that is *not* representative of the population. It is possible that a SRS of size 1,000 could have 600 females and 400 males. Assuming the population is evenly divided among females and males, this sample is not representative of that population. The stratified sampling procedure prevents this from happening.

Definition

A **stratified random sample** is obtained by dividing the population into groups called strata, and then selecting a SRS from each stratum.

Stratification can take the form of prestratification or poststratification. Prestratification stratifies the population prior to selection of the sample. If a sample of 1,000 voters is desired, one could use prestratification and take two random samples by randomly selecting 500 females and randomly selecting 500 males from a sampling frame. If prestratification is not possible, the sample could be stratified after it is selected; this is poststratification. For example, suppose that 20% of the national population has a college education. If education is not recorded on any sampling frame, then prestratification cannot be used. One must select the random sample, determine the education level of each subject in the sample, and then reduce the sample size so that 20% of the sample have a college education and 80% do not.

The strata in stratification can be geographical regions, religious preference, race, income bracket, political party, and sex, as mentioned above. The stratification can be based on any variable relevant to the survey. Stratification will guarantee that the sample is similar to the population in those characteristics on which one chooses to stratify.

EXAMPLE 1.12

Suppose a bank wishes to sample its savings accounts. Suppose, furthermore, that they know that 5% of their accounts are over $50,000, 20% are between $10,000 and $50,000, 25% are between $5,000 and $10,000, and 50% are below $5,000. Using stratification they could assure themselves that a sample of 100 accounts would have accounts representing all possible categories.

Often, stratified random sampling is accompanied by proportional sample size. This means that if, for example, stratum 1 is twice as big as stratum 2, then the random sample from stratum 1 will be twice as large as the sample from stratum 2. That is, the sample size for a stratum is proportional to the population size in that stratum.

EXAMPLE 1.13

Using proportional sampling in Example 1.12, the bank would sample so as to have 5 of the desired 100 accounts from the over $50,000 category. Similarly, 20 accounts would be from the $10,000 to $50,000 category, 25 from the $5,000 to $10,000 category, and the remaining 50 accounts from the below $5,000 category. So, not only does the sample consist of accounts from all possible categories, but the sample sizes from each category are proportional to the stratum sizes.

To use proportional sampling it is necessary that the stratum sizes be known. This information can only be obtained from a census. Thus, it is important that up-to-date census information be available.

Cluster Sample   Cluster sampling is an economical way of selecting a sample for a survey.

Definition

Cluster sampling consists of selecting clusters of units in a population and then performing a census of each cluster. The selection of clusters could be based on some desired feature of the population or could be a simple random sample of clusters in the population.

Suppose that a sample of residents of a large city is desired. Using cluster sampling, the entire city could be divided by a city map into blocks (clusters) of households. Then a simple random sample of clusters could be obtained and each resident of each selected cluster would be interviewed. Clearly, this is a more cost-effective method than traveling around the city looking for individuals in a simple random sample. On the other hand, cluster sampling may be susceptible to sampling bias. A certain ethnic group may not be represented in the sample simply because its cluster is not selected. Also, the population units in a cluster usually are very homogeneous (similar age, income, educational background, recreational interests, etc.) and thus a cluster sample is less informative than a simple random sample of the same size. With cluster sampling we may be unable to get a true cross section of the population.

EXERCISES 1.3

1.22 Explain why a sample is preferable to a census in the following situations:

(a) Testing the effectiveness of an experimental drug, which is supposed to reduce pain

(b) Determining the potential market for a new product by distributing free samples

(c) Estimating the amount of timber in a forest

(d) Estimating the life of a flashlight battery

1.23 To estimate the murder rates in the major U.S. cities, a law enforcement agency determined the rates in the 12 largest cities in 1981. What is the target population? What is the sampled population?

1.24 In a house-to-house survey of a small city, the interviewer selected every corner house and asked the head of the household how he or she intended to vote on a municipal bond for upgrading the sewer system. Would there be sampling bias in the results? Why?

1.25 Use Table B1 in Appendix B to select a random sample of size 5 from your class.

1.26 Identify the sampling technique used in the following situations:

(a) Every fiftieth person is selected from a list of registered voters.

(b) The winning number is selected from a revolving barrel.

(c) Five taxpayers are randomly selected from each county in the state.

(d) A college administrator uses a freshman class for a study she is conducting.

1.27 How would you select a sample of 100 students from your student body in order to get representatives of each class? What type of sampling technique would you use?

1.28 To assess the possible success of a new shopping center, a questionnaire was mailed to a sample of people obtained from the mailing list of the local chamber of commerce. The questionnaire addressed the issue of whether or not the respondent felt that the availability of merchandise was adequate in the city. Distinguish between the target population and the sampled population.

1.4 Experimental Design

Generally the sample survey is used to describe the characteristics of an existing population. Experimentation is concerned with investigating a population that has been altered for study. Unlike survey sampling, the experimenter actively intervenes by controlling the environment and administering a *treatment* to the subjects in order to study its effect on some other variable. One of the simplest forms of an experiment is one in which two homogeneous groups of subjects that are treated alike except that one group receives the treatment and the other does not. Any observed difference would be due to the treatment under study.

The basic idea of experimentation is to study the effect that a change in one variable has on another variable. The variable that is manipulated is called the *independent variable* and the other is called the *response* or *dependent variable*. The different values through which the independent variable is manipulated are called the *levels* of the variable.

Definitions

An **experimental unit** is the basic element on which the experiment is conducted. It could also be called a **subject.**

An **independent variable** is one that is controlled or manipulated by the experimenter. In an experiment it is often referred to as a **factor.**

A **level** is one specific value assumed by a factor.

A **dependent variable** is a variable that is measured on each experimental unit to determine whether its value is affected by the independent variable.

A **treatment** is a specific combination of the various levels of the experimental factors.

EXAMPLE 1.14

Suppose that we wish to study the effect of jogging on lung capacity. We could take a group of homogeneous subjects (subjects whose physical characteristics are very similar) and randomly assign them to several groups. One group jogs 1 mile each day and another group jogs 5 miles each day. To further control the experiment we have a *control* group that does not jog. Ideally we would require that all subjects be treated the same (diet, sleep, emotional stress, etc.) except for the distance that they jog. After a period of time, we would measure the change in lung capacity. The independent variable is jogging—its levels are 1 mile, 5 miles, and no miles. The dependent variable is the change in lung capacity.

Design

▲ A possible modification to this design would be to take observations both before and after the treatment.

The design of an experiment describes how the treatments are applied to the experimental units. The experiment can be relatively simple or somewhat complicated. The simplest would be a *one factor design* in which a single treatment is applied to the experimental units and then observations are made.

The aim of such an experiment is to determine whether the treatment has any effect on the experimental units. For this design to be valid, however, we must make sure that no outside or *extraneous* variables affect the units.

Definition

An independent variable and an extraneous variable are **confounded** when their effects on the dependent variable cannot be distinguished from each other.

EXAMPLE 1.15

To study the effects of unemployment on the divorce rate, a sociologist selected a random sample of unemployed people from the local unemployment agency and a sample of employed people from his graduation class. He then compared the divorce rate for the two groups.

Here, educational level (extraneous variable) possibly would have an effect on divorce rate (dependent variable) that could not be separated from the effect of employment status (independent variable) on divorce rate. They are confounded on the dependent variable.

EXAMPLE 1.16

A group of senior citizens was randomly divided into two groups. One group was given daily doses of vitamin C and the other group was given no treatment. After the winter the vitamin C group reported fewer colds than the group with no treatment. The experimenter concluded that vitamin C helps prevent colds.

It is quite possible that the vitamin C group in the preceding example experienced the *placebo effect*—a psychological effect to a treatment in which the subject has confidence. The mind is a very powerful tool, and the subjects who received vitamin C *knew* they should have fewer colds.

To combat the placebo effect, experimenters give dummy pills (sugar tablets), which are called *placebos*. In a similar study, patients were given a placebo, but were told it was vitamin C. They had fewer colds than those given vitamin C, but thought it was a placebo. ("Is Vitamin C Really Good for Colds?" *Consumer Reports,* Feb., 1976, pp. 68–70.)

Comparative Experiment   Unless the experiment is well designed, the placebo effect often is confounded with the treatment effect. The *comparative experiment* is used to avoid confounding of experimental with extraneous factors. In the *comparative experiment* we randomly divide the subjects into two equivalent groups, with one group receiving the treatment and the other serving as a *control group*. A configuration of the design is

Subjects randomly
assigned

↓   ↓

Experimental   Control
group   group

↓   ↓

Observations   Observations

The control group is treated the same as the experimental group except that they do not receive the treatment; then, any difference observed between the two groups should be due to the treatment. Extraneous factors should affect both groups the same and thus nullify their effect. Often the control group will receive a placebo. Or if we were comparing two drugs, for example, we could assign drug A to one group and drug B to the other group and not have a control group. Again, any observed difference in the two groups should be due to the effects of the different drugs. This design also lends itself to the before-after type experiment in which subjects are measured both before and after receiving the treatment.

The simple comparative experiment can be extended to what is called the *completely randomized design* with multiple groups:

Subjects randomly
assigned

↓	↓	↓	↓
Treatment group 1	Treatment group 2	Treatment group 3	Treatment group 4
↓	↓	↓	↓
Observations	Observations	Observations	Observations

Of course one of the four treatments could be a control group with no treatment. More will be said about the completely randomized design in Chapter 13.

Reducing Bias

▲ The random assignment of subjects to groups could be done with a table of random digits.

To reduce possible bias in the comparative experiment, subjects are randomly assigned to the groups. We hope *randomization* will create equivalent groups prior to the experiment. For example, to study the effect of alcohol on reaction time we certainly would not want our subjects to choose which group (control or experimental) they would enter. Even if the subjects are homogeneous (alike in all characteristics relevant to the experiment), they do have some differences. Random allocation tends to distribute these differences evenly between the two groups.

Another means of reducing bias in an experiment is the *double-blind* experiment. In this experiment, neither the subject nor the one administering the treatment knows who is receiving the actual treatment and who is receiving the placebo.

The double-blind procedure is used in most studies involving medical trials. Subconsciously, the diagnosing physician may be influenced by knowledge of what treatment the subject receives. For this reason, he is kept ignorant of who is receiving the treatment and who is receiving the placebo. For reasons discussed previously (placebo effect) the subject does not know if he or she is receiving the treatment or the placebo. Only the director of the study knows who is receiving the treatment and who is receiving the nontreatment.

Randomized Block Design

Randomization tends to average out the differences and produce nearly equivalent groups for comparison. However, there is no guarantee that it will. For example, suppose that there are 20 subjects to be randomly assigned to two experimental groups of 10 each. It is possible that a completely randomized procedure would put 10 subjects of high intelligence in one group and 10 subjects of low intelligence in the other group. If it is important that the two groups be equivalent as far as intelligence is concerned, then the randomization procedure has failed. The problem can be avoided by assigning subjects according to the *randomized block design*.

Definition

A **randomized block design** is the random assignment of subjects within a block to treatments, with each treatment appearing exactly once in every block.

In the preceding example the 20 subjects can be divided into 10 intelligence groups (blocks) of 2 subjects each. The 2 subjects in each block are then randomly allocated to the two experimental groups. The end result would be that each of the groups would have subjects from each of the 10 intelligence blocks, and thus would be equivalent as far as the intelligence of the subject is concerned. A configuration of the design is

Block 1	Block 2	. . .	Block 10
↓	↓		↓
Random assignment	Random assignment	. . .	Random assignment
↓ ↓	↓ ↓		↓ ↓
Group 1 Group 2	Group 1 Group 2	. . .	Group 1 Group 2

The randomized block design can be generalized to b blocks and g groups and the blocks can represent any variable that has a confounding effect on the dependent variable. We also see that the random assignment to groups is no longer applied to all subjects as in a completely randomized design, but is restricted to the subjects within each block.

A special case of the randomized block design is the *matched pair design* in which each block consists of two matched subjects that are randomly assigned to one of two treatment groups. Each pair should be as much alike as possible prior to the experiment. Any differences observed then would be due to the treatment.

EXAMPLE 1.17

A clothing manufacturer would like to compare the durability of a newly designed line of children's clothing to that of their existing line of clothing. Ten sets of identical twin children are chosen for the experiment. One of each set of twins is randomly selected to wear the new clothing and the other will wear the old type. After a period of time the durability of the two types of clothing will be evaluated. In this example the sets of twins make up the blocks and the new and old lines of clothing are the two groups. A configuration of the design would be

Twin set (blocks)	1	2	3	4	5	6	7	8	9	10
Clothing type A	x	x	x	x	x	x	x	x	x	x
Clothing type B	y	y	y	y	y	y	y	y	y	y

The value of x denotes a measure of durability for type A and the value of y denotes the measure of durability for type B. Identical twins are used as blocks because it is assumed that they will give approximately the same treatment to the two sets of clothing.

Observational Study

In many situations it is not possible to allocate subjects randomly to groups in order to study the effects of a treatment. Yet, it is desirable to study the effects of the treatment. For example, if we wish to study the effects of income on one's spending habits, it would not be possible to assign subjects randomly to different income groups. However, we can observe the spending habits of people who already fall in different income groups, and then make statistical generalizations about how income affects one's spending habits.

Definition

An **observational study** (also called a quasi-experiment) is an experiment in which one observes how a treatment has already affected the subjects.

The most obvious observational study that comes to mind is the smoking/nonsmoking issue. Clearly we *cannot* randomly allocate subjects to smoking and nonsmoking groups. Yet, we can compare the incidence of lung cancer for those who smoke and those who do not smoke. The treatment (smoking) is not manipulated by the experimenter but has already been manipulated by, let's say, natural circumstances. The smoking and nonsmoking groups already exist and we are able to compare their rates of lung cancer with an observational study.

All observational studies must deal with the problem of confounding. Their effects cannot be reduced by randomization.

EXAMPLE 1.18

A university would like to compare the salaries for male and female faculty members. It is not possible to assign the subjects randomly to a sex. Therefore, the two groups (male and female) will have to be compared, as they are, in an observational study.

Suppose that a difference is observed in the salaries of men and women. Does this mean that the university should be accused of sex discrimination? Possibly not. It is possible, and very likely, that the years of experience differ for the two groups, which could account for the discrepancy in the salaries. We would say that the years of experience are confounded with sex on the salaries of the professors.

In the preceding example, it is possible to control for the confounding variable by comparing the salaries of males and females with similar years of experience. In other examples, extraneous variables are so confounded with the treatment that it is impossible to compare groups.

EXAMPLE 1.19

In the 1950s, the School of Hygiene and Public Health at Johns Hopkins University conducted a study of the effects of public housing on such issues as health and social attitudes. The treatment group consisted of 600 families assigned to a public housing project, and the control group was made up of 300 families who lived in the slums. After three years, it was concluded that those in the treatment group (public housing) were happier than those in the control group (slums) and that the mortality rate in the control group was higher. Does this mean, for example, that the treatment is effective in reducing the mortality rate? After a closer investigation it was learned that families were not randomly selected from the slums for the housing project. There were thousands who applied for admission. The Housing Authority selected the tenants, then the treatment group was selected from the accepted applicants and the control group from those not accepted. Thus, the treatment group consisted of "desirable" tenants and the control did not. So the difference in the two groups could be due to the selection process rather than the public housing. It is not possible to separate the two; the extraneous variable, selection process, is confounded with the treatment, public housing, and thus it is not possible to determine whether the change in mortality rate is due to the public housing or the selection process (Freedman et al, 1979).

Often, it is possible to isolate an extraneous variable that is confounded. In those cases it may be that the bias due to the uncontrolled extraneous source can be reduced with regression analysis. In Chapter 11 we will take a closer look at this very powerful tool.

Human Experimentation

Serious ethical questions should be raised when experiments are conducted on human subjects. For example,

1. If a specific treatment has the potential of curing a disease, should it be withheld from some victims for the sake of having a control group?
2. Should an experimenter deliberately expose humans to a substance suspected of being detrimental to their health?

The subject should be made aware of the possibility of consequences of the experiment. In fact, the federal government requires that the *informed consent* of the subjects be obtained in order to use them in an experiment. Subjects are told of the possible risks and benefits and asked if they would consent to random assignment to the experimental and control groups.

One of the largest experiments ever conducted on humans occurred in 1954 when the Public Health Service experimented with a polio vaccine developed by Jonas Salk (Tanur et al, 1978). The study involved almost 2 million children in grades 1, 2, and 3. The study was done in two parts, one of which involved 750,000 children. In this experiment 400,000 consented to treatment and 350,000 refused treatment. The 400,000 were then randomly assigned to two groups, a treatment group to get the Salk vaccine and a control group to get a salt-water injection. The subjects did not know whether they were receiving the vaccine or the salt-water solution. Many forms of polio are difficult to diagnose, and in a borderline case the diagnosis could easily be affected by knowledge of whether or not the subject received the treatment. For this reason, the technician was also not told who received the treatment. We can sum up by saying that the study was a double-blind randomized controlled experiment with informed consent.

The second part of the study, involving over a million students, was conducted by the National Foundation for Infantile Paralysis (NFIP). For their study, they used all second-grade students with parental consent as their treatment group (Salk vaccine) and first- and third-grade students as their control group (salt solution). Even though the sample size was larger, the results in this experiment were less reliable than in the previously described experiment. First, polio is a contagious disease so the incidence in the second grade could have been much higher (or lower) than in the control group of first and third graders. Thus, the study possibly was biased. Second, parental consent was required in the treatment group but not in the control group. It was known that parents with higher incomes would tend to consent more readily than the lower-income parents. Moreover, children of higher-income parents were more likely to contract polio than children of lower-income parents. This seems the opposite of what one would believe, but remember that polio is a disease of hygiene. Children who live in less sanitary conditions develop antibodies early in childhood that protect them from polio later in life. Again, the study was biased against the vaccine, because the subjects in the treatment and control groups had different family backgrounds.

Both studies showed that the Salk vaccine was effective (see Table 1.2) in preventing polio, but the difference in the randomized double-blind experiment was more pronounced.

Considering that the vaccine was effective we might ask whether the vaccine should have been given to all children. Prior to the study it was not known whether the vaccine would be an effective treatment for polio and, in fact, it might well have had an adverse

TABLE 1.2
Salk vaccine experiment
of 1954
(rate is number of polio cases
per 100,000)

	Randomized double-blind experiment		
	Treatment	Control	No consent
Sample Size	200,000	200,000	350,000
Rate/100,000	28	71	46
	NFIP design		
	Vaccine (Grade 2)	Control (Grades 1&3)	No consent (Grade 2)
Sample Size	225,000	725,000	125,000
Rate/100,000	25	54	44

SOURCE: Data from Thomas Francis, Jr. *Am. J. of Public Health* 45:5, 1955, pp. 1–63.

effect on the subjects. So, the vaccine should not have been given to all children. Moreover, to get a valid evaluation of the vaccine a treatment and a control group were necessary. In addition, it would have been unethical if it were known that the vaccine (or any treatment for that matter) would be harmful to one's health. In human experimentation we should not use a treatment if we know it will cause harm to the subjects. Historically, harmful treatments have been given to subjects, but federal regulations now prohibit such studies. Most research institutions have review boards that screen all research involving human subjects.

Animal Experimentation

Experiments conducted on animals are less restricted than those conducted on humans. We may purposely expose rats to a substance that is suspected of causing cancer, in order to learn if it indeed causes cancer, and then to infer the same for the human population. But can we generalize to the human population? Critics of animal experimentation say not. They contend that the results are not valid because a laboratory animal reacts differently to a drug than a human. Moreover, to test a substance, the lab animal is frequently given a massive overdose so as to reduce the duration of the experiment. The drawback is that the animal may develop cancer as a result of the massive overdose. Even so, many feel that if a substance causes cancer in a lab animal (regardless of the dose), it should not be given to humans. In fact, the Delaney Amendment of 1958 requires that the Food and Drug Administration (FDA) outlaw additives that are found to produce cancer in humans *or animals*.

EXAMPLE 1.20

To evaluate the sugar substitute saccharin, rats were fed a diet consisting of 5% saccharin. A second group of like rats was fed and treated the same except that it got no saccharin. The results of the experiment showed that the treatment group had significantly more bladder tumors than did the control group. Saccharin was then classified as a carcinogen—a substance that produces cancer in laboratory animals.

Some respected scientists criticized the saccharin experiment because of the overdose given. They contend that the large amounts of saccharin irritated the bladders of the rats, causing tumors. The effect of saccharin on humans has not been determined as yet.

Causation

Experimentation and survey sampling both can be used to study the relationship between independent and dependent variables. The difference, however, is that survey sampling can only show a relation between variables, whereas experimentation (in principle) can show causation. If the subjects are treated alike except for manipulation of the independent variable, then any change in the dependent variable is caused by the change in the independent variable. Practically speaking, though, it is difficult to show causation. It is possible that both the independent and the dependent variables are related through a third extraneous variable. The change observed in the dependent variable may be due to the change in the extraneous variable.

In 1964 the Surgeon General reported that smoking can be dangerous to one's health. This conclusion was *not* a result of the Surgeon General establishing a cause-and-effect relationship between smoking and lung cancer. To show that smoking *causes* lung cancer an experiment would have to be designed in which a homogeneous group of adults is divided into two groups, with one group required to smoke and the other group not allowed to smoke. Furthermore, the two groups would have to be exposed to the same diets, the same environmental conditions, the same working conditions, and so on. After several years the incidence of lung cancer for the two groups would be compared. Clearly, such an experiment would be impossible to conduct.

It is possible that a genetic or heredity factor causes cancer-prone people to smoke. The association seen between smoking and lung cancer is through this third genetic factor. These people would be more susceptible to lung cancer regardless of whether they smoked or not. There are data to support the genetic third-factor theory; however, there is more evidence to the contrary. For example, more women are smoking and the rate of lung cancer in women has increased dramatically, discounting the genetic theory. Many independent studies have established strong associations between smoking and lung cancer. Other experiments have shown that cigarette smoke causes lesions in the skin of laboratory animals, suggesting that smoke could cause lesions in the human lungs. Although it has not been proven, the evidence that smoking is a cause of lung cancer is about as strong as it can be without conducting the comparative experiment on humans described earlier.

Caution. An association between two variables does not establish causation. For example, test results show that the average IQ of black students is lower than the average IQ of white students. But this does *not* mean that if you are black you are predisposed to a lower IQ. Many other factors such as quality of education and social environment contribute to the difference in mean IQ scores. The effects of these other variables are confounded with the effects of race on IQ. So the association between race and IQ is simply through these other variables. The observed difference in IQ's would disappear if blacks and whites had the same quality of education and social environment.

EXERCISES 1.4

1.29 What is meant by a double-blind experiment?

1.30 What is meant by the confounding of a variable with the dependent variable?

1.31 How can confounding be avoided or at least reduced?

1.32 Describe the placebo effect.

1.33 Suppose we wish to compare two different methods of teaching first graders to read. At the beginning of the school year we will have our two first-grade teachers choose their students for their respective reading instructions. At the end of the school year all students will be given a reading test and then the two methods evaluated.

(a) What is an experimental unit?

(b) What is the independent variable? How many levels?

(c) What is the dependent variable?

(d) Are any extraneous variables confounded with the independent variable on the dependent variable?

(e) Is there possible bias in this experiment?

(f) Has randomization been used to select the samples?

(g) Is there a control group? Is one needed?

1.34 A study of 31,604 pregnancies reported in the *Journal of the American Medical Association* that babies whose mothers had one or two alcoholic drinks daily weighed an average of about 3 ounces less than those of nondrinking mothers.

(a) Is this an observational study or a randomized experiment?

(b) Are there any potential confounding variables in this study?

1.35 To study the effects of TV violence on children an experimenter randomly assigned a group of 3-year-old children to two groups. One group viewed a half-hour TV program filled with violent acts. The other group viewed a half-hour segment of the "Mr. Rogers" show. The number of violent acts by the children in the hour following the viewing were recorded for the two groups.

(a) Is this an observational study or a randomized experiment?

(b) Are there any potential confounding variables in this study?

1.36 Explain why a control group would be advisable in the following studies:

(a) Seventy percent of those who took pain reliever A experienced relief from pain within 3 hours after taking the pill.

(b) Farmers who used a new crop fertilizer experienced a 10% increase in crop yield over the yield of the preceding year.

1.37 Classify each of the following as either a sample survey, a randomized experiment, or an observational study.

(a) Two-thousand employees of a large corporation were asked whether they preferred the new health insurance plan or the old one.

(b) A medical lab compared the incidence of lung cancer in people who smoke with those who do not smoke.

(c) A building contractor used insulation type A in half the houses he built and insulation type B in the other half. Afterward the energy efficiency of each house was measured.

(d) The birth weights of newborn babies whose mothers drank alcohol were compared with the birth weights of babies of nondrinking mothers.

(e) Students in a tenth-grade class were divided into two groups and taught two different methods of typing. A typing test was then given to compare the two methods.

1.38 Of the examples listed in Exercise 1.37, which do you think would possibly have a biased sample?

Statistical Computer Packages (optional)

The use of statistics has grown enormously in the last few decades. The concepts and methods are currently used in all fields of human endeavor in which numerical and graphical descriptions of data are necessary. The computer has greatly improved the techniques for exhibiting and analyzing the collected data and has contributed significantly to the growth of statistics as a discipline.

Among the many statistical computer packages available for summarizing data are:

1. BMDP® (UCLA Biomedical Statistical Package)
2. SPSS-X® (Statistical Package for the Social Sciences)
3. SAS®
4. Minitab®

(References given at end of this section)*

Originally these packages were accessed through a mainframe computer housed in a computer center located centrally to a university or business. More recently, versions of all four statistical packages were made available to a select group of microcomputers. The statistical analysis required by most statisticians can be done on the microcomputer.

You may never be called upon to analyze your own data set; however, as a *user* of statistics, you will need to become familiar with the terminology and be able to interpret statistical data. Therefore, you should become familiar with the procedures of at least one of the major statistical packages listed above. Each package has its own set of manuals that explains in detail how the programs operate and what information you should expect to receive from them. To access any of the packages you will need the job control language (JCL) that is unique to your particular computer. Having obtained the JCL from your local computer center, you are then ready to use the statistical package.

The BMDP, SPSS-X, and SAS software packages are extensive programs that provide techniques ranging from analysis of simple data sets to advanced statistical procedures. Some computer experience is required for their use. Minitab is not as versatile as the other three packages, but it is capable of handling all the analysis required in this text. It is described as an easy-to-use package intended for those who have had no previous experience with computers. For this reason we will devote our attention to Minitab. Commands to perform the analysis discussed in each chapter will be described at the end of each chapter.

The Minitab "worksheet" consists of columns identified as C1, C2, C3, and so on. In Figure 1.4 an empty array of the worksheet is shown before any data are entered.

Approximately 150 commands operate on the data stored in the worksheet. The commands are given in English almost as you would tell someone to do the calculations by hand.

---

*BMDP is the registered trademark of BMDP Statistical Software, Inc.

SPSS-X is the registered trademark of SPSS, Inc. for its proprietary computer software.

SAS is the registered trademark of SAS Institute Inc.

Minitab is the registered trademark of Minitab Inc.

FIGURE 1.4
Minitab worksheet

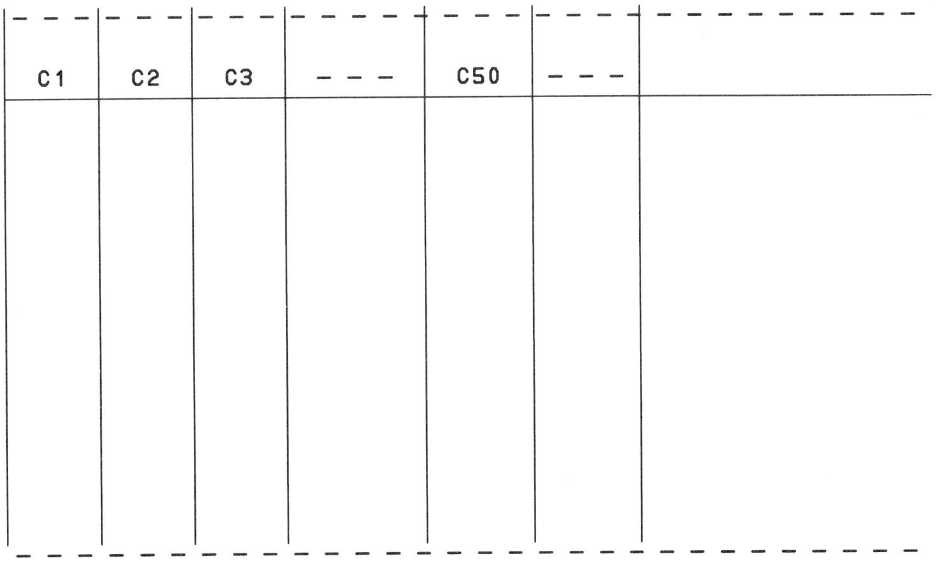

C1	C2	C3	- - -	C50	- - -

READ **Command**

To read data into column 1 of the worksheet the command is:

```
READ the data into C1
```

followed by the data. In this command, READ is the command name that is looked up in the list of 150 command names; C1 is called the argument of the command. After looking up the command name and detecting no errors, it then knows to look for a column number, C1 in this case. After a little practice you will see that in a command you only need the command name and its arguments. Thus the previous command could be abbreviated as

```
READ C1
```

The READ command can have several arguments. For example,

```
READ C1, C2, C3, C4
```

says to fill the first four columns of the worksheet with data.

EXAMPLE 1.21

Suppose we have five students and wish to store in the computer their grades on the first three tests of a course and then print back the results. We will also identify their sex as 0 for male and 1 for female. The collection of commands (program) would be as follows:

```
READ sex in C1, first grade in C2, second grade in C3, third in C4
1   65   72   80
1   92   95   94
0   38   69   54
0   75   82   81
1   79   76   80
END
PRINT C1, C2, C3, C4
STOP
```

We see that the first student is female with grades of 65, 72, and 80. The third student is male with grades of 38, 69, and 54, and so on. The END command signals the end of the data. Once the data are stored we ask the computer to print back the four columns of data with the PRINT command. Minitab can be used interactively, meaning that after each command the analysis is performed and the answer given. The four columns will be printed out immediately after the PRINT command. The STOP command indicates the end of the Minitab session.

The printout is as shown in Figure 1.5.

FIGURE 1.5

Minitab output for Example 1.21

```
MTB > READ SEX IN C1,FIRST GRADE IN C2,SECOND IN C3,THIRD IN C4
DATA> 1   65   72   80
DATA> 1   92   95   94
DATA> 0   38   69   54
DATA> 0   75   82   81
DATA> 1   79   76   80
DATA> END

      5 ROWS READ

MTB > PRINT C1,C2,C3,C4

  ROW   C1    C2    C3    C4

    1    1    65    72    80
    2    1    92    95    94
    3    0    38    69    54
    4    0    75    82    81
    5    1    79    76    80

MTB > STOP
```

Additional Features

Other interesting points about Minitab are:

1. Only the first four letters of each command name are necessary; for example, PRINT could have been PRIN.

2. Data values can be separated with blanks or commas, thus,

```
1   65   72   80
```

could have been

```
1,65,72,80
```

3. Each command must start on a separate line.

4. Minitab will only accept numerical data. M and F for male and female would not be acceptable.

5. When the session is completed, give the command STOP.

6. The dash can be used between two arguments to include all values between the two, for example,

```
READ C1-C4
```

is equivalent to

```
READ C1, C2, C3, C4
```

7. Any column may be given a name. The form of the command is

```
NAME C1 'name', C2 'name'
```

Each name must be less than eight characters long and it must be enclosed by single quotes when used in a command.

8. At any point the command HELP can be given for assistance.

9. Two manuals can be used to learn more about Minitab. They are the *Minitab Reference Manual* and the *Minitab Handbook* (second edition). They are referenced at the end of this section.

SET Command

The command

```
READ C1
```

requires that each observation be read one line at a time as follows:

```
READ IQ in C1
 98
106
110
 95
115
100
105
END
PRINT C1
STOP
```

The same data can be stored in C1 in a more efficient way with the SET command as is shown here.

```
SET IQ in C1
98,106,110,95,115,100,105
END
PRINT C1
STOP
```

Note that only one line is needed for the data as opposed to seven lines for the READ command. The READ command can have multiple arguments, however, whereas the SET command can have only one (one column at a time) argument. Figure 1.6 is the printout.

FIGURE 1.6

Minitab output

```
MTB > SET IQ IN C1
DATA> 98, 106, 110, 95, 115, 100, 105
DATA> END

MTB > PRINT C1
C1
      98     106     110      95     115     100     105

MTB > STOP
```

In each of the following chapters new Minitab commands that relate to the material in the chapter will be given.

EXERCISES 1.5

Although the following exercises are intended for a Minitab solution, they can be solved with any of the other statistical packages. The commands would be different but the end result would be the same.

1.39 Enter the following data into column 1 of the worksheet and then print it back out.

38, 62, 26, 44, 51, 73, 35, 48

1.40 Following are the sex, height, and weight of five adults. Enter sex, height, and weight of each individual and then print it back out.

	Male	Male	Female	Male	Female
Height (inches)	72	74	64	69	68
Weight (pounds)	195	214	105	172	137

1.41 Suppose you were asked to keep the records of a scholastic fraternity at your university. Identify five different quantities that you think are important to keep track of (e.g., the sex of the member). Write a short Minitab program to keep track of the records for the fraternity.

Sources for Computer Packages

1. Minitab, Inc.: 3081 Enterprise Drive, State College, Pa. 16801.
2. *Minitab Student Handbook,* 2d ed.: Ryan, Joiner, and Ryan, PWS Publishers, Boston, Mass.
3. *Minitab Reference Manual:* Ryan, Joiner, and Ryan, Minitab, Inc., 3081 Enterprise Drive, State College, Pa. 16801.
4. SAS® Institute Inc.: Box 8000 SAS Circle, Cary, N.C. 27511-8000.
5. *SAS® User's Guide: Basics,* Version 5 Edition: Available from SAS Institute Inc.
6. *SAS® User's Guide: Statistics,* Version 5 Edition: Available from SAS Institute Inc.
7. SPSS Inc.: 444 North Michigan Ave., Chicago, Ill. 60611.
8. *SPSS-X User's Guide 2d ed.:* Available from SPSS Inc.
9. BMDP Statistical Software, Inc.: 1440 Sepulveda Boulevard, Los Angeles, Calif. 90025.
10. *BMDP Statistical Software Manual,* 1985 reprinting: W. J. Dixon et al. (1985), University of California Press, 2120 Berkeley Way, Berkeley, Calif. 94720.

☑ The topics of statistics considered in this book are collecting data, descriptive statistics, and inferential statistics. *Descriptive statistics* is the means by which the collected data are organized and summarized. *Inferential statistics* is drawing conclusions about a population from the data in a sample.

☑ A *variable* is a characteristic associated with each unit in a population. An *observation* is a single value assumed by a variable.

☑ The study of statistics can be generalized to two areas: sample surveys and experiments. A *survey* is for studying the opinions of the subjects in an existing population. An *experiment* is concerned with a population that has been altered for study. *Precision* and *bias* are important considerations in both the survey and the experiment.

☑ Things to consider when studying a survey are

1. population
2. method of contact
3. response rate
4. questions
5. timing
6. sample design

☑ The *target population* and the *sampled population* may differ, but it is hoped that their characteristics of interest are similar.

☑ The goal of sampling is to obtain a sample that is representative of the population. The different types of sampling techniques are

1. convenience sample
2. simple random sample
3. systematic sample
4. stratified sample
5. cluster sample

☑ An *experiment* attempts to establish a cause-and-effect relationship between two variables. The *independent variable* is the variable that is controlled or manipulated by the experimenter. The *dependent variable* is the variable whose change, if any, is caused by the independent variable. An *extraneous variable* is a variable, outside of the experiment, whose effect might be confounded with the independent variable on the dependent variable. A *treatment* is a specific combination of the levels of the independent variables. A *control group* is the group that does not get the treatment.

☑ The *comparative experiment* is used to avoid confounding. The *randomized block design* is used to produce groups that are equivalent in regard to the blocking variable. An *observational study* is the study of the effect that a treatment has on the subjects when random assignment is not possible.

☑ An association between two variables does *not* establish *causation*.

Learning Goals

Having completed this chapter you should be able to:

1. Understand the concept of a population. *Section 1.1*
2. Understand the concept of a sample. *Section 1.1*
3. Identify a variable and understand how it is used in a statistical study. *Section 1.1*
4. Understand the concept of precision and bias as it relates to a survey or an experiment. *Section 1.1*
5. Understand the details of a sample survey and how it is used to describe a population. *Section 1.2*
6. Distinguish between the target population and the sampled population. *Section 1.3*
7. Know what a census is and why a sample is usually more practical. *Section 1.3*
8. Identify the various sampling techniques. *Section 1.3*
9. Understand the details of an experiment and how the experiment is used to study the effects of a treatment on the subjects. *Section 1.4*

Review Questions

To test your skills, answer the following questions. Multiple choice—circle the correct letter.

1.42 In 1936 the *Literary Digest* (a popular magazine of the 1920s and 1930s) conducted a poll and predicted that Alf Landon would defeat Franklin D. Roosevelt by a 3-to-2 margin. FDR won the election. What happened?

(a) Landon was caught robbing a bank the day before the election.

(b) The sampling frame, made up of *Literary Digest* subscribers and those who owned telephones, was not representative of the population.

(c) The *Digest* only polled residents of major cities, and therefore did not get a true reflection of how the nation felt about the election.

(d) The poll was conducted weeks before the actual election at which time Landon was leading; however, he lost ground and subsequently lost the election.

1.43 Also in the Landon–Roosevelt election 10 million sample ballots were mailed out, of which 2.3 million were returned. Mailed or voluntary questionnaires are

(a) a very reliable source of data in a survey

(b) easy to administer and almost always represent the population

(c) the most widely used method of collecting survey data

(d) usually biased, because angry people have a high response rate

1.44 The most reliable method of obtaining a simple random sample is

(a) with a table of random digits

(b) with a physically mixing bowl

(c) with a telephone book

(d) with a stratification technique

1.45 To use a table of random digits

(a) you always start at the top left-hand corner;

(b) you need not enumerate the elements of the population;

(c) you need a sampling frame of the population;

(d) you need a computer.

1.46 In 1977, a Gallup poll of 1,500 people stated that 83% of the American people opposed preferential treatment for women and minorities in college admission and job placement. This figure is

(a) inaccurate because only 1,500 people were asked;

(b) biased because there is no guarantee of women and minorities being in the sample;

(c) meaningless because college admission and job placement are unrelated;

(d) none of the above.

1.47 A census is

(a) an attempt to sample the whole population

(b) rarely practical

(c) sometimes less reliable than a scientific sample

(d) all of the above

(e) none of the above

1.48 A psychology teacher uses his class as a sample of university students for a study he is conducting. The sampling technique used is
(a) stratified sampling
(b) systematic sampling
(c) convenience sampling
(d) lottery sampling

1.49 Every fiftieth person is selected from a list of registered voters. The sampling technique used is
(a) stratified sampling
(b) systematic sampling
(c) convenience sampling
(d) lottery sampling

1.50 The winning number is selected from a revolving barrel. The sampling technique used is
(a) stratified sampling
(b) systematic sampling
(c) convenience sampling
(d) lottery sampling

1.51 Five taxpayers are randomly selected from each county in the state. The sampling technique used is
(a) stratified sampling
(b) systematic sampling
(c) convenience sampling
(d) lottery sampling

Fill in the blanks (1.52–1.54).

1.52 In a well-designed experiment the control group is given _____ so that the response is to the treatment rather than to the idea of a treatment.

1.53 A _____ experiment is one in which neither the subject nor the person measuring the response knows who is receiving the treatment.

1.54 Two variables are _____ when the effects of the two cannot be separated.

1.55 To study the effects of exercise on the risk of heart disease, an investigator wishes to compare the incidence of heart disease in bus drivers and pedestrian police in New York City. He selects his subjects so that the ages of the two groups are similar, and each subject must have been on the job for at least 10 years.
(a) What is the independent variable?
(b) What is the dependent variable?
(c) What is a unit?
(d) Name two possible confounding variables.
(e) Is this an observational study or a randomized experiment?

1.56 Answer true or false.
(a) An experiment can establish causation.
(b) A survey can establish causation.
(c) An association between two variables means one causes the other to happen.
(d) A placebo is always required for a comparative experiment.
(e) The double-blind experiment should increase bias.

SUPPLEMENTARY EXERCISES FOR CHAPTER 1

1.57 Earlybird Airlines is interested in opening a new route between Charlotte and Dallas. A survey concerning the issue was sent to 2,000 of their past customers. Describe
(a) the population of interest
(b) the sample
(c) a variable of interest

1.58 A stock market investor is interested in oil stocks. She collects last year's price/earnings ratio on 10 selected oil stocks. Describe
(a) the population of interest
(b) the sample
(c) a variable of interest

1.59 A newspaper article reported the number of farmers in each state who went bankrupt in 1984; for example, Texas had 428 and North Carolina had 230 go out of business. Is this a meaningful measure to compare the number of bankrupt farmers in the various states? Can you think of a more valid measure for comparison?

1.60 Of the methods of contact in a sample survey, which method
(a) has the lowest response rate?
(b) has the highest response rate?
(c) is the most reliable?
(d) is the least reliable?
(e) is the most expensive to conduct?
(f) is the least expensive to conduct?
(g) is the most often used?

1.61 In a sample survey, how might response rate be increased?

1.62 Give two reasons why a sample would be preferable to a census.

1.63 Identify the sampling technique used in the following:

(a) Twenty classes are randomly selected across campus and each student in the class is given a survey to complete.

(b) Samples of size 100, 80, 70, and 60 are randomly selected from the freshman, sophomore, junior, and senior classes, respectively.

(c) An interviewer selects a person every 10 minutes as he or she leaves the student union building.

1.64 Give a reason why the Nielsen ratings might be biased.

1.65 To obtain the opinions of the residents of an agriculture state on a decision by the President to phase out all farm supports, a survey was sent to a random sample of 1,000 registered voters in the state by the Department of Agriculture.

(a) What is the target population?

(b) What is the sampled population?

(c) Is this a survey or an experiment?

(d) Do you think the results will be reliable?

1.66 The following question appeared on the survey referred to in Exercise 1.65.

"Don't you agree that a farmer should be allowed to raise as much of any crop he chooses without any restrictions or supports from the federal government?"

Should this question be revised so as not to lead the respondent? If so, how would you revise it?

1.67 The National Women and Stress Survey was designed by *9 to 5,* the National Association of Working Women, and appeared simultaneously in *Glamour, Ms.,* and *Working Woman* in 1984. At the same time *Essence* magazine urged its readers to "Speak Out on Stress!" and respond to the survey. Because *Essence* (a magazine geared to black women) encouraged its readers to respond to the survey, is it possible that the results of the survey are biased? If so, in what way would it be biased?

1.68 In reference to Exercise 1.67, it was reported that nearly 7,000 *Essence* readers responded to the survey. Of the black women who responded, they

(i) were an average age of 30.4 years (younger than black women as a whole)

(ii) had an average of 14.7 years of education (better educated than black women as a whole)

(iii) earned an average of $18,189 before taxes (higher than black women as a whole)

(iv) and 94% were employed (more than black women as a whole)

On the question,

"How would you rate your job overall?"

their responses were

"Very stressful": 30%

"Somewhat stressful": 61%

"Not at all stressful": 9%

(a) Can these results be generalized to the general population of black women?

(b) To what population would these results pertain?

1.69 In reference to Exercise 1.67, if *Essence* did not encourage its readers to respond to the survey, is it possible that the results would still be biased? If so, why?

1.70 Does the following question lead the respondent?

"The existence of the textile industry in the United States is threatened by foreign imports. Don't you agree that the government should limit foreign imports of textile materials?"

If so, how should it be reworded so as not to lead the subject?

1.71 A sample of 500 males and 500 females were asked how often they experience depression. Is this a sample survey or an experiment?

1.72 Sixty students were randomly divided into two groups to study the effects of alcohol on reaction time. Each member of one group consumed a specified amount of alcohol and members of the other group had a nonalcoholic beverage. The reaction times of both groups were measured before and after the beverage.

(a) What is the independent variable?

(b) What is the dependent variable?

(c) What is an experimental unit?

(d) What is the purpose of one group having a nonalcoholic beverage?

(e) Are there any potential confounding variables in this study?

1.73 Is the study in Exercise 1.72 an observational study or a randomized experiment?

1.74 In Exercise 1.72, explain how to construct a randomized block design with blocks based on sex.

1.75 A Latin teacher claims that studying Latin increases one's verbal skills, because in their school the average SAT verbal score for Latin students is 532 and for those not taking Latin it is only 489.

(a) Is this an observational study or a randomized experiment?

(b) Are there any confounding variables in this case?

(c) How reliable is this comparison?

1.76 A survey on pets and people was conducted by *Psychology Today* (August 1984, pp. 52–57). Twelve percent of the 13,000 who responded have no pet, but most expect to get one eventually, when it is more convenient or when they move to a place that allows them. In comparing the demographics of the pet owners and nonowners, it was learned that

(i) 34% of the pet owners had incomes of more than $40,000, compared with 25% for the nonowners.

(ii) 49% of the owners were married, compared with 33% of the nonowners.

Psychology Today pointed out that the pet owners who responded were more satisfied with their lives than the nonowners. Does this mean that owning a pet increases one's quality of life? Explain.

1.77 Which method of contact for a survey sample would you recommend in the following situations?

(a) The local radio station wants to know what percent of the county residents listen to the farm report.

(b) The state medical examiner wants to know what percent of the public is aware of the Poison Control Center.

(c) A U.S. senator wants to know the opinions of his constituency on several national issues.

(d) The city planner wants to know the opinions of the local restaurant owners on a new sign ordinance.

1.78 A 1975 study determined that education has a significant effect on one's health. The study concluded that additional education extends your life expectancy. Give a variable you think might be confounded with the effect of education on health.

1.79 Identify the sampling technique in the following:

(a) To obtain a sample of books in the library the librarian randomly selects a book from the first rack in the library. He then selects a book from that same position from each rack in the library.

(b) To obtain a sample of voters from the county, ten voters are randomly selected from each precinct.

(c) A local TV station asks people to send in their name on a postcard for a free trip to Disney World. The postcards are placed in a large cage, which is rotated several times before the winning card is selected.

(d) To investigate the attitudes of the general population on the subject of nuclear weapons, *Time* magazine takes a random sample of its readers.

1.80 A study was designed to determine whether SAT scores can be used to predict success in college.

(a) What is the independent variable?

(b) What is the dependent variable?

(c) What is a unit?

(d) Give a variable you think might be confounded with SAT on success in college.

1.81 A large department store has its charge accounts listed alphabetically by the customer's last name. The store wishes to estimate the total amount of unpaid balances. Should systematic, stratified, or simple random sampling be employed? Discuss the advantages of your choice.

2

Organizing and Displaying Data

CONTENTS

The survey sample and the experiment, which we considered in Chapter 1, are two general areas of statistics. In a survey, sample data are used to draw conclusions about a population. In an experiment, data are used to investigate the relationship between a treatment and the dependent variable. In each, the data are the important quantities. The data must be collected, organized, and summarized.

Having completed Chapter 1, you should have a good understanding of how the data are collected. In this chapter we will study the different types of data and see how they are organized. In Chapter 3 we will look at ways the data are numerically summarized.

Tuition Costs Rising

According to the U.S. Department of Education, expenses at public colleges and universities are rising faster than the rate of inflation (*Watauga Democrat,* July 3, 1985). In the 1984–1985 school year an education cost an average of $4,522 per student. This represents an 11.1% increase from the 1983–1984 school year. Of the $4,522, $1,055 is paid by the student in tuition and the remaining $3,467 is paid by taxpayers.

As might be expected the appropriations vary from state to state, with some states picking up more of the tab than others. Following are the state-by-state tuition costs per student and state appropriations per student.

State	Tuition per Student	Appropriations per Student
Ala.	$ 1,158	$ 3,541
Alaska	1,099	10,584
Ariz.	956	3,120
Ark.	1,001	3,660
Calif.	500	3,885
Colo.	1,622	2,424
Conn.	1,197	3,314
Del.	2,335	3,607
D.C.	805	8,128
Fla.	734	3,197
Ga.	1,075	3,710
Hawaii	943	6,055
Idaho	655	3,705
Ill.	923	2,899
Ind.	1,571	3,054
Iowa	1,398	3,352
Kans.	923	3,173
Ky.	1,105	3,305
La.	903	3,159
Maine	1,499	3,186
Md.	1,294	3,033
Mass.	902	4,014
Mich.	1,681	2,994
Minn.	1,169	3,553

State	Tuition per Student	Appropriations per Student
Miss.	991	2,611
Mo.	1,178	2,959
Mont.	820	3,841
Nebr.	1,006	2,646
Nev.	1,098	3,058
N.H.	2,836	1,856
N.J.	1,374	4,051
N. Mex.	590	4,111
N.Y.	952	5,008
N.C.	507	3,207
N. Dak.	1,039	2,801
Ohio	2,049	2,857
Okla.	650	2,839
Oreg.	1,203	3,082
Pa.	2,009	3,287
R.I.	1,425	3,845
S.C.	1,464	4,202
S. Dak.	1,065	2,084
Tenn.	1,081	3,476
Tex.	580	3,498
Utah	948	3,617
Vt.	3,545	1,905
Va.	1,246	3,046
Wash.	887	3,286
W. Va.	1,060	2,557
Wis.	1,226	3,220
Wyo.	627	5,195

As the data are presented here, it is possible to determine the tuition cost for any particular state; to make general statements, however, the data would have to be reorganized and *displayed* in different ways.

In this chapter several methods of displaying data graphically will be presented. The objective will be to present the data in a manner that will convey as much information as possible. We will look first at the different types of data.

41

Types of Data

If we are to display data properly, we must first understand the different types of data. Consider the data set in Table 2.1, which lists the sex, major, classification, grade point average (GPA), and Scholastic Aptitude Test (SAT Math and SAT Verbal) scores for a group of undergraduate students.

TABLE 2.1

Sex	Major	Class	GPA	SATM	SATV
M	Physics	Junior	3.2	620	460
M	Political Sci.	Soph.	2.8	410	550
F	Psychology	Junior	3.4	460	620
M	Psychology	Soph.	2.2	400	530
F	Sociology	Senior	3.5	560	670

The column headings, Sex, Major, Class, GPA, SATM, and SATV represent different variables measured on each subject. The various descriptive words or numbers that they assume for the subjects are the *observations* or the values of the variable. The data set is a *multivariate* data set because several variables have been measured on each unit.

Definitions

A **univariate data set** is a data set in which one measurement (variable) has been made on each experimental unit.

A **bivariate data set** is a data set in which two measurements (variables) have been made on each experimental unit.

A **multivariate data set** is a data set in which several measurements (variables) have been made on each experimental unit.

We notice that there is a basic difference in the values of the first three variables in Table 2.1 (Sex, Major, Class) and the values of the last three variables (GPA, SATM, SATV). The last three are numerical and the first three are not.

Definition

A **quantitative variable** is a variable whose values are numerical in nature.

Examples of quantitative variables include

1. Grade point average
2. Weight
3. Age
4. Family income
5. Birth rate
6. Number of children in a family
7. Number of suicides each year

8. Number of voters who favor a certain issue
9. Number of successfully treated patients
10. Number of A's in a class

Quantitative variables are further classified as being either discrete or continuous.

Definition

A **discrete variable** is a quantitative variable that can assume a finite number or at most a countable number of values. (Countable means you can associate the values with the counting numbers 1, 2, 3, . . . ; that is, the values can be counted.)

Normally, discrete variables are the result of some type of count. Example variables 6 through 10 just given are all discrete. The number of children in a family, for example, is represented as a whole number (0, 1, 2, 3, . . .), as opposed to all possible numbers in a real number line interval.

Definition

A **continuous variable** is a quantitative variable that can assume an infinite number of values associated with the numbers on a line interval.

Normally continuous variables are the result of some measurement process. Example variables 1 through 5 are all continuous. For example, grade point average is a continuous variable because it could assume any value in the line interval from 0.0 to 4.0.

As observed previously, not all variables necessarily represent quantities.

Definition

A **qualitative variable** is a variable whose values are classifications or categories and are not subject to a quantitative interpretation.

Examples of qualitative variables include

1. Sex
2. Major
3. Classification
4. Political party affiliation
5. Occupation
6. Religious preference
7. Marital status
8. Employment status

Although qualitative variables are nonnumerical, we can assign a number code to the values the variable could assume. For example, for the variable Sex, we could code Male as 0 and Female as 1; or, the variable Classification could be coded

Freshman-1   Sophomore-2   Junior-3   Senior-4

▲ Assigning a numerical code to a qualitative variable does not make the variable quantitative.

The coding of qualitative variables is very useful when we wish to store the data in a computer, but remember that it is only a code, and we should never perform any arithmetic (such as average) on the values.

EXAMPLE 2.1

Classify the following as either qualitative or quantitative. If quantitative, further classify as discrete or continuous.

(a) The divorce rate

(b) An opinion on a political issue

(c) The number of hospitals that have a trauma center

SOLUTION

(a) Divorce rate is measured in different ways. It can be given as the ratio of divorces to marriages in a given year. For example, in 1983 there were 2,444,000 marriages and 1,179,000 divorces, so the divorce rate was 48% of the 1983 marriages. An alternate method of measuring the divorce rate is to give the number of divorces per 1,000 people. In 1983, there were 5.0 divorces per 1,000. At the same time the marriage rate was 10.5 per 1,000 people, which is consistent with the ratio of divorces to marriages being approximately 48%. Regardless of the way it is measured, the fact remains that it is measured and hence is a continuous quantitative variable. Generally all rates (death rate, abortion rate, etc.) are continuous.

(b) Opinion is not a measurement but rather a classification such as for or against, hence it is qualitative. This is not to be confused with the variable that gives the *number* of those who favor an issue. The number of voters who favor an issue is a discrete quantitative variable.

(c) This is a *count of* variable and thus is a discrete quantitative variable.

EXERCISES 2.1

For Exercises 2.1 through 2.4:

(a) Describe the population of interest.

(b) Describe how you might obtain a random sample.

(c) Give three qualitative variables that would be of interest.

(d) Give three quantitative variables that would be of interest, and classify them as either continuous or discrete.

2.1 Suppose you are interested in studying the welfare recipients of a state.

2.2 Suppose you are interested in evaluating a new treatment for heart disease.

2.3 Suppose you are interested in studying the effectiveness of computers as a learning tool for third graders.

2.4 Suppose you are interested in studying the effect of alcohol on teenage crime.

2.5 Classify the following as qualitative or quantitative. Further classify any that are quantitative as being discrete or continuous.

(a) Age of freshmen U.S. senators

(b) Faculty rank

(c) Weight of newborn babies

(d) Per capita income of residents of a state

(e) Murder rate in a major city

(f) Number of students in a classroom

(g) Brand of television set

2.6 A doctor is keeping a record of patients who are in need of surgery. For each patient she records their sex, age, marital status, occupation, type of surgery needed, and how urgently they need surgery. Classify each of these variables as qualitative or quantitative.

2.7 A university placement office keeps track of past graduates of the university. For each former student they record their sex,

major, grade point average at graduation, number of times they used the placement service, and the type of employment at graduation. Classify each of these variables as qualitative or quantitative.

2.8 Give a continuous and a discrete variable that might be of interest in a survey of city residents on the issue of gun control.

2.2   Scales of Measurement for Qualitative and Quantitative Variables

After a variable has been defined, data must be collected and analyzed. Before we analyze the data, however, we should determine the appropriate scale of measurement that applies to the variable. This will determine the amount of information contained in the data and indicate possible procedures for further analysis.

The Nominal Scale

Consider, for example, the marital status of the employees of a large company. Typically, the categories of marital status are

Single, Married, Divorced, Widowed

These are only categories and there is neither numerical quantity nor order implied. The scale of measurement for the qualitative variable marital status is called the *nominal scale.*

Definition

A **nominal scale** simply identifies categories of the variable. The various categories are called the **classes** or **levels** of the scale.

Other variables using the nominal scale are

1. Sex
2. Religious preference
3. Political party affiliation
4. Occupation

The statistical analysis of a variable measured on a nominal scale is somewhat limited. As stated previously, numerical codes can be assigned to the different levels of a qualitative variable such as sex (Male—0, Female—1), but no arithmetic analysis can be performed on those numerical codes. What is found is the percent of observations falling in the various categories. For example, the marital status of the employees of the large company mentioned previously might break down as follows:

Marital status	Code	Percent
Single	1	22%
Married	2	38%
Divorced	3	33%
Widowed	4	7%

The Ordinal Scale

As implied by the name, the *ordinal scale* is for variables whose levels can be ordered from smallest to largest. For example, there is a definite ranking of the levels of the classification of students in a university—freshman, sophomore, junior, and senior.

Definition

An **ordinal scale** includes all the properties of a nominal scale with the additional property that observations can be ranked from the smallest to the largest.

Other variables measured on an ordinal scale include the outcome of a beauty contest or the response to a questionnaire that attempts to record degrees of feeling or opinion. For example, using the ordinal scale, you can rate your teacher's performance as poor, fair, average, good, or excellent.

The Interval Scale

Even though we can rank observations on an ordinal scale, we cannot assign a distance between ranks. The *interval scale* is for variables in which it is possible to specify a numerical distance between the values.

Definition

An **interval scale** includes all the properties of an ordinal scale with the additional property that distance between observations is meaningful.

It is generally believed that IQ is measured on an interval scale; however, some will argue that IQ simply places an order on intelligence (hence is measured on the ordinal scale). In this sense, one who scores 100 on an IQ exam is perceived as more intelligent than one who scores 90, but the 10-point difference is not relevant. That is, a 10-point spread from 90 to 100 does not indicate necessarily the same difference in intelligence as a 10-point spread from, say, 100 to 110. We will leave this argument up to the psychologist.

Although distance between observations is meaningful on the interval scale, ratios are not meaningful. We can say the difference between temperatures of 40 and 50 degrees is the same as the difference between temperatures of 80 and 90 degrees, namely, 10 degrees. But we *cannot* say a temperature of 80 degrees is *twice* as hot as a temperature of 40 degrees. Thus temperature is measured on the interval scale. The key to the interval scale is that zero is not uniquely defined. For example, zero degrees does not mean the absence of temperature; in fact, zero appears at different temperatures on the two scales we use. When there is no uniquely defined zero, ratios will not be meaningful.

The Ratio Scale

The *ratio scale* is for variables that have a uniquely defined zero, and thus ratios of observations on this scale are relevant.

Definition

The **ratio scale** includes all the properties of an interval scale with the additional property that ratios of observations are meaningful.

Examples of variables measured on the ratio scale are

1. Weight
2. Height
3. Age
4. Birth rate
5. Grade point average
6. Income

As long as the variable is the result of some measurement process and the zero is uniquely defined, we can say the variable is measured on the ratio scale. A uniquely defined zero allows for ratios to make sense. When we look at a number line with zero fixed, we see, for example, that 200 is twice 100, and thus the ratio is 2. A weight of 200 pounds is twice the weight of 100 pounds because 0 pounds means the absence of weight.

We see that the amount of information conveyed by the different measurement scales is somewhat varied. The nominal scale is the least informative and the ratio scale is the most informative. It is important that we understand the difference in the four scales because some of the statistical procedures we wish to employ may apply only to one of the more informative scales. That is, some procedures have been developed for interval or ratio data and should not be applied to nominal or ordinal data. Generally, however, interval and ratio can be lumped into one scale called the *interval/ratio scale*, because both are the result of a measurement process. Any statistical procedure that applies to one also applies to the other. The only difference is that one has a uniquely defined zero and the other does not. Therefore, in the exercises we will distinguish between the three scales—nominal, ordinal, and interval/ratio. It should be pointed out that a variable measured on one scale automatically satisfies any of the less informative scales. That is, interval/ratio data can be described on an ordinal or nominal scale, and ordinal data can be described on a nominal scale. For instance, we could convert grade point average (interval/ratio data) to a letter grade of A, B, C, D, F, which is ordinal data. This will ignore some information, but sometimes it will be convenient to do so.

In regard to the distinction between qualitative and quantitative variables, we can say that qualitative variables (also called *categorical variables*) are generally measured on either a nominal or ordinal scale. Quantitative variables (also called *measurement variables*) are generally measured on an interval/ratio scale.

EXAMPLE 2.2

Identify the most informative scale of measurement that applies to the following variables.

(a) The length (in hours) of a baseball game

(b) Colors of paint in the inventory of a paint company

(c) Ranks of personnel in the military

SOLUTION

(a) A 4-hour game is twice as long as a 2-hour game; thus the most informative scale of measurement is the ratio scale. Time of any sort will always be measured on the ratio scale. Because we are combining the interval and ratio scales, our answer is interval/ratio.

(b) Color is simply an attribute. One color is not ranked above another. The nominal scale is the most informative.

(c) Rank in the military is not numerical, but it does exhibit a definite ordering; the most informative scale is the ordinal scale.

EXERCISES 2.2

2.9 Discuss the difference between qualitative and quantitative data.

2.10 Identify each of the following as either qualitative or quantitative, and then identify the most informative scale of measurement that applies.
(a) Phone number
(b) Social Security number
(c) Age
(d) Street address
(e) Gross annual income
(f) Number of years of education

2.11 Identify each of the following as either continuous or discrete variables.
(a) Number of hours children watch TV
(b) Suicide rate
(c) Number of violent crimes
(d) Housing cost
(e) Number of adults over 65
(f) Unemployment rate

2.12 Consider the following data:

1983 Auto Theft Rate (per 100,000 people)

City	Rate
Asheville, N.C.	154.1
Boston, Mass.	1,245.5
Cleveland, Ohio	919.8
Detroit, Mich.	1,324.2
Fresno, Calif.	454.3
Knoxville, Tenn.	380.4
Madison, Wis.	164.4

SOURCE: *Uniform Crime Reports for the United States, 1983*, U.S. Department of Justice, Washington, D.C.

(a) What scale of measurement is appropriate for the auto theft rate?

(b) What scale of measurement is appropriate for the city variable?

(c) Reclassify the auto theft rate as Low (less than 200), Medium (between 200 and 1,000), and High (greater than 1,000). What scale of measurement is appropriate now?

(d) Is auto theft rate a continuous or discrete variable?

2.13 A psychologist wishes to investigate the threshold reaction time (in seconds) for persons subjected to emotional stress.

(a) What scale of measurement is appropriate for threshold reaction time?

(b) Is the variable continuous or discrete?

2.14 A sociologist wishes to study the occupations and salaries of Vietnam veterans discharged prior to 1974.
(a) What scale of measurement is appropriate for occupations?
(b) What scale of measurement is appropriate for salaries?

2.3 Displaying Nominal and Ordinal Data

It is important to know how to organize and display data. We will now look at several ways of displaying data measured on a nominal or ordinal scale.

The Frequency Table

Data are often organized and displayed in the form of tables and graphs. For nominal or ordinal data the table or graph usually takes the form of a frequency count of each of the categories.

Definition

A table listing the different categories of the nominal or ordinal data and the associated frequency of occurrence is called a **frequency table.**

Suppose we ask the following question of a sample of 1,200 university students:

"Do you believe that the possession of a small amount of marijuana is a criminal offense?"

Yes_____   No_____   No opinion_____

After obtaining the responses of the 1,200 students, we tabulate the results in Table 2.2.

TABLE 2.2
A survey of 1,200 university students on the question—"Do you believe that the possession of a small amount of marijuana is a criminal offense?"

Response	Frequency
Yes	516
No	648
No opinion	36
Total	1200

Note that the *frequency* is simply a count of the number of subjects falling in the different categories.

EXAMPLE 2.3

A small company, XYZ, has 400 employees of which 88 are single, 152 are married, 132 are divorced, and 28 are widowed. Summarize these data in a frequency table.

SOLUTION   The following is a frequency table of the marital status of the employees of Company XYZ.

Marital status	Frequency
Single	88
Married	152
Divorced	132
Widowed	28
Total	400

The frequencies may be divided by the total to obtain *relative frequencies*. From Table 2.2 we have the following relative frequencies:

$$516/1200 = .43 \text{ or } 43\% \text{ answered yes}$$
$$648/1200 = .54 \text{ or } 54\% \text{ answered no}$$
$$36/1200 = .03 \text{ or } 3\% \text{ had no opinion}$$

In completed form we have Table 2.3.

TABLE 2.3

A survey of 1,200 university students on the question—"Do you believe that the possession of a small amount of marijuana is a criminal offense?"

Response	Frequency	Relative frequency
Yes	516	.43
No	648	.54
No opinion	36	.03
Total	1200	1.00

Relative frequency is important in that its value is independent of the size of the data set. This is necessary when comparing two (or more) data sets of different size.

EXAMPLE 2.4

Suppose the sample of 1,200 university students comprised 700 males and 500 females. Suppose also that of the 516 who responded yes, 294 were male; of the 648 who responded no, 385 were male; and of the 36 with no opinion, 21 were male. Summarize these data in a frequency table, including relative frequencies for both males and females.

SOLUTION

A survey of 1,200 university students on the question—"Do you believe that the possession of a small amount of marijuana is a criminal offense?"

Response	Frequency		Relative frequency	
	Male	Female	Male	Female
Yes	294	222	.420	.444
No	385	263	.550	.526
No opinion	21	15	.030	.030
Total	700	500	1.000	1.000

Although fewer females (222) than males (294) responded yes, the relative frequencies indicate that a greater percent of females (44.4%) than males (42.0%) responded yes.

The frequency table is a common way of summarizing not only nominal data but ordinal data as well.

EXAMPLE 2.5

A sample of 1,000 university students living in dormitories is surveyed in regard to their housing on campus. They were asked to rate their current housing as

1: very desirable
2: desirable
3: sufficient
4: livable
5: undesirable

The results showed that 120 chose category 1, 180 chose category 2, 360 chose category 3, 240 chose category 4, and 100 chose category 5. Summarize these data in a frequency table to include relative frequency.

SOLUTION   A relative frequency table giving student opinion on dormitory living conditions follows.

▲ The relative frequency may be stated as a decimal or as a percent. It is a simple matter of moving the decimal point to get from one to the other.

Opinion	Frequency	Relative frequency
Very desirable	120	12%
Desirable	180	18%
Sufficient	360	36%
Livable	240	24%
Undesirable	100	10%
Total	1000	100%

Thus far we have used the frequency table to display only nominal and ordinal data; however, it can be used to display interval/ratio data. We will see in the next section that interval/ratio data can be displayed in a number of ways.

The data in the above examples are fictitious, so no source was given. But if it is known, be sure to give the source of the data, as in Table 2.4, so that they may be verified if so desired.

TABLE 2.4

Estimated rate (per 100,000 persons 12 years of age or older) of personal victimization, United States, 1979

Type of victimization	Rate per 100,000
Rape and attempted rape	145
Robbery	1709
Assault	5490
Personal larceny with contact	783
Personal larceny without contact	17185

SOURCE: *Sourcebook of Criminal Justice Statistics—1981*, p. 251, U.S. Department of Justice, Bureau of Justice Statistics, Washington, D.C.

Technically speaking, the data in Table 2.4 are not exactly in the form of a frequency table because the data are given as a rate per 100,000 instead of as a frequency count; however, it is a way of displaying nominal and ordinal data of this type.

The Bar Graph

Tables of numbers are sometimes difficult to interpret. In those cases a picture might better illustrate the point. A *bar graph* presents a graphical illustration of the data from a frequency table.

Definition

A **bar graph** is a picture consisting of horizontal and vertical axes with rectangles representing the frequency (or relative frequency) of the data. The values of the variable are listed along one axis and the frequency (or relative frequency) along the other.

FIGURE 2.1

Marital status of the employees of Company XYZ

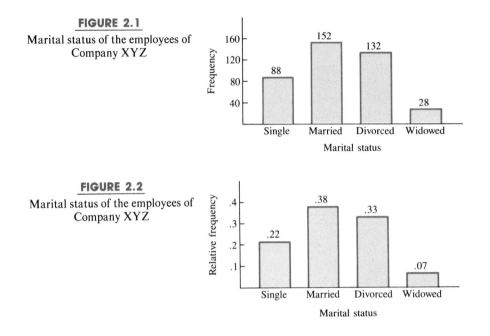

FIGURE 2.2

Marital status of the employees of Company XYZ

FIGURE 2.3

Distribution of oil in countries other than the Middle East

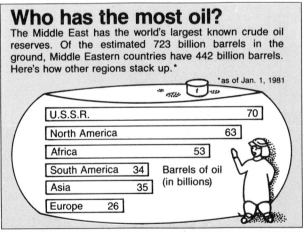

USA SNAPSHOTS

A look at statistics that shape your finances

Who has the most oil?

The Middle East has the world's largest known crude oil reserves. Of the estimated 723 billion barrels in the ground, Middle Eastern countries have 442 billion barrels. Here's how other regions stack up.*

*as of Jan. 1, 1981

U.S.S.R. 70
North America 63
Africa 53
South America 34
Asia 35
Europe 26

Barrels of oil (in billions)

by Elys A. McLean, USA TODAY

SOURCE: *USA Today*, November 26, 1984. Copyright, 1984 *USA Today*. Reprinted with permission. Data from U.S. Department of the Interior Geological Survey.

Figure 2.1 is a bar graph showing the frequency associated with the marital status of the employees of company XYZ given in Example 2.3. The height of the bars can be measured in either frequency or relative frequency, as in Figure 2.2.

Observe that the frequency (or relative frequency) is listed within each bar; this is a practice that helps to describe the data. The bars of the bar graph can be either vertical or horizontal (as in Figure 2.3).

▲ The bars should always have the same width so that the difference in frequencies is reflected only in the height of the bars.

Pie Chart

The pie chart is another useful way to exhibit nominal and ordinal data. The pie is divided into pieces corresponding to the categories of the variable, so that the size (angle) of the slice is proportional to the relative frequency of the category. Normally the relative frequency (as a percent) is shown in each slice.

EXAMPLE 2.6

Figure 2.4 shows pie charts comparing the national budget for 1962 and 1982.

FIGURE 2.4

Comparison of the national budget (figures in billions of dollars)

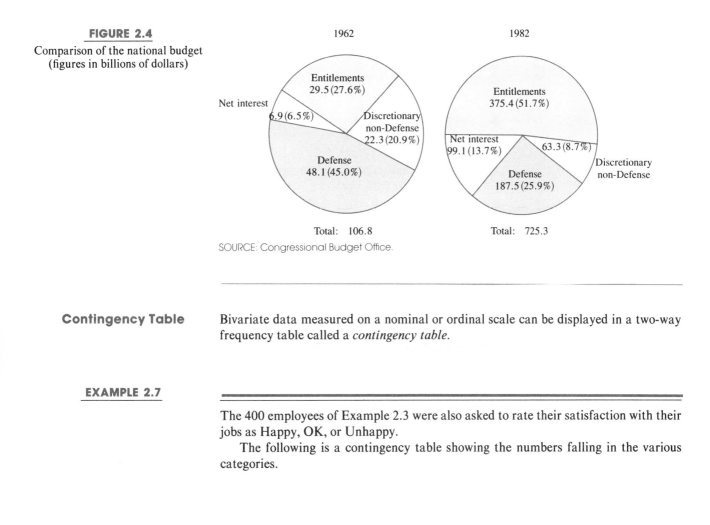

SOURCE: Congressional Budget Office.

Contingency Table

Bivariate data measured on a nominal or ordinal scale can be displayed in a two-way frequency table called a *contingency table*.

EXAMPLE 2.7

The 400 employees of Example 2.3 were also asked to rate their satisfaction with their jobs as Happy, OK, or Unhappy.

The following is a contingency table showing the numbers falling in the various categories.

Job satisfaction	Single	Married	Divorced	Widowed	Total
		Marital status			
Happy	46	46	28	5	125
OK	30	64	42	18	154
Unhappy	12	42	62	5	121
Total	88	152	132	28	400

▲ Observe that one variable (marital status) is nominal and the other (job satisfaction) is ordinal. It is not uncommon to have the various measurement scales mixed when bivariate data are exhibited.

Various questions can be asked about this bivariate data. For example,

1. What percent of the married are happy with their jobs? Answer: $46/152 = .303 = 30.3\%$.

2. What percent of those who are happy with their jobs are married? Answer: $46/125 = .368 = 36.8\%$.

3. What percent of the employees are married and happy with their jobs? Answer: $46/400 = .115 = 11.5\%$.

4. Which of the four marital groups is most happy with its jobs? Answer: Single—$46/88 = .523$; Married—$46/152 = .303$; Divorced—$28/132 = .212$; Widowed—$5/28 = .179$. Thus, the Single group has the highest percent happy with their jobs.

The data in a contingency table will be statistically analyzed in Chapter 12. Bivariate data can also be displayed in a bar graph.

EXAMPLE 2.8

A local election is to be held to determine whether the sale of alcoholic beverages within the city limits should be legalized. A survey of city residents yielded the results given in the following table.

Should the sale of alcoholic beverages within the city limits be legalized?

	Yes	No	No opinion	Total
Male	55	16	4	75
Female	53	26	9	88
Total	108	42	13	163

From the contingency table, display the data in a bar graph.

SOLUTION   See Figure 2.5.

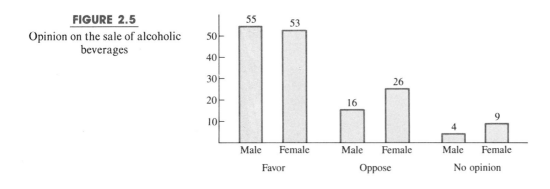

FIGURE 2.5

Opinion on the sale of alcoholic beverages

EXERCISES 2.3

2.15 In a random sample of college students, it was found that 124 had blue eyes, 150 had brown eyes, 15 had green eyes, and 103 had hazel eyes. Display this data in a frequency table, including relative frequencies.

2.16 Draw a bar graph of the data in Exercise 2.15.

2.17 Construct a pie chart for the data in Exercise 2.15.

2.18 A sample of 120 employed people is classified according to sex and occupation group. There were 48 women of which 9 were blue-collar workers, 23 were white-collar workers, 4 were farm workers, and the rest were classified as "other." Of the men, 31 were blue-collar workers, 22 were white-collar workers, 15 were farm workers, and the rest were classified as "other." Arrange these data in a bivariate table.

(a) What percent are women?

(b) What percent of the men are blue-collar workers?

(c) What percent are female blue-collar workers?

2.19 Draw a bar graph depicting the frequency of men and women for each occupation group for the data in Exercise 2.18.

2.20 The data in the table are from a poll by George Gallup.

(a) Using only the levels of the sex variable, draw a relative frequency bar graph illustrating the opinions on the issue of the possession of marijuana.

(b) Using only the education level, draw a relative frequency bar graph showing the opinions on the issue of the possession of marijuana.

(c) Using only the age level, draw a relative frequency bar graph illustrating the opinions on the issue of the possession of marijuana.

TABLE FOR EXERCISE 2.20

Attitude toward the treatment of possession of small amounts of marijuana as a criminal offense, United States, 1980.
Question: "Do you think the possession of small amounts of marijuana should or should not be treated as a criminal offense?"

	Should be treated as a criminal offense	Should not be treated as a criminal offense	No opinion
National	43%	52%	5%
Sex			
Male	42%	53%	5%
Female	44%	51%	5%
Education			
College	30%	67%	3%
High school	45%	50%	5%
Grade school	58%	33%	9%
Age			
18–24	27%	67%	6%
25–29	26%	70%	4%
30–49	45%	52%	3%
50–older	54%	39%	7%
Religion			
Protestant	49%	47%	4%
Catholic	39%	55%	6%

SOURCE: George H. Gallup, *The Gallup Opinion Index Report No. 179*, Princeton, N.J.: The Gallup Poll, July 1980, p. 15. Used with permission.

2.21 One hundred psychology majors were classified according to sex and class level. Ten were lower division women, 20 were upper division women, 40 were lower division men, and 30 were upper division men. Arrange the data in a bivariate table.

2.22 The SAT scores for 900 female and 800 male college freshmen were classified as being low, medium, or high as follows:

Range	Sex	
	Men	Women
High	190	250
Med.	430	520
Low	180	130
Total	800	900

(a) Draw a bar graph illustrating the frequency of the SAT scores for males.

(b) Draw a bar graph illustrating the frequency of the SAT scores for females.

(c) Can these two bar graphs be compared as drawn?

2.23 A survey of 1,000 males and 1,000 females revealed that 100 of the males and 500 of the females had no job. Further, 650 of the males and 400 of the females had one job. The rest had more than one job.

(a) Arrange this information in a bivariate table.

(b) Construct an employment bar graph for the males.

(c) Construct an employment bar graph for the females.

(d) Can these two bar graphs be compared as drawn?

2.24 A survey of 500 students, 100 faculty members, and 30 administrators revealed that 360 students oppose the present parking policy, 60 faculty oppose it, and only 5 administrators oppose it. Arrange the data in a bivariate table by attitude (favor, oppose) and status (student, faculty, administrator).

(a) What percent of those sampled oppose the present parking policy?

(b) What percent of the students oppose the policy?

(c) What percent are faculty and oppose the policy?

(d) Give the appropriate bar graphs in order to compare the opinions of students, faculty, and administrators.

2.25 In 1979 there were 587 murders reported in North Carolina (a 1.18% decrease from 1978). Of the 587 murders, 422 were by firearms, 89 by knife, 30 by a blunt object, 22 by hands, fist, or feet, and 24 by other means. Arrange these data in a frequency table to include relative frequencies.

(a) What was the weapon used most often and what percent of all murders were committed with that weapon?

(b) What percent of all murders were committed by a weapon other than a firearm?

2.4 Displaying Interval/Ratio Data

In the previous section methods of displaying data measured on a nominal or ordinal scale were presented. We now look at ways of exhibiting data measured on the interval/ratio scale.

A First Look at the Stem and Leaf Plot

The **stem and leaf plot** is a quick and easy way to display quantitative data measured on the interval/ratio scale. It is extremely useful for arranging the observations from smallest to largest so that specific locations within the data set can be found. To illustrate the construction of a stem and leaf plot, consider Example 2.9.

EXAMPLE 2.9

A test is given to a class of 20 students in a beginning typing class. Their scores were:

84	59	82	78	74	96	44	76	85	66
77	91	62	54	72	65	84	38	76	70

Construct a stem and leaf plot of the scores.

SOLUTION By observation we see that the scores are two-digit numbers and range from the thirties to the nineties. To construct the display each observation will be divided into a stem and a leaf. In this example, the digits in the tens place of the numbers will become the stems, and the digits in the units place will become the leaves of the stem and leaf plot. A vertical line is drawn to separate the stems from the leaves.

Figure 2.6(a) shows only the stem and the first observation (84) plotted. Note that only the 4 of 84 is plotted on the 8 stem. Figure 2.6(b) shows the completed stem and leaf plot after all 20 leaves (corresponding to the 20 observations) are properly plotted.

FIGURE 2.6

```
3 |                    3 | 8
4 |                    4 | 4
5 |                    5 | 9 4
6 |                    6 | 6 2 5
7 |                    7 | 8 4 6 7 2 6 0
8 | 4                  8 | 4 2 5 4
9 |                    9 | 6 1

    (a)                    (b)
```

Rotating the stem and leaf 90 degrees counterclockwise as in Figure 2.7, we see that the leaves form a graphical picture, which shows how the test scores are *distributed* along the stem values.

FIGURE 2.7

```
                    0
                    6
                    2
                    7 4
                5 6 5
              4 2 4 2 1
          8 4 9 6 8 4 6
          ─────────────
          3 4 5 6 7 8 9
```

Frequency Histogram

The **histogram** is a graphical alternative for presenting quantitative data. It can be constructed from the stem and leaf plot. Each stem of a stem and leaf plot defines an interval of values called a *class*. The *class limits* are the smallest and largest possible values for the leaves of that stem. Thus the class limits for the stem value 3 are 30 and 39, and for stem value 4 are 40 and 49, and so on. Once the class limits are determined, the data can be presented in a *grouped frequency table*(grouped in the sense that the data have been grouped into the various classes), such as the one for Example 2.9 given in Table 2.5.

TABLE 2.5

Test scores for a beginning typing class

Class limits	Class boundaries	Frequency	Relative frequency
30–39	29.5–39.5	1	1/20
40–49	39.5–49.5	1	1/20
50–59	49.5–59.5	2	2/20
60–69	59.5–69.5	3	3/20
70–79	69.5–79.5	7	7/20
80–89	79.5–89.5	4	4/20
90–99	89.5–99.5	2	2/20
		Total 20	

The *class boundaries* are obtained by lowering the lower class limits by .5 and raising the upper class limits by .5; thus, class limits 30 and 39 become class boundaries 29.5 and 39.5, respectively. The class boundaries are given so that the classes are continuous; that is, the first class runs from 29.5 to 39.5 and immediately the second class picks up at 39.5 and runs to 49.5, then the third class starts at 49.5, and so on.

The histogram can now be constructed from the grouped frequency table. The class boundaries are scaled off on the horizontal axis. Bars are constructed over each class boundary so that their height is given by the frequency of the class, which is marked off on the vertical axis. Figure 2.8 is the completed histogram for the data of Example 2.9.

FIGURE 2.8

Test scores for a beginning typing class

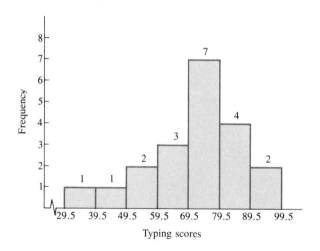

The histogram, also called a *frequency distribution,* gives a picture of the distribution of the data. It is very similar to the bar graph presented in Section 2.3 for nominal and ordinal data. The basic difference is that the histogram is for interval/ratio data, which are continuous and, consequently, the bars are joined at the class boundaries.

Relative Frequency Histogram

Just as in the case of the bar graph, the vertical axis of the histogram can be scaled in relative frequency as depicted in Figure 2.9.

The *relative frequency histogram* is important for two reasons. First, it removes the dependence of the frequency distribution on the size of the data set. This is important

FIGURE 2.9

Test scores for a beginning
typing class

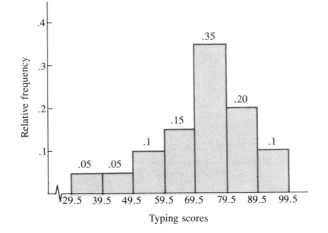

when we compare two or more frequency distributions. For example, do women score higher than men in introductory typing classes? If considerably more women take typing courses than men, it would not be realistic to compare their frequencies. On the other hand, the relative frequencies of the two groups are independent of their sizes, and thus a fair comparison can be made.

The relative frequency distribution can also be used in a *probability* sense. From Table 2.5 (or Figure 2.9) we see that the relative frequency of the class with boundaries from 79.5 to 89.5 is 4/20. Thus the percent of students who scored somewhere between 79.5 and 89.5 is 4/20 = .2 = 20%. The percent that scored somewhere between 69.5 and 89.5 is 7/20 + 4/20 = 11/20 = .55 = 55%. Later we will see that the relative frequency distribution obtained from a sample is used to approximate the distribution of the entire population of scores. If the relative frequency distribution given in Figure 2.9 accurately describes the population distribution, we could say that 55% of the population would score somewhere between 69.5 and 89.5. Alternatively we could say that the chance or probability of a randomly selected individual scoring between 69.5 and 89.5 is 55%.

As previously mentioned, the stem and leaf plot also gives a picture of the frequency distribution. Moreover, the stem and leaf plot has preserved the data and is quickly constructed. This does not mean that we should not use a histogram. The histogram is certainly appropriate when presenting the data in a completed report or when the data set is very large. But for preliminary analysis and exploratory analysis we should use the stem and leaf plot. We might say that the histogram is the formal presentation of the data, and the stem and leaf plot is the informal presentation of the data.

Frequency Polygon

The **frequency polygon** is another useful tool in describing interval/ratio data. It is a graph of connected line segments that correspond to the frequencies of the various classes. The histogram can be easily converted to a frequency polygon by connecting the midpoints of the tops of the bars by straight line segments. On the end bars the midpoints are connected down to the horizontal axis to the midpoints of the classes that would have been adjacent to those end classes. Figure 2.10 is a frequency polygon constructed from the histogram given in Figure 2.8.

▲ To construct a frequency polygon it is *not* necessary to construct a histogram first.

FIGURE 2.10

Test scores for a beginning
typing class

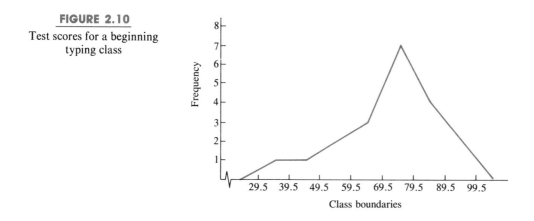

The frequency polygon is useful in comparing frequency distributions. For example, to compare test scores for males and females, frequency polygons for both males and females could be graphed on the same coordinate axis system.

Frequency Curve

When the number of observations is large, a frequency polygon would take on the appearance of a continuous curve similar to what might be expected if the entire population of values were graphed. Usually the entire population is not available; however, a stem and leaf plot, histogram, or frequency polygon obtained from a representative sample should closely approximate the shape of the *population frequency curve*. The frequency polygon in Figure 2.10 might suggest the population frequency curve in Figure 2.11.

FIGURE 2.11

Population scores for a beginning
typing class

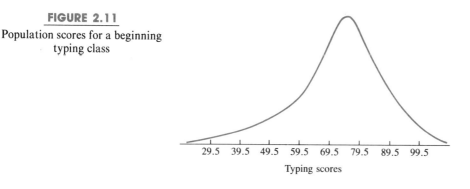

The distributional shape of a population is described by its frequency curve. In Chapter 6 we will discuss some of the distributional shapes commonly encountered in the study of statistics.

Scatterplot

We saw in Example 2.6 that the contingency table can be used to display bivariate data where each of the two variables is measured on the nominal and/or ordinal scales. The joint display of bivariate data in which the variables are interval/ratio is accomplished with a *scatterplot*, which is illustrated in Example 2.10.

EXAMPLE 2.10

It has been suggested that there is a relationship between sleep deprivation and the ability to solve simple tasks. To evaluate this hypothesis, 12 subjects were asked to solve simple tasks after having gone without sleep for 15, 18, 21, and 24 hours. The number of tasks completed in 10 minutes after having been ·deprived of sleep was recorded for the 12 subjects, three at each of the four deprivation levels. The data are recorded as follows:

Subject		1	2	3	4	5	6	7	8	9	10	11	12
No. hours without sleep		15	15	15	18	18	18	21	21	21	24	24	24
No. tasks completed		13	9	15	8	12	10	5	8	7	3	5	4

To construct a scatterplot, a rectangular coordinate system is drawn with the number of hours without sleep recorded on the horizontal scale (*x*-axis) and the number of tasks completed on the vertical scale (*y*-axis). Then each of the pairs of observations is plotted in the *x-y* plane. Figure 2.12 is a scatterplot of the preceding data.

FIGURE 2.12

Scatterplot for data in
Example 2.10

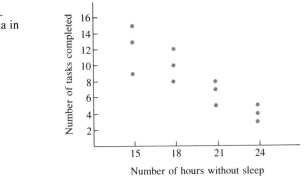

The scatterplot is useful in detecting trends in the data or in revealing an association between the two variables of the bivariate data. It appears from the preceding data that there is a downward linear (straight line) trend between sleep deprivation and the number of tasks performed. In Chapter 3 we will measure the association between the variables, and in Chapter 11 we will find the equation of the straight line and evaluate the association.

A variation of the scatterplot can be used to display bivariate data in which one variable is quantitative and the other is qualitative. This is illustrated in Example 2.11.

EXAMPLE 2.11

At the end of a semester, a university professor asked her students to rate her teaching on a scale from 0 to 10. She also asked them to record their present course grade as A, B, C, D, or F. Graph the data that were obtained:

Course grade	B	C	C	A	B	D	C	C	C	F	A	C	D	B	B	F
Rating	8	6	9	10	9	3	8	10	5	1	9	8	7	6	10	4

SOLUTION   To graph this data, a coordinate system is drawn with the course grades listed on the horizontal scale and the ratings on the vertical scale (see Figure 2.13). As in the scatterplot, the data are plotted by placing a dot at the intersection of the grade and the rating for each of the 16 students.

FIGURE 2.13

▲ The numeral 2 in the scatterplot means that two data points fall in the same spot.

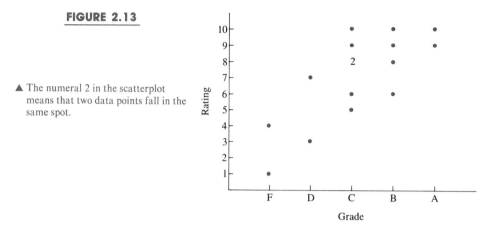

Note that, as might be expected, there is an upward trend—the higher the grade the higher the rating. To use this technique of graphing, it is necessary that the qualitative variable be measured on an ordinal scale.

EXERCISES 2.4

2.26 Construct a stem and leaf plot of the following data.

24	31	54	62	36
28	37	55	18	27
58	32	37	41	55
39	56	42	29	35

2.27 Construct a frequency histogram from the stem and leaf plot found in Exercise 2.26.

2.28 Construct a frequency polygon of the following data.

87	79	94	60	75	94
77	83	68	74	82	73
63	75	77	83	92	57
64	53	53	82	73	90
55	68	72	88	65	78

2.29 A random sample of 24 high school seniors was given a college entrance exam, which has a maximum score of 100. Construct a stem and leaf plot of the following scores.

64	75	81	43	69	75
86	58	63	66	82	62
79	91	83	55	68	74
48	66	84	77	73	59

2.30 Construct a frequency histogram of the data in Exercise 2.29.

2.31 Construct a frequency polygon of the data in Exercise 2.29.

2.32 To estimate the number of trees on a tree farm, a farmer divided the farm into 1,000 small grids. He then randomly selected 20 grids and counted the number of trees with the results that follow. Construct a stem and leaf plot and a frequency histogram of the data.

81	96	87	83	99
64	77	63	93	84
102	68	94	81	70
84	92	109	74	86

2.33 Construct a relative frequency histogram from the following stem and leaf plot.

2		3 4 8
3		2 4 3 8 4 7
4		1 8 9 2 6 7 3 0
5		4 5 2 7 5
6		7 3 0 8
7		5 2 7
8		3

2.34 A study of the association between reading ability and IQ scores was conducted by a reading coordinator in a large public school system. A random sample of 14 eighth grade students was given a reading achievement test and an IQ test. The scores are recorded as follows:

Reading score	42	35	61	28	48	46	59
IQ score	105	110	122	92	112	100	120
Reading score	21	47	29	65	37	35	53
IQ score	85	125	96	130	90	107	120

Organize these data in a scatterplot.

2.35 In order to understand some of the problems facing the American farmer, construct a scatterplot of the following data corresponding to the price of a bushel of wheat and the national farm debt recorded in billions of dollars.

Year	1980	1981	1982	1983	1984
Wheat price	$3.91	3.65	3.55	3.54	3.36
Farm debt	$165.8	182.0	201.7	216.3	214.7

2.36 Construct a scatterplot of the following data corresponding to private pay increase for salaried employees and the inflation rate.

Year	1976	1977	1978	1979	1980	1981
Raise	8.1%	7.9	8.1	7.7	9.2	9.8
Inflation	4.6%	6.5	9.0	13.2	12.2	8.8
Year	1982	1983	1984			
Raise	9.0	6.9	6.6			
Inflation	3.9	3.8	4.0			

2.5 Other Graphs and Descriptive Techniques (optional)

Time Series

Business and economic data, such as the daily Dow Jones Industrial Average or the quarterly earnings of a company, are often observed at regular time intervals. Any set of data recorded in time intervals is called *time series* data. The data can be presented graphically in a number of ways. Figure 2.14 is a bar graph of the time series corresponding to the world population from 1950 through 1980 with estimates for 1990 and 2000.

FIGURE 2.14

Bar graph of time series for the world population from 1950 through 1980

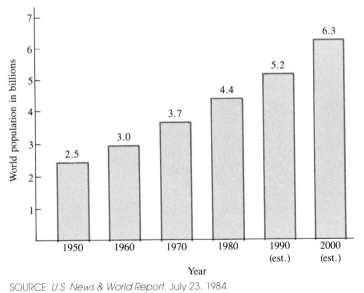

SOURCE: *U.S. News & World Report*, July 23, 1984.

FIGURE 2.15

USA SNAPSHOTS

A look at statistics that shape the nation

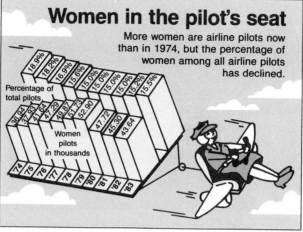

By John Sherlock, USA TODAY

SOURCE: *USA Today*, August 6, 1984. Copyright, 1984 *USA Today*.
Reprinted with permission. Data from Aircraft Owners and Pilots
Association.

Figures 2.15 and 2.16 are also graphs of time series data, which are presented in an eye-catching manner. This is acceptable (if done correctly) and seems to be the trend today.

The graphs are informative because they give the actual numerical values, but it does take time to understand the graph. The text in Figure 2.16 points out that women are gaining slightly on men, in that the women's average wage was 65.6% of men's average wage in 1983 as opposed to 62.4% in 1979. The graph does not show this,

FIGURE 2.16

USA SNAPSHOTS

A look at statistics that shape the nation

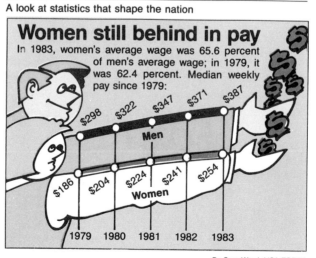

By Sam Ward, USA TODAY

SOURCE: *USA Today*, July 30, 1984. Copyright, 1984 *USA Today*.
Reprinted with permission. Data from U.S. Department of Labor.

however, because no baseline is given. In fact, the difference in pay grows *larger* each year. In 1979 the difference in median weekly pay (median will be defined in Chapter 3) was $112, whereas in 1983 it was $133. Without giving the baseline it appears that women are dropping further behind. The graph would be more descriptive if the data were presented in bar graph form as in Figure 2.17, with the baseline being 0 dollars at the horizontal scale.

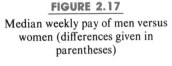

FIGURE 2.17

Median weekly pay of men versus women (differences given in parentheses)

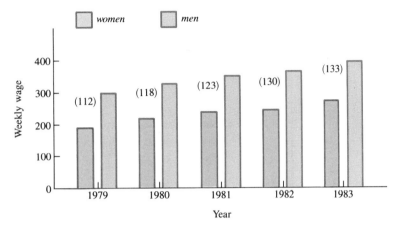

▲ A baseline is usually needed when you are comparing two or more graphs on the same coordinate system.

Comparing the bars for men and women, we see that in 1983 the women's wages are a greater percent of men's wages than they were in 1979.

Time series data are frequently presented in the form of a curve as in Figure 2.18. This is preferable to a bar graph when the time increments are close together as they are here.

Note that this graph does not begin at 0 on the vertical axis. This is acceptable here, because giving the rest of the graph (from 0 to 800) would not be informative and would simply take up space.

FIGURE 2.18

Dow Jones Industrial Average

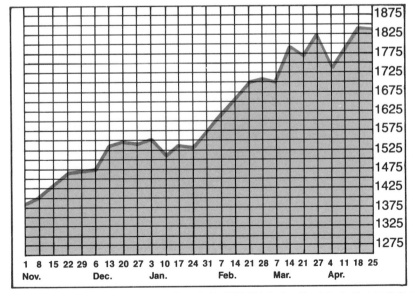

SOURCE: *USA Today*, April 28, 1986. Copyright, 1986 *USA Today*. Reprinted with permission.

FIGURE 2.19

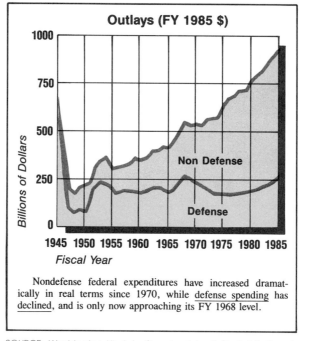

Nondefense federal expenditures have increased dramatically in real terms since 1970, while defense spending has declined, and is only now approaching its FY 1968 level.

SOURCE: *Washington Update*, Senator John P. East, U.S. Senate, 716 Hart Building, Washington, D.C. Data from U.S. Department of Defense.

Frequently multiple time series are graphed on the same coordinate system. Figure 2.19 compares the federal expenditures for defense and nondefense from 1945 through 1985.

Figure 2.20 is an interesting time series. Can you explain the decline in 20-game winners?

FIGURE 2.20

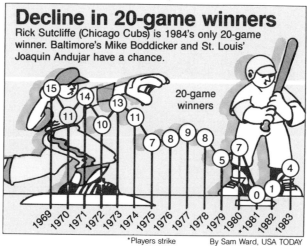

SOURCE: *USA Today*, September 28, 1984. Copyright, 1984 *USA Today*. Reprinted with permission. Data from *The Sporting News*.

The computer age has greatly enhanced the graphic presentation of data. Computers can easily graph data that previously had to be done by hand.

The pie chart in Figure 2.21 was produced by the GCHART procedure of SAS/GRAPH® software.*

The bar graph can also be produced by the GCHART procedure in a simple fashion as in Figure 2.22 (a) or a more enhanced version as in Figure 2.22 (b).

---

*SAS/GRAPH is a registered trademark of SAS Institute Inc., Cary, N.C., USA.

FIGURE 2.21

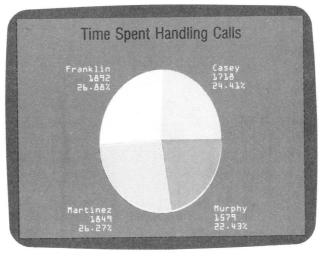

SOURCE: *SAS/GRAPH User's Guide,* Version 5 Edition, p. 215. Cary, N.C.: SAS Institute Inc., 1985. Used with permission.

FIGURE 2.22

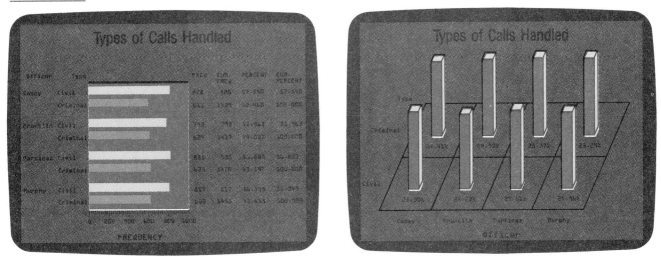

(a)   (b)

SOURCE: *SAS/GRAPH User's Guide,* Version 5 Edition, p. 208. Cary, N.C.: SAS Institute Inc., 1985. Used with permission.

The GMAP procedure of SAS/GRAPH software can produce interesting statistical maps as in Figure 2.23.

FIGURE 2.23

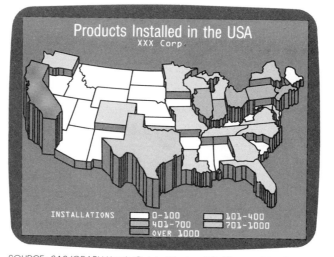

SOURCE: *SAS/GRAPH User's Guide*, Version 5 Edition, p. 246. Cary, N.C.: SAS Institute Inc., 1985. Used with permission.

Programs have been written specifically to display stock market data as is presented by the Dow Jones Market Analyzer in Figure 2.24.

It seems that there is no limit to the various ways of exhibiting data graphically. Clearly, the computer has revolutionized the art of graphical data display.

FIGURE 2.24

Stock—IBM

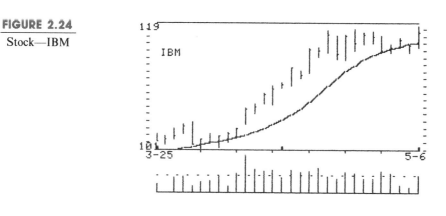

DAILY QUOTES WITH MOVING AVERAGE

SOURCE: The Dow Jones Market Analyzer, Dow Jones & Company/RTR Software, Inc., El Paso, Texas. Used with permission.

2.37 Following are the daily highs, lows, and closing prices of the Dow Jones Industrial Average for a 2-week period ending on February 12, 1985.

Date	Wed.	Thu.	Fri.	Mon.	Tue.
High	1305	1293	1286	1295	1301
Close	1288	1287	1278	1290	1285
Low	1279	1273	1270	1269	1279
Date	Wed.	Thu.	Fri.	Mon.	Tue.
High	1295	1297	1297	1292	1283
Close	1281	1290	1290	1276	1277
Low	1275	1280	1281	1269	1266

Construct time series graphs for all three prices (high, low, close) on the same coordinate system similar to Figure 2.19.

2.38 Divorce rate may be recorded as the number of divorces for every 1,000 people. According to the 1985 *World Almanac and Book of Facts* (New York: Newspaper Enterprise Association, Inc.), in 1960 the rate was 2.2, in 1970 it was 3.5, in 1980 it was 5.2, and in 1983 it was 5.0. Illustrate these time series data in a bar graph.

2.39 The following table gives the median age of men and women at their first marriage.

Year	1970	1971	1972	1973	1974
Median age men	23.2	23.0	23.3	23.2	23.1
Median age women	20.8	20.9	20.9	21.0	21.2
Year	1975	1976	1977	1978	1979
Median age men	23.5	23.8	24.0	24.2	24.4
Median age women	21.1	21.3	21.5	21.8	22.1

Year	1980	1981	1982	1983
Median age men	24.7	24.9	25.2	25.4
Median age women	22.0	22.4	22.5	22.8

SOURCE: U.S. Bureau of the Census.

Compare these time series by graphing both on the same coordinate system.

2.40 The number of heroin addicts in Great Britain has increased greatly since 1981. The increase is attributed to the ease with which drugs enter the country. Following is the number of heroin addicts getting approved treatment in Britain. The actual number of addicts is much higher, estimated by some officials to be well over 100,000.

Year	1979	1980	1981	1982	1983	1984
No. of addicts treated	1,200	1,300	1,300	2,200	3,700	4,300

SOURCE: British Home Office, and *USA Today*, September 21, 1984.

Illustrate the data in a time series graph.

2.41 The *g*ross *n*ational *p*roduct (GNP) is a measure of the national productivity. Following is the annual growth of the GNP from one quarter to the next for the period from fourth quarter 1982 through third quarter 1984.

Year	1982	1983				1984		
Quarter	4	1	2	3	4	1	2	3
GNP annual growth	.8	3.7	9.3	6.8	6.0	10.0	7.1	3.6

SOURCE: *USA Today*, September 21, 1984.

Illustrate these time series data with a bar graph.

2.6 Computer Session (optional)

In Chapter 1 we were introduced to the Minitab computer package. There we learned how to use the READ and SET commands to enter data into the computer. We also learned how to use the PRINT command to print out the data in any specified column. Finally we learned that the STOP command is necessary when we complete our session.

Other Minitab commands that relate to the material in this chapter are: HISTOGRAM, STEM-AND-LEAF, and PLOT.

HISTOGRAM

Once data are stored in a column of the Minitab worksheet, the HISTOGRAM command can be given to print out a histogram of the data. The form of the command is

```
HISTOGRAM OF DATA IN C, C,...,C
```

A separate histogram is constructed for the data in each of the specified columns. Minitab chooses intervals (classes) with rounded midpoints, then prints the midpoint of each interval, the number of observations in the interval, and a graph. The graph, unlike the histogram we learned to draw, prints a * for each frequency count. The output of HISTOGRAM may be easily transformed by hand into the histogram of rectangles, with which we are familiar.

EXAMPLE 2.12

Following are the test grades in a beginning statistics class:

76	73	81	65	83	90	77	60	67	76
84	72	57	64	71	78	87	76	92	75
58	65	79	86	95	81	71	68	74	

Construct a histogram of the test scores.

SOLUTION First the scores are read into the computer with the SET command, and then the command HISTOGRAM is given. The printout (Figure 2.25) gives the results.

FIGURE 2.25

Minitab output for Example 2.12

```
MTB > SET SCORES IN C1
DATA> 76 73 81 65 83 90 77 60 67 76 84 72 57 64
DATA> 71 78 87 76 92 75 58 65 79 86 95 81 71 68 74
DATA> END

MTB > HISTOGRAM C1

Histogram of C1   N = 29

Midpoint    Count
      55       1    *
      60       2    **
      65       4    ****
      70       4    ****
      75       7    *******
      80       4    ****
      85       4    ****
      90       2    **
      95       1    *
```

You may choose your own class intervals for the histogram by specifying the subcommands INCREMENT and START.

```
HISTOGRAM OF C,C,...,C;
  INCREMENT = K; (K = width of each interval)
  START with midpoint at K.
```

Note: If a command has subcommands, then the command and all subcommands except the last one must end in a semicolon. The final subcommand ends in a period.

To specify the class intervals for the data in Example 2.12, you might try the following command:

```
HISTOGRAM C1;
  INCREMENT =10;
  START = 54.5.
```

STEM-AND-LEAF

This command is used in the same way as HISTOGRAM. The end result, however, is a stem and leaf plot of the data. The command is

```
STEM-AND-LEAF OF DATA IN C, C,...,C
```

A stem and leaf plot is constructed for the data in each of the specified columns. The first column of the constructed stem and leaf plot is a cumulative frequency count of the data, until the line containing the middle observation is reached. The line containing the middle observation has a frequency count just for that line, and the lines below that have a cumulative frequency count of the data from the bottom of the display.

EXAMPLE 2.13

Construct a stem and leaf plot of the data given in Example 2.12.

SOLUTION   Assuming the data have been read into column 1 of the worksheet, we give the command STEM-AND-LEAF OF DATA IN C1 (Figure 2.26).

FIGURE 2.26
Minitab output for Example 2.13

```
MTB > STEM-AND-LEAF OF DATA IN C1;
SUBC> INCREMENT 10.

Stem-and-leaf of C1        N  = 29
Leaf Unit = 1.0

     2    5 78
     8    6 045578
   (12)   7 112345666789
     9    8 113467
     3    9 025
```

PLOT

The following command gives a scatterplot of the data:

```
PLOT C VERSUS C
```

The first column specified is put on the vertical axis, and the second column is put on the horizontal axis. A ∗ is plotted at each point of the scatterplot, unless more than one observation falls at that point; in that case, a count is plotted. Minitab automatically chooses a scale; however, you may specify your own scale with the following subcommands:

```
XINCREMENT = K;
XSTART AT K [GO TO K];
YINCREMENT = K;
YSTART AT K [GO TO K].
```

Time series data can be plotted with the PLOT command by putting the time increments in one column and the series data in the other column.

FIGURE 2.27

Minitab output for Example 2.14

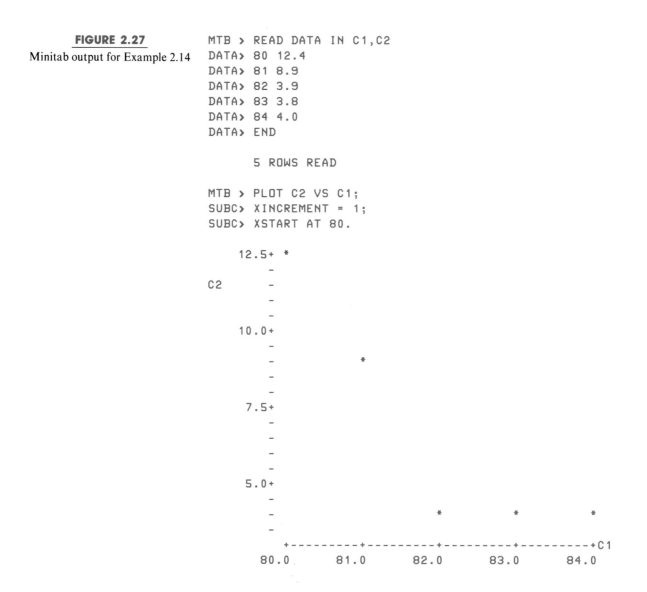

```
MTB > READ DATA IN C1,C2
DATA> 80 12.4
DATA> 81 8.9
DATA> 82 3.9
DATA> 83 3.8
DATA> 84 4.0
DATA> END

      5 ROWS READ

MTB > PLOT C2 VS C1;
SUBC> XINCREMENT = 1;
SUBC> XSTART AT 80.


        12.5+  *
            -
 C2         -
            -
            -
        10.0+
            -
            -              *
            -
            -
         7.5+
            -
            -
            -
            -
         5.0+
            -
            -              *        *        *
            -
            +---------+---------+---------+---------+C1
          80.0      81.0      82.0      83.0      84.0
```

EXAMPLE 2.14

Following is the Consumer Price Index for 1980 through 1984:

Year	1980	1981	1982	1983	1984
CPI	12.4%	8.9%	3.9%	3.8%	4.0%

SOURCE: U.S. Treasury Department.

Graph the time series with the PLOT command.

SOLUTION   See Figure 2.27 on the preceding page.

EXERCISES 2.6

2.42 Following is the Gross National Product annual growth for the years 1980 through 1984:

Year	1980	1981	1982	1983	1984
GNP annual growth	−0.3%	2.5%	−2.1%	3.7%	6.8%

SOURCE: U.S. Treasury Department.

Graph the data with the PLOT command.

2.43 From the following data, using the PLOT command, compare the farm debt with the price the farmer gets for a bushel of wheat.

Year	1980	1981	1982	1983	1984
Debt (in billions)	$165.8	182.0	201.7	216.3	214.7
Price bushel wht.	$3.91	3.65	3.55	3.54	3.36

SOURCE: *USA Today*, January 24, 1985, page 1A.

2.44 Analyze the salary data for executives (given in Table A1 of Appendix A) by using the HISTOGRAM and STEM-AND-LEAF commands.

(a) Set the salary in C1.

(b) Name C1 'SALARY'.

(c) Give a histogram of C1 and comment on the shape of the distribution.

(d) Give a stem and leaf plot of C1.

2.45 Using the tuition data in the Statistical Insight problem at the beginning of the chapter, complete the following:

(a) Read the tuition per student in C1 and the appropriations per student in C2.

(b) Name C1 and C2.

(c) Give stem and leaf plots of C1 and C2.

(d) Plot C1 versus C2.

(e) Is there a trend-line relationship beween tuition and appropriations?

2.7 Summary and Review

Key Concepts

☑ A *variable* is a quantity measured on each subject. When two variables are measured on each subject, the data set is called *bivariate*. When several variables are measured on each subject, the data set is called *multivariate*. Variables are classified as either *quantitative* or *qualitative*. Quantitative variables are classified as *discrete* or *continuous*. Four scales of measurement apply to variables—the *nominal scale*, the *ordinal scale*, the *interval scale*, and the *ratio scale*. The ratio scale is the most informative and the nominal scale is the least informative.

☑ Nominal and ordinal data can be listed in a table called a *frequency table*. The *relative frequency* for a category is obtained by dividing the frequency of the category by the total frequency. Bivariate data in which both variables are qualitative is displayed in a *contingency table*. Quantitative bivariate data can be displayed in a *scatterplot*. The frequency table for nominal or ordinal data can be graphed in a *bar graph* or *pie chart*.

☑ Interval and ratio data can be displayed in a *stem and leaf plot*. The data in a stem and leaf plot can be easily transformed into a *histogram* or a *frequency polygon*. The advantage of the stem and leaf plot is that the data are left intact.

☑ Data recorded in time are called *time series* data.

☑ A variety of graphs and charts can be produced by the computer.

Learning Goals

Having completed this chapter you should be able to:

1. Distinguish between a quantitative and qualitative variable.   *Section 2.1*

2. Distinguish between a discrete and continuous variable.   *Section 2.1*

3. Identify the appropriate scale of measurement that applies to a given variable.   *Section 2.2*

4. Present data in the form of a frequency table.   *Section 2.3*

5. Construct bar graphs.   *Section 2.3*

6. Organize data in a contingency table.   *Section 2.3*

7. Construct stem and leaf plots.   *Section 2.4*

8. Draw a histogram from a stem and leaf plot.   *Section 2.4*

9. Construct a frequency polygon from a histogram.   *Section 2.4*

10. Construct a scatterplot.   *Section 2.4*

11. Construct a time series graph.   *Section 2.5*

Review Questions

To test your skills, answer the following questions:

2.46 Discuss the difference between continuous and discrete variables, and give two examples of each.

2.47 The personnel office of a corporation records each of the following variables on all of its employees. Identify the most informative scale of measurement (N for nominal, O for ordinal, or IR for interval/ratio) for each of the variables.

(a) Name _____   (b) Sex _____

(c) Age _____   (d) Race _____

(e) Marital status _____   (f) Appearance _____

(g) Years of education _____ (h) Years of service _____

(i) Current salary _____ (j) Job difficulty _____

2.48 Mark the following as QN for quantitative or QL for qualitative. Also give the most informative scale of measurement.

	QN or QL	Scale of measurement
(a) Race	_____	_____
(b) IQ	_____	_____
(c) Divorce rate	_____	_____
(d) Hair color	_____	_____
(e) Soc. Sec. #	_____	_____
(f) Lawyer's fee	_____	_____

2.49 Two hundred people were interviewed to determine whether they feel children of working mothers are adequately cared for.

	Men	Working mothers	Nonworking mothers
Yes	28	44	18
No	32	14	64

(a) What percent of the sample are women? _____

(b) What percent believe the children are adequately cared for? _____

(c) What percent of the working mothers feel that the children are adequately cared for? _____

(d) Of those who do not agree, what percent are women? _____

2.50 Of 2,898 robberies involving weapons in Atlanta in 1974, there were 1,611 with a gun, 857 with a knife, 356 with some other weapon, and in 74 cases it was not determined what weapon was used. Organize these data in a frequency table, and then construct a bar graph.

2.51 For the following data, construct a stem and leaf plot that corresponds to the ages of 25 executives.

```
35  45  63  42  59
45  50  62  36  64
50  26  51  54  45
59  57  64  28  38
48  42  61  54  60
```

2.52 Construct a frequency table and draw a histogram of the data in Exercise 2.51.

SUPPLEMENTARY EXERCISES FOR CHAPTER 2

2.53 Identify the following variables as either qualitative (QL) or quantitative (QN). If quantitative, then further identify as being continuous (C) or discrete (D). Also give the most informative scale of measurement for each variable.

Variable	QL or QN	C or D	Scale
(a) Amount of tar in a cigarette	_____	_____	_____
(b) SAT score for entering freshmen	_____	_____	_____
(c) Brand of breakfast cereal	_____	_____	_____
(d) Number of gold medals won by U.S. athletes	_____	_____	_____
(e) One's ratings of his or her favorite TV shows	_____	_____	_____
(f) Number of defective computer parts	_____	_____	_____

2.54 A group of students selected randomly from the student body was asked to identify their majors. There were 58 women, of which 9 were science majors, 15 were business majors, 22 were education majors, and the rest were liberal arts majors. There were 44 men, of which 17 were science majors, 15 were business majors, 5 were education majors, and the rest were liberal arts majors.

(a) Arrange the data in a bivariate frequency table.

(b) What percent are science majors?

(c) What percent of the women are education majors?

(d) What percent of the education majors are women?

2.55 In a class there are 10 psychology majors, 3 sociology, 7 criminal justice, 2 planning, 3 computer science, and 5 classified as other. Organize the data in a frequency table and draw a bar graph.

2.56 Suppose you are studying the welfare recipients in your state.

(a) Describe the population of interest.

(b) Describe how you might obtain a random sample.

(c) Give three qualitative variables that would be of interest.

(d) Give three quantitative variables that would be of interest and classify them as either continuous or discrete.

2.57 Identify each of the following variables as quantitative or qualitative.

(a) Number of violent crimes

(b) Social Security number

(c) Number of cities with a population greater than 100,000

(d) Employment status

(e) Reaction time to a stimulus

2.58 Identify each of the following quantitative variables as either continuous or discrete.

(a) Blood pressure

(b) Suicide rate

(c) Number of apartments in a city

(d) Gross National Product (GNP)

(e) Number of trials to complete a task

2.59 Following are the number of inmates per 100,000 population in the prisons across the United States in 1983. Some states were omitted, because the data were not available.

Ala.	113	Mont.	50
Alaska	8	Nebr.	53
Ariz.	99	Nev.	105
Ark.	69	N.H.	50
Calif.	166	N.J.	80
Colo.	88	N. Mex.	96
Conn.	N/A	N.Y.	91
Del.	N/A	N.C.	57
D.C.	456	N. Dak.	36
Fla.	137	Ohio	66
Ga.	178	Okla.	67
Hawaii	N/A	Oreg.	87
Idaho	61	Pa.	85
Ill.	77	R.I.	N/A
Ind.	66	S.C.	82
Iowa	29	S. Dak.	45
Kans.	55	Tenn.	128
Ky.	100	Tex.	97
La.	192	Utah	56
Maine	49	Vt.	N/A
Md.	107	Va.	103
Mass.	57	Wash.	84
Mich.	84	W. Va.	52
Minn.	47	Wis.	64
Miss.	97	Wyo.	66
Mo.	76	Total	98

SOURCE: Bureau of Justice Statistics.

Construct a stem and leaf plot of the available data.

2.60 The 251,012 poisons reported to 16 poison control centers, which served 11% of the U.S. population in 1983, were distributed as follows:

Type poison	No. reported
Drugs	150,857
Cleaning agent	22,347
Plants	22,326
Cosmetics	13,192
Insecticides	8,438
Alcohol	9,201
Other	24,651

Organize these data in a relative frequency bar graph.

2.61 The CLEP Subject Examination in Calculus was given to a group of college students. Their scores were:

```
42  45  39  40  46  46  46  39
52  44  42  41  44  46  60  45
60  38  43  63  45  46  50  41
40  40  41  44  42  44  50  48
57  47  46  41  42  42  38  47
50  40  50  34  44  52  44
```

Construct a stem and leaf plot of the scores.

2.62 A group of college students diagnosed as having dyslexia was asked to attend a reading workshop. The students were given a reading test, which was scored as the number of words read per minute. The following information was also recorded for each student:

Words/min.	165	201	75	124	105	143	126	92
Age	21	18	19	19	20	18	19	20
Sex	M	F	F	M	F	M	M	M
L/R(handed)	L	R	R	R	L	R	L	R
Wt. (lb.)	165	115	138	187	100	210	178	155
Ht. (in.)	70	66	65	72	61	71	69	68
No. children in family	2	1	4	3	2	1	1	3

(a) Identify each variable as being either quantitative or qualitative.

(b) Identify the quantitative variables as being either continuous or discrete.

(c) Give the most informative scale of measurement for each variable.

2.63 A survey by the American Council of Life Insurance of 516 adults on the issue of birth control gave the following results:

Birth control	Percent
Completely for	50%
Somewhat for	28%
Somewhat against	10%
Completely against	11%
Undecided	1%

Illustrate these data in a bar graph.

2.64 The money that Americans spend on alcoholic beverages is distributed as follows:

Beverage	Percent
Beer	52%
Wine	23%
Liquor	25%

Illustrate these data in a pie chart.

2.65 Suppose the following display represents 500 automobile accidents that occurred in a certain city:

	No fatalities	At least one fatality
Involved alcohol	68	142
No alcohol	194	96

(a) Fill in the marginal totals.

(b) What percent of the accidents involved alcohol?

(c) What percent of the accidents with at least one fatality involved alcohol?

(d) What percent of the nonalcohol-related accidents had at least one fatality?

(e) What percent of the accidents involved alcohol and had at least one fatality?

2.66 Following are the weekly rentals for 45 apartments in a large metropolitan area:

$100	130	130	305	175	155	150	95	295
210	80	270	135	130	335	230	235	75
90	285	65	345	110	135	185	300	70
250	125	180	150	305	170	95	90	145
90	160	130	80	490	235	75	60	425

Organize the data in a grouped frequency table, and construct a frequency histogram.

2.67 Classify the following as qualitative or quantitative. If quantitative, further classify as discrete or continuous.

(a) Number of vehicles owned by a family

(b) Homicide rate in a major city

(c) Socioeconomic status

(d) Concentration of PCB in a chemical spill

(e) Duration of a kidney transplant operation

2.68 Suppose you were studying the characteristics of automobile accidents in your city.

(a) Describe the population of interest.

(b) Give three relevant qualitative variables.

(c) Give three relevant quantitative variables, and classify them as discrete or continuous.

(d) Give the most informative scale of measurement for the variables listed in parts b and c.

2.69 A market analyst is studying the percent of the computer market shared by the major manufacturers. She records the corporation name, the area of the market they are active in (e.g., personal computers, software, etc.), the assets of the company, and the percent of the market they control.

(a) What scale of measurement is appropriate for corporate name?

(b) What scale of measurement is appropriate for market area?

(c) What scale of measurement is appropriate for assets?

(d) What scale of measurement is appropriate for market percent?

2.70 A survey of 1,000 adults (conducted by R. H. Bruskin Associates for *USA Today,* February 18, 1985) found that 720 thought the legal drinking age should be 21, 50 said 20 years old, 60 said 19 years old, 110 said 18 years old, and 60 were undecided.

(a) Organize these data in a relative frequency table.

(b) Display the data in a bar graph.

(c) Display the data in a pie chart.

2.71 On December 2, 1982, Barney Clark became the first human to receive an artificial heart. He was in surgery for 7 hours. On February 17, 1985, Murray Haydon, the third artificial heart recipient, was in surgery for $3\frac{1}{2}$ hours. (William Schroeder was the second recipient on November 25, 1984 and was in surgery for $6\frac{1}{2}$ hours.) Suppose you were to keep records on the artificial heart recipients. What variables would be of interest?

(a) List five relevant variables.

(b) Classify them as qualitative or quantitative.

(c) Give the most informative scale of measurement for each variable.

2.72 Following are the birth places, ages at inauguration, and ages at death of the first 40 Presidents of the United States:

President	Birth	Inaug. age	Age at death
G. Washington	Va.	57	67
J. Adams	Mass.	61	90
T. Jefferson	Va.	57	83
J. Madison	Va.	57	85
J. Monroe	Va.	58	73
J. Q. Adams	Mass.	57	80
A. Jackson	S.C.	61	78
M. Van Buren	N.Y.	54	79
W. Harrison	Va.	68	68
J. Tyler	Va.	51	71
J. Polk	N.C.	49	53
Z. Taylor	Va.	64	65
M. Fillmore	N.Y.	50	74
F. Pierce	N.H.	48	64
J. Buchanan	Pa.	65	77
A. Lincoln	Ky.	52	56
A. Johnson	N.C.	56	66
U. S. Grant	Ohio	46	63
R. Hayes	Ohio	54	70
J. Garfield	Ohio	49	49
C. Arthur	Vt.	50	57
G. Cleveland	N.J.	47	71
B. Harrison	Ohio	55	67
G. Cleveland	N.J.	55	71
W. McKinley	Ohio	54	58
T. Roosevelt	N.Y.	42	60
W. Taft	Ohio	51	72
W. Wilson	Va.	56	67
W. Harding	Ohio	55	57

President	Birth	Inaug. age	Age at death
C. Coolidge	Vt.	51	60
H. Hoover	Iowa	54	90
F. D. Roosevelt	N.Y.	51	63
H. Truman	Mo.	60	88
D. Eisenhower	Tex.	62	78
J. Kennedy	Mass.	43	46
L. Johnson	Tex.	55	64
R. Nixon	Calif.	56	
G. Ford	Nebr.	61	
J. Carter	Ga.	52	
R. Reagan	Calif.	69	

(a) Construct a bar graph of the birth states of the Presidents.

(b) Construct back-to-back stem and leaf plots (stem and leaf plots with a common stem) of the inaugural age and age at death.

2.73 A sample of 1,100 people was asked to give their favorite sport. The responses were:

Sport	Number
Baseball	324
Basketball	149
Football	347
Golf	29
Ice hockey	68
Soccer	92
Tennis	45
Other	46

(a) Organize the data in a relative frequency table.

(b) Construct a bar graph of the data.

3

Numerically Describing Data

Suppose you are asked to determine whether there is a relationship between the birth weights of children and their adult heights. From a sample of adults you could determine their heights, and from their birth certificates you could determine their birth weights. Thus for each adult you would have two measurements, their birth weight and their adult height.

After organizing the data in a scatterplot you might detect an association between the variables. But how do you report the degree of association? Clearly some variables are more closely related than others. What is needed is a single number, which, in a sense, measures the amount of association between two variables. That number, called the *correlation,* is a summary measure computed from the collected data. Correlation is just one of the many summary measures used to describe data that are presented in this chapter.

Are Corporate Executives Overpaid?

A private study conducted by the Democracy Project, a New York-based research institute headed by a former aide to consumer activist Ralph Nader, suggests that corporate executives are overpaid for the services they provide (*Winston-Salem Journal,* April 24, 1984). The study, entitled "The Trouble With Executive Compensation," was based on 60 interviews with executives and compensation consultants. The report points out that there has been a marked increase in executive salaries over the last five to seven years despite White House efforts to restrain such pay increases. It is not clear whether the stockholders are getting what they are paying for. The study suggests that there should be some sort of merit pay system for executives similar to that suggested for schoolteachers; that is, reward those who perform.

To study this issue further, the reported salaries of the 100 highest-paid executives of major corporations across the United States and their companies' financial reports (from *U.S. News & World Report*) are recorded in Appendix A1. The salaries are reported in thousands, and the assets, sales, market value, and net profit are reported in millions. The ranks are the 1984 rankings given to the top 500 corporations (Forbes 500) by *Forbes* magazine. For example, Gulf and Western's assets are ranked 233 out of the top 500 corporations in the United States. Note that the rank is not the rank within the data set of 100 corporations, but is the rank in the top 500 corporations. For example, the assets of Gannett did not make the Forbes 500 and therefore no rank is given.

By examining the data we can see that the largest reported salary is $2,122,000 and the smallest is $806,000, but little else can be determined about the general distribution of the salaries across the corporations. The same can be said about the assets, sales, market value, and net profits of the corporations. What is a typical salary? How many executives earn over $1 million? Is there any correlation between the salaries and the sizes of the companies? These are just a few of the many questions that can be asked about a data set. In this chapter we will investigate ways to summarize the data so that questions such as these can be answered.

3.1 Summary and Measures

The objective of statistics is to draw generalizations about a population based on the collected data in a sample. In most cases it is difficult to work with the complete distribution of values; hence summary measures are introduced to help answer our statistical questions. For example, the comparison of the effects of a certain drug on the reaction time of women to that of men may simplify to a comparison of two numerical values that correspond to the centers of the two distributions of reaction times. These summary measures can apply to either a sample or to the entire population.

Definition

A **parameter** is a numerical summary measure of a population distribution.

Typically the values of the parameters associated with a population are unknown. Thus we must select a random sample from the population and use summary measures of the sample to estimate the unknown population parameter values.

Definition

A **statistic** is a numerical quantity calculated from the observations in a sample.

Corresponding to each parameter of a population there is a statistic that is its sample counterpart. If the summary measure refers to the entire population, it is called a parameter; if it is obtained from information in the sample, it is called a statistic.

Among the summary measures are the *measures of location* and *measures of variability*. Many of these statistics are found from the data that have been arranged in a stem and leaf plot. In the next section we will take a closer look at the stem and leaf plot.

3.2 A Second Look at Stem and Leaf Plots

As stated in Chapter 2, the stem and leaf plot preserves the data for future calculations. Some of those calculations involve finding specific locations in the data; this would only be relevant if the data were ordered from smallest to largest. The stem and leaf plot allows us to accomplish this easily. All that is necessary is that we reorder the leaves of each stem from smallest to largest. The resulting stem and leaf plot is called an *ordered* stem and leaf plot.

EXAMPLE 3.1

A psychologist wishes to test a new method to improve rote memorization by college students. A sample of 20 college students is taught the new technique and then asked to memorize a list of 100 word phrases. Following is a list of the number of correct word phrases for each of the 20 students:

60	72	81	93	72	66	71	94	81	76
84	81	82	99	89	78	90	89	79	74

Construct an ordered stem and leaf plot for the data.

SOLUTION   From the data we get the unordered stem and leaf plot

```
6 | 0  6
7 | 2  2  1  6  8  9  4
8 | 1  1  4  1  2  9  9
9 | 3  4  9  0
```

By ordering each stem we get the ordered stem and leaf plot

```
6 | 0  6
7 | 1  2  2  4  6  8  9
8 | 1  1  1  2  4  9  9
9 | 0  3  4  9
```

Following are some examples that illustrate how the stem and leaf plot might be modified to accommodate data that might not immediately yield itself to a stem and leaf interpretation.

EXAMPLE 3.2

An ecologist wishes to investigate the level of mercury pollution in a major lake. He catches 25 lake trout and measures the concentration of mercury (measured in parts per million) in each fish. From the following data, construct an ordered stem and leaf plot.

2.2	3.4	3.0	2.6	3.8
1.8	2.8	3.2	3.7	1.4
2.7	3.6	1.9	2.2	3.0
3.3	2.3	1.7	2.6	3.5
3.0	2.9	3.4	3.1	2.4

SOLUTION   When confronted with a decimal point, we ignore it while constructing the stem and leaf plot, but understand that the numbers do actually have decimals. Thus in this problem we treat the numbers as two-digit numbers that range from 14 to 38. Using stem values of 1, 2, and 3, we get the following stem and leaf plot.

```
1 | 8  4  9  7
2 | 2  6  8  7  2  3  6  9  4
3 | 4  0  8  2  7  6  0  3  5  0  4  1
```

Realizing that the plot is somewhat compact, we can spread it out by constructing a *double-stem* stem and leaf plot. This is accomplished by breaking up each stem into two stems called ∗ and −. The ∗ is used to denote leaves 0 through 4 and the − denotes leaves 5 through 9. We have the following double-stem stem and leaf plot:

```
1∗ | 4
1− | 8  9  7
2∗ | 2  2  3  4
2− | 6  8  7  6  9
3∗ | 4  0  2  0  3  0  4  1
3− | 8  7  6  5
```

Ordering each stem, we have an ordered double-stem stem and leaf plot.

```
1*  │  4
1−  │  7   8   9
2*  │  2   2   3   4
2−  │  6   6   7   8   9
3*  │  0   0   0   1   2   3   4   4
3−  │  5   6   7   8
```

each value × 10⁻¹

▲ Each value × 10⁻¹ means to multiply each number in the stem and leaf plot by 1/10 in order to reinsert the decimal point.

EXAMPLE 3.3

Murder rates for major cities are measured as the number of murders per 100,000 inhabitants. Following are the 1983 murder rates for 35 northern metropolitan statistical areas, rounded to the nearest whole number.

City	Rate	City	Rate	City	Rate
Albany, N.Y.	3	Allentown, N.J.	2	Atlantic City, N.J.	4
Baltimore, Md.	11	Boston, Mass.	4	Buffalo, N.Y.	6
Chicago, Ill.	14	Cleveland, Ohio	10	Danbury, Conn.	1
Dayton, Ohio	9	Detroit, Mich.	16	Erie, Pa.	2
Fort Wayne, Ind.	5	Harrisburg, Pa.	5	Hartford, Conn.	4
Indianapolis, Ind.	6	Johnstown, Pa.	2	Manchester, N.H.	5
Middletown, Conn.	5	Milwaukee, Wis.	4	Newark, N.J.	8
New York, N.Y.	19	Niagara Falls, N.Y.	4	Philadelphia, Pa.	9
Pittsburgh, Pa.	4	Portland, Maine	1	Providence, R.I.	3
Rochester, N.Y.	4	South Bend, Ind.	4	Stamford, Conn.	3
Toledo, Ohio	7	Trenton, N.J.	9	Washington, D.C.	9
York, Pa.	2	Youngstown, Ohio	6		

SOURCE: *Uniform Crime Reports for the United States, 1983*, U.S. Department of Justice, Washington, D.C., pp. 353–382.

Construct a stem and leaf plot for these data.

SOLUTION   The only stems available are 0 and 1, in which case the stem and leaf plot takes the following form:

```
0  │  3  2  4  4  6  1  9  2  5  5  4  6  2  5  5  4  8  4  9  4  1  3  4  4  3  7  9  9  2  6
1  │  1  4  0  6  9
```

Certainly this is too condensed; a double-stem stem and leaf plot takes the following form:

```
0*  │  3  2  4  4  1  2  4  2  4  4  4  1  3  4  4  3  2
0−  │  6  9  5  5  6  5  5  8  9  7  9  9  6
1*  │  1  4  0
1−  │  6  9
```

Again this seems too condensed, so we try a *five-stem* stem and leaf plot. Now * corresponds to leaves 0–1, t corresponds to leaves 2–3, f corresponds to leaves 4–5, s corresponds to leaves 6–7 and, finally, – corresponds to leaves 8–9. The final stem and leaf plot takes the form

```
0* | 1  1
0t | 3  2  2  2  3  3  2
0f | 4  4  4  4  4  4  4  4  5  5  5  5
0s | 6  6  7  6
0– | 9  8  9  9  9
1* | 1  0
1t |
1f | 4
1s | 6
1– | 9
```

▲ Note that 2 and 3 start with t, 4 and 5 start with f, and 6 and 7 start with s; hence the stems t, f, and s.

Clearly this stem and leaf plot is more informative than the other attempts.

Occasionally the data will have negative values as is illustrated in the following example.

EXAMPLE 3.4

The nicotine content of 16 different brands of cigarettes is measured and recorded as above or below an acceptable norm. Construct a stem and leaf plot of the following data recorded as a percent above (+) or below (−) the norm:

− 5	+ 5	− 25	+ 22	+ 10	+ 9	− 32	− 26
+ 24	− 14	− 6	+ 17	− 8	− 25	+ 31	− 12

SOLUTION   We see that the data run from the negative 30s to the positive 30s, which indicates stem values of

$$-3, \quad -2, \quad -1, \quad -0, \quad +0, \quad +1, \quad +2, \quad +3$$

so the stem and leaf plot becomes

```
−3 | 2
−2 | 5  6  5
−1 | 4  2
−0 | 5  6  8
+0 | 5  9
+1 | 0  7
+2 | 2  4
+3 | 1
```

▲ By convention, we would place a score of 0 in the +0 stem.

Observe that we need a −0 stem as well as a +0 stem to distinguish between, say − 5 and + 5.

The following example illustrates the possibility of multidigit leaves.

EXAMPLE 3.5

Following are the yearly per bed rental costs incurred by a random sample of 35 halfway houses:

472	303	280	282	417	400	257
205	384	264	317	76	643	480
136	250	100	732	317	264	384
750	402	422	373	325	313	749
791	196	891	283	52	186	693

Construct a stem and leaf plot of these data.

SOLUTION The obvious stems are the values corresponding to the hundreds place of the number, and therefore the leaves will be the remaining two digits of the number.

```
0 | 76, 52
1 | 36, 00, 96, 86
2 | 80, 82, 57, 05, 64, 50, 64, 83
3 | 03, 84, 17, 17, 84, 73, 25, 13
4 | 72, 17, 00, 80, 02, 22
5 |
6 | 43, 93
7 | 32, 50, 49, 91
8 | 91
```

▲ The leaves are separated by commas so that, for example, 76 is not interpreted as a 7 and a 6.

Occasionally one or two scores may be far removed from the rest of the data, in which case it is not realistic to continue the stems all the way down to those values. These extreme values, called *outliers*, are handled by putting them in a class by themselves with stem value either HI (for high) or LO (for low), depending on whether it is on the high side or the low side of the data. For example, suppose that in Example 3.4 there had been a $+80$ and a $+95$. Instead of putting in stem values of $+4$, $+5$, $+6$, $+7$, $+8$, and $+9$, we would construct the stem and leaf plot as follows:

```
-3 | 2
-2 | 5 6 5
-1 | 4 2
-0 | 5 6 8
+0 | 5 9
+1 | 0 7
+2 | 2 4
+3 | 1
-------------
HI | +80, +95
```

Note that it is necessary to give the entire number for the outliers, because it would not be clear which stem value applies.

Certainly there are other possibilities we have not considered, but, with a little practice, you will be able to come up with a meaningful stem and leaf plot for almost any data set.

3.1 Following are 20 scores on the Miller Personality Test:

22	21	16	26	22	23	31	25	20	25
33	17	27	29	25	22	30	18	23	25

Arrange the data in an ordered double-stem stem and leaf plot.

3.2 Fifty families were interviewed and the number of dependent children recorded.

3	2	2	4	1	1	2	3	4	1
2	0	1	2	1	0	4	2	1	0
0	1	3	0	3	2	2	3	0	3
2	5	0	1	2	1	4	3	0	5
2	0	1	1	2	6	1	2	1	5

Arrange the data in an ordered stem and leaf plot.

3.3 The number of defective items produced by the 20 employees in a small business were:

17	26	10	19	18	27	17	36	22	18
20	31	30	18	16	24	10	23	12	13

Arrange the data in an ordered stem and leaf plot.

3.4 The number of days that the 20 employees in the small business in Exercise 3.3 were absent during the year are listed below:

1	0	4	3	2	0	0	2	0	2
10	0	0	2	3	4	5	0	2	1

Arrange the data in a stem and leaf plot.

3.5 A group of 23 students participated in a psychology experiment, and their scores were recorded based on the number of correct responses. The scores were:

12	14	18	7	11	15	8	15
10	14	19	14	6	13	14	12
16	10	9	12	15	8	17	

Arrange the data in a stem and leaf plot.

3.6 Following are the daily profits of 20 newsstands. Construct an ordered double-stem stem and leaf plot.

$81.32	61.47	64.90	70.88	76.02
75.06	76.73	64.21	74.92	77.56
58.01	68.05	73.37	75.41	59.41
65.43	74.76	76.51	65.10	76.02

3.7 An exercise program had 30 members who did exercises 5 days a week for 1 month. Following are the weight losses of the 30 members (a negative indicates that the person gained weight). Construct an ordered stem and leaf plot of the weight losses.

5	15	3	−4	8	7	5	10
−3	2	−5	9	5	−4	10	6
−2	7	4	3	−12	−6	11	4
8	−10	9	5	5	2		

3.8 The price of regular unleaded gasoline was obtained from 25 service stations within a major city. Construct an ordered stem and leaf plot of the following data:

$1.16	1.22	1.20	1.22	1.18
1.20	1.20	1.19	1.18	1.20
1.19	1.20	1.19	1.25	1.19
1.20	1.22	1.19	1.21	1.20
1.19	1.20	1.18	1.21	1.19

3.9 Following are the weights of 25 soccer players. Arrange the data in a five-stem stem and leaf plot. Construct a frequency histogram from the stem and leaf plot with interval widths of 20 and the first lower class limit at 100.

144	162	197	173	183
129	209	190	117	160
179	177	154	132	151
159	175	154	148	166
184	157	162	150	136

3.3   Measures of Location

A measure of location is a quantity that locates a particular position in the frequency distribution. We will first consider measures of the center of the distribution.

Center

The most common measure of center is the *mean* that locates the balance point of the distribution.

Definitions

The **population mean,** denoted by the Greek letter μ, is the numerical value that locates the *balance point,* also called the *center of mass,* of the population distribution.

If a sample consists of observations $y_1, y_2, y_3, \ldots, y_n$, then the **sample mean** is

$$\bar{y} = (y_1 + y_2 + \cdots + y_n)/n = \Sigma y_i/n$$

Figure 3.1 illustrates that μ is the center of gravity of the frequency curve, the point about which the curve would balance.

FIGURE 3.1

Location of the mean

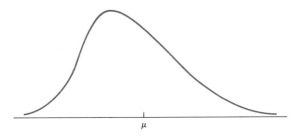

If the entire population were available and finite, the population mean μ would be calculated the same way as the sample mean. The entire population is rarely available, however, and thus the value of μ remains unknown. Only $\bar{y}$ is calculated (because the sample will be available), and its value serves as an approximate value of μ.

EXAMPLE 3.6

A sample of 30 observations is selected at random from a population whose mean is unknown. The sample consists of

```
0 | 4
1 | 3  4  6  2
2 | 5  8  3  2  1  6  2  7
3 | 4  8  2  1  6  3  8
4 | 3  1  2  6
5 | 8  4  2
6 | 2  5
7 | 1
```

Calculate the sample mean.

SOLUTION   The sample mean is calculated by first finding the total of all observations

$$4 + 13 + 14 + \cdots + 65 + 71 = 1{,}029$$

and then dividing by n, the number of observations. We have

$$\bar{y} = 1{,}029/30 = 34.3$$

which is an estimate of the unknown population mean.

EXAMPLE 3.7

Following are the 1984 flat monthly phone rates for the 50 states and the District of Columbia.

Ala.	$18.05	Ky.	$17.65	N. Dak.	$11.35
Alaska	12.25	La.	12.43	Ohio	12.95
Ariz.	9.21	Maine	10.29	Okla.	6.55
Ark.	14.00	Md.	16.49	Oreg.	13.30
Calif.	7.00	Mass.	12.10	Pa.	9.46
Colo.	7.28	Mich.	10.65	R.I.	15.80
Conn.	10.05	Minn.	13.65	S.C.	16.20
Del.	11.17	Miss.	19.01	S. Dak.	13.50
D.C.	12.49	Mo.	9.55	Tenn.	13.15
Fla.	12.20	Mont.	8.37	Tex.	10.75
Ga.	14.90	Nebr.	10.63	Utah	16.44
Hawaii	11.40	N.J.	7.75	Vt.	11.00
Idaho	5.05	N.H.	13.72	Va.	12.55
Ill.	31.84	N. Mex.	10.70	Wash.	12.50
Ind.	13.84	N.Y.	15.65	W. Va.	17.77
Iowa	11.15	Nev.	9.20	Wis.	13.00
Kans.	11.65	N.C.	14.95	Wyo.	9.10

SOURCE: Local Bell and GTE operating companies.

Organize the data in a stem and leaf plot and calculate the sample mean.

SOLUTION   The stem and leaf plot, with the exception of the $31.84 rate for Illinois, appears to be bell-shaped and centered around $12.00.

5	05
6	55
7	00, 28, 75
8	37
9	10, 20, 21, 46, 55
10	05, 29, 63, 65, 70, 75
11	00, 15, 17, 35, 40, 65
12	10, 20, 25, 43, 49, 50, 55, 95
13	00, 15, 30, 50, 65, 72, 84
14	00, 90, 95
15	65, 80
16	20, 44, 49
17	65, 77
18	05
19	01
HI	31.84

Averaging the 51 observations, we find

$$\bar{y} = \$12.582$$

However, it seems only natural that $\bar{y}$ is adversely affected by the large rate in Illinois.

In reference to Example 3.7, an obvious question is, should we always use the mean to represent the center of a distribution? Are there alternatives to $\bar{y}$ that would tend to be better measures of the center of a distribution? Consider for the moment the distribution in Figure 3.2, which has significantly long tails.

FIGURE 3.2

A long-tailed distribution

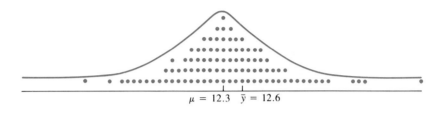

A distribution with long tails tends occasionally to produce an extreme observation as is indicated by the far right observation in the graph. Such outliers (observations that are remote to or stand apart from the remainder of the data) tend to pull the value of $\bar{y}$ toward them and thus give a distorted measure of the center of the distribution. A solution to this problem is a *trimmed mean*.

Definitions

An $\alpha\%$ **trimmed population mean** denoted by

$$\mu_{T.\alpha}$$

is obtained by trimming $\alpha\%$ of the distribution off both ends and finding the mean of the remaining distribution.

An $\alpha\%$ **trimmed sample mean** denoted by

$$\bar{y}_{T.\alpha}$$

is found by trimming $\alpha\%$ off both ends of an ordered stem and leaf plot and calculating the mean of the remaining observations. If $\alpha\%$ is not a whole number, round to the next whole number.

EXAMPLE 3.8

Find the 10% trimmed mean for the phone rate data given in Example 3.7.

SOLUTION   Ten percent of 51 observations is 5.1; thus we trim six observations off each end of the ordered stem and leaf plot. We trim 5.05, 6.55, 7.00, 7.28, 7.75, and 8.37 off the low end; we trim 31.84, 19.01, 18.05, 17.77, 17.65, and 16.49 off the high end of the ordered stem and leaf plot to obtain the trimmed stem and leaf plot.

9	10,	20,	21,	46,	55		
10	05,	29,	63,	65,	70,	75	
11	00,	15,	17,	35,	40,	65	
12	10,	20,	25,	43,	49,	50,	55,	95
13	00,	15,	30,	50,	65,	72,	84
14	00,	90,	95				
15	65,	80					
16	20,	44					

▲ Remember to order the data before trimming.

The average of the remaining 39 observations is

$$\bar{y}_{T.10} = 12.279$$

which is some 30 cents lower than $\bar{y}$ and might possibly be a more representative measure of the center of the distribution.

Clearly trimming minimizes the effect of outliers. In Example 3.8, the 31.84 is definitely an outlier and has a measurable impact on the ordinary sample mean. Once it is trimmed off, it has no effect on the trimmed mean, $\bar{y}_{T.10}$. In fact, the 10% trimmed mean is not affected by the lower or upper 10% of the observations.

Another measure of location that is not affected by outliers is the *median*.

Definitions

The **population median,** denoted by θ, is the numerical value that divides the population distribution in half.

If the sample observations $y_1, y_2, \ldots, y_n$ are arranged in order from smallest to largest, the **sample median,** denoted by M, is the middle observation if n is odd, or the average of the two middle observations if n is even.

Figure 3.3 illustrates that the population median is the value that separates the lower 50% of the distribution from the upper 50% of the distribution.

FIGURE 3.3
Location of the population median

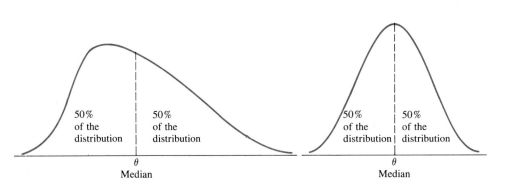

The sample median is located in the center of the sample after it has been ordered from smallest to largest. It will be convenient to find the sample median from an ordered stem and leaf plot if we first find its location.

> The *location of the median* in a sample that has been ordered is
>
> $$l(M) = (n + 1)/2$$
>
> where n is the size of the sample.

If n is odd, then $l(M)$ will be a whole number. In this case, count the leaves in the ordered stem and leaf plot, in order, until you get to the observation in the $l(M)$ position. That observation is the sample median.

EXAMPLE 3.9

Find the sample median of the following data that are presented in an *ordered* stem and leaf plot:

```
0 | 1 4 6
1 | 3 6 7 8 9 9
2 | 2 4 6 8 9
3 | 4 6 8
4 | 2 5
5 | 3 6
6 | 4
7 | 3
8 | 1
9 | 4
```

SOLUTION   By counting the leaves we see that $n = 25$; therefore the location of the median is

$$l(M) = (n + 1)/2 = (25 + 1)/2 = 26/2 = 13$$

Counting leaves until we get to the thirteenth observation, we find that the sample median is

$$M = 28$$

If n is even, then $l(M)$ will not be a whole number but will have a .5 decimal part. For example, if $n = 26$, then

$$l(M) = (26 + 1)/2 = 27/2 = 13.5$$

The sample median is then the average of the two observations adjacent to the median location. Thus if $l(M) = 13.5$, then the median is the average of the thirteenth and fourteenth observations in an ordered stem and leaf plot.

EXAMPLE 3.10

Find the sample median for the following data that are presented in an ordered stem and leaf plot:

```
 0 | 1  4  6
 1 | 3  6  7  8  9  9
 2 | 2  4  6  8  9
 3 | 4  6  8
 4 | 2  5
 5 | 3  6
 6 | 4
 7 | 3
 8 | 1
 9 | 4
10 | 4
```

SOLUTION Because $n = 26$, the location of the median is

$$l(M) = (26 + 1)/2 = 27/2 = 13.5$$

Counting down the leaves of the ordered stem and leaf plot, we find that the thirteenth observation is 28 and the fourteenth observation is 29, so the median is

$$M = (28 + 29)/2 = 28.5$$

In comparison, by averaging the data, we find

$$\bar{y} = 37$$

▲ If the right tail of the distribution is longer than the left tail, then the mean will be larger than the median. The opposite is true if the left tail is longer than the right tail; that is, the mean will be smaller than the median.

which is considerably larger. Because the data have extreme observations on one tail, the median of 28.5 seems to be a more representative measure of the center of the data.

Example 3.10 illustrates that the sample mean may give a distorted measure of the center of a distribution that has outliers. It is pulled toward the extreme observations, whereas the sample median is not affected by the outliers. Because the median is not affected by the extreme observations, we say it is *resistant* to the influence of outliers.

Quantiles

Just as the population median divides the population distribution in half, the population *quartiles* divide it into quarters.

Definitions

The **first quartile,** denoted by θ_1, is the numerical value that divides the lower half of the population distribution in half. The **third quartile,** denoted by θ_3, is the value that divides the upper half of the population distribution in half.

The **first** and **third sample quartiles,** Q_1 and Q_3, are similarly defined for samples. The median is the **second quartile,** Q_2.

Figure 3.4 illustrates that the first quartile, the median, and the third quartile divide the population distribution into four quarters.

FIGURE 3.4

Location of median and first and
third quartiles

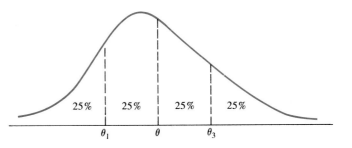

Having ordered the stem and leaf plot, the sample quartiles can be found just as easily
as the sample median was found.

The location of each quartile can be found by

$$l(Q) = ([l(M)] + 1)/2$$

where $[l(M)]$ is the integer part of $l(M)$, meaning that we drop off the .5 decimal
of $l(M)$ if it is present.

The first sample quartile, Q_1, is found by counting the leaves from the *lower* end of the
ordered stem and leaf plot until we get to the observation in the $l(Q)$ position. The third
sample quartile, Q_3, is found by counting the leaves from the *higher* end of the ordered
stem and leaf plot until we get to the observation in the $l(Q)$ position. Again if $l(Q)$ has
a decimal part of .5, we average the two observations adjacent to the $l(Q)$ position.

EXAMPLE 3.11

Find Q_1 and Q_3 for the data in Example 3.10. The stem and leaf plot is repeated here
for convenience.

```
 0 | 1  4  6
 1 | 3  6  7  8  9  9
 2 | 2  4  6  8  9
 3 | 4  6  8
 4 | 2  5
 5 | 3  6
 6 | 4
 7 | 3
 8 | 1
 9 | 4
10 | 4
```

SOLUTION   Recall that $l(M) = 13.5$. The integer part is $[l(M)] = 13$, thus

$$l(Q) = (13 + 1)/2 = 14/2 = 7$$

So Q_1 is the seventh observation from the lower end and Q_3 is the seventh observation
from the higher end of the ordered stem and leaf plot. We find

$$Q_1 = 18 \quad \text{and} \quad Q_3 = 53$$

We can further divide the data into eighths by first finding the location of the eighth by

$$l(E) = ([l(Q)] + 1)/2$$

and counting the leaves of an ordered stem and leaf plot from the lower end to find E_1 and counting from the higher end to find E_7.

EXAMPLE 3.12

Find E_1 and E_7 for the data in Example 3.10.

SOLUTION   From Example 3.11 we found $l(Q) = 7$, thus

$$l(E) = (7 + 1)/2 = 4$$

So E_1 is the fourth observation from the lower end, or $E_1 = 13$; and E_7 is the fourth observation from the higher end, or $E_7 = 73$.

Two additional measures of location we shall consider are the smallest and largest observations. They are called the *range* scores and are denoted as L and H for low and high.

Finally, the measures of location are summarized in a *quantile* summary diagram as follows:

▲ The word *quantile* is used here instead of quartile, because we are referring to quantities other than quarters.

	n	
M (location)	M	
Q (location)	Q_1	Q_3
E (location)	E_1	E_7
R	L	H

The information obtained in Examples 3.10 through 3.12 is presented here in a quantile summary diagram:

	26	
M (13.5)	28.5	
Q (7)	18	53
E (4)	13	73
R	1	104

EXAMPLE 3.13

The reading scores for a group of fifth graders are given here in an ordered stem and leaf plot. Summarize the data in a quantile summary diagram.

4	8
5	
6	7
7	3
8	1 3 6
9	2 3 4 4 4 5 7 8 8 9
10	0 1 2 2 3 5 6 7 8
11	5 7
12	3
13	4
14	9

SOLUTION   There are $n = 30$ scores, which gives

$$l(M) = (30 + 1)/2 = 15.5$$

Thus

$$M = (98 + 99)/2 = 98.5$$

We then have

$$l(Q) = (15 + 1)/2 = 8$$

Thus

$$Q_1 = 93 \quad \text{and} \quad Q_3 = 106$$

Finally, we have

$$l(E) = (8 + 1)/2 = 4.5$$

Thus

$$E_1 = (81 + 83)/2 = 82 \quad \text{and} \quad E_7 = (115 + 117)/2 = 116$$

Summarizing in the quantile summary diagram, we have:

	30	
$M(15.5)$	98.5	
$Q(8)$	93	106
$E(4.5)$	82	116
R	48	149

The information in the quantile summary diagram can be used to obtain other measures of the center of the data. These are called the *midsummary* statistics. The average of Q_1 and Q_3 is called the *midquarter* or midQ. The average of E_1 and E_7 is called the *mideighth* or midE. The average of the two extremes, low and high, is called the *midrange* or midR. The median is already a midsummary. These midsummaries can be easily exhibited next to the quantile summary diagram as illustrated:

	n		Midsummaries
$M(\text{loc})$	M		M
$Q(\text{loc})$	Q_1	Q_3	midQ
$E(\text{loc})$	E_1	E_7	midE
R	L	H	midR

Here the midsummaries are presented with the quantile summary diagram found in Example 3.13.

		30	Midsummaries
$M(15.5)$		98.5	98.5
$Q(8)$	93	106	99.5
$E(4.5)$	82	116	99
R	48	149	98.5

In Chapter 6 the midsummaries will be used to analyze the shape of the distribution from which the sample came.

EXERCISES 3.3

3.10 Find the mean and a 10% trimmed mean for the following data:

19, 22, 34, 28, 18, 24, 16, 25, 27, 31

3.11 Find the mean and a 20% trimmed mean for the following data:

215	247	169	210	242
198	184	275	170	211

3.12 The aggressive tendency scores for a group of teenage boys are as follows:

38	27	44	39	41	26	35
45	39	28	16	37	11	36
33	46	42	37	40	19	24
29	32	34	31	30	32	43

Find the mean and 10% trimmed mean for the data.

3.13 Construct a quantile summary diagram for the data given in Exercise 3.12.

3.14 Calculate the midsummaries for the data given in Exercise 3.12.

3.15 The Self Criticism scores on the Tennessee Self Concept Scale (a test to measure one's self confidence) for the group of teenagers in Exercise 3.12 are as follows:

26	19	23	27	24	33	25
29	14	30	20	25	5	18
7	28	31	37	28	3	20
25	45	29	22	41	34	22

Find the mean and 20% trimmed mean.

3.16 Construct a quantile summary diagram for the data given in Exercise 3.15.

3.17 Calculate the midsummaries for the data given in Exercise 3.15.

3.4 Measures of Variability

Measures of central location are important because they describe a "typical" observation in the data set. Not all observations are typical, however; some are atypical in that they deviate from the center. The amount of deviation from the center is an important consideration when we are investigating the properties of a data set. For example, suppose that two patients in a hospital have their heart rates taken every 4 hours. Their heart rates over a 24-hour period were

Patient A	68	70	69	70	71	72
Patient B	65	85	90	65	55	60

In each case, the average heart rate is 70, but there is a difference in the heart rates of the two patients. Patient A's rate is stable, but Patient B's rate probably needs to be monitored more closely.

EXAMPLE 3.14

Suppose that we give a 10-point quiz to two classes of students. Their scores were:

Class A	4	6	3	5	5	4	7	5	6	4	6	6	4	3	7
Class B	6	3	8	5	9	7	0	1	7	6	5	9	2	4	3

Using a hand-held calculator we can find the mean of each data set to be

$$\bar{y}_A = 5 \quad \bar{y}_B = 5$$

Both classes, each consisting of 15 students, have the same mean of 5. However, back-to-back stem and leaf plots reveal a difference in the two data sets.

Data set A		Data set B
	0*	0 1
3 3	0t	2 3 3
4 4 4 4 5 5 5	0f	4 5 5
6 6 6 6 7 7	0s	6 6 7 7
	0–	8 9 9

Clearly the difference is that the observations in data set B are more *spread* out from the mean than those in data set A. This was also the case for the heart rates of the two patients.

This concept of variability is extremely important in studying the characteristics of a data set. The simplest measure of variability is the *range*.

Definition

The **range** is the highest measurement minus the lowest measurement, H – L.

The range in heart rate for Patient A is $72 - 68 = 4$ and for Patient B is $90 - 55 = 35$.

The population range applies to the entire population, and the sample range applies to the sample. On the 10-point quiz in Example 3.14, the maximum possible is 10 and the minimum possible is 0; thus the population range is $10 - 0 = 10$. On the other hand, from the stem and leaf plots we find that the two sample ranges are

Range Class A = 7 – 3 = 4
Range Class B = 9 – 0 = 9

Note that it is a very simple matter to find the sample range from an ordered stem and leaf plot.

The range is a meaningful measure of variability for small data sets, but a more sensitive measure of variability is needed for large data sets. The most widely accepted measure of dispersion is the *standard deviation*. To obtain the standard deviation we must first find the *variance*.

Definitions

The **population variance,** σ^2, is the average squared distance of all measurements from the population mean.

The **sample variance,** s^2, is an average squared distance of the sample values from the sample mean. It is calculated by the formula

$$s^2 = \frac{\Sigma(y_i - \bar{y})^2}{n - 1}$$

The expression $(y_i - \bar{y})$ in the preceding formula for s^2 is called a deviation from the mean and represents how far the observation y_i is from $\bar{y}$. Because $\bar{y}$ is the mean of all the measurements, these deviations will sum to zero. That is, the negative deviations will cancel the positive deviations. The *average* of the squared deviations, $(y_i - \bar{y})^2$, will not be zero and yields the sample variance. We find the average by dividing by $n - 1$ rather than n, because with a denominator of n the sample variance tends to underestimate the population variance; that is, it has a negative bias. Dividing by $n - 1$ eliminates this bias. More will be said about this in Chapter 8.

Because the sample variance is a sum of squares, it will have units that are the square of the original units of measurement. That is, if we originally measured our data in miles, then s^2 will be in miles squared. To overcome this problem we define the *standard deviation* as the square root of the variance.

Definition

In either case, population or sample, the positive square root of the variance is called the **standard deviation.** The population standard deviation is denoted by σ. The sample standard deviation is given by

$$s = \sqrt{\frac{\Sigma(y_i - \bar{y})^2}{n - 1}}$$

EXAMPLE 3.15

The two data sets corresponding to the 10-point quiz in Example 3.14 are presented here.

Class A	4	6	3	5	5	4	7	5	6	4	6	6	4	3	7
Class B	6	3	8	5	9	7	0	1	7	6	5	9	2	4	3

Calculate the sample standard deviation of each class.

SOLUTION   To find the standard deviation we must first find the sample variance.

Solution for Data Set A: For ease of computation it is suggested that the data be put in tabular form as follows:

y_i	$y_i - \bar{y}$	$(y_i - \bar{y})^2$
4	$4 - 5 = -1$	1
6	$6 - 5 = +1$	1
3	$3 - 5 = -2$	4
5	$5 - 5 = 0$	0
5	$5 - 5 = 0$	0
4	$4 - 5 = -1$	1
7	$7 - 5 = +2$	4
5	$5 - 5 = 0$	0
6	$6 - 5 = +1$	1
4	$4 - 5 = -1$	1
6	$6 - 5 = +1$	1
6	$6 - 5 = +1$	1
4	$4 - 5 = -1$	1
3	$3 - 5 = -2$	4
7	$7 - 5 = +2$	4

$\bar{y} = 75/15 = 5$ $\qquad\qquad\qquad$ $s^2 = 24/14 = 1.714$

Thus $s = \sqrt{1.714} = 1.309$

Solution for Data Set B: Again, to find the sample variance using the definition, we use the following tabular form:

y_i	$y_i - \bar{y}$	$(y_i - \bar{y})^2$
6	$6 - 5 = +1$	1
3	$3 - 5 = -2$	4
8	$8 - 5 = +3$	9
5	$5 - 5 = 0$	0
9	$9 - 5 = +4$	16
7	$7 - 5 = +2$	4
0	$0 - 5 = -5$	25
1	$1 - 5 = -4$	16
7	$7 - 5 = +2$	4
6	$6 - 5 = +1$	1
5	$5 - 5 = 0$	0
9	$9 - 5 = +4$	16
2	$2 - 5 = -3$	9
4	$4 - 5 = -1$	1
3	$3 - 5 = -2$	4

$\bar{y} = 75/15 = 5$ $\qquad\qquad\qquad$ $s^2 = 110/14 = 7.857$

Thus $s = \sqrt{7.857} = 2.803$

We see that the sample standard deviation for data set A is not as large as the standard deviation for data set B. Therefore we say that data set B is more variable.

An alternative expression for s, called the *computational form,* will often make the calculations easier.

▲ Many hand-held calculators will calculate both $\bar{y}$ and s by keying in the data and pressing the appropriate buttons. All statistics students should have such a calculator.

It should be pointed out that the computational form is often easier to use with a hand-held calculator. Several significant digits should be carried in the calculation, especially when $\bar{y}$ is very large in comparison with the value of s.

EXAMPLE 3.16

To illustrate the computational form, calculate s for data set B in Example 3.15.

SOLUTION   To use the computational form, we need only two columns, both of which are totaled.

y_i	y_i^2
6	36
3	9
8	64
5	25
9	81
7	49
0	0
1	1
7	49
6	36
5	25
9	81
2	4
4	16
3	9
75	485

Therefore

$$s^2 = [485 - (75)^2/15]/14$$
$$= 110/14 = 7.857$$

Thus $s = \sqrt{7.857} = 2.803$, which is the same as that obtained by the definitional form.

Interpreting the Standard Deviation

An interpretation of s is not as easy as, say, the interpretation of the mean as a balance point of a distribution. As we saw in Example 3.15, when we compare two data sets, a larger value of s in one reflects greater variation of the observations from the mean than in the other. But how do we explain variability for a single set of data?

One interpretation is to indicate the percent of the data that is within a specified number of standard deviations of the mean. For example, what percent of the distribution is within one standard deviation of the mean? within two standard deviations of the mean? The answer, of course, depends on the shape of the distribution. If the frequency distribution is mound-shaped, then the *Empirical rule* gives an approximate answer to the percent of the distribution within one, two, and three standard deviations of the mean. Theoretically speaking, the Empirical rule applies to a population distribution and its mean, μ, and standard deviation, σ. However, it gives a reasonable approximation when applied to mound-shaped data. We will state it as it applies to sample data.

Empirical Rule Applied to Sample Data

If a stem and leaf plot, histogram, or a similar descriptive tool has a "bell"-shaped appearance, then

1. Approximately 68% of the measurements will fall within one standard deviation of the mean.

   The boundaries are $\bar{y} \pm s$.

2. Approximately 95% of the measurements will fall within two standard deviations of the mean.

   The boundaries are $\bar{y} \pm 2s$.

3. Essentially all the measurements will fall within three standard deviations of the mean.

   The boundaries are $\bar{y} \pm 3s$.

For distributions that are not necessarily bell shaped, in fact, for any shaped distribution, we have Chebyshev's rule.

Chebyshev's Rule

Regardless of the shape of the distribution we have

1. At least three-fourths of the measurements will fall within two standard deviations of the mean.

   The boundaries are $\bar{y} \pm 2s$.

2. At least eight-ninths of the measurements will fall within three standard deviations of the mean.

   The boundaries are $\bar{y} \pm 3s$.

EXAMPLE 3.17

A standardized math test was given to 30 students in the tenth grade. Their scores were

| 44 | 49 | 62 | 45 | 51 | 59 | 57 | 55 | 70 | 64 | 54 | 58 | 65 | 75 | 43 |
| 42 | 67 | 63 | 71 | 54 | 60 | 53 | 40 | 49 | 52 | 50 | 54 | 61 | 42 | 38 |

Interpret the amount of variability in the data set using either the Empirical rule or Chebyshev's rule.

SOLUTION   With a hand-held calculator we find

$$\bar{y} = 54.9 \quad \text{and} \quad s = 9.75$$

and drawing a stem and leaf plot we have

```
3 | 8
4 | 0 2 2 3 4 5 9 9
5 | 0 1 2 3 4 4 4 5 7 8 9
6 | 0 1 2 3 4 5 7
7 | 0 1 5
```

From the stem and leaf plot of the data, it appears that the data are somewhat bell shaped, so we apply the Empirical rule, which says that approximately 68% of the measurements should be between

$$\bar{y} - s \quad \text{and} \quad \bar{y} + s$$

which is between

$$54.9 - 9.75 = 45.15 \quad \text{and} \quad 54.9 + 9.75 = 64.65$$

The actual percent is $18/30 = .60 = 60\%$.

The Empirical rule further says that approximately 95% of the measurements should be between

$$\bar{y} - 2s \quad \text{and} \quad \bar{y} + 2s$$

which yields

$$54.9 - 2(9.75) = 35.4 \quad \text{and} \quad 54.9 + 2(9.75) = 74.4$$

The actual percent is $29/30 = .967 = 96.7\%$.

Finally, all measurements are between

$$\bar{y} - 3s \quad \text{and} \quad \bar{y} + 3s$$

or between

$$54.9 - 3(9.75) = 25.65 \quad \text{and} \quad 54.9 + 3(9.75) = 84.15$$

If for some reason we thought that the distribution was not bell shaped, we could apply Chebyshev's rule, which says at least three-fourths of the data is between 35.4 and 74.4, and at least eight-ninths of the data is between 25.65 and 84.15. Clearly, both of these conditions are satisfied by our data.

We saw in Section 3.3 that outliers tend to have an adverse effect on the value of $\bar{y}$ (its value is pulled toward the long tail). The standard deviation is also extremely sensitive to outliers. Example 3.18 shows that if the distribution has symmetrical long tails, then the sample mean is not distorted (the two long tails tend to cancel each other's effect), but the standard deviation, s, is severely affected by the long tails.

▲ Symmetrical means that the left half is a mirror image of the right half.

EXAMPLE 3.18

Calculate $\bar{y}$ and s for the following data. Then recalculate $\bar{y}$ and s after having trimmed three observations from each end of the sample.

Ordered stem and leaf plot

```
 4 | 8
 5 |
 6 | 7
 7 | 3
 8 | 1  3  6
 9 | 2  3  4  4  4  5  7  8  8  9
10 | 0  1  2  2  3  5  6  7  8
11 | 5  7
12 | 3
13 | 4
14 | 9
```

Trimmed stem and leaf plot

```
 8 | 1  3  6
 9 | 2  3  4  4  4  5  7  8  8  9
10 | 0  1  2  2  3  5  6  7  8
11 | 5  7
```

SOLUTION   Using the complete data set and a calculator, we find

$$\bar{y} = 98.8 \quad \text{and} \quad s = 18.92$$

But for the trimmed data, we find

$$\bar{y} = 98.75 \quad \text{and} \quad s = 8.79$$

The long tails do not affect the value of $\bar{y}$, but we see that s is more than doubled.

(Note: In Chapter 8 we will see that if we wish to calculate a "trimmed" standard deviation, it is not enough simply to calculate the standard deviation of the trimmed sample as has been done here. An alternative to the trimmed sample, called the Winsorized sample, will be found.)

Other measures of variability, some of which are *resistant* to the influence of outliers, can be obtained from the quantiles found in the quantile summary diagram.

Definitions

The **Q-spread** is the distance between the first and third sample quartiles, $Q_3 - Q_1$.

The **E-spread** is the distance between the first and seventh sample eighths, $E_7 - E_1$.

The corresponding population spreads are similarly defined using the population quantiles in place of the sample quantiles.

EXAMPLE 3.19

From the following quantile summary diagram of the data that appeared in Example 3.13, find the Q-spread, the E-spread, and the range, R.

	30	
M(15.5)	98.5	
Q(8)	93	106
E(4.5)	82	116
R	48	149

The Q-spread is the difference between Q_3 and Q_1, so

$$Q\text{-spread} = Q_3 - Q_1 = 106 - 93 = 13$$

The E-spread is the difference between E_7 and E_1, so

▲ Note that the range is *not* resistant to outliers, but rather is highly influenced by them.

$$E\text{-spread} = E_7 - E_1 = 116 - 82 = 34$$

The range is the difference between H and L, so

$$\text{Range} = H - L = 149 - 48 = 101$$

This information is conveniently placed next to the midsummaries on the quantile summary diagram. Thus, we have

	30		Midsummaries	Spreads
M(15.5)	98.5		98.5	
Q(8)	93	106	99.5	13
E(4.5)	82	116	99	34
R	48	149	98.5	101

EXERCISES 3.4

3.18 For the following data

7, 12, 10, 9, 22

(a) Check that the deviations from the mean sum to zero.
(b) Calculate the standard deviation.

3.19 For the following data

6.8, 5.7, 9.2, 8.4, 7.4

(a) Check that the deviations from the mean sum to zero.
(b) Calculate the standard deviation.

3.20 Calculate s^2 by the computational formula for the data:

30, 18, 35, 22, 36, 18, 21, 31, 28

3.21 A social scientist examines the records at several hospitals and finds the following sample of women's ages at the birth of their first child:

30	18	35	22	23	22	36	24
23	28	19	23	25	24	33	21
24	19	33	23	19	32	21	18
36	21	25	17	21	24	39	22
23	18	22	28	18	15	25	21
23	26	38	24	20	36	27	21
28	26	22	28	33	18	17	21
15	20	16	21	23	15	20	38
16	24	42	22	24	24	20	17
26	39	22	21	28	20	29	14
25	20	19	17	21	24	26	

Construct a quantile summary diagram to include midsummaries and spreads.

3.22 Following are test scores for two beginning statistics classes:

Class 1							
81	73	86	90	75	80	75	81
85	87	83	75	70	65	80	76
64	74	86	80	83	67	82	78
76	83	71	90	77	81	82	

Class 2							
87	77	66	75	78	82	82	71
79	73	91	97	89	92	75	89
75	95	84	75	82	74	77	87
69	96	65					

Construct back-to-back stem and leaf plots and compare the standard deviations of the two classes.

3.5   Correlation

If two variables are related in such a way that the points of a scatterplot tend to fall in a straight line, then we say that there is an association between the variables and that they are linearly correlated. The most common measure of the strength of the association between the variables is the *Pearson correlation coefficient*.

Definition

The **Pearson correlation coefficient,** denoted by r, is given by the following expression:

$$r = \frac{SS_{xy}}{\sqrt{SS_{xx}SS_{yy}}}$$

where

$$SS_{xy} = \Sigma x_i y_i - (\Sigma x_i)(\Sigma y_i)/n$$
$$SS_{xx} = \Sigma x_i^2 - (\Sigma x_i)^2/n$$
$$SS_{yy} = \Sigma y_i^2 - (\Sigma y_i)^2/n$$

The population correlation coefficient is denoted by the Greek letter ρ (rho).

EXAMPLE 3.20

A laboratory wishes to study the relationship between the dose of a growth stimulant and weight gain for laboratory animals. Seven animals of the same sex, age, and size are selected and randomly assigned to one of seven dosage levels of the growth stimulant.

Dosage	0.0	1.0	2.0	3.0	4.0	5.0	6.0
Weight gain	1.0	1.2	2.0	2.4	3.4	4.9	5.1

Plot the data in a scatterplot and calculate the correlation between the two variables.

SOLUTION   To plot the data we place the growth stimulant on the *x*-axis and the weight gain on the *y*-axis. The data in Figure 3.5 appear to fall close to a straight line, which suggests a strong correlation. The value of r should bear this out.

FIGURE 3.5

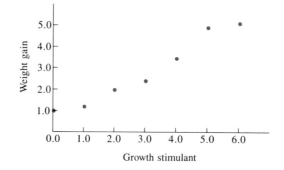

The computations for r are simplified by placing the data in the following table:

x_i	y_i	x_i^2	y_i^2	$x_i y_i$
0.0	1.0	0.0	1.00	0.0
1.0	1.2	1.0	1.44	1.2
2.0	2.0	4.0	4.00	4.0
3.0	2.4	9.0	5.76	7.2
4.0	3.4	16.0	11.56	13.6
5.0	4.9	25.0	24.01	24.5
6.0	5.1	36.0	26.01	30.6
Totals 21.0	20.0	91.0	73.78	81.1

Therefore

$$SS_{xy} = 81.1 - (21)(20.0)/7 = 21.1$$
$$SS_{xx} = 91.0 - (21)^2/7 = 28.0$$
$$SS_{yy} = 73.78 - (20.0)^2/7 = 16.64$$

Thus we have

$$r = 21.1/\sqrt{(28.0)(16.64)} = .9775$$

which is a very high correlation.

Interpretation of r

▲ Caution—there could be a nonlinear relationship between x and y, and r could still be near 0.

The value of r will always be between -1 and $+1$. The closer it is to either -1 or $+1$ the stronger the *linear* relationship between x and y. If r is 0, then x and y are not linearly related. Figure 3.6 illustrates three scatterplots of data for various values of r. If r is positive, then there is a positive linear relationship between the two variables, meaning that as one increases the other also increases. If r is negative, then there is a negative linear relationship, meaning that as one variable increases the other decreases.

Another useful interpretation of the correlation coefficient involves its square, r^2. Note that SS_{yy} is the numerator of the sample variance for the y data, and thus in a sense, is a measure of the *total* variability of the y values about their mean. Then treating y as the *dependent variable* and x as the *independent variable,* we have that r^2 is the amount of the variability in y explained by the x variable. The quantity r^2 is referred to as the *coefficient of determination*.

FIGURE 3.6

Scatterplots for various values of *r*

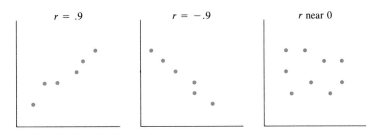

Definition

The **coefficient of determination** is given by

$$r^2 = \frac{(SS_{xy})^2}{SS_{xx}SS_{yy}}$$

and can be interpreted as the percent of the variability in the dependent variable that is explained by the independent variable.

EXAMPLE 3.21

Calculate the coefficient of determination for the data in Example 3.20 and give an interpretation.

SOLUTION   In Example 3.20 we found that $r = .9775$, and thus the coefficient of determination is

$$r^2 = .9555$$

Thus the growth stimulant explains 95.6% of the variability in weight gain in the laboratory animals.

Spearman's Rank Correlation

A measure of correlation for ordinal data is given by *Spearman's rank correlation* coefficient, r_S. Given the sample $(x_1, y_1), (x_2, y_2), \ldots, (x_n, y_n)$, ranks of 1 to *n* are assigned to the *x*'s and to the *y*'s separately. For tied observations, the average of the ranks that would have been assigned had there been no ties is given to each tied observation. The formula for r_S is the same as the formula for Pearson's correlation coefficient except that it is applied to the ranks of the data.

EXAMPLE 3.22

A married couple who have children were asked to rank 10 points in raising children from the most important (10) to the least important (1). From the collected data compute Spearman's rank correlation coefficient.

Husband's ranking x_i	Wife's ranking y_i
1	6	6
2	3	3
3	1	2
4	7	9
5	2	1
6	8	7
7	4	5
8	9	8
9	5	4
10	10	10

SOLUTION The data are already in the form of ranks, and therefore we apply Pearson's correlation formula to the ranks.

In tabular form we have

	x_i	y_i	x_i^2	y_i^2	$x_i y_i$
	6	6	36	36	36
	3	3	9	9	9
	1	2	1	4	2
	7	9	49	81	63
	2	1	4	1	2
	8	7	64	49	56
	4	5	16	25	20
	9	8	81	64	72
	5	4	25	16	20
	10	10	100	100	100
Totals	55	55	385	385	380

▲ The sum of the ranks from 1 to n is always the sum of the first n integers, which can be found by the formula $n(n + 1)/2$. The sum of the squares of the integers is $n(n + 1)(2n + 1)/6$.

Therefore

$$\text{SS}_{xy} = 380 - (55)(55)/10 = 77.5$$
$$\text{SS}_{xx} = 385 - (55)^2/10 = 82.5$$
$$\text{SS}_{yy} = 385 - (55)^2/10 = 82.5$$

Thus we have

$$r = 77.5/\sqrt{(82.5)(82.5)} = .9394$$

Spearman's rank correlation can also be applied to interval/ratio data. All that is required is that the data for each variable be ranked from smallest to largest, and then the formula is applied to the ranks. Again, if there are tied observations (have the same value), then average ranks are assigned.

Like the median, Spearman's rank correlation is resistant to the influence of outliers. It treats all observations equally, so that outliers have no more effect on the outcome than any other observation. Additionally, Spearman's correlation identifies both linear and nonlinear relationships. For example, a weight lifter makes great strides on the amount he can lift early in his training career. Later in his training, however, he may be able to add only ounces to the amount he can lift from week to

FIGURE 3.7

Lift of a weight lifter as it relates to the length of training

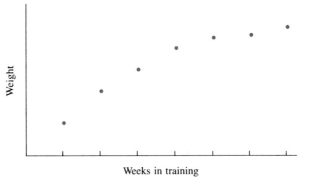

week. Figure 3.7 illustrates that this is not a linear relationship; it is called a *monotonic* relationship.

Unlike Pearson's correlation coefficient, Spearman's correlation coefficient will measure nonlinear trend as above.

EXAMPLE 3.23

A math teacher suspects that her students' algebra test scores are related to the amount of test anxiety exhibited prior to the test. With an anxiety scale that ranges from 0 to 100 (the higher the score the more anxiety exhibited), she records each student's anxiety score and algebra score. From the data, calculate (a) Pearson's correlation and (b) Spearman's correlation; (c) comment on the two correlations.

Anxiety score	Anxiety rank	Math score	Math rank
61	5	76	5
47	2	84	8
82	9	51	2
65	6	79	6
72	7	71	4
84	10	28	1
55	4	87	9
49	3	82	7
75	8	60	3
12	1	98	10

SOLUTION   (a) From the scores we find

$$\Sigma x_i = 602 \quad \Sigma x_i^2 = 40{,}314 \quad \Sigma y_i = 716 \quad \Sigma y_i^2 = 54{,}996 \quad \Sigma x_i y_i = 39{,}844$$

so that

$$SS_{xy} = 39{,}844 - (602)(716)/10 = -3{,}259.2$$
$$SS_{xx} = 40{,}314 - (602)^2/10 = 4{,}073.6$$
$$SS_{yy} = 54{,}996 - (716)^2/10 = 3{,}730.4$$

Thus we have

$$r = -3{,}259.2/\sqrt{(4{,}073.6)(3{,}730.4)} = -.836$$

(b) From the ranks we find

$$\Sigma x_i = 55 \qquad \Sigma x_i^2 = 385 \qquad \Sigma y_i = 55 \qquad \Sigma y_i^2 = 385 \qquad \Sigma x_i y_i = 224$$

so that

$$SS_{xy} = 224 - (55)(55)/10 = -78.5$$
$$SS_{xx} = 385 - (55)^2/10 = 82.5$$
$$SS_{yy} = 385 - (55)^2/10 = 82.5$$

Thus we have

$$r_S = -78.5/\sqrt{(82.5)(82.5)} = -.9515$$

(c) The last anxiety score of 12 is an unusually low score (an outlier), which decreases the Pearson correlation. The exact numerical value of the score is not important; what is important is that it is the lowest anxiety score, and we would expect the student to score high on the algebra test. Spearman's correlation ignores the fact that the score is a 12 and simply considers it as the lowest score, a rank of 1, and is not affected by its magnitude. It is for this reason that Spearman's correlation of $-.9515$ is the more representative value of the association between the anxiety scores and the algebra scores.

EXERCISES 3.5

3.23 For the following data sets construct a scatterplot and calculate the Pearson's correlation coefficient between x and y:

(a)

x	2	1	3	2	3	1
y	4	0	6	6	8	2

(b)

x	0	1	2	3	4	5	6
y	10	8	7	6	5	5	3

3.24 Calculate Pearson's correlation coefficient for the following data:

x	42	61	12	71	52	48	74	65	53
y	75	49	95	64	83	84	38	58	81

x	63	55	94	19
y	47	78	51	93

3.25 Rank the data in Exercise 3.24 and calculate Spearman's correlation coefficient.

3.26 Calculate Pearson's correlation coefficient for the following data:

x	85	84	79	82	86	81	13	91
y	7.4	6.3	4.9	5.3	8.1	5.1	4.2	9.8

3.27 Rank the data in Exercise 3.26 and calculate Spearman's correlation coefficient.

3.28 Recent studies suggest that smoking during pregnancy has an effect on the birth weights of newborn infants. To study this issue further a sample of 16 women smokers was asked to estimate the average number of cigarettes smoked per day. Following are their estimates and the birth weights (in pounds) of their children.

No. of cigarettes	22	16	4	19	42	8	12	30
Birth weight	6.4	7.2	8.1	6.9	6.1	8.4	7.6	6.5
No. of cigarettes	14	16	5	20	32	2	15	48
Birth weight	8.4	8.1	8.5	6.6	6.0	7.9	7.1	5.5

Calculate Pearson's correlation coefficient.

3.29 From the data in Exercise 3.28, calculate the coefficient of determination and give an interpretation.

3.30 A group of 15 football recruits was ranked by two coaches as to the potential value of each recruit to the team. From the following ranks compute Spearman's correlation coefficient.

	Ranks									
Player	1	2	3	4	5	6	7	8	9	10
Coach A	3	12	9	1	5	14	8	10	2	13 (cont.)
Coach B	5	10	7	1	8	15	9	12	3	13

		Ranks			
Player	11	12	13	14	15
Coach A	6	4	11	15	7
Coach B	4	2	11	14	6

3.31 A professor ranked her history class at the beginning of the course based on each student's performance in a group discus-

sion. Following are her rankings together with the final grades for the class:

Student	1	2	3	4	5	6	7	8	9
Ranking	6	3	2	8	4	9	1	5	7
Grade	72	91	90	68	89	55	95	84	62

Correlate the rankings with the grades received in the class.

3.6   Distorting the Truth (optional)

Whenever you are presented with data and asked to draw a conclusion from it, you should be concerned with the origin of the data. You should also be concerned with the reliability of the source, with the technique used to collect the data, and with the way in which the data are presented. For example, suppose it is reported that four out of five dentists recommend brand A toothpaste. Was this a scientific poll of dentists across the United States by an independent agency, or did the manufacturer of brand A toothpaste poll five preselected dentists? These are considerations that must be addressed when you are evaluating the presented data. You should not be skeptical of all data presented to you, but you should learn to challenge statistical data that you read, see, and hear.

We are speaking of the reliability of the source of the data, which should not be confused with the "statistical" reliability. Statistical reliability is concerned with the ideas of accuracy, precision, and magnitude of the error in an estimation problem and will be discussed later in the book. Here we are concerned with the reliability or confidence we have in the source and the usefulness of the data.

Some questions that might be useful in determining the validity of the data and the reliability of their sources are:

1. Is there reason for the source to be biased in reporting the results?
2. What procedure was used to collect the data?
3. Are the data reasonable; would you normally expect such results?
4. Are the data useful, relevant, and reported properly?
5. Are the graphs misleading?

Is the Source Biased?

The field of advertising is extremely competitive. Manufacturers are constantly dreaming up schemes to market their products. A very effective method of advertising is to quote statistical data similar to the "Four out of five dentists recommend" statement. When the toothpaste manufacturer declares its product superior, we must question the motive. If a competitor did the study, would the results be the same?

EXAMPLE 3.24

In a 1976 Pepsi television commercial, Coke drinkers were given a glass of Coke labeled "Q," and a glass of Pepsi labeled "M." Over half chose the brand labeled as

"M" (Moore, 1979, p. 89). Can Pepsi be accused of deceptive advertising? Coca-Cola thought so. Why?

SOLUTION The drinks are very similar, and many people would have trouble distinguishing between the two. In those cases, the subject most likely would choose the more appealing letter "M."

The person reporting the statistics may not be aware of the fact that his or her source is unreliable, as noted in the following example.

EXAMPLE 3.25

A newspaper columnist once reported that 500,000 Vietnam war veterans had attempted suicide. This figure seems farfetched. The columnist later admitted that the figure had come from an article in *Penthouse* magazine, which had obtained it from a radical veterans' group that presumably made it up (Haack, 1979, p. 20).

Were the Data Collected Properly?

Were the data obtained by an independent party? Was a scientific procedure used to collect the data? Did the data represent the intended population? Was the sample size adequate? Were the data contrived from other information? Each of these questions must be answered when you are examining the origin of the data.

We hope that the data are collected by a party that will not gain monetarily by the results. A scientific procedure should always be used in designing an experiment or conducting a survey sample.

The following advertisement has appeared in many magazines, newspapers, and on TV.

"Duracell Batteries Last 5 Times Longer"

Longer than what? If we read closely, we see in fine print that "Duracell batteries last *up to five* times longer than regular *carbon* batteries." Were the Duracell batteries also carbon? We all know that alkaline batteries last longer than carbon batteries. Were the tested batteries of comparable price? We are left to guess at the answers to these questions.

We have already seen (1936 presidential election) that, even with good intentions, it is possible to collect a sample that does not represent the intended population. Many organizations conduct their own polls and try to generalize the results. Magazines such as *People, Glamour,* and *Ladies' Home Journal* conduct polls of their readers. Do their results reflect the views of all Americans or just the views of their readers?

Are the Data Reasonable?

It has been reported that 80% of Miami's economy depends on drugs. First, this is an extremely high figure and, second, how can it be verified? Dealing in drugs is illegal and much of it goes undetected, so there is no way of knowing what percent of the economy is dependent on drugs.

Also, we should be aware of variability in measurements. Two people using the same device to measure a patient's blood pressure will probably get different readings. However, the readings obtained by each would be valid measurements of the blood pressure if the procedure were carried out properly. In fact, if the measurements were exactly the same, then we would begin to worry, as is illustrated in Example 3.26.

Cyril Burt was an English psychologist who studied the IQ scores of identical twins who were raised apart. A high correlation between the IQ's of the separated twins indicates that heredity is a factor in IQ. In 1955, Burt reported a correlation of .771 for 21 pairs of twins. In 1958, he reported a correlation of .771 for "over 30" pairs of twins. In 1966, he reported a correlation of .771 for a sample of 53 pairs of twins. History has discounted Burt's research as useless and perhaps fraudulent (Wade, 1976).

Often we are only given summary information. The "raw data" (original data) are not supplied. It is possible that only favorable results are reported and unfavorable results are ignored. Also, if only summary information is supplied, we should be concerned with misinterpretation.

EXAMPLE 3.27

In discussing energy sources, a national columnist reported that New England supplied only 9.1% of its own energy needs. This figure seems extremely low. In fact, the columnist later corrected himself by stating that New England supplied 9.1% of its energy needs with natural gas.

It is a widely held opinion that housing costs today are high because labor and materials are so costly. To the contrary, the National Association of Home Builders reported that in 1949 labor and materials accounted for 69% of the cost of a new single-family home, but by 1969 the figure had declined to 55%, and by 1982 it was only 45% of the cost of a new home. Also, today's materials are better and homes are more efficient. The real cost of homes today is in financing and land. In 1949 financing was 5% and land was 11% of the total cost. In 1982 financing costs had risen to 15% and land to 24% of the total costs (Watauga *Democrat,* August 3, 1984). Confusion also exists about the prices of new homes. Today the average price is over $100,000, whereas the median price is somewhere between $75,000 and $80,000. Later we will discuss which of these two measures we should use to describe the cost of a "typical house."

Are the Data Useful, Relevant, and Reported Properly?

Figure 3.8 is the front page headline of a major regional newspaper of October 1982. This headline is in obvious error. It would lead one to believe that the Dow Jones Industrial Average dropped to the level of the 1929 market. On October 28, 1929, the beginning of the Great Crash, the Dow Jones Industrial Average of 30 industrials dropped 38.33 points in 5 hours to a value in the 200–300 range. The headline suggests that the Dow dropped to that same level in 1982. The headline intended to point out that the drop of 36.33 points on October 25, 1982 was the largest point drop since the 38.33 drop in 1929. Less dramatic is the fact that on October 25, 1982 the Dow dropped 36.33 to 995.13, a 3.522% drop; whereas the 38.33 drop on October 28, 1929 accounted for a 12.8% drop.

▲ The Dow Jones Industrial Average dropped 86 points on September 11, 1986 when it was at the 1,900 level.

FIGURE 3.8

Stock Market Dips To Lowest Since '29

As reported in the August 6, 1984, issue of *USA Today,* "Robbers steer clear of Montana's banks." (See Figure 3.9.) Could it be that Montana has no banks? The data point out that Montana had no bank robberies from July 1982 through June 1983. Obviously Montana has fewer banks than, say, California, which had 2,383 bank robberies in the same time period. Knowing the number of banks in each state would help in interpreting this data.

The data in Table 3.1 suggest that of the five southwestern states listed, California and Texas are getting all of the scholarships given in that region in the thirtieth annual Merit Scholarship Program.

FIGURE 3.9

Robbers steer clear of Montana's banks

From July 1982 through June 1983 there were 6,094 bank robberies: the use of violence or threat to wrongfully take property. There were also 444 bank burglaries: entering a building unlawfully with the intent to commit a crime.

State	Bank robberies*	Bank burglaries*	State	Bank robberies*	Bank burglaries*
Ala.	38	9	Neb.	21	3
Alaska	22	1	Nev.	81	5
Ariz.	111	6	N.H.	8	1
Ark.	6	6	N.J.	152	20
Calif.	2,383	63	N.M.	18	3
Colo.	97	4	N.Y.	520	25
Conn.	49	10	N.C.	84	7
Del.	8	1	N.D.	7	1
D.C.	64	0	Ohio	159	9
Fla.	279	38	Okla.	29	14
Ga.	65	12	Ore.	170	2
Hawaii	30	1	Pa.	211	9
Idaho	12	0	R.I.	4	2
Ill.	63	14	S.C.	28	9
Ind.	79	8	S.D.	4	0
Iowa	16	2	Tenn.	56	9
Kan.	36	1	Texas	237	27
Ky.	35	14	Utah	45	0
La.	91	17	Vt.	1	2
Maine	10	3	Va.	149	16
Md.	116	6	Wash.	127	5
Mass.	96	2	W.Va.	6	3
Mich.	97	18	Wis.	41	7
Minn.	56	13	Wyo.	7	0
Miss.	22	8	**Total**	**6,094**	**444**
Mo.	48	7			
Mont.	0	1	*Federally chartered banks		

SOURCE: *USA Today,* August 6, 1984. Copyright, 1984 *USA Today.* Reprinted with permission. Data from 1984 FBI crime statistics.

TABLE 3.1

Number of semifinalists in the
thirtieth annual Merit Scholarship
program for five selected
southwestern states

State	No. of scholarships	Rel. freq.
Arizona	148	.0597
California	1,294	.5220
Nevada	49	.0198
New Mexico	95	.0383
Texas	893	.3602
Total	2,479	1.0000

SOURCE: National Merit Scholarship Corp.

The truth is that they should get most of the scholarships. If we also consider the populations of the five states, we see that California and Texas comprise approximately 89% of the total population. Thus they should have approximately 89% of the scholarships. If we consider the rate of scholarships per 100,000 residents, as presented in Table 3.2, we see that California has the lowest rate and New Mexico and Nevada the highest.

TABLE 3.2

Scholarship semifinalists in the
thirtieth annual Merit Scholarship
program for five selected
southwestern states

State	No. of scholarships	Rate per 100,000 residents
Arizona	148	62.87
California	1,294	58.04
Nevada	49	77.77
New Mexico	95	78.38
Texas	893	68.62
Total	2,479	

Misleading Graphs

The histogram and bar graph are popular devices used to exhibit data. We should be aware, however, that they can give false impressions. The frequency or relative frequency of the various classes should be reflected only in the height of the bars. Figure 3.10 is an example of a bar graph where the bars have been replaced with

FIGURE 3.10

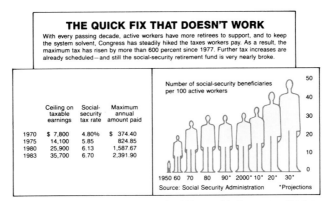

THE QUICK FIX THAT DOESN'T WORK

With every passing decade, active workers have more retirees to support, and to keep the system solvent, Congress has steadily hiked the taxes workers pay. As a result, the maximum tax has risen by more than 600 percent since 1977. Further tax increases are already scheduled—and still the social-security retirement fund is very nearly broke.

	Ceiling on taxable earnings	Social-security tax rate	Maximum annual amount paid
1970	$ 7,800	4.80%	$ 374.40
1975	14,100	5.85	824.85
1980	25,900	6.13	1,587.67
1983	35,700	6.70	2,391.90

Number of social-security beneficiaries per 100 active workers

1950 60 70 80 90* 2000* 10* 20* 30*
Source: Social Security Administration   *Projections

SOURCE: *Newsweek*, September 24, 1984, p. 33. Copyright 1984, by *Newsweek*, Inc. All Rights Reserved. (*Newsweek*—Christopher Blumrich) Reprinted with permission.

pictures. This is acceptable; however, these data are misleading because the widths of the "so-called" bars are not the same. By changing both the width and height of the bars, we get a distorted view of the actual frequencies because our eyes respond to the difference in areas.

The same is true of the pictogram shown in Figure 3.11. Here the bags are intended to illustrate that per capita income in the United States doubled from $4,000 in 1970 to $8,000 in 1978. Again the picture does not convey a "fair" impression of the actual change. Not only is the difference reflected in the heights of the two bags, but also is reflected in the areas covered by the bags.

In the next example we see that the scale used for either the vertical or horizontal axes can conceal or reveal information. Table 3.3 gives the world population in 1950, 1960, 1970, 1980, and projections for the years 1990 and 2000.

FIGURE 3.11

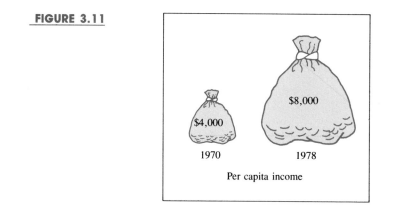

Per capita income

TABLE 3.3

Year	World population (in billions)
1950	2.5
1960	3.0
1970	3.7
1980	4.4
1990 (est.)	5.2
2000 (est.)	6.3

SOURCE: Data from *U.S. News and World Report,* July 23, 1984.

These data can be used to show a moderate or a high growth by manipulating the vertical and the horizontal scales. To show a moderate growth of the world population over this time period, we could shrink down the vertical scale of a bar graph or frequency polygon as has been done in Figure 3.12. You can see that the population growth is barely apparent.

FIGURE 3.12
World population in billions

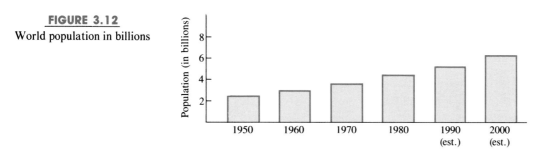

Using the same data, we could make the population growth appear large by stretching out the vertical axis as has been done in Figure 3.13. Then it appears that the world population is growing without bounds.

To properly exhibit the data, the vertical scale should cover about the same distance as the horizontal scale as in Figure 3.14. (This bar graph also appeared in Figure 2.15.)

Figure 3.15 gives a deceptive view of the trade deficit with Japan. To dramatize the decline, the graph has been rotated. A rotation in the opposite direction would make the graph appear to *increase* rather than decrease.

FIGURE 3.13
World population in billions

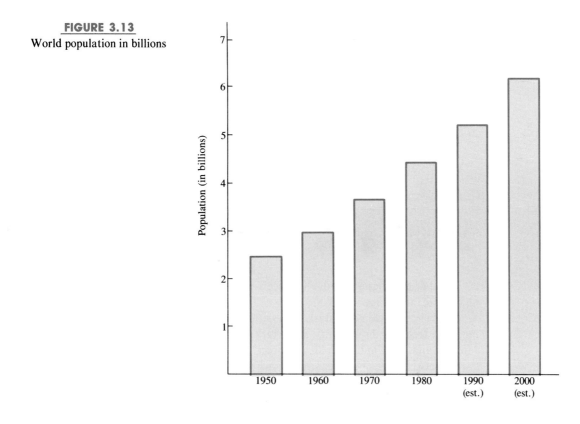

FIGURE 3.14
World population in billions

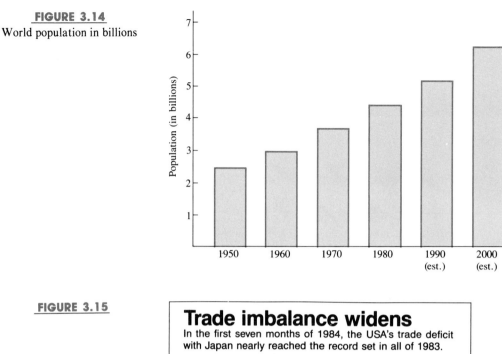

FIGURE 3.15

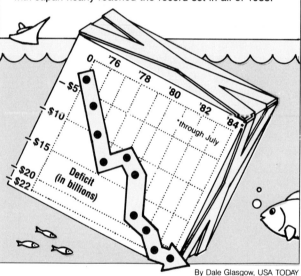

SOURCE: *USA Today*, October 9, 1984. Copyright, 1984 *USA Today*. Reprinted with permission. Data from Data Resources Inc.

Another technique that can be deceiving is to start the vertical scale at some point above zero, as in Figure 3.16.

Here the vertical scale starts at $39,000, giving the appearance that Manufacturing, Research and engineering, and Marketing managers earn significantly more than Finance and legal and Human resources and public relations managers. In fact, the graph would have us believe that Finance and legal managers earn very little. Starting the vertical scale at zero, as in Figure 3.17, we see only a slight difference in the salaries of the various managers.

FIGURE 3.16

USA SNAPSHOTS

A look at statistics that shape your finances

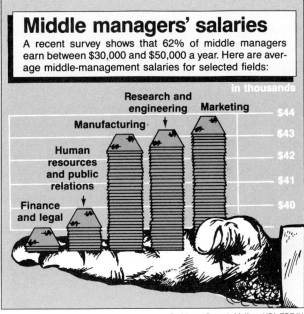

Middle managers' salaries

A recent survey shows that 62% of middle managers earn between $30,000 and $50,000 a year. Here are average middle-management salaries for selected fields:

By Marcy Eckroth Mullins, USA TODAY

SOURCE: *USA Today*, September 14, 1984. Copyright, 1984 *USA Today*. Reprinted with permission. Data from the Wyatt Co.

FIGURE 3.17

Middle managers' salaries

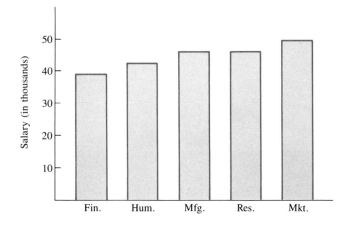

Graphs are the most widely used technique of exhibiting statistical data, and they can be very informative; however, we have seen that they can be misleading. To get the most out of a graph we must take a close look at the scales of measurement on both axes and ask ourselves whether the data have been reported properly.

3.32 The American Cancer Society and the tobacco industry disagree on the results of research concerning the tobacco–lung cancer issue. Clearly the tobacco industry has something at stake. What interest does the American Cancer Society have in this controversy?

3.33 The mayor of a major city reported that 95% of the police force has never taken a bribe. Is this figure reasonable? Can it be verified?

FIGURE FOR EXERCISE 3.34

Hawaiians got fewest speeding tickets in '83

During fiscal 1983, more than 8 million motorists on the USA's highways received tickets for exceeding the 55 mph speed limit. **(See story, 1A)**

	Tickets issued		Tickets issued
Alabama	103,875	Nebraska	77,755
Alaska	11,200	Nevada	59,885
Arizona	197,638	New Hampshire	31,472
Arkansas	67,170	New Jersey	164,426
California	1,022,180	New Mexico	150,602
Colorado	101,019	New York	247,444
Connecticut	67,130	North Carolina	226,517
Delaware	24,752	North Dakota	45,106
Florida	325,054	Ohio	399,636
Georgia	194,781	Oklahoma	141,495
Hawaii	4,759	Oregon	99,287
Idaho	30,781	Pennsylvania	218,087
Illinois	222,491	Rhode Island	27,975
Indiana	97,438	South Carolina	194,284
Iowa	135,180	South Dakota	33,084
Kansas	205,639	Tennessee	147,845
Kentucky	73,758	Texas	881,673
Louisiana	171,358	Utah	108,606
Maine	19,212	Vermont	32,900
Maryland	129,051	Virginia	212,583
Massachusetts	208,007	Washington	143,179
Michigan	202,535	West Virginia	54,553
Minnesota	110,745	Wisconsin	95,524
Mississippi	180,160	Wyoming	56,020
Missouri	184,284	Puerto Rico	46,505
Montana	96,802	**Total**	**8,083,443**

*Source: Federal Highway Administration

USA TODAY

SOURCE: *USA Today*, September 19, 1984. Copyright, 1984 *USA Today*. Reprinted with permission. Data from Federal Highway Administration.

3.34 From the data presented in the accompanying figure, supplied by the Federal Highway Administration and reported by *USA Today* (September 19, 1984), is it reasonable that Hawaii would have the fewest speeding tickets? Why would California and Texas have the most speeding tickets? What other information would help in interpreting these data? Can you think of a better way of reporting the information?

3.35 The accompanying figure illustrates the number of reports of sexual abuse of children for the years 1977 through 1982. The figures are staggering. Can you think of a reason why there was such a dramatic increase over this period of time? Does the number of cases reported give any indication of how many actual cases there were? The sizes of the children inside the rectangles are misleading to a certain extent. Can you explain why?

FIGURE FOR EXERCISE 3.35

By Julie Stacey, USA TODAY

SOURCE: *USA Today*, September 17, 1984. Copyright, 1984 *USA Today*. Reprinted with permission. Data from the American Humane Association, Denver.

3.36 On November 4, 1984, advice columnist Ann Landers asked her estimated 70 million readers: "Would you be content to be held close and treated tenderly and forget about 'the act'?"

Of the more than 90,000 who responded, 72% said they would rather be hugged. Is this a scientific poll? How reliable is this figure? Can it be generalized to the general population?

Computer Session (optional)

In Section 3.2 methods were given to modify a stem and leaf plot so that almost any data set could be exhibited in an informative manner. The STEM-AND-LEAF command (discussed in Section 2.6) in Minitab automatically orders the data and chooses the most desirable form for the plot. Useful subcommands that can be used with the STEM-AND-LEAF command (as well as the HISTOGRAM and DOT-PLOT commands) are SAME and BY C.

The SAME subcommand is used when several columns are to be compared. It uses the same scale for the plots of all columns specified on the main command. If SAME is not given, then each column listed on the main command is scaled separately. A separate plot is given for each value in C when the BY C subcommand is given. Also all plots are on the same scale.

The following Minitab commands, which relate to the material in this chapter, are self-explanatory:

```
MEAN     of the data in C [put value in C]
STDEV    of the data in C [put value in C]
MEDIAN   of the data in C [put value in C]
MINIMUM  of the data in C [put value in C]
MAXIMUM  of the data in C [put value in C]
RANK        the data in C put ranks in C
```

Two additional commands are DESCRIBE and CORRELATION.

DESCRIBE   This command prints a table of summary numbers, which includes

```
N
MEAN
MEDIAN
TRMEAN (a 5% trimmed mean)
STDEV
SEMEAN (standard error of the mean to
        be discussed in Chapter 7)
MAX
MIN
Q1
Q3
```

The form of the command is

```
DESCRIBE C, C,...,C
```

EXAMPLE 3.28

Following are the weight gains for a herd of steers after 60 days in a feedlot:

203	189	193	185	216	229	194	197	162	184
204	210	179	183	199	172	206	225	197	184

Summarize the data with the DESCRIBE command.

SOLUTION   Having stored the data in C1, the command is executed as in Figure 3.18.

FIGURE 3.18
Minitab output for Example 3.28

```
MTB > DESCRIBE C1

              N      MEAN    MEDIAN    TRMEAN     STDEV    SEMEAN
C1           20    195.55    195.50    195.55     16.87      3.77

            MIN       MAX        Q1        Q3
C1       162.00    229.00    184.00    205.00
```

CORRELATION

This command computes Pearson's correlation for the data in two specified columns. The form of the command is

```
CORRELATION coefficient between C, C,...,C
```

If two columns are specified, then the correlation between the data in those two columns is computed. If several columns are specified, then the correlation between all pairs of columns is computed.

EXAMPLE 3.29

Following are the grade point averages, SAT math scores, and final exam grade in College Algebra for a group of 20 sophomores. Determine the correlations between all pairs of variables.

Student	GPA	SAT	Final exam
1	2.6	510	84
2	2.1	460	77
3	3.5	680	94
4	1.7	390	45
5	2.2	420	71
6	2.9	510	89
7	2.3	370	65
8	3.2	550	90
9	2.3	420	82
10	2.8	470	87
11	3.9	700	99
12	2.2	450	83
13	1.4	380	63
14	2.7	440	75
15	2.8	460	79
16	3.0	520	85
17	1.8	430	70
18	2.0	410	72
19	3.6	650	98
20	2.5	500	75

SOLUTION   See Figure 3.19.

FIGURE 3.19
Minitab output for Example 3.29

```
MTB > READ DATA IN C1-C3
DATA> 2.6   510   84
DATA> 2.1   460   77
DATA> 3.5   680   94
DATA> 1.7   390   45
DATA> 2.2   420   71
DATA> 2.9   510   89
DATA> 2.3   370   65
DATA> 3.2   550   90
DATA> 2.3   420   82
DATA> 2.8   470   87
DATA> 3.9   700   99
DATA> 2.2   450   83
DATA> 1.4   380   63
DATA> 2.7   440   75
DATA> 2.8   460   79
DATA> 3.0   520   85
DATA> 1.8   430   70
DATA> 2.0   410   72
DATA> 3.6   650   98
DATA> 2.5   500   75
DATA> END

     20 ROWS READ

MTB > CORRELATION C1-C3

            C1       C2
C2       0.901
C3       0.861    0.825
```

Spearman's correlation can be computed by using the RANK command followed by the CORRELATION command, because Spearman's correlation is Pearson's correlation applied to the ranks of the data.

EXAMPLE 3.30

Find Spearman's correlation between the grade point averages and final exam scores from the data in Example 3.29.

SOLUTION   See Figure 3.20.

FIGURE 3.20
Minitab output for Example 3.30

```
MTB > RANK C1 PUT IN C4
MTB > RANK C2 PUT IN C5
MTB > RANK C3 PUT IN C6

MTB > PRINT C4-C6
```

```
ROW       C4       C5       C6

  1      11.0     14.5     13.0
  2       5.0     10.5      9.0
  3      18.0     19.0     18.0
  4       2.0      3.0      1.0
  5       6.5      5.5      5.0
  6      15.0     14.5     16.0
  7       8.5      1.0      3.0
  8      17.0     17.0     17.0
  9       8.5      5.5     11.0
 10      13.5     12.0     15.0
 11      20.0     20.0     20.0
 12       6.5      9.0     12.0
 13       1.0      2.0      2.0
 14      12.0      8.0      7.5
 15      13.5     10.5     10.0
 16      16.0     16.0     14.0
 17       3.0      7.0      4.0
 18       4.0      4.0      6.0
 19      19.0     18.0     19.0
 20      10.0     13.0      7.5


MTB > CORRELATION C4-C6

                C4       C5
     C5       0.871
     C6       0.893    0.915
```

EXERCISES 3.7

3.37 Following are the anxiety scores before a major math test for a group of tenth-grade students:

12	16	6	22	17	14	8	27	18	19
11	9	21	3	15	17	22	19	13	5

Store the data in a column of Minitab and compute the mean and standard deviation of the data, using the two separate commands. Next use the DESCRIBE command to compute the summary measures associated with it.

3.38 Following are the math test scores for the tenth-grade students whose anxiety scores are listed in Exercise 3.37. The math scores are listed in the same order as the anxiety scores.

75	56	91	48	69	73	88	65	72	65
88	94	48	99	78	65	58	52	70	90

Using Minitab, compute the correlation between the anxiety scores and the test scores.

3.39 In 1984, the U.S. Department of Education ranked the states in several categories related to public schools. A scoring method was devised by the Gannett News Service based on the Department of Education's rankings of such quantities as pupil-teacher ratios, dropout rates, per capita income spent on education, and teachers' salaries. Following are the composite scores together with the reported 1984 SAT scores for each of the 50 states and the District of Columbia.

State	Score	Verbal SAT	Math SAT
Ala.	16.0	467	503
Alaska	88.3	443	471
Ariz.	18.6	469	509
Ark.	15.8	482	521
Calif.	36.3	421	476
Colo.	38.9	465	514
Conn.	54.7	436	468
Del.	60.0	433	469
D.C.	59.5	397	426
Fla.	19.7	423	467
Ga.	13.5	392	430
Hawaii	50.7	395	474
Idaho	21.6	480	512
Ill.	25.1	463	518
Ind.	14.2	410	454
Iowa	19.0	519	570
Kans.	38.0	502	549
Ky.	14.7	479	518
La.	18.0	472	508
Maine	20.8	429	463
Md.	46.8	429	468
Mass.	42.9	429	467
Mich.	53.1	461	515
Minn.	52.8	481	539
Miss.	19.5	480	512
Mo.	16.9	469	512
Mont.	71.1	469	512
Nebr.	57.0	493	548
Nev.	17.9	442	489

State	Score	Verbal SAT	Math SAT
N.H.	33.4	448	483
N.J.	60.4	418	458
N.M.	29.3	487	527
N.Y.	26.6	424	470
N.C.	16.3	395	432
N. Dak.	35.6	500	554
Ohio	26.2	460	508
Okla.	18.7	484	525
Oreg.	61.7	435	472
Pa.	45.9	425	462
R.I.	66.6	424	461
S.C.	17.0	384	419
S. Dak.	50.6	520	566
Tenn.	14.5	486	523
Tex.	15.4	413	453
Utah	34.6	503	542
Vt.	67.4	437	470
Va.	21.8	428	466
Wash.	26.7	463	505
W. Va.	32.0	466	510
Wis.	57.0	475	532
Wyo.	93.4	489	545

SOURCE: *USA Today*, January 2, 1985 and The College Board.

Use Minitab to correlate the composite score with each of the SAT scores. Because there is no uniform evaluation system from state to state, the composite score is not totally reliable in rating the states. For this reason it is suggested that Spearman's correlation be computed.

Key Concepts

☑ Populations and samples are characterized by summary measures. A *parameter* is a summary measure associated with a population. A *statistic* is a summary measure associated with a sample. The stem and leaf plot is used to organize sample data so that the statistics can be calculated more easily.

☑ Summary measures are often classified as measures of location and measures of variability. Among the measures of location are those that measure the center of the distribution. They include the *mean*, the *trimmed mean*, and the *median*. Other measures of location are the *quartiles*, the *eighths*, and the *high* and the *low*. From these, other measures of center can be found. They are called midsummaries and include the *midQ*, the *midE*, and the *midrange*.

☑ Measures of variability include the *range*, the *standard deviation*, the *Q-spread*, and the *E-spread*. Standard deviations can be interpreted with the Empirical rule and Chebyshev's rule.

☑ A *correlation* is a summary measure that measures the degree of association between two variables. Two correlations considered in this text are *Pearson's correlation coefficient* and *Spearman's correlation coefficient*. Pearson's correlation is applicable to interval/ratio data. Spearman's correlation is more applicable to ordinal data, yet can be applied to interval/ratio.

☑ When studying data, we should be concerned with

1. The source
2. The collection procedures
3. The usefulness of the data
4. Whether it is presented properly

Learning Goals

Having completed this chapter you should be able to:

1. Decide whether a summary measure is a parameter or a statistic.   *Section 3.1*

2. Organize any data set in a stem and leaf plot.   *Section 3.2*

3. Compute the measures of location described in this chapter.   *Section 3.3*

4. Construct a quantile summary diagram.   *Section 3.3*

5. Compute the measures of variability described in this chapter.   *Section 3.4*

6. Give an interpretation of standard deviation using the Empirical rule or Chebyshev's rule.   *Section 3.4*

7. Compute Pearson's and Spearman's correlation coefficients.   *Section 3.5*

To test your skills, answer the following questions.

3.40 The military would like to know whether the black soldier feels he or she is treated fairly. After all, 60% of the military is black. A sample of 2,000 black soldiers was interviewed, and 11% of those interviewed by whites said they were treated unfairly and 35% of those interviewed by blacks said they were treated unfairly. Answer True or False.

(a) The 60% is a statistic.

(b) The 2,000 is a parameter.

(c) The 11% is a statistic.

(d) The 35% is a statistic.

(e) The 35% should be discarded because it is not realistic.

(f) The sample size of 2,000 is much too small to determine anything about the issue.

3.41

(a) Name three statistics that measure the center of a data set.

(b) Name three statistics that measure the variability of a data set.

3.42 Following are the education levels of nine workers at a plant.

8 8 1 12 14 9 12 14 12

(a) Find the mean and the standard deviation.

(b) Find the 10% trimmed mean.

3.43 Following are the numbers of reported serious reactions per million doses of vaccines in 11 southern states:

Alabama	3.9
Arkansas	27.8
Florida	5.0
Georgia	73.5
Louisiana	24.8
Mississippi	12.1
North Carolina	54.9
Oklahoma	35.3
South Carolina	11.2
Tennessee	138.8
Texas	9.2

SOURCE: Centers for Disease Control, Atlanta, Georgia.

Organize these data in an ordered stem and leaf plot and construct a quantile summary diagram.

3.44 Calculate the correlation between the following two variables.

x	3	5	4	7	9	2	6	2	5
y	15	11	10	8	4	16	9	14	10

SUPPLEMENTARY EXERCISES FOR CHAPTER 3

3.45 A government study showed that of all burglaries, 13% took place when someone was home. Is 13% a parameter or a statistic?

3.46 A sample of 10,000 homes revealed that the average family income in 1982 was $26,259. Is $26,259 a parameter or a statistic?

3.47 In 1983, more than 44 million women were in the U.S. work force. Assuming we are interested only in the 1983 data, is 44 million a parameter or a statistic?

3.48 The U.S. fertility rate is 1.8 children per woman. Limiting ourselves to the United States, is 1.8 a parameter or a statistic?

3.49 Display the following data in an *ordered* stem and leaf plot.

1.45	1.38	4.37	2.97	1.06	.44	2.20	.33
1.25	2.23	3.96	9.12	3.49	5.64	1.74	
1.69	4.33	.28	2.43	1.46	3.57	2.03	
1.39	3.35	4.33	2.68	2.48	1.47	3.47	

3.50 Construct a frequency table and histogram for the data in Exercise 3.49.

3.51 Following are the per capita incomes for 20 randomly selected counties in North Carolina:

7,300	5,700	8,200	6,500	8,100
9,400	7,400	5,300	7,000	10,100
8,400	11,000	6,600	5,500	8,200
9,200	7,700	6,400	6,800	6,900

(a) Organize the data in an ordered stem and leaf plot.

(b) Graph the data in a frequency histogram.

3.52 The 1982 Census of Retail Sales reported the following per resident sales for groceries for the fifty states:

Ala.	$ 889.96	Calif.	1,041.24
Alaska	1,503.37	Colo.	1,158.22
Ariz.	1,121.08	Conn.	1,020.07
Ark.	861.05	Del.	1,005.58

Fla.	1,103.52	N.J.	1,000.89		La.	13.8	Ohio	16.4
Ga.	929.16	N. Mex.	1,030.74		Maine	18.0	Okla.	16.4
Hawaii	1,008.88	N.Y.	856.79		Md.	13.8	Oreg.	17.6
Idaho	1,023.89	N.C.	961.20		Mass.	17.6	Pa.	18.8
Ill.	844.76	N. Dak.	788.22		Mich.	15.5	R.I.	18.9
Ind.	902.64	Ohio	960.65		Minn.	17.1	S.C.	14.1
Iowa	973.15	Okla.	1,125.49		Miss.	17.1	S. Dak.	19.5
Kans.	942.26	Oreg.	951.77		Mo.	18.6	Tenn.	16.4
Ky.	917.41	Pa.	896.59		Mont.	16.3	Tex.	13.2
La.	1,082.77	R.I.	826.27		Nebr.	18.6	Utah	12.3
Maine	1,057.85	S.C.	929.84		Nev.	12.7	Vt.	16.2
Md.	989.76	S. Dak.	826.93		N.H.	15.9	Va.	13.7
Mass.	914.01	Tenn.	938.84		N.J.	16.8	Wash.	15.4
Mich.	856.63	Tex.	1,154.98		N. Mex.	13.7	W.Va.	18.0
Minn.	866.88	Utah	921.90		N.Y.	16.9	Wis.	17.6
Miss.	865.05	Vt.	1,099.43		N.C.	15.2	Wyo.	11.5
Mo.	913.57	Va.	1,014.82		N. Dak.	17.8		
Mont.	1,113.82	Wash.	1,074.46					
Nebr.	850.29	W.Va.	979.60					
Nev.	1,306.81	Wis.	887.16					
N.H.	1,246.98	Wyo.	1,189.96					

SOURCE: U.S. Bureau of the Census.

SOURCE: U.S. Department of Commerce.

Construct a stem and leaf plot of the per resident sales of groceries in 1982.

Construct a stem and leaf plot of the percent of residents in the states over 65 years of age.

3.53 Construct a quantile summary diagram for the data in Exercise 3.52.

3.59 Construct a quantile summary diagram for the data given in Exercise 3.58.

3.54 Calculate the midsummaries for the data given in Exercise 3.52.

3.60 Calculate the midsummaries for the data given in Exercise 3.58.

3.55 Suppose 15 artificial heart transplants were conducted and the duration (in hours) was recorded as follows:

7.0	6.5	3.5	3.8	3.1
2.8	2.5	2.6	2.4	2.1
1.8	2.3	3.1	3.0	2.5

Organize these data in an ordered stem and leaf plot.

3.61 Calculate the mean and standard deviation for the following sample of nine measurements:

8, 12, 18, 16, 9, 10, 2, 8, 7

3.62 The following represents the degree of urbanization of 10 localities as measured on a 100-point scale.

63, 56, 32, 56, 48, 45, 45, 96, 57, 72

Find the mean and standard deviation.

3.56 Construct a quantile summary diagram for the data in Exercise 3.55.

3.57 Calculate the midsummaries for the data given in Exercise 3.55.

3.63 Seventy people were asked to do a certain task, after which the number of trials necessary to complete the task was recorded:

3.58 Following is the percent of the population of each state and the District of Columbia that is over the age of 65:

Ala.	16.8	Fla.	22.8
Alaska	4.5	Ga.	13.9
Ariz.	17.1	Hawaii	12.5
Ark.	19.8	Idaho	16.0
Calif.	14.2	Ill.	16.2
Colo.	11.9	Ind.	16.1
Conn.	17.1	Iowa	19.4
Del.	14.8	Kans.	18.1
D.C.	15.3	Ky.	16.3

No. of trials	Frequency
18	2
17	4
16	9
15	7
14	5
13	7
12	8
11	10
10	12
9	6
Total	70

Calculate the mean, median, standard deviation, and Q-spread for the number of trials to complete the task.

3.64 The "fasting-blood-sugar value" was taken from each of nine apparently normal individuals:

102, 103, 117, 101, 101, 93, 107, 87, 89

Calculate the mean and standard deviation.

3.65 Find the mean, 10% trimmed mean, and standard deviation for the following data:

11.7	12.8	17.2	14.5	17.3	15.9	12.3	13.6
11.5	19.0	15.8	14.7	11.7	19.6	16.9	13.8
13.5	13.9	17.6	13.7	12.6	13.3	15.0	14.6
15.7	13.7	18.0	13.4	13.5	12.8	20.9	14.4
16.4	18.5	16.8	12.1	19.5	14.9	14.8	21.6
13.9							

3.66 Construct an ordered stem and leaf plot for the data in Exercise 3.65.

3.67 Complete a quantile summary diagram to include mid-summaries and spreads for the data in Exercise 3.65. Use the stem and leaf plot constructed in Exercise 3.66.

3.68 The fertility rate is the number of births a woman can expect in her childbearing years. An average rate of 2.12 is needed to keep the population constant. Following are the fertility rates in all states, including the District of Columbia:

Ala.	1.9	Mont.	2.1
Alaska	2.3	Nebr.	2.0
Ariz.	2.1	Nev.	1.8
Ark.	2.0	N.H.	1.7
Calif.	1.9	N.J.	1.6
Colo.	1.8	N. Mex.	2.2
Conn.	1.5	N.Y.	1.6
Del.	1.8	N.C.	1.6
D.C.	1.5	N. Dak.	2.1
Fla.	1.7	Ohio	1.8
Ga.	1.9	Okla.	2.0
Hawaii	2.1	Oreg.	1.8
Idaho	2.5	Pa.	1.6
Ill.	1.9	R.I.	1.5
Ind.	1.8	S.C.	1.8
Iowa	2.0	S. Dak.	2.4
Kans.	2.0	Tenn.	1.7
Ky.	1.9	Tex.	2.1
La.	2.2	Utah	3.2
Maine	1.7	Vt.	1.7
Md.	1.6	Va.	1.6
Mass.	1.5	Wash.	1.8
Mich.	1.8	W.Va.	1.8
Minn.	1.9	Wis.	1.9
Miss.	2.2	Wyo.	2.4
Mo.	1.9		

SOURCE: Population Reference Bureau.

Find the mean, median, and 10% trimmed mean for the fertility rate data.

3.69 A lottery agent is one who sells lottery tickets in states that have legalized lotteries. The amount the agent earns depends on the number of tickets he or she sells. Following are the numbers of lottery agents and the average annual sales commissions (1983) paid lottery agents in the 18 states with lotteries:

State	No. of lottery agents	Average sales commission
Arizona	2,100	$ 2,228
Colorado	2,500	4,880
Connecticut	3,000	3,200
District of Columbia	850	3,986
Delaware	324	5,247
Illinois	7,710	3,320
Maine	1,629	490
Maryland	1,239	20,399
Massachusetts	3,600	5,279
Michigan	7,293	4,758
New Hampshire	1,150	574
New Jersey	4,000	12,000
New York	12,449	4,281
Ohio	5,400	3,704
Pennsylvania	7,700	8,013
Rhode Island	883	7,667
Vermont	750	307
Washington	4,711	1,804

SOURCE: Gaming and Wagering Business and *USA Today*, January 15, 1985.

Calculate the mean, median, 10% trimmed mean, and standard deviation for the sales commissions.

3.70 From an annual income of $20,000, the average single householder spends $3,456 on food each year (according to the Bureau of Labor Statistics). Following are the annual food expenditures for a random sample of 40 single households in the state of Ohio:

$2,845	3,170	2,352	4,978	3,820	2,475	3,160
5,780	2,175	2,648	2,872	4,250	3,970	2,534
6,870	2,734	2,847	4,670	5,176	3,640	2,765
1,180	3,679	3,320	7,580	2,416	3,743	2,830
3,127	3,249	2,648	1,976	2,784	3,869	2,086
5,587	3,420	2,645	8,147	4,367		

Calculate the mean, 10% trimmed mean, and standard deviation for the annual food expenditures.

3.71 Construct an ordered stem and leaf plot for the data given in Exercise 3.70.

3.72 Complete a quantile summary diagram to include mid-summaries and spreads for the data given in Exercise 3.70. Use the stem and leaf plot constructed in Exercise 3.71.

3.73 Modern technology, regulations, malpractice premiums, and physicians' fees all contribute to the spiraling cost of health care. Following are the fees for a coronary bypass for a random sample of 20 physicians:

$3,820	3,540	2,800	4,260	3,920
3,400	4,370	3,890	4,420	3,860
3,970	4,250	5,200	3,270	3,950
4,180	4,470	3,920	4,840	4,650

Calculate the mean, 10% trimmed mean, and standard deviation for the above physician fees.

3.74 Construct an ordered stem and leaf plot of the data given in Exercise 3.73.

3.75 Complete a quantile summary diagram to include mid-summaries and spreads for the data given in Exercise 3.73. Use the stem and leaf plot constructed in Exercise 3.74.

3.76 Following are the 1973 and 1984 rankings of the nation's favorite breeds of dogs.

	Ranking	
Dog	1973	1984
Poodle	1	1
Spaniel	9	2
Labrador	10	3
Shepherd	2	4
Doberman	7	5
Beagle	3	6

SOURCE: *Dog World* magazine.

Correlate the 1973 and 1984 rankings.

3.77 In 1985 Americans spent $24 per capita per week on groceries. It is suspected that the more people there are in a family the less the cost per person would be. To evaluate this a marketing institute randomly sampled 20 families and found the following:

Number in family	Cost of food per person
2	35
2	28
1	38
3	26
4	22
3	24
2	29
4	19
1	29
3	25
5	18
2	32
2	28
3	22
4	26
1	40
2	34
6	17
3	27
2	29

Correlate the number in the family with the cost per person for food.

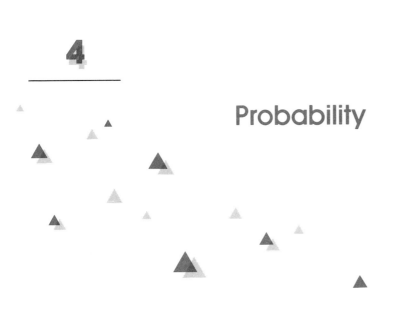

4

Probability

Y ou will recall that it is usually impractical to investigate a population in its entirety. Thus we resort to a sample and infer from the sample to the population. Because the decisions or predictions made about a population are based on sample information, a degree of uncertainty is involved. That uncertainty is measured by probability.

In this chapter you will be introduced to the vocabulary and laws of probability and be exposed to problems that illustrate how probabilities are determined. In later chapters you will see how probability relates to statistical inference.

Delivery Room Birthdays

Quincy, Mass. (AP)

When Justine Lee Mitchell was born last week, she wasn't the only one in the operating room celebrating a birthday.

Her mother turned 18, the obstetrician turned 37 and the nurse turned 32, all on Saturday.

"It's mind boggling," the obstetrician, Dr. Richard R. Adams, said in an interview published today.

While Lauralyn Mitchell was delivering Justine at Quincy City Hospital, Adams said he boasted to registered nurse Beth MacLeod, Ms. Mitchell was "going to have a great kid because we have the same birthday."

Ms. McLeod responded that it was her birthday, too, and then the new mother announced it was her birthday. Adams said he demanded a look at her driver's license to be sure, and it was confirmed, she was telling the truth.

Justine was born at 5:30 a.m. and weighed 8 pounds, 3 ounces.

SOURCE: Watauga *Democrat*, July 11, 1984.

As one might expect, the chance of having four randomly selected people (random as far as their birthdays are concerned) with the same birthday in a room is extremely rare, approximately 1 chance out of 50 million! Yet it did happen as is documented in the accompanying newspaper article. Even two randomly selected people in a room having the same birthday is somewhat rare, with less than 3 chances out of 1,000. On the other hand, if there are 50 people in a room, it is almost certain that at least 2 will have the same birthday. In this chapter we will develop the tools necessary to understand this so-called birthday problem, as well as other interesting probability problems.

There is evidence that the concept of probability or chance has existed since the beginning of recorded history. Games of chance played by tossing bones have been documented as early as 3000 B.C. This apparently evolved into our modern-day games involving the six-sided die. Gambling with dice has been popular since Roman times, but was not studied mathematically until the sixteenth and seventeenth centuries.

As early as the fourteenth century, insurance companies insured the contents of cargo ships. In order to determine the premiums (some 12% to 15% of the value of the cargo) the insurance companies had to determine the probability that they might have to pay a claim. Other uses of probability centered around the recording of vital statistics such as births, deaths, and marriages. The *Bill of Mortality* in London in the early 1500s involved the calculation of the probability of death at a certain age. Such was the beginning of life insurance.

It is generally believed that the mathematical theory of probability was started by the French mathematicians, Blaise Pascal (1623–1662) and Pierre Fermat (1601–1665). Their solutions to certain gambling problems involving dice posed by the gambler and French nobleman, Chevalier de Méré, are considered to be among the first contributions to probability theory. As it relates to statistics, significant contributors to probability theory were Jacob Bernoulli (1654–1705), Abraham de Moivre (1667–1754), Pierre-Simon de Laplace (1749–1827), and Carl Friedrich Gauss (1777–1855). A significant contributor to modern-day probability theory is the Russian mathematician, A. N. Kolmogorov, who in 1933 presented a consistent axiomatic approach to the study of probability.

Today, probability is a useful tool in almost all fields of study. Moreover its applications are much broader than the original gambling problems. For instance, probability has even entered the courtroom. A black male and his blond female companion were convicted of a crime simply because they were together in a yellow Trans Am in the vicinity of a crime scene. The prosecutor convinced the jury that the chance of another couple of that description in a yellow Trans Am was so unlikely that they had to be the two who committed the crime. The Supreme Court of the state reversed the conviction on the basis that the laws of probability were not used properly (*Time* magazine, April 26, 1968). At the end of Section 4.4 we will see how those laws were improperly used.

Another instance of the use of probability in the courts occurred on August 4, 1983, when a judge ruled that the Educational Testing Service (ETS) acted fairly in invalidating the College Board Scores of four high school classmates accused of cheating. ETS cited a statistical analysis that showed the odds were as high as 300 billion to 1 in some parts of the exam that the four would have honestly given the same incorrect answers (Charlotte *Observer*, August 5, 1983).

One of the most important goals of statistics is to make inferences about a population from the information contained in a sample. Once an inference is made, however, we must then specify a measure of the reliability of that inference. Probability is the tool that will allow us to do so.

What Is Probability?

We all have some idea of what is meant by the term *probability* or *chance*. We say that he "guessed" on the true/false problem so he has a 50–50 chance of getting it right. Or there is a 10% chance that a controversial bill will pass Congress. A baseball player has a .324 batting average; thus each time at bat he has a 32.4% chance of getting a hit.

Intuitively we think of probability as a numerical value associated with some event, which indicates how likely it is that the event will occur. We say that a small probability indicates that the event is not likely to occur and a large probability indicates that it is likely that the event will occur. The controversial bill has a 10% chance of passing—the 10% is small, indicating that the event is not very likely to happen. But if the probability were 90%, it has a very good chance of passing.

The probability of an event can be described as the *relative frequency* with which the event will occur if we repeated the experiment a large number of times. For example, if we tossed a fair coin, we could say the probability of the event of getting a head on the coin is 50% or 1/2, because half of the time it should land on a head and half of the time it should land on a tail. This does not mean that out of 10 tosses, 5 should be heads and 5 should be tails; but rather, if we tossed it 10,000 times, then the number of heads should be near 5,000 (i.e., the relative frequency of heads should be near .5).

In the preceding examples we have used four different expressions of a probability (50–50, 10%, .324, 1/2), each of which can be transformed to any of the other forms. For example, 50–50 on the true/false question means half of the time the answer is right and half of the time it is wrong. This could be written as 50% or .50 or 1/2, all meaning the same thing.

In any event, we can interpret a probability as a number between 0 and 1; the closer it is to 0 the less likely the event is to occur and the closer it is to 1 the more likely the event is to occur. In fact, if its probability is 0, then the event cannot happen; if its probability is 1, then the event is certain to occur.

EXAMPLE 4.1

Suppose you are playing bridge (all 52 cards are dealt to four players) and you are dealt four aces. What is the probability that one of the other three players has an ace?

SOLUTION   Because there are only four aces in the deck and you have all four, then no one else can have an ace. Therefore the probability is zero.

EXAMPLE 4.2

A group of 10 people, consisting of 4 men and 6 women, is in a meeting and, at random, they select a person to preside. What is the probability that the person selected is a woman?

SOLUTION   Because 6 of the 10 people are women and each person is equally likely to be selected, the probability that a woman is selected is 6 out of 10 or .6.

EXAMPLE 4.3

A man has two indistinguishable keys and has two doors to open in the dark. At each door, he selects a key at random. What is the probability that both attempts are successful?

SOLUTION   A *tree diagram* will help list the possibilities.

Door 1	Door 2	Result

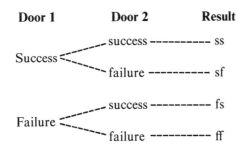

We see that if the first try were successful, then the second try could still be either successful or not. Or if the first try were unsuccessful, the second try could be successful or not. Hence there are four possibilities—ss, sf, fs, ff—and because he was selecting at random, each of the four possibilities is equally likely. Thus the probability that both attempts are successful (ss) is 1 out of 4 or .25.

EXERCISES 4.1

4.1 Roll a fair die. What is the probability that you observe a number greater than 4?

4.2 There are 200 names in a bowl and 1 is drawn as the winner. What is the probability that you win, assuming that your name is in the bowl?

4.3 Suppose that there are 3 doctors in a group of 10 people. If we select a person at random, what is the probability that he or she is a doctor?

4.4 A card is drawn from an ordinary deck of 52 cards. What is the probability that it is red?

4.5 Two cards are drawn from a deck of cards, one after the other without replacement. The first is red. What is the probability that the second card is also red?

4.6 Suppose you are playing blackjack and you score blackjack. What is the probability that you have an ace?

4.7 Suppose in a game of bridge you are dealt three aces. What is the probability that your partner has the other ace?

4.8 A man has three suits from which to select—a brown one, a gray one, and a blue one. If he chooses one at random, what is the probability that he selects the gray one?

4.9 A basketball player has a 50% chance of making a free throw. Out of two attempts what is the probability that he misses both shots?

4.10 A family has two children. Assuming a boy or girl is equally likely, what is the probability of two girls?

4.11 Roulette is played by spinning a ball on a round table that is divided into 38 slots of equal size. The slots are numbered 00, 0, 1, 2, 3, . . . , 35, 36. The slots are also colored as follows:

Red	1	3	5	7	9	12	14	16	18
	19	21	23	25	27	30	32	34	36
Black	2	4	6	8	10	11	13	15	17
	20	22	24	26	28	29	31	33	35
Green	00	0							

Find the probability of a black number on a single spin. Find the probability that it is neither red nor black on a single spin.

4.12 A group of people consists of two children under 12 years of age, three teenagers, and five adults. If a person is selected at random, what is the probability that the person is an adult? What is the probability that the person selected is over 12 years old?

To study probability further we need to introduce some terminology. The first is that of an experiment.

Definition

An **experiment** is the process of making an observation or taking a measurement.

All experiments result in a certain collection of possible outcomes. For example, the roll of a die results in

$$1, \quad 2, \quad 3, \quad 4, \quad 5, \quad \text{or} \quad 6$$

The simple experiment of tossing a coin results in one of two possible outcomes—a head or a tail. The experiment of weighing a person results in the weight of the person. Thus we can say that any activity that results in outcomes is an experiment. Other examples include recording voters' opinions on a certain issue, recording SAT scores for entering college students, measuring the dissolved oxygen content of a river, and observing the diameter of trees in a forest.

Once an experiment is defined we wish to have a list of all the possible outcomes—this will be the sample space.

Definition

The collection of all possible outcomes of an experiment is called the **sample space, S.**

For the die experiment, the sample space would be

$$S = \{1, 2, 3, 4, 5, 6\}$$

For the coin experiment, the sample space would be

$$S = \{\text{head, tail}\}$$

For the experiment of weighing a person, the sample space (assuming no one weighs over 600 pounds) would be

$$S = \{x \mid x \text{ is a real number between 0 and 600}\}$$

Definition

Any subset of the sample space is called an **event**. Those subsets containing a single outcome are called **simple events**.

In the die rolling experiment, we might consider the event A that an odd number occurs. Thus

$$A = \{1, 3, 5\}$$

There are six simple events—$\{1\}, \{2\}, \{3\}, \{4\}, \{5\}, \{6\}$.

Definition

An event is said to **have occurred** if any one of its elements happens when the experiment is conducted.

The event *A* just mentioned occurs if, when the die is rolled, any one of the three possibilities, 1, 3, or 5 comes up.

A more complicated experiment might consist of tossing a coin three times. A tree diagram will help in listing the possible outcomes. The first toss is either heads or tails. If the first toss is heads, the second can be either heads or tails, or if the first toss is tails, the second can be either heads or tails. So after the second toss there are four possibilities. Then there are two possibilities for each of those four possibilities, resulting in eight different possible outcomes. The tree diagram is

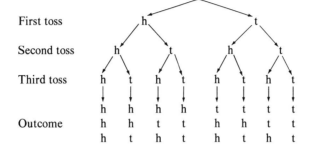

The sample space is

$$S = \{(h, h, h), (h, h, t), (h, t, h), (t, h, h), (h, t, t),$$
$$(t, h, t), (t, t, h), (t, t, t)\}$$

where, for example, (h, h, t) means the first two coins were heads and the third was a tail.

An event *B*, which might be of interest, is that exactly two of the three coins turn up heads. In this case event *B* is

$$B = \{(h, h, t),(h, t, h),(t, h, h)\}$$

If any of the three possibilities comes up, then we say *B* did occur.

EXAMPLE 4.4

A company gives an exam to its potential employees. There is only a 30% pass rate for the exam. Suppose three potential employees take the exam. List the elements of the sample space and the elements that make up the following events:

> Event *A*—Two of the three pass the exam.
> Event *B*—Only one passes the exam.
> Event *C*—All three pass the exam.
> Event *D*—At least one of the three passes the exam.

SOLUTION   If, for example, we let the simple event (p, f, p) represent the fact that the first and third person passed and the second failed, then the sample space of possible outcomes will be

$$S = \{(p, p, p), (p, p, f), (p, f, p), (f, p, p), (p, f, f), (f, p, f), (f, f, p), (f, f, f)\}$$

From these elements we can list the elements of the events just mentioned. We have

$$A = \{(p, p, f), (p, f, p), (f, p, p)\}$$
$$B = \{(p, f, f), (f, p, f), (f, f, p)\}$$
$$C = \{(p, p, p)\}$$
$$D = \{(p, f, f), (f, p, f), (f, f, p), (p, p, f), (p, f, p), (f, p, p), (p, p, p)\}$$

Note that the 30% pass rate is not used to list the sample space or events. Later we will need it to assign probabilities.

EXAMPLE 4.5

Suppose that the results of the exam in the previous example were: Jane passed, Bob passed, Sally failed. List which of the four events A, B, C, or D occurred and which did not occur.

SOLUTION   Assuming that Jane is the first person, Bob is the second, and Sally is the third, the element we observed is (p, p, f), which is in events A and D. Thus we can say events A and D occurred, but events B and C did not.

The occurrence or nonoccurrence of an event depends on its *probability*. First we will look at the probability of a simple event. If probabilities can be assigned to the simple events, then we can find the probability of other events that are comprised of the simple events.

Definition

The **probability of a simple event** is a number between 0 and 1 that measures the likelihood that it will occur when the experiment is performed.

EXAMPLE 4.6

In rolling a fair die there are six simple events

$$\{1\}, \quad \{2\}, \quad \{3\}, \quad \{4\}, \quad \{5\}, \quad \{6\}$$

Because the die is fair we can say the six simple events are equally likely to occur. Therefore we assign a probability of 1/6 to each simple event because there is one chance out of six of it coming up. If we add the probabilities of all simple events, we get $6/6 = 1$, which says that one of the simple events will occur when the experiment is conducted. Certainly when a die is rolled, one of the six sides will turn up.

Assigning Probabilities

In assigning probabilities to the simple events of a sample space two conditions must be satisfied:

1. The probability of each simple event must be between 0 and 1.

2. The probabilities of all simple events in the sample space must sum to 1.

EXAMPLE 4.7

In Example 4.3 a man had two indistinguishable keys and had to open two doors in the dark. On each attempt he was either successful or unsuccessful. We saw that the sample space of possible outcomes was

$$S = \{ss, sf, fs, ff\}$$

Assign probabilities to each of the simple events.

SOLUTION   As observed in Example 4.3, the four outcomes (simple events) are equally likely, hence each would be assigned a probability of $1/4$.

EXAMPLE 4.8

Suppose a die is unfair in such a way that the even numbers are twice as likely to turn up as the odd. Assign probabilities to the simple events.

SOLUTION   There are still the six simple events

$$\{1\}, \quad \{2\}, \quad \{3\}, \quad \{4\}, \quad \{5\}, \quad \{6\}$$

Suppose the probabilities of $\{1\}$, $\{3\}$, and $\{5\}$ are each a value called p. Then the probabilities of $\{2\}$, $\{4\}$, and $\{6\}$ are each $2p$. The sum of all the probabilities must be 1, so we have

$$1 = p + 2p + p + 2p + p + 2p = 9p$$

Because $9p = 1$, we have that $p = 1/9$ and $2p = 2/9$. That is, the probabilities of each of the odds are $1/9$ and the probabilities of each of the evens are $2/9$.

Once probabilities are assigned to the simple events, we can find the probability of any other event.

Definition

The **probability of an event** A is the sum of the probabilities of the simple events in A. We write it as $P(A)$.

We saw in Example 4.6 that if a die is fair, we can assign probabilities $1/6$ to each of the six simple events. Then the probability of event A, that an odd number occurs, is

$$\begin{aligned}
P(A) &= P(\{1\}) + P(\{3\}) + P(\{5\}) \\
&= 1/6 + 1/6 + 1/6 \\
&= 3/6 \\
&= 1/2
\end{aligned}$$

On the other hand, for the unfair die of Example 4.8, the probability that an odd number occurs is

$$\begin{aligned}
P(A) &= P(\{1\}) + P(\{3\}) + P(\{5\}) \\
&= 1/9 + 1/9 + 1/9 \\
&= 3/9 \\
&= 1/3
\end{aligned}$$

EXAMPLE 4.9

Calculate the probability of observing exactly two heads in the toss of three fair coins.

SOLUTION   In tossing three coins, the sample space is

$$S = \{(h, h, h), (h, h, t), (h, t, h), (t, h, h), (h, t, t), (t, h, t), (t, t, h), (t, t, t)\}$$

Because all coins are fair, a head is just as likely as a tail. Thus for three coins, (h, h, h) is just as likely as (h, h, t), which is just as likely as (h, t, h), and so on. We have that the eight simple events are equally likely and thus each has a probability of 1/8. The event of interest is that exactly two heads appear, which is

$$A = \{(h, h, t), (h, t, h), (t, h, h)\}$$

Adding the probabilities of the simple events in A, we have

$$P(A) = 1/8 + 1/8 + 1/8 = 3/8$$

EXAMPLE 4.10

In the game of craps the "shooter" wins on the first roll of a pair of fair dice if the sum of the two dice is 7 or 11. Calculate the probability he or she wins on the first roll.

SOLUTION   The sample space for the roll of a pair of dice is

$$
\begin{aligned}
S = \{&(1, 1), (1, 2), (1, 3), (1, 4), (1, 5), (1, 6), \\
&(2, 1), (2, 2), (2, 3), (2, 4), (2, 5), (2, 6), \\
&(3, 1), (3, 2), (3, 3), (3, 4), (3, 5), (3, 6), \\
&(4, 1), (4, 2), (4, 3), (4, 4), (4, 5), (4, 6), \\
&(5, 1), (5, 2), (5, 3), (5, 4), (5, 5), (5, 6), \\
&(6, 1), (6, 2), (6, 3), (6, 4), (6, 5), (6, 6)\}
\end{aligned}
$$

Assuming both dice are fair we have that the 36 simple events are equally likely, that is, all have probability 1/36. The event of interest is that a total of 7 or 11 turns up. In set notation the event of interest is

$$A = \{(1, 6), (2, 5), (3, 4), (4, 3), (5, 2), (6, 1), (5, 6), (6, 5)\}$$

Adding the probabilities of the simple events in A, we have

$$P(A) = 8(1/36) = 8/36$$

We can summarize the steps in calculating the probability of any event as follows:

Calculating the Probability of an Event

1. Define the experiment and list the sample space.

2. Assign probabilities to the simple events such that each is between 0 and 1 and they add to 1.

3. List the elements of the event of concern.

4. Sum the probabilities of the simple events that are in the event of concern.

Thus far, listing the simple events in a sample space has been fairly easy. Assigning probabilities to the simple events has not been a difficult task either. However, more difficult experiments do exist. In some situations, listing the sample space may be straightforward, but assigning probabilities may not be obvious. In other situations the sample space may be so large that listing it is impractical. For example, we could not attempt to list all possible five-card poker hands. If we knew how many possible hands there were, however, we could assign an equal probability to each one (the hands are equally likely because you are just as likely to get any 5 of the 52 possible cards) and then find the probability of any type of hand. Combinatorial mathematics can be used to count the number of possible poker hands as well as to determine the number of elements in other sample spaces. Also in the next section we will learn of some laws of probability that will be helpful in determining the probability of events.

EXERCISES 4.2

4.13 A single die is tossed. List the elements in the following events:

A: Observe a 3.
B: Observe a number less than 3.
C: Observe a number greater than or equal to 4.
D: Observe a number between 2 and 5 exclusive.

4.14 Three coins are tossed. List the elements in the following events:

A: Two coins are heads.
B: No more than one coin is heads.
C: At least two coins are heads.
D: No coins are heads.

4.15 A single card is drawn from a deck of cards, and we are concerned only with the color. List the sample space of the experiment.

4.16 A person conducting a poll randomly chooses one of two houses. He then randomly chooses either a male or female from the chosen house. List the sample space of possibilities.

4.17 Two cards are drawn from a deck of cards and we are concerned only with the color. List the sample space of the experiment.

4.18 In Exercise 4.17 we are concerned with event R that both cards are red. One of the cards is the two of spades. Did event R occur?

4.19 In the roll of a pair of dice we are concerned with the event W that a total of 7 is rolled. One of the dice turns up on 4. What must the other die be in order for event W to occur?

4.20 Consider the sample space $S = \{s_1, s_2, s_3, s_4\}$ in which

$$P(s_1) = .4, P(s_2) = .2, P(s_3) = .1$$

What is $P(s_4)$?

4.21 A man has a blue suit that he wears 30% of the time. Let B correspond to his wearing the blue suit and N indicate that he does not. For two consecutive days, list the elements of the sample space. Are the simple events equally likely?

4.22 Student loan applications are either approved or denied. If three students apply, list the sample space of possible outcomes. If 60% of the applications are approved, are the simple events equally likely? If 50% of the applications are approved, are the simple events equally likely? Assuming 50% are approved, assign probabilities to the simple events.

4.23 There are three traffic lights on your way home. As you arrive at each light assume that it is either red (R) or green (G) and that it is green with probability .7. List the elements of the sample space. Are the simple events equally likely?

4.24 Assume that 60% of the student body is female. Three students are selected at random. List the sample space of possible sexes of the three students. Are the simple events equally likely? Assign probabilities to the simple events if 50% of the student body is female.

Most events of interest are the result of combining other events. For example, in the roll of a pair of dice we are not interested in whether we roll a (6, 1) or a (1, 6); we are interested in whether or not the total is 7. Similarly if one tosses three coins, he or she is not concerned with which of the three coins turns up heads, but, rather, the total number of heads. Events that are formed from other events are called *compound events*. Compound events are formed by taking intersections and unions of other events.

Definition

The **union** of two events A and B is the event containing all simple events in A or B or both. It is denoted as $A \cup B$.

The **intersection** of two events A and B is the event containing all simple events that are in both A and B. It is denoted as $A \cap B$.

Figure 4.1 is called a *Venn diagram*; here it illustrates the union of A and B. Any element in the shaded area is in the union of A and B. The shaded area in the Venn diagram in Figure 4.2 depicts the intersection of A and B.

FIGURE 4.1
$A \cup B$

FIGURE 4.2
$A \cap B$

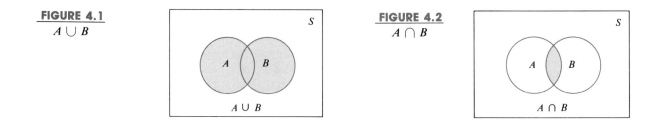

EXAMPLE 4.11

Consider the roll of a die. Define events:

       A: An even number turns up.
       B: A number less than 3 turns up.

Describe events $A \cup B$ and $A \cap B$.

SOLUTION   The sample space is

$$S = \{1, 2, 3, 4, 5, 6\}$$

and events A and B are

$$A = \{2, 4, 6\} \quad B = \{1, 2\}$$

Hence we have that

$$A \cup B = \{1, 2, 4, 6\} \quad \text{and} \quad A \cap B = \{2\}$$

The concept of unions and intersections can be extended to more than two events. For example, the union of events A, B, and C is the event consisting of all simple events in A or B or C. The intersection of A, B, and C is the event consisting of all simple events common to A, B, and C.

EXAMPLE 4.12

Consider the experiment of tossing three coins. Let events A, B, and C be defined as follows:

> A: The first coin is heads.
> B: The second coin is heads.
> C: The third coin is heads.

Describe events $A \cup B \cup C$ and $A \cap B \cap C$.

SOLUTION   We have that

$$A = \{(h, h, h), (h, h, t), (h, t, h), (h, t, t)\}$$
$$B = \{(h, h, h), (h, h, t), (t, h, h), (t, h, t)\}$$
$$C = \{(h, h, h), (h, t, h), (t, h, h), (t, t, h)\}$$

which gives us that

$$A \cup B \cup C = \{(h, h, h), (h, h, t), (h, t, h), (h, t, t), (t, h, h), (t, h, t), (t, t, h)\}$$

and

$$A \cap B \cap C = \{(h, h, h)\}$$

Often there are relations between events that will help in determining probabilities. Two relationships we will consider are complementary events and mutually exclusive events.

Definitions

The **complement of event** A is the collection of all simple events not in A. It is denoted as A'.

Events A and B are said to be **mutually exclusive** if they have no intersection. That is, $A \cap B = \varnothing$, the empty set.

EXAMPLE 4.13

Suppose A is the event that an even turns up on the roll of a single die. The complement of A, A', is the event that an odd turns up. In set notation,

$$A = \{2, 4, 6\} \quad \text{and} \quad A' = \{1, 3, 5\}$$

Because $A \cap A' = \varnothing$, they are mutually exclusive.

The Venn diagrams in Figure 4.3 illustrate the complement of an event and mutually exclusive events.

FIGURE 4.3

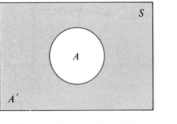

The complement of event A

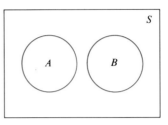

Mutually exclusive events

Because A' is all simple events not in A, the two events make up the entire sample space. Accordingly we have

---

The Complement Law

Let A be an event with probability denoted by $P(A)$. Then

$$P(A') = 1 - P(A)$$

Thus if the probability of an event is known, then the probability of its complement is also known.

EXAMPLE 4.14

A fair coin is tossed six times. What is the probability that at least one head turns up?

SOLUTION   We recall that when 3 coins are tossed, there are 8 different outcomes. More precisely,

$$S = \{(h, h, h), (h, h, t), (h, t, h), (t, h, h), (h, t, t), (t, h, t), (t, t, h), (t, t, t)\}$$

If 4 coins are tossed, each outcome will be of the form (h, h, h, h), (h, h, h, t), and so on, and there are 16 of them. A tree diagram will help in listing all 16.

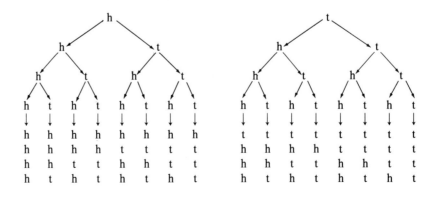

If we toss 5 coins, there will be 32 different outcomes of the form (h, h, h, h, h), (h, h, h, h, t), and so on. We see that the number of outcomes goes up by powers of 2 because there are 2 possibilities on each toss, namely, heads and tails. Consequently for 6 coins there are 64 different outcomes of the form (h, h, h, h, h, h), (h, h, h, h, h, t), and so on. Assuming that the coins are fair, each sequence of heads and tails has the same chance of occurring, and therefore all the simple events are equally likely. Because there are 64 of them, each has probability 1/64.

The event of interest, A, is that at least one head turns up. Event A contains 63 of the 64 simple events in the sample space. Certainly we do not want to list all 63. However the complement of A, A', contains only one simple event

$$A' = \{(t, t, t, t, t, t)\} \quad \text{and} \quad P(A') = 1/64$$

Using the complement law, we have

$$P(A) = 1 - P(A') = 1 - 1/64 = 63/64$$

Another useful law in determining probabilities of events is the additive law.

The Additive Law

Let A and B be two events, then

$$P(A \cup B) = P(A) + P(B) - P(A \cap B)$$

If A and B happen to be mutually exclusive, then

$$P(A \cup B) = P(A) + P(B)$$

EXAMPLE 4.15

Again consider the roll of a fair die and events A and B as defined earlier in Example 4.11. Namely

> A: An even number turns up.
> B: A number less than 3 turns up.

Find the probability of $A \cup B$.

SOLUTION   As before

$$A = \{2, 4, 6\} \quad \text{and} \quad B = \{1, 2\}$$

Therefore $A \cap B = \{2\}$.

Because the die is fair, each of the simple events has probability 1/6. Consequently

$$P(A) = P(\{2\}) + P(\{4\}) + P(\{6\}) = 1/6 + 1/6 + 1/6 = 3/6$$

and

$$P(B) = P(\{1\}) + P(\{2\}) = 1/6 + 1/6 = 2/6$$

and

$$P(A \cap B) = 1/6$$

Using the additive law, we have

$$
\begin{aligned}
P(A \cup B) &= P(A) + P(B) - P(A \cap B) \\
&= 3/6 + 2/6 - 1/6 \\
&= 4/6 \\
&= 2/3
\end{aligned}
$$

Note that in adding $P(A)$ and $P(B)$ we are adding in $P(\{2\})$ twice; but the simple event $\{2\}$ appears only once in $A \cup B = \{1, 2, 4, 6\}$. Thus we subtract it out one time by subtracting $P(A \cap B) = P(\{2\})$. In general, when finding $P(A \cup B)$, we must subtract $P(A \cap B)$ because it has been added in twice. Thus the formula

$$P(A \cup B) = P(A) + P(B) - P(A \cap B)$$

EXERCISES 4.3

4.25 Suppose $S = \{2, 3, 4, 5, 6, 7, 8\}$, $A = \{3, 5, 7\}$, $B = \{4, 5, 6\}$, and $C = \{3, 4, 5, 6, 7\}$. Find

(a) $A \cup B$ (b) $A \cap B$

(c) $A \cap C$ (d) C'

(e) $A' \cap C$ (f) $A \cup B \cup C$

(g) $A \cap B \cap C$ (h) Are A and B mutually exclusive? Why?

4.26 Suppose events A and B are such that $P(A) = .5$, $P(B) = .4$, and $P(A \cap B) = .3$. Find

(a) $P(A')$

(b) $P(A \cup B)$

(c) Are A and B mutually exclusive? Why?

4.27 Suppose events A and B are such that $P(A) = .5$, $P(B) = .4$, and $P(A \cup B) = .8$. Find

(a) $P(B')$

(b) $P(A \cap B)$

(c) Are A and B mutually exclusive? Why?

4.28 Suppose that a pop quiz is given in class A 30% of the time and in class B 20% of the time. If they occur in both 6% of the time, what is the probability of a pop quiz in either class A or class B?

4.29 Assign probabilities to the simple events in the sample space and calculate the probabilities of the events given in Exercise 4.13.

4.30 Suppose a card is drawn from a standard deck of 52 cards.

(a) What is the probability that the card is a club or a face card?

(b) What is the probability that the card is not a face card?

(c) Are the events of drawing a club and drawing a face card mutually exclusive?

4.31 A manufacturer of golf clubs has plants in Phoenix, Denver, and Memphis. The Memphis plant produces 60% of the company's clubs, and the remaining 40% is evenly divided between the Phoenix and Denver plants. A set of clubs is randomly selected from a K mart in Amarillo, Texas.

(a) Are the events: produced in Memphis, produced in Denver, and produced in Phoenix mutually exclusive?

(b) What is the probability that the clubs were produced by either the Memphis or Denver plants?

(c) What is the probability that the clubs were not produced by the Memphis plant?

4.32 A survey of teenagers 16 years of age and older showed that 35% feel that the legal drinking age should be 18, 20% feel that marijuana should be legalized, and 15% agree with both. What percent feel that the legal drinking age should be 18 or that marijuana should be legalized?

In our earlier discussion of probability we determined that the probability of rolling a 1 on the toss of a single fair die is $1/6$. Suppose you are told that it landed on an odd number. Now what is the probability that you rolled a 1? This is the notion of *conditional probability*. We are given additional information about the event that might affect its probability.

If no special conditions about the event are given, then its probability is called an *unconditional probability*. On the other hand, if certain conditions are given, then the probability of the event is called a conditional probability. Thus, we wish to find the conditional probability of rolling a 1 given that the die landed on an odd number. Let

A be the event that the die lands on a one

and

B be the event that the die lands on an odd number

Then we denote the conditional probability of A given that B occurred as

$$P(A \mid B)$$

Knowing that the outcome is odd, the sample space of possible outcomes can be reduced to $\{1, 3, 5\}$, the elements in B.

Therefore

$$P(A \mid B) = 1/3$$

that is, assuming the simple events are equally likely, there is one favorable outcome, event $\{1\}$, out of three possible outcomes, event $\{1, 3, 5\}$. We see that the numerator consists of simple events common to both events A and B, and the denominator consists of simple events in B. This suggests a possible definition of a conditional probability.

Definition

The **conditional probability** of event A given that event B has occurred is found by dividing the probability that both A and B occur by the probability that B occurs. That is,

$$P(A \mid B) = \frac{P(A \cap B)}{P(B)}$$

provided that $P(B)$ is not zero.

In the preceding die example we have

$$A = \{1\} \quad \text{and} \quad B = \{1, 3, 5\}$$

from which we find

$$A \cap B = \{1\}$$

Because all simple events are equally likely, we have

$$P(A \cap B) = P(\{1\}) = 1/6$$

and
$$P(B) = P(\{1, 3, 5\}) = 3/6$$
so that
$$P(A \mid B) = \frac{P(A \cap B)}{P(B)} = \frac{1/6}{3/6} = 1/3$$

EXAMPLE 4.16

Suppose that at a large university 40% of the students take English, 25% take math, and 12% take both. If a student is randomly selected from an English class, what is the probability that he or she is also taking math?

SOLUTION   This is a conditional probability problem because we are given that the student is taking English and are asked to find the probability he or she is taking math. Let

   M be the event that the student is taking math

and

   E be the event that the student is taking English

We have
$$P(M) = .25, P(E) = .40, \text{ and } P(M \cap E) = .12$$
Therefore
$$P(M \mid E) = \frac{P(M \cap E)}{P(E)} = \frac{.12}{.40} = .30$$

The conditional probability formula can be rewritten so that
$$P(A \cap B) = P(A \mid B) P(B)$$

In this form, the probability that both A and B happen is the product of the conditional probability of A given B and the unconditional probability of B. For this reason it is called the multiplication law of probability.

The Multiplication Law

Let A and B be two events with nonzero probability, then
$$P(A \cap B) = P(A \mid B) P(B)$$

EXAMPLE 4.17

Bowl A contains three red balls and five white balls. Bowl B contains four red and two white balls. One of the bowls is selected at random and then a ball is drawn. What is the probability that the ball is red?

SOLUTION Let

R be the event that the selected ball is red
A be the event that bowl A is selected
B be the event that bowl B is selected

Then

$$R \cap A$$

is the event that bowl A is chosen *and* the selected ball is red, and

$$R \cap B$$

is the event that bowl B is chosen *and* the selected ball is red.

Clearly $R \cap A$ and $R \cap B$ are mutually exclusive because both bowls cannot be selected. Moreover event R occurs with

$$R \cap A \quad \text{or} \quad R \cap B$$

Thus from the additive law we have

$$P(R) = P(R \cap A) + P(R \cap B)$$

Using the multiplication law, we have

$$P(R) = P(R \mid A) P(A) + P(R \mid B) P(B)$$

Because the bowls are selected at random, we have

$$P(A) = P(B) = 1/2$$

Knowing the number of balls in each bowl, we have

$$P(R \mid A) = 3/8 \quad \text{and} \quad P(R \mid B) = 4/6$$

Consequently

$$
\begin{aligned}
P(R) &= (3/8)(1/2) + (4/6)(1/2) \\
&= 3/16 + 4/12 \\
&= (9 + 16)/48 \\
&= 25/48
\end{aligned}
$$

In some situations, the occurrence of event B will have no effect on the probability of A. That is,

$$P(A \mid B) = P(A)$$

In those cases we say that events A and B are *independent*.

Definitions

Events A and B are said to be **independent** if the occurrence of one does not affect the probability of the occurrence of the other, that is,

$$P(A \mid B) = P(A) \quad \text{and} \quad P(B \mid A) = P(B)$$

Otherwise A and B are **dependent**.

If events A and B are independent, then we have from the multiplication law that

$$P(A \cap B) = P(A)P(B)$$

which is often given as the definition of independent events. The equation is also used to check the independence of two events.

The Multiplication Law for Independent Events

Let A and B be two independent events, then

$$P(A \cap B) = P(A)P(B)$$

EXAMPLE 4.18

Toss two fair coins and define the following events:

> A: A head appears on the first coin.
> B: A head appears on the second coin.

Show that A and B are independent.

SOLUTION   The sample space is

$$S = \{(h, h), (h, t), (t, h), (t, t)\}$$

Because the coins are fair the four simple events each have probability $1/4$. The simple events in A and B are

$$A = \{(h, h), (h, t)\} \quad \text{and} \quad B = \{(h, h), (t, h)\}$$

We then have that

$$A \cap B = \{(h, h)\}$$

Calculating probabilities, we have

$$P(A \cap B) = P(h, h) = 1/4 = (1/2)(1/2) = P(A)P(B)$$

Thus events A and B are independent. Our intuition would say that the outcome of the first coin is independent of the outcome of the second coin. This result could be extended to three or more coins and establish that the outcome on any one coin is independent of the outcome of any other coin.

Note that A and B are *not* mutually exclusive because their intersection is not empty.

EXAMPLE 4.19

In Example 4.4 a company gives its potential employees an exam that has a 30% pass rate. If three potential employees take the exam, the sample space of possible outcomes is

$$S = \{(p, p, p), (p, p, f), (p, f, p), (f, p, p), (p, f, f), (f, p, f), (f, f, p), (f, f, f)\}$$

Assign probabilities to the eight simple events of this sample space.

SOLUTION   Event $\{(p, p, p)\}$ means all three employees pass and each passes with a 30% probability. One employee passing is independent of any other employee passing, so

$$P(\{(p, p, p)\}) = P(\text{1st passes})P(\text{2nd passes})P(\text{3rd passes})$$
$$= (.3)(.3)(.3)$$
$$= .027$$

In a similar manner, event $\{(p, p, f)\}$ should be assigned probability $(.3)(.3)(.7) = .063$, as should events $\{(p, f, p)\}$ and $\{(f, p, p)\}$, because each consists of two people passing and one failing.

Events $\{(p, f, f)\}$, $\{(f, p, f)\}$, and $\{(f, f, p)\}$ should each be assigned probabilities $(.3)(.7)(.7) = .147$.

Finally, event $\{(f, f, f)\}$ should be assigned probability $(.7)(.7)(.7) = .343$.

Thus we have the following table of probabilities.

Simple event	Probability
$\{(p, p, p)\}$	.027
$\{(p, p, f)\}$	.063
$\{(p, f, p)\}$	.063
$\{(f, p, p)\}$	.063
$\{(p, f, f)\}$	.147
$\{(f, p, f)\}$	.147
$\{(f, f, p)\}$	.147
$\{(f, f, f)\}$	.343
Total	1.000

EXAMPLE 4.20

In Example 4.4 events A, B, C, and D were defined as follows:

> A: Two of the three pass the exam.
> B: Only one passes the exam.
> C: All three pass the exam.
> D: At least one of the three passes the exam.

Determine the probabilities of events A, B, C, and D.

SOLUTION   Event A consists of simple events

$$\{(p, p, f)\}, \{(p, f, p)\}, \quad \text{and} \quad \{(f, p, p)\}$$

From Example 4.19 each has probability .063. Thus

$$P(A) = .063 + .063 + .063 = .189$$

Event B consists of simple events

$$\{(p, f, f)\}, \{(f, p, f)\}, \quad \text{and} \quad \{(f, f, p)\}$$

each of which has probability .147. Thus

$$P(B) = .147 + .147 + .147 = .441$$

Event C consists of only

$$C = \{(p, p, p)\}$$

Thus

$$P(C) = .027$$

Finally event D consists of all simple events *except* (f, f, f), which has probability .343. Because all the probabilities must add to 1.000, we know that

$$P(D) = 1.000 - .343 = .657$$

Probability in the Courtroom

An elderly woman was mugged in the suburb of a large city. A jury convicted a couple of the crime on the basis of circumstantial evidence presented by the prosecution. The prosecution used the multiplication law on the following events:

Event	Assumed probability
Driving a yellow Trans Am	1/10
Interracial couple	1/1,000
Blond woman	1/4
Woman wears ponytail	1/10
Man has beard	1/10
Man black	1/3

These probabilities were multiplied in order to find the probability that a couple of this description would be in the vicinity of the crime. The claimed probability was 1/12,000,000. The jury proceeded to convict the defendants, because the likelihood of another couple of that description would be almost impossible. We recall, however, that in order to use the multiplication law the events must be independent. Clearly, for example, interracial couple and black man are not independent. The defense lawyer pointed this out in the appeal and, consequently, the Supreme Court of the state reversed the conviction (*Time* magazine, April 26, 1968, p. 41).

EXERCISES 4.4

4.33 Suppose events A and B are such that $P(A) = .3$, $P(B) = .5$, and $P(A \cap B) = .2$. Find

(a) $P(A')$

(b) $P(A \cup B)$

(c) $P(A \mid B)$

(d) Are A and B independent? Why?

4.34 Suppose events A and B are such that $P(A) = .4$, $P(B) = .7$, and $P(A \cup B) = .8$. Find

(a) $P(B')$

(b) $P(A \cap B)$

(c) $P(B \mid A)$

(d) Are A and B independent? Why?

4.35 In Exercise 4.21, a man wears his blue suit 30% of the time. Assign probabilities to the simple events listed in the sample space found in that exercise.

4.36 In Exercise 4.22, assume 60% of the student loans are approved. Assign probabilities to the simple events listed in the sample space found in that exercise.

4.37 In Exercise 4.36, what is the probability that at least two loans are approved?

4.38 The three traffic lights of Exercise 4.23 are green with probability .7 and red with probability .3. Assign probabilities to the simple events listed in the sample space found in that exercise.

4.39 In Exercise 4.38, what is the probability that you stop no more than one time?

4.40 Suppose 3 of the 12 bottles in a case of wine are bad. If you randomly select 2 bottles, what is the probability that
(a) both are good?
(b) both are bad?
(c) one is good and one is bad?

4.41 Suppose two fair dice are rolled, and we are concerned with events:

A: The sum on the two up faces is even.
B: The sum on the two up faces is less than 6.

Are events *A* and *B* independent? Are they mutually exclusive?

4.42 Suppose events *A* and *B* are such that

$P(A) = .4$ and $P(B) = .3$.
(a) If *A* and *B* are independent, find $P(A \cup B)$.
(b) If *A* and *B* are mutually exclusive, find $P(A \cup B)$.

4.43 Two rescue teams set out to find a lost hiker in the Grand Canyon. Team *A* has a 30% chance of finding the hiker, and team *B* has a 40% chance of finding the hiker. What is the probability that the hiker will be rescued?

4.5   The Birthday Problem (optional)

On July 7, 1984, a rare event occurred when four people in a hospital delivery room discovered that they were all born on July 7. Using the laws of probability we can determine the probability of this and other similar events.

First suppose there are two people in a room. To determine the probability that both have the same birthday, let b_1 denote the birthday of one of the two and b_2 denote the birthday of the other. Both b_1 and b_2 can be any one of 365 possible days of the year (assuming it is not leap year). The sample space of possibilities is

$$S = \{(b_1, b_2) \mid b_1 = 1, 2, 3, \ldots, 365 \text{ and } b_2 = 1, 2, 3, \ldots, 365\}$$

Clearly the number of elements (simple events) in *S* is

$$365 \times 365 = 365^2$$

and they are equally likely.

Let *B* be the event that $b_1 = b_2$; that is, the two people have the same birthday. How many elements of the sample space are there in event *B*? There is no restriction on b_1, so it could be any one of 365 days. However b_2 must equal b_1, so there is only one possibility for b_2. Hence there are $365 \times 1 = 365$ simple events in event *B*.

Listing the elements of *B*, we have

$$B = \{(1, 1), (2, 2), (3, 3), \ldots, (365, 365)\}$$

and we see that there are indeed 365 simple events.

Because the simple events are equally likely, we have

$$P(B) = 365/(365)^2 = 1/365 = .00274$$

Next suppose there are three people in a room. The sample space of possible birthdays is

$$S = \{(b_1, b_2, b_3) \mid b_i = 1, 2, 3, \ldots, 365; i = 1, 2, 3\}$$

The number of simple events in S is 365^3, and they are equally likely.

Let B be the event such that $b_1 = b_2 = b_3$ (the three have the same birthday), then we have

$$B = \{(1, 1, 1), (2, 2, 2), (3, 3, 3), \ldots, (365, 365, 365)\}$$

Then, as in the preceding, the number of simple events in B is 365. So

$$P(B) = 365/(365)^3 = 1/(365)^2$$
$$= .0000075$$

Now consider the delivery room problem. There were four people (one was the newborn) in the room. The sample space of possible birthdays is

$$S = \{(b_1, b_2, b_3, b_4) \mid b_i = 1, 2, 3, \ldots, 365; i = 1, 2, 3, 4\}$$

As before, the number of elements in S is $(365)^4$ and they are equally likely. The event B that they all have the same birthday is the event where $b_1 = b_2 = b_3 = b_4$. That is, we have that

$$B = \{(1, 1, 1, 1), (2, 2, 2, 2), (3, 3, 3, 3), \ldots, (365, 365, 365, 365)\}$$

and the number of simple events in B is 365. Because the simple events are equally likely, we have that

$$P(B) = 365/(365)^4$$
$$= 1/(365)^3$$
$$= .00000002056$$
$$= 1 \text{ out of } 48,627,125$$

or about 1 out of 50 million.

Let's now look at a slightly different problem. Suppose that there are 10 people in a room. What is the probability at least 2 have the same birthday? Let $b_1, b_2, b_3, \ldots, b_{10}$ denote the birthdays of the 10 people. Then the sample space of possible birthdays is

$$S = \{(b_1, b_2, b_3, \ldots, b_{10}) \mid b_i = 1, 2, 3, \ldots, 365; i = 1, 2, \ldots, 10\}$$

As before, the number of elements in S is $(365)^{10}$. We are interested in the event A that at least 2 of the 10 have the same birthday. However the complement of A, A', would be easier to work with. A' would be the event that no two have the same birthday; that is,

$$b_1 \neq b_2 \neq b_3 \neq \cdots \neq b_{10}$$

There would be 365 possibilities for b_1, but because $b_2 \neq b_1$ there are only 364 possibilities for b_2. Similarly there are 363 possibilities for b_3, and so on. We can see that the number of elements in A' is $365 \times 364 \times 363 \times \cdots \times 356$. Therefore we have that

$$P(A') = (365 \times 364 \times 363 \times \cdots \times 356)/(365)^{10}$$
$$= .883$$

Using the complement law, we have

$$P(\text{at least two have the same birthday}) = P(A)$$
$$= 1 - P(A')$$
$$= 1 - .883$$
$$= .117$$

Table 4.1 gives the probability that at least 2 people will have the same birthday when the number of people ranges from 10 to 50.

TABLE 4.1

Number of people	Probability at least two have the same birthday
10	.117
15	.253
20	.411
21	.444
22	.476
23	.507
24	.538
25	.569
30	.706
40	.891
50	.970

We observe that in a group of 23 people there is better than a 50% chance that at least 2 will have the same birthday. Among 50 people it is almost certain that at least 2 will have the same birthday.

EXERCISES 4.5

4.44 Determine whether two people in your class have the same birthday. Determine the number of students in the classroom and compare the results with the probabilities in Table 4.1.

4.45 A company packages pictures of 40 former U.S. Presidents in its product. Assuming an equal number of pictures of each President are distributed, what is the probability that out of five packages you will get five different pictures?

4.46 Compute the probability that, out of five people, at least two have the same birth month.

RANDOM

Two Minitab subcommands relating to the material in this chapter are described in this section. Each subcommand is executed from the Minitab command RANDOM.

RANDOM simulates observations from a number of probability distributions, among which are the uniform and Bernoulli distributions. For a complete list of all distributions, see the *Minitab Handbook*. The uniform distribution simply means that all values have an equal chance of occurring. The Bernoulli distribution, to be discussed in Section 5.4, is a distribution in which only two possible outcomes are possible. To simulate observations from a uniform distribution we use the Minitab command RANDOM with the subcommand INTEGER.

INTEGER

RANDOM, with the subcommand INTEGER, randomly generates independent integers and stores them in the specified columns of the worksheet. The command is

```
RANDOM K OBSERVATIONS INTO EACH OF C, C,...,C;
INTEGERS BETWEEN K AND K.
```

INTEGER can be used to simulate the roll of a fair die. For example, if we wish to simulate 18 rolls of a fair die, the command would be

```
RANDOM 18 OBSERVATIONS, PUT IN C1;
INTEGERS BETWEEN 1 AND 6.
```

EXAMPLE 4.21

Simulate 18 rolls of a fair die and repeat it 10 times. Ideally we would expect each of the 6 possible integers to turn up 3 times in 18 rolls. For each simulation determine the frequency of occurrence of each possible outcome.

SOLUTION The data will be simulated with the RANDOM command with the INTEGER subcommand (Figure 4.4). After printing out the simulated data, a summary will be given. The TALLY command can be used to give the frequency table for each column.

A summary of all 10 simulations is given here:

| | Counts | | | | | | | | | |
Observation	C1	C2	C3	C4	C5	C6	C7	C8	C9	C10
1	3	4	4	0	1	4	4	3	2	2
2	2	2	1	5	4	3	1	3	2	2
3	5	1	6	4	3	2	0	2	1	3
4	2	3	4	4	3	5	3	2	4	4
5	4	5	2	3	3	2	2	4	6	5
6	2	3	1	2	4	2	8	4	3	2

As stated, we would have expected each of the six possible values to have come up three times each. We see that in simulation number 7, however, the 3 did not appear, whereas the 6 appeared eight times. On the other hand, simulation number 8 is close to what we would expect.

FIGURE 4.4

Minitab output for Example 4.21

```
MTB > RANDOM 18 OBSERVATIONS INTO C1-C10;
SUBC> INTEGERS BETWEEN 1 AND 6.

MTB > PRINT C1-C10
```

ROW	C1	C2	C3	C4	C5	C6	C7	C8	C9	C10
1	4	3	5	2	2	2	5	4	6	5
2	6	5	3	2	5	5	4	2	4	4
3	1	1	4	4	3	4	1	1	1	1
4	5	4	5	3	3	1	6	2	3	3
5	1	6	1	4	2	3	1	6	5	4
6	2	5	2	5	5	4	4	2	2	4
7	3	1	3	3	6	2	1	1	5	3
8	3	4	4	4	4	6	6	6	5	5
9	3	5	1	6	6	5	6	6	4	2
10	1	5	1	5	5	3	4	5	5	5
11	3	4	4	4	3	1	6	4	4	4
12	5	5	3	3	2	4	6	1	5	2
13	6	1	4	5	6	1	2	5	5	5
14	3	6	3	2	2	4	5	6	6	3
15	5	2	3	6	6	1	6	5	1	1
16	5	2	3	3	1	6	6	3	4	6
17	4	1	6	2	4	2	1	5	2	6
18	2	6	1	2	4	4	6	3	6	5

BERNOULLI

In many experiments there are only two possible outcomes. For example, in the toss of a coin there is the possibility of heads or tails, or a treatment is either successful or not. We will learn in Section 5.4 that an outcome of such an experiment is called a Bernoulli trial. BERNOULLI will simulate the outcomes of a series of Bernoulli trials. The command is

```
RANDOM K OBSERVATIONS INTO C, C,...,C;
BERNOULLI WITH P = K.
```

The outcome is coded as a 0 or 1. The probability of a 1 on any given trial is P. Finally the results are stored in the specified columns.

To simulate the toss of a fair coin 10 times, we use the command

```
RANDOM 10 OBSERVATIONS IN C1;
BERNOULLI WITH P = 0.5.
```

EXAMPLE 4.22

Suppose a field-goal kicker has an 80% success rate inside the 35-yard line. Simulate 8 kicks inside the 35 during a game. Determine the number of successes. Simulate 8 kicks inside the 35 for 10 successive games. Determine his success rate in each game and his success rate for the 10-game season.

SOLUTION   Figure 4.5 simulates 8 kicks with a success rate of 80%:

FIGURE 4.5
Minitab output for Example 4.22

```
MTB > RANDOM 8 OBSERVATIONS IN C1;
SUBC> BERNOULLI TRIALS WITH P = 0.800.

MTB > PRINT C1
C1
    1     0     0     1     1     1     0     1

MTB > TALLY C1

    C1   COUNT
    0      3
    1      5
   N=      8
```

▲ The TALLY command gives a frequency count of each of the values in the column.

Out of 8 kicks he missed 3, which were the second, third, and seventh kick. Figure 4.6 gives the simulations for the 10 games with their summary:

FIGURE 4.6
Minitab output for Example 4.22

```
MTB > RANDOM 8 OBSERVATIONS IN C1-C10;
SUBC> BERNOULLI WITH P= .8000.

MTB > PRINT C1-C10
```

ROW	C1	C2	C3	C4	C5	C6	C7	C8	C9	C10
1	0	1	1	1	1	1	1	1	0	1
2	0	1	0	1	0	0	1	1	1	1
3	1	1	1	1	1	1	1	1	1	1
4	1	1	1	1	0	1	1	0	1	0
5	1	1	0	0	1	1	1	1	1	1
6	1	1	1	1	1	1	1	0	1	1
7	1	1	1	1	1	1	1	1	1	1
8	1	1	1	1	1	1	1	1	1	1

In game 1 he missed the first two tries and made the rest. He made all 8 in game 2, and so on. Following is a summary with his success rate in each game:

Game	No. hits	No. misses	Success rate
1	6	2	75%
2	8	0	100%
3	6	2	75%
4	7	1	87.5%
5	6	2	75%
6	7	1	87.5%
7	8	0	100%
8	6	2	75%
9	7	1	87.5%
10	7	1	87.5%
Total	68	12	85%

4.47 Simulate the roll of a die 60 times. How often would one expect each of the six integers to turn up? Determine the frequency of occurrence of each possible outcome in the 60 rolls. Repeat the process five times and compare the results.

4.48 INTEGER can be used to simulate birthdays if we associate each birthday with its corresponding day of the year. For example, January 23 is day 23, February 19 is day 50 (31 + 19 = 50), until we get to December 31, which is day 365. We saw in Section 4.5 that in a sample of 23 people the chance of at least two having the same birthday is greater than 50%. Simulate 23 birthdays with INTEGER and see if there are any matches. In a sample of 50 people it is almost certain that at least 2 will have the same birthday. Simulate 50 birthdays and see if there are any matches.

4.49 Approximately 55% of entering freshmen eventually graduate from a 4-year college. Simulate the success or failure to graduate for 50 entering freshmen. How many graduated? Does the number seem reasonable?

4.50 A component for a spacecraft launch fails 5% of the time. Suppose five such components are connected in series for a launch. Simulate the success or failure of the five components in a launch. Did all work? Simulate 10 different launches. How many got off the ground?

4.7 Summary and Review

Key Concepts

☑ An *experiment* is an activity involving making an observation or taking a measurement, which leads to a collection of outcomes associated with the experiment. The set of all outcomes is called the *sample space*. Any subset of the sample space is called an *event*. A *simple event* is an event containing a single element.

☑ *Probability* is a number between 0 and 1 that describes the likelihood with which an event is to occur. If probabilities can be assigned to the simple events in a sample space, then the probability of any other event A is the sum of the probabilities that have been assigned to the simple events in event A.

☑ Events that are made up of other events are called *compound events*. The *union* of two events is a compound event containing all simple events in either of the two events. The *intersection* of two events is the event containing all simple events that are common to the two events.

☑ Several laws of probability apply to compound events. The *additive law* is

$$P(A \cup B) = P(A) + P(B) - P(A \cap B)$$

☑ For *mutually exclusive* events A and B, $P(A \cap B) = 0$.

☑ The *complement law* says that

$$P(A') = 1 - P(A)$$

☑ $P(A \mid B)$ is the *conditional probability of A given B* and is given by the following formula:

$$P(A \mid B) = \frac{P(A \cap B)}{P(B)}$$

☑ The *multiplicative law* says that

$$P(A \cap B) = P(A \mid B) P(B)$$

☑ Events A and B are independent if $P(A \mid B) = P(A)$ or if $P(B \mid A) = P(B)$.

☑ The terms *mutually exclusive* and *independent* are sometimes confused. Mutually exclusive means that if one event occurs the other one cannot occur. Independence means that if one event occurs it will not affect the probability of the other one occurring.

Learning Goals

Having completed this chapter you should be able to:

1. Discuss the intuitive concepts of probability. *Section 4.1*
2. Calculate probabilities for simple experiments. *Section 4.2*
3. List the elements of the sample space for certain experiments. *Section 4.2*
4. Understand the concept of a simple event and determine the simple events that comprise other events. *Section 4.2*
5. Assign probabilities to the simple events and determine the probability of other events. *Section 4.2*

6. Be familiar with the laws of probability and use them to determine probabilities of events. *Sections 4.3 and 4.4*

7. Understand the concept of a conditional probability and decide whether two events are independent. *Section 4.4*

Review Questions

To test your skills answer the following questions:

4.51 A man has three keys, one of which will open a door. If he randomly picks one key after another until he finds the one to open the door, what is the probability that he will open the door on the first try? What is the probability that it will take more than three tries to open the door?

4.52 A room has five women and five men. Two are selected at random. What is the probability both are women?

4.53 A tetrahedron (regular four-sided polyhedron) has four sides numbered 1, 2, 3, 4. A pair of fair tetrahedra is tossed. How many outcomes are possible? _____ Are they equally likely? _____ A variable is defined to be the sum of the two numbers on the two tetrahedra. What are the possible outcomes of the variable? _____ Are they equally likely? _____ We win in a game if we roll a 5 or a 7 on the first roll. What is the probability that we win on the first roll? _____

4.54 A circular game board is divided into five slots numbered 1, 2, 3, 4, and 5. The odd numbers are colored red and the even numbers are colored black. A marble is spun around the board and randomly falls in one of the slots.

(a) If the slots are equally likely, what is the probability that it lands on red?

(b) Suppose the slots are such that an even number is twice as likely as an odd number. What is the probability that the marble lands on red?

4.55 Suppose 30% of the new employees hired by a computer firm are female. Three new employees are selected at random.

(a) List the sample space of possible sexes of the three employees.

(b) Assign a probability to each of the simple events.

(c) Suppose A is the event that exactly two of the three selected are women. List the elements that make up A.

(d) What is $P(A)$?

SUPPLEMENTARY EXERCISES FOR CHAPTER 4

4.56 Roll a fair die. What is the probability it turns up odd and greater than 4?

4.57 A card is drawn from a standard deck of cards. What is the probability that the card is red and a face card?

4.58 Roll a pair of fair dice. What is the probability that the sum of the two up faces is 7 or 11?

4.59 The big wheel on "The Price Is Right" game show is marked off in 20 equal increments from 5¢ to $1.00. The contestant gets two tries to spin a total as close as possible to $1.00 without going over. Suppose the contestant gets 60¢ on the first spin and decides to spin again.

(a) What is the probability she gets a total of $1.00 on her next spin?

(b) What is the probability she goes over $1.00?

4.60 There are eight books on a bookshelf, of which three are fiction.

(a) Supposing you select a book at random, what is the probability that it is nonfiction?

(b) A second book is selected without replacing the first. If the first one was fiction, what is the probability the second is also fiction?

4.61 Complete the following:

(a) A test has two true-false questions on it. If a student guesses, what is the probability he or she gets both right?

(b) A test has two multiple-choice questions, and each question has four choices. If a student guesses, what is the probability he or she gets both right?

4.62 Suppose your skill at playing a particular video game is such that you stand a 40% chance of winning each time you play. Suppose you play three games and either win (W) or lose (L).

(a) List the sample space of outcomes.

(b) Assign probabilities to the simple events.

(c) What is the probability you win at least two of the three games?

4.63 In the game of craps, the "shooter" loses on the first roll of a pair of fair dice if the two dice are double ones or double sixes.

(a) What is the probability that he or she loses on the first roll?

(b) What is the probability that he or she neither loses nor wins on the first roll? (Recall that the shooter wins on the first roll if the sum is 7 or 11.)

4.64 Suppose 40% of the employees of a company are female. Three employees are selected at random to participate in a discussion of salaries.

(a) List the sample space of possible sexes of the three employees.

(b) Assign probabilities to each of the simple events.

(c) Suppose A is the event that exactly two of the three selected are women. List the elements that make up A.

(d) What is $P(A)$?

4.65 During the month of February, school is held only 80% of the time, because of the weather. Let H stand for school being held on a given day and N stand for school not being held. Assume (perhaps erroneously) that school being held on one day is independent of it being held on another day.

(a) List the sample space of possibilities of school being held the first three days of February.

(b) Assign probabilities to each of the simple events.

(c) Suppose B is the event that school is held on exactly one day out of the first three days. List the elements that make up B.

(d) What is $P(B)$?

4.66 At the end of the television season a show is either renewed or canceled. One producer has three shows on the air.

(a) List the sample space of possibilities of his shows being canceled or renewed.

(b) If the probability a show is canceled is 60%, assign probabilities to the simple events in the sample space.

(c) What is the probability at least two of his shows will be renewed?

4.67 Three-fourths (75%) of the women who wear Andrea Acrylic Shield Nail Color have no significant chipping or peeling after 3 days. Let C denote chipping and N denote no significant chipping. For three women who try the nail color:

(a) List the sample space of possibilities of chipping and no chipping.

(b) Assign probabilities to the simple events in the sample space.

(c) What is the probability no more than one experiences chipping?

4.68 Suppose a man has a certain trait that will be passed on to his offspring with 30% probability. His wife has the same trait and she passes it to her offspring with 20% probability. Assuming that the trait can be passed on by either parent, what is the probability that their child will have the trait?

4.69 In the game of roulette (see Exercise 4.11), suppose you play black on three consecutive plays. What is the probability you win on all three plays?

4.70 In the game "three strikes you're out" on "The Price Is Right" game show, seven circular chips are placed in a bag. Four of the chips are white and have numbers on them corresponding to the digits in the price of an automobile. The other three chips are red and have X's on them corresponding to strikes. If the contestant draws the four digits of the price of the car and places them in order before drawing the three strikes, then he or she wins the car. On the first draw, what is the probability that the contestant will get a white chip? If he or she gets a white chip and guesses, what is the probability that he or she will select the right position of it in the price of the car?

4.71 According to the state Center for Health Statistics, 116,564 pregnancies were reported in North Carolina in 1983. Of those 116,564 pregnancies, 31,892 ended in abortions. The report indicated that it was the first abortion for 23,217 of the women, the second for 6,680 women, the third for 1,561 women, and the fourth or more for 434 women.

(a) What percent of the pregnancies ended in abortion?

(b) Of those who had an abortion, what percent had previously had an abortion?

4.72 An urn contains a red, a green, and a black marble. Two marbles are drawn from the urn with replacement. List the elements in the sample space and the following events:

A: Both marbles are red.
B: None is red.
C: One is red.

4.73 Drivers can select any one of three pumps at a gas station. If two drivers enter the station at the same time (obviously they cannot use the same pump), list the sample space of possible selections for the two drivers. Are the simple events equally likely?

4.74 A box contains a nickel, a dime, and a quarter. Two coins are selected without replacement. List the elements of the sample space and the elements of the following events:

A: The selection has a quarter.
B: The selection totals less than 25 cents.

Are the simple events equally likely?

4.75 A fair die is rolled. If it is even, the die is rolled again; if it is odd, a fair coin is tossed. List the sample space and assign probabilities to each of the simple events.

4.76 Given the experiment described in Exercise 4.75, what is the probability of a head on the coin?

4.77 There are three checkout lines in a department store. Suppose that three customers randomly pick a line. Assuming that all three could choose the same line, list the elements of the sample space. Assign probabilities to the simple events.

4.78 A beer drinker is asked to rank three unmarked glasses of beer according to taste. List the elements of the sample space. What must be true in order to assume that the simple events are equally likely? Assuming that they are, assign probabilities to them.

4.79 A shopper wishes to buy two pairs of shoes but cannot decide among four different pairs. List the elements of the sample space of possibilities. If she randomly chooses the two pairs, assign probabilities to the simple events.

4.80 A trial lawyer claims that she wins 80% of her cases. Out of two independent cases, what is the probability that she wins both, assuming that her claim is true? What is the probability that she loses both? What is the probability that she wins only one?

4.81 Eighty students who are candidates for an honor society are classified according to sex and academic standing in the following table.

	Men	Women	Total
Freshman	16	14	30
Sophomore	24	26	50
Total	40	40	80

Are the events Sophomore and Women independent?

4.82 What is the probability that the next four babies born at a hospital will be girls? Assume a boy and girl are equally likely.

4.83 In 1982, 39% of all business/management bachelor's degrees were awarded to women. This compares to only 8.1% in 1971 (*USA Today,* October 23, 1984). Suppose that three business/management students are selected at random from the 1982 graduating class. What is the probability that all are women?

4.84 Jan Stenerud of the Minnesota Vikings has been kicking field goals since 1967, when he joined the Kansas City Chiefs. His season's best was in 1981 with the Green Bay Packers. That year he set an NFL record by hitting 22 of 24 attempts for a 91.7% success rate. His lifetime record is 66.2% (*USA Today,* October 12, 1984).

Suppose he has three attempts in a game. Assuming his attempts are independent, what is the probability that

(a) he will hit all three?

(b) he will hit two of the three?

(c) he misses all three?

5

Random Variables and Probability Distributions

Suppose you are interested in studying the age (in months) at which children begin to walk. You know, of course, that all children do not begin walking at the same age. In fact, if you measured this variable on a large group of children, you would get a large number of different values. The collection of all possible measurements represents the observed values of the *random variable,* age. In other words, a random variable represents the possible values of the measurement that you are taking.

You will see, in this chapter, that all random variables have a distribution. The distribution simply describes the probabilities associated with the different values that the variable assumes. Two important random variables considered in this chapter are the binomial random variable and the normal random variable.

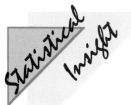

Does Acid Rain Cause Lopsided Tomatoes?

In 1984 scientists at North Carolina State University, funded by the U.S. Department of Agriculture, began a 3-year study of the effects of acid rain on a selected group of crops (Watauga *Democrat,* August 8, 1984). The study was designed to determine which crop varieties are at risk and to answer questions regarding the effects of the acid rain on the development of individual plants. Specifically does acid rain damage leaves when it falls directly on them, or does it hinder pollination by making it more difficult for pollen to stick to the plant in the flowering stage? Improper pollination prevents complete fertilization, which might result in lopsided tomatoes, smaller cucumbers, or cause only half the kernels on an ear of corn to develop.

Acid rain is described as any precipitation that results when sulfur dioxides and nitrogen oxides react with other chemical compounds in the atmosphere to produce sulfuric or nitric acid. The acidity is measured by the pH scale in which normal rain has a pH of 5.6 and lemon juice has a pH of 2.2.

In the research projects, plants are exposed to simulated acid rain concentrations ranging from a pH of 2.4 to 5.6. Suppose there are 20 plants of a certain variety that are exposed to acid rain with a pH of 4.4. Out of the 20 plants, it is possible that none will be damaged, all 20 will be damaged, or any number less than 20 will be damaged. It is not possible to determine in advance how many will be damaged. Under certain assumptions, the number damaged is called a binomial random variable. In Section 5.4 we will take a close look at the binomial random variable.

Suppose that any rainfall with a pH below 4.4 causes serious problems with the yields of crops. Then one might ask, what percent of the time are there rainfalls with a pH below 4.4? Also what percent of the time is the pH between 4.4 and 4.9? The pH level of the rainfall is also a random variable, which might possibly be what we call a normal random variable. In Section 5.5 we will study the normal random variable. But first, we will begin with a general discussion of random variables.

5.1 Random Variables

In Example 4.14, the experiment of tossing a coin six times resulted in a sample space of 64 different outcomes of the form

(h, h, h, h, h, h), (h, h, h, h, h, t), (h, h, h, h, t, h), . . . , etc.

It is not an easy matter to list all 64 possibilities; moreover we may not want to list all 64 possibilities. In this problem (and similar problems) we are not interested in the distinct outcomes but rather in the number of heads that appear in the six tosses.

Instead of concerning ourselves with the sample space consisting of elements of the form (h, h, h, h, h, h), (h, h, h, h, h, t), (h, h, h, h, t, h), etc., we need concern ourselves only with the numerical values 0, 1, 2, 3, 4, 5, and 6, which clearly is much simpler.

It appears that for the sample space, we need a rule that will assign a numerical quantity to each of the outcomes in the sample space. This rule is called a *random variable*.

Definition

A **random variable,** which is denoted by a letter such as x or y, is a rule that represents the potential numerical values associated with the outcomes of an experiment. The list of values will be called the *range* of the random variable.

In the preceding example, the random variable, x, denotes the number of heads in six tosses of a coin. The range for x is

$$R_x = \{0, 1, 2, 3, 4, 5, 6\}$$

As another example, let x represent the potential outcome of the roll of a die. The possible values that x could assume are

$$R_x = \{1, 2, 3, 4, 5, 6\}$$

EXAMPLE 5.1

Four students are asked whether they believe that marijuana should be legalized. Each will respond with either Yes or No. Define a random variable on the sample space of possible outcomes.

SOLUTION   Several possible random variables could be defined on this sample space, but it seems reasonable that we will be interested only in the number of students who favor the issue. Thus we define the random variable

y = number who believe marijuana should be legalized

Clearly, out of four students, the range of y would be

$$R_y = \{0, 1, 2, 3, 4\}$$

EXAMPLE 5.2

An experiment consists of tossing a coin until a head appears. Define a random variable on this experiment.

SOLUTION   Because the head might appear on the first toss or perhaps the second toss or maybe not until the seventeenth toss, the sample space will consist of an infinite number of simple events:

$$S = \{h, th, tth, ttth, tttth, \ldots\}$$
("..." means it continues forever)

On this sample space we can let the random variable x be the number of tosses until the head appears, in which case the range for x will be

$$R_x = \{1, 2, 3, 4, \ldots\}$$

We observe in the preceding examples that the range of each random variable consists of a finite number of values such as $\{0, 1, 2, 3, 4\}$, or at most a countably infinite number of values such as $\{1, 2, 3, 4, \ldots\}$. We recall from Section 2.1 that variables of this type are called *discrete*.

Continuous variables were defined in Section 2.1 to be variables that could assume all values in a line interval.

EXAMPLE 5.3

In 1981 the average life expectancy in the United States reached an all-time high of 74.2 years. Obviously this does not mean that each of us can expect to live to an age of 74.2 years. In fact, if you are female, you can expect to live over 7 years longer than men (77.9 years for women and 70.4 years for men). Also blacks have a significantly shorter life expectancy than whites (68.7 years for blacks and 74.8 for whites). Your occupation also has an effect on your life expectancy. Define a random variable related to life expectancy.

SOLUTION   Insurance companies that insure specific groups of people must study the life expectancy of that group. For example, companies that insure schoolteachers are very familiar with the random variable, call it y, that represents the life expectancy of schoolteachers. Theoretically this random variable can assume any value greater than 0. Practically, however, the life expectancy of all teachers should range from about 20 to 120 years. Most teachers can expect to live to about the age of 75, but a few will die very young and a few will live longer than 75 years. In any case, the random variable

$$y = \text{life expectancy of a schoolteacher}$$

can assume any value in the interval from 20 to 120 years, and thus is a continuous random variable.

Typically we think of a discrete random variable as one corresponding to a count or "the number of," and a continuous random variable as one corresponding to some measurement process. Examples of discrete random variables include:

1. The number of penalties incurred by a football team in a single game ($x = 0, 1, 2, 3, \ldots$)
2. The number of students passing a psychology test out of a class of 30 ($x = 0, 1, 2, 3, \ldots, 30$)

3. The number of students from the student body favoring beer being sold on campus ($x = 0, 1, 2, 3, \ldots, 10{,}000$, assuming the size of the student body is 10,000)

4. The number of violent crimes committed per month in a certain section of the city ($x = 0, 1, 2, 3, \ldots$)

5. The number of computer terminals that are in use during a certain period of the day ($x = 0, 1, 2, 3, \ldots, n$, where n is the number of terminals)

Examples of continuous random variables include:

1. The length of a prison term for possession of marijuana
2. The concentration of DDT in a sample of milk
3. The potential IQ of a randomly selected third grader
4. The amount of a hospital bill for 2 weeks in coronary care
5. The amount of cola in a 12-ounce can
6. The weight of a box of oranges

EXERCISES 5.1

5.1 The experiment consists of tossing five fair coins. Let x be the random variable that is the number of heads in the five tosses. List the range of values for x. Is x discrete or continuous?

5.2 The experiment consists of rolling a pair of fair dice. Let the random variable y be the total of the faces of the two dice. List the range of values for y. Is y discrete or continuous?

5.3 Let w be the random variable that indicates the amount of time it takes a person to drive to work. List the range of values for w. Is w discrete or continuous?

5.4 The Red Cross is searching for AIDS carriers who have donated blood. They will continue testing subjects until they find a carrier. Let

y = no. of subjects tested until a carrier is found

List the range of values for y. Is y discrete or continuous?

5.5 The Red Cross has 15 blood donors and is searching for type B blood. Let

b = no. who have type B blood out of the 15

List the range of values for b. Is b discrete or continuous?

5.6 Let x be the SAT Math score of a student selected at random from the student body of your school. List the range of values for x. Is x discrete or continuous?

5.7 Classify the following as either discrete or continuous:

(a) The self-concept score on the Tennessee Self Concept Scale

(b) The number of alcohol-related traffic accidents per week in your state

(c) The number of people applying for food stamps per week in your city

(d) The length of time of a strike by steel workers

(e) The potential life expectancy of a 40-year-old man applying for an insurance policy

Probability Distribution of a Discrete Random Variable

As stated in the previous section, a discrete random variable assumes only a discrete set of values. Thus we might ask, "What are the chances or probabilities associated with each possible value of the random variable?" For example, what is the probability of getting a 4 on the roll of a fair die; or the probability of two heads in three tosses of a fair coin?

Definition

A table or function that lists the possible values of a discrete random variable and their associated probabilities is called the **probability distribution** of the random variable.

Table 5.1, for example, is the probability distribution for the random variable y associated with the outcomes of the roll of a fair die.

TABLE 5.1

y	$p(y)$
1	1/6
2	1/6
3	1/6
4	1/6
5	1/6
6	1/6
Total	1.00

Note that the table lists all six possible values and each occurs with a probability of 1/6. Observe that the probabilities are all between 0 and 1. Further they total 1 (or 100%), because the table lists all possible values of the random variable. If any entry in the table falls outside the interval from 0 to 1, or if the probabilities do not add to 1, then it is *not* a probability distribution. Again this is a discrete random variable, because it can assume only the discrete values

$$R_y = \{1, 2, 3, 4, 5, 6\}$$

To illustrate how the probabilities are *distributed* over the values we will graph the distribution of the random variable. Figure 5.1(a) is such a graph of the distribution with line segments of height 1/6 drawn at each value of the random variable. Figure 5.1(b) is the same distribution drawn in the form of a histogram. Both presentations are correct. Figure 5.1(a) illustrates the discrete nature of the random variable by showing that the probabilities are concentrated at the values 1, 2, 3, 4, 5, and 6. An advantage of the histogram form [Figure 5.1(b)] is that the areas of the rectangles correspond to the probabilities associated with the various numerical values. Because this concept of area carries over to continuous random variables we shall graph the distributions in the form of a histogram.

FIGURE 5.1

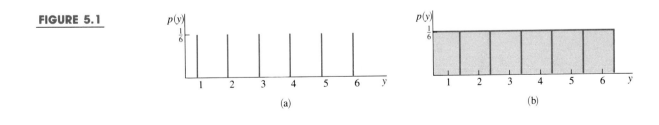

(a) (b)

EXAMPLE 5.4

Consider families with three children. Let the variable w be the number of boys in a family. Find the probability distribution for w and graph the distribution.

SOLUTION   The possibilities for the three children, that is, the sample space for the experiment is

$$S = \{(g, g, g), (g, g, b), (g, b, g), (b, g, g), (b, b, g), (b, g, b), (g, b, b), (b, b, b)\}$$

We see there are eight possibilities, and assuming a boy is just as likely as a girl, each of the eight possibilities is equally likely. Of the eight only one corresponds to zero boys, that is, $w = 0$, thus the probability of zero boys is one-eighth. Three of the eight possibilities correspond to $w = 1$ (one boy) and therefore the probability is three-eighths. Continuing in this fashion we have the completed probability distribution of w in Table 5.2.

TABLE 5.2

w	0	1	2	3
$p(w)$	1/8	3/8	3/8	1/8

Observe that there could be either 0, 1, 2, or 3 boys in the family, and the table gives the probability of each.

The graph of this distribution is given in Figure 5.2.

FIGURE 5.2

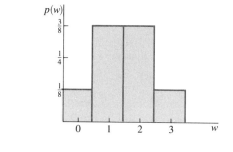

Note that in Table 5.2 the probabilities add to 1, which is always the case with a probability distribution table.

Also if you observe closely, this is exactly the same as the distribution for the random variable

x = no. of heads in three tosses of a fair coin

Recall that the sample space for this experiment is

$$S = \{(h, h, h), (h, h, t), (h, t, h), (t, h, h), (h, t, t), (t, h, t), (t, t, h), (t, t, t)\}$$

which is completely analogous to the sample space just given for the possibilities of the three children. And because a head or tail turns up half the time, just as we assumed a boy is born half the time, the probability distribution for x is exactly the same as the probability distribution for w.

So we see that a genetics problem can be viewed as a simple coin toss experiment. The coin toss problem serves as a model for many problems confronting us in the real world.

Had the problem specified six children, the probability distribution for the number of boys would be as given in Table 5.3.

TABLE 5.3

w	0	1	2	3	4	5	6
$p(w)$	1/64	6/64	15/64	20/64	15/64	6/64	1/64

Because each of the six children can be either a boy or a girl, there are $2^6 = 64$ possibilities in the sample space. Only one, (g, g, g, g, g, g), corresponds to $w = 0$, hence $p(0) = 1/64$. Six of the possibilities correspond to $w = 1$, hence $p(1) = 6/64$. Continuing we could soon come up with the probabilities in Table 5.3. The graph of this distribution is given in Figure 5.3.

FIGURE 5.3

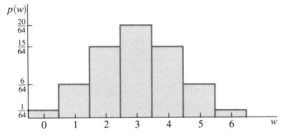

Once a probability distribution is obtained, we can then use it to obtain probabilities of other events.

EXAMPLE 5.5

The probability distribution given in Table 5.3 corresponds to the random variable w, which gives the number of boys in a family of six children. What is the probability of at least four boys?

SOLUTION   The probability of at least four boys out of six children can be interpreted as

$$P(w \geq 4)$$

The event $w \geq 4$ can happen only if $w = 4$, 5, or 6; thus we simply add the probabilities corresponding to 4, 5, and 6 and get

$$\begin{aligned} P(w \geq 4) &= p(4) + p(5) + p(6) \\ &= 15/64 + 6/64 + 1/64 \\ &= 22/64 \\ &= 11/32 \end{aligned}$$

Combinatorial Method

The remainder of this section is *optional* and can be omitted without a loss of continuity.

In order to determine the probabilities associated with each value of w given in Table 5.3, we must be able to count the number of simple events associated with each value of w. This is accomplished with the *combination formula*.

Definition

The number of **combinations** of k objects taken from n objects is given by

$$\binom{n}{k} = \frac{n!}{k!(n-k)!}$$

where $n!$ is read "n factorial" and is given by

$$n! = n(n-1)(n-2)(n-3)\cdots(2)(1)$$

and

$$0! = 1$$

EXAMPLE 5.6

Suppose that there are four vacant seats at a stadium and two people arrive. How many possible seating arrangements are there?

SOLUTION   Label the seats as S_1, S_2, S_3, and S_4. In the following configuration we mark an X for an occupied seat. We list the possibilities as follows:

	S_1	S_2	S_3	S_4
Arrangement 1	X	X		
Arrangement 2	X		X	
Arrangement 3	X			X
Arrangement 4		X	X	
Arrangement 5		X		X
Arrangement 6			X	X

There are six different possible arrangements. Without listing the possibilities we can use the combination formula to find out how many there are. We have

$$\binom{4}{2} = \frac{4!}{2!(4-2)!} = \frac{4 \times 3 \times 2 \times 1}{2 \times 1 \times 2 \times 1} = 3 \times 2 = 6$$

The probabilities in Table 5.3 can be obtained in the following manner.

1. Recall from Example 4.14 that in the coin toss experiment (which is analogous to this experiment) there were $2^6 = 64$ possibilities in the sample space. In this experiment each simple event consists of a sequence of six g's or b's. For example, (g, g, b, g, b, g) means the first, second, fourth, and sixth children were girls and the third and fifth were boys. Because a boy is just as likely as a girl, the 64 simple events are equally likely and have probability $1/64$.

2. There is $\binom{6}{0} = \dfrac{6!}{0!6!} = 1$

outcome that results in no boys ($w = 0$); it is

$$(g, g, g, g, g, g)$$

hence $P(w = 0) = 1/64$.

3. There are $\binom{6}{1} = \dfrac{6!}{1!5!} = 6$

possibilities that result in one boy ($w = 1$); they are

$$(b, g, g, g, g, g), (g, b, g, g, g, g), (g, g, b, g, g, g),$$
$$(g, g, g, b, g, g), (g, g, g, g, b, g), (g, g, g, g, g, b)$$

hence $P(w = 1) = 6/64$.

4. In a similar manner there are $\binom{6}{2} = \dfrac{6!}{2!4!} = 15$

ways that $w = 2$; hence $P(w = 2) = 15/64$.

5. Continuing with the combination formula, we come up with Table 5.3.

We will learn in Section 5.4 that this is an example of a binomial random variable, and the corresponding probabilities can be found in the binomial tables given in Appendix B.

EXERCISES 5.2

5.8 For random variable x in Exercise 5.1, give the probability distribution table and graph it.

5.9 For random variable y in Exercise 5.2, give the probability distribution table and graph it.

5.10 Suppose x is a discrete random variable with probability distribution table as follows:

x	1	3	5	7	9
p(x)	1/15	2/15	3/15	4/15	5/15

(a) What value of x is most likely to come up?
(b) What is the probability that x will come up even?
(c) What is the probability that x will be less than 7?

5.11 Given the function

$$P(x = k) = (k - 1)/6 \quad \text{for} \quad k = 2, 3, 4$$

find $P(x = 2)$, $P(x = 3)$, and $P(x = 4)$. Does the equation define a probability distribution?

5.12 Given the function

$$P(x = k) = (k - 2)/10 \quad \text{for} \quad k = 3, 4, 5, 6$$

find $P(x = 3)$ and $P(x = 6)$. Does the equation define a probability distribution?

5.13 Suppose y is a discrete random variable with probability distribution table as follows:

y	1	2	3	4
p(y)	.111	.222	.333	.334

(a) What value of y is most likely to come up?
(b) What is the probability that y will come up even? odd?
(c) Graph this distribution.

5.3   Expected Value and Variance

In the previous section we saw how the probabilities associated with the values of a discrete random variable are distributed over a number line. In this section the discussion of discrete random variables continues as we find the average value (expected value) of the variable and a measure of the spread associated with the distribution.

Expected Value of a Discrete Random Variable

Consider the probability distribution table for the random variable w (number of boys in a family of six children) given in Table 5.3. Theoretically, this distribution table means that out of 64 families 1 family should have no boys, 6 should have 1 boy, 15 should have 2 boys, 20 should have 3 boys, 15 should have 4 boys, 6 should have 5 boys, and 1 should have 6 boys. Practically speaking, however, there is no guarantee that each group of 64 families will be distributed perfectly as just mentioned. But to study the statistical properties of w we will consider the ideal case, namely, the theoretical distribution of w. To find the *mean* number of boys per family we expect that

> 0 boys should occur 1/64 of the time
>
> 1 boy should occur 6/64 of the time
>
> 2 boys should occur 15/64 of the time
>
> 3 boys should occur 20/64 of the time
>
> 4 boys should occur 15/64 of the time
>
> 5 boys should occur 6/64 of the time
>
> 6 boys should occur 1/64 of the time

Therefore we have

$$
\begin{aligned}
\text{mean of } w &= 0(1/64) + 1(6/64) + 2(15/64) + 3(20/64) \\
&\quad + 4(15/64) + 5(6/64) + 6(1/64) \\
&= 6/64 + 30/64 + 60/64 + 60/64 + 30/64 + 6/64 \\
&= 192/64 \\
&= 3
\end{aligned}
$$

Figure 5.4 is a graph of the probability distribution of w with the mean marked off. Note that the mean of w is at a point that would balance the distribution. We see that, to get the mean of w, we multiply each possible value of w by its probability, $p(w)$, and then sum over all possible values of w. The mean of w is also called the *expected value*

FIGURE 5.4

Identifying the mean of random variable w

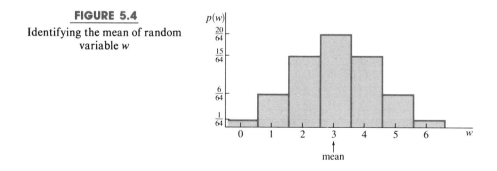

of w, and is denoted as $E(w)$. It is the value of *w* we would expect in the long run; it is identified with the mean of a population, and thus is denoted by the Greek letter μ.

Definition

The **expected value** of a discrete random variable *y* is

$$\mu = E(y) = \Sigma y p(y)$$

(μ is the Greek letter equivalent to *m*.)

EXAMPLE 5.7

In Table 5.1, we have the probability distribution of the random variable *y*, which is the outcome of the roll of a fair die. Find $E(y)$, and locate it on a graph of the probability distribution.

SOLUTION   The expected value of *y* is found by multiplying each value of *y* by its probability and summing. We have

$$E(y) = 1(1/6) + 2(1/6) + 3(1/6) + 4(1/6) + 5(1/6) + 6(1/6)$$
$$= 1/6 + 2/6 + 3/6 + 4/6 + 5/6 + 6/6$$
$$= 21/6$$
$$= 3.5$$

A graph of the distribution is given in Figure 5.5. The expected value of *y*, μ, which is located at 3.5, would balance the distribution.

FIGURE 5.5

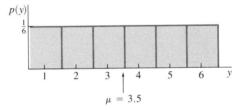

EXAMPLE 5.8

A new drug, Adriamycin, may be the most effective treatment available for breast cancer (Watauga *Democrat,* September 19, 1984). Many doctors are reluctant to prescribe it, however, because of its side effects. When given in high doses, it is reported that Adriamycin causes heart damage in 10% of all patients.

Suppose five patients are given high doses of Adriamycin for the treatment of breast cancer. The probability distribution for the number *y* of patients experiencing heart damage is given as follows:

y	0	1	2	3	4	5
p(y)	.59049	.32805	.07290	.00810	.00045	.00001

Find $E(y)$ and locate it on a graph of the probability distribution.

SOLUTION Whenever a distribution table is available, it is helpful to find the expected value in tabular form as follows:

y	$p(y)$	$yp(y)$
0	.59049	0
1	.32805	.32805
2	.07290	.14580
3	.00810	.02430
4	.00045	.00180
5	.00001	.00005
	Sum =	.50000

The last column, the product of y and $p(y)$, is summed to get the expected value; hence we have

$$\mu = E(y) = .5$$

Thus out of 5 patients the expected number to experience heart damage is $1/2$. Certainly $1/2$ of a patient does not make sense, but this means that, in the long run, out of every 5 patients we would expect, on the average, $1/2$ of a patient to experience heart damage. So, if there were 10 patients, we would expect 1 to experience heart damage after having had high doses of Adriamycin. A graph of the distribution is shown in Figure 5.6. The mean of y is located at .5, the value that would balance the distribution.

FIGURE 5.6

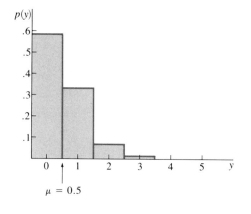

$\mu = 0.5$

Variance of a Discrete Random Variable

We have seen that the mean or expected value of a random variable is an important characteristic of the random variable. An equally important characteristic of a random variable is the *variance,* denoted as $V(y)$. It is a measure of the variability of the variable.

Definition

The **variance** of a discrete random variable y is

$$V(y) = E[(y - \mu)^2]$$
$$= \Sigma\, (y - \mu)^2 p(y)$$

The variance is obtained by subtracting μ from each possible y, squaring the difference, multiplying by $p(y)$, and summing over all possible y.

EXAMPLE 5.9

Find the variance of the random variable y, which is the outcome of the roll of a fair die given in Example 5.7.

SOLUTION   In Example 5.7, the mean of y was found to be 3.5. To find the variance we subtract 3.5 from each possible value of y, square the difference, multiply by the probability, and then sum. We have

$$
\begin{aligned}
V(y) &= (1 - 3.5)^2(1/6) + (2 - 3.5)^2(1/6) \\
&\quad + \cdots + (6 - 3.5)^2(1/6) \\
&= (1/6)[(-2.5)^2 + (-1.5)^2 + \cdots + (+2.5)^2] \\
&= (1/6)(6.25 + 2.25 + .25 + .25 + 2.25 + 6.25) \\
&= (1/6)(17.5) \\
&= 2.9167
\end{aligned}
$$

The variance of a random variable can also be found by the following computational formula.

> **Computational Formula for Variance**
>
> $$V(y) = \Sigma\, y^2 p(y) - \mu^2$$

EXAMPLE 5.10

Use the computational formula to find the variance of the random variable y in Example 5.8.

SOLUTION   Here we square each value of y and multiply by its probability and sum. In tabular form we have:

y	$p(y)$	y^2	$y^2 p(y)$
0	.59049	0	0
1	.32805	1	.32805
2	.07290	4	.29160
3	.00810	9	.07290
4	.00045	16	.00720
5	.00001	25	.00025
			Sum = .70000

From the sum we subtract the square of the mean and get

$$V(y) = .7 - (.5)^2 = .7 - .25 = .45$$

The variance of a random variable is in squared units of the variable. To get a measure of variability in the original unit of measurement, we take the square root of the variance and call it the *standard deviation*. It is denoted by the Greek letter σ.

Definition

The **standard deviation** of a random variable y is

$$\sigma = \sqrt{V(y)}$$

(σ is the Greek letter for the standard deviation of a population and is equivalent to s.)

EXAMPLE 5.11

Find the standard deviation of the random variable

w = no. of boys in a family of six children

whose distribution was given in Table 5.3.

SOLUTION   In tabular form we have

w	$p(w)$	$wp(w)$	$w - \mu$	$(w - \mu)^2$	$(w - \mu)^2 p(w)$
0	1/64	0	$0 - 3 = -3$	9	9/64
1	6/64	6/64	$1 - 3 = -2$	4	24/64
2	15/64	30/64	$2 - 3 = -1$	1	15/64
3	20/64	60/64	$3 - 3 = \ \ 0$	0	0
4	15/64	60/64	$4 - 3 = \ \ 1$	1	15/64
5	6/64	30/64	$5 - 3 = \ \ 2$	4	24/64
6	1/64	6/64	$6 - 3 = \ \ 3$	9	9/64

$$E(w) = 192/64$$
$$= 3$$

$$V(w) = 96/64$$
$$= 3/2 = 1.5$$

The standard deviation is

$$\sigma = \sqrt{1.5} = 1.225$$

Interpretation of the Standard Deviation

The interpretation of the standard deviation of a sample given in Chapter 3 applies to population variables as well. That is, one can determine the percent of the distribution that lies within one or two standard deviations of the mean.

In Example 5.11, we found that a family of six children should expect three boys with a standard deviation of 1.225. One standard deviation below the mean and one standard deviation above the mean would give the bounds

$$3 - 1.225 = 1.775 \quad \text{and} \quad 3 + 1.225 = 4.225$$

From the distribution table given in the example we find that

$$15/64 + 20/64 + 15/64 = 50/64 = 78\%$$

of the distribution lies between 1.775 and 4.225.

Two standard deviations above and below the mean ($\mu \pm 2\sigma$) will give bounds

$$3 - 2(1.225) = 3 - 2.45 = .55 \quad \text{and}$$
$$3 + 2(1.225) = 3 + 2.45 = 5.45$$

From the distribution table we have

$$6/64 + 15/64 + 20/64 + 15/64 + 6/64 = 62/64 = 97\%$$

of the distribution lies between .55 and 5.45, that is, within two standard deviations of the mean.

Generally for bell-shaped distributions the Empirical rule applies.

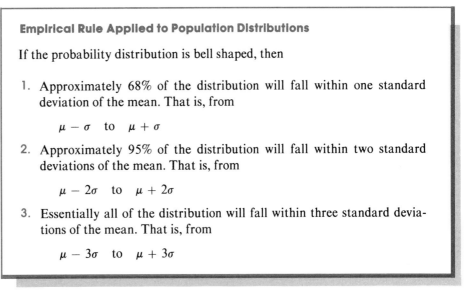

Empirical Rule Applied to Population Distributions

If the probability distribution is bell shaped, then

1. Approximately 68% of the distribution will fall within one standard deviation of the mean. That is, from

   $$\mu - \sigma \quad \text{to} \quad \mu + \sigma$$

2. Approximately 95% of the distribution will fall within two standard deviations of the mean. That is, from

   $$\mu - 2\sigma \quad \text{to} \quad \mu + 2\sigma$$

3. Essentially all of the distribution will fall within three standard deviations of the mean. That is, from

   $$\mu - 3\sigma \quad \text{to} \quad \mu + 3\sigma$$

For distributions that are not necessarily bell shaped, in fact for any shaped distribution, we have Chebyshev's rule due to the Russian mathematician P. L. Chebyshev (1821–1894).

Chebyshev's Rule

Regardless of the shape of the distribution we have

1. At least 3/4 of the distribution will fall within two standard deviations of the mean. That is, from

   $$\mu - 2\sigma \quad \text{to} \quad \mu + 2\sigma$$

2. At least 8/9 of the distribution will fall within three standard deviations of the mean. That is, from

   $$\mu - 3\sigma \quad \text{to} \quad \mu + 3\sigma$$

EXAMPLE 5.12

It has been claimed that TV violence leads to aggressiveness and violent conduct among children. Data have shown that in children's weekend shows, the average number of violent incidents has reached a record high of 30.3 per hour (*Watauga Democrat*, October 12, 1984). If the standard deviation for the number of violent incidents per hour is 6.4, obtain an interval that will contain the number of violent incidents per hour for most weekend shows.

SOLUTION If we can assume that the number of violent incidents has a bell-shaped distribution, then we can apply the Empirical rule, which states that, with 95% confidence, the number of violent incidents per hour will be within two standard deviations of the mean. This will give

$$30.3 - 2(6.4) = 30.3 - 12.8 = 17.5$$

and

$$30.3 + 2(6.4) = 30.3 + 12.8 = 43.1$$

Thus most weekend children's shows will have between 17.5 and 43.1 violent incidents per hour.

If we could not assume the bell distribution, then we would have to apply Chebyshev's rule, which states that at least 75% of the children's shows would have from 17.5 to 43.1 violent incidents per hour.

EXERCISES 5.3

5.14 Suppose y is a discrete random variable with probability distribution as follows:

y	1	3	5	7	9
$p(y)$	1/15	2/15	3/15	4/15	5/15

Find $E(y)$ and $V(y)$. Sketch a graph of the probability distribution and locate the mean μ. Construct the interval $\mu \pm 2\sigma$ and give an interpretation using either the Empirical rule or Chebyshev's rule.

5.15 The probability distribution for the random variable y is given in the following table:

y	-1	0	1	2	3
$p(y)$	.05	.1	.15	.4	.3

Find $E(y)$ and $V(y)$. Sketch a graph of the probability distribution and locate the mean μ. Construct the interval $\mu \pm 2\sigma$ and give an interpretation using either the Empirical rule or Chebyshev's rule.

5.16 The probability distribution for the number of daily requests for assistance from a poison control center is

No. requests	0	1	2	3	4	5	6
Probability	.1	.1	.2	.3	.1	.1	.1

Find the expected number of requests and the standard deviation. Construct the interval $\mu \pm 2\sigma$ and give an interpretation using either the Empirical rule or Chebyshev's rule.

5.17 An oil well drilling company generally drills four wells per month. The probability distribution for the number of successful attempts out of the four wells is given by

No. successes	0	1	2	3	4
Probability	.1	.4	.3	.1	.1

Find the expected number of successful wells and the standard deviation. Construct the interval $\mu \pm 2\sigma$ and give an interpretation using either the Empirical rule or Chebyshev's rule.

5.18 One fourth of all pregnancies end in abortion. Suppose that three pregnant women are randomly selected. Let y be the number, out of the three, who get an abortion. The probability distribution of y is given by

y	0	1	2	3
$p(y)$	.422	.422	.141	.015

Find the mean and standard deviation of y. Construct the interval $\mu \pm 2\sigma$ and give an interpretation using either the Empirical rule or Chebyshev's rule.

5.19 Suppose that 60% of all flights by On-Time Air are delayed at least 20 minutes. For four randomly selected flights, let y be the number of flights delayed at least 20 minutes. The probability distribution for y is

y	0	1	2	3	4
$p(y)$	.026	.154	.345	.345	.130

Find the mean and standard deviation of y. Construct the interval $\mu \pm 2\sigma$ and give an interpretation using either the Empirical rule or Chebyshev's rule.

5.4 Binomial Distribution

The two distributions given in Tables 5.2 and 5.3 are examples of the binomial distribution, which is associated with a Bernoulli population. In each case, the outcome of the experiment resulted in one of two possibilities—a boy or a girl. Many other experiments share this same characteristic; namely, the outcome is only one of two possibilities. This is what we commonly call a Bernoulli population after Jacob Bernoulli (1654–1705), the first of a number of gifted mathematicians of the Bernoulli family of Antwerp, Belgium.

Definition

A **Bernoulli population** is a population in which each element is one of two possibilities. The two possibilities are usually designated as either success or failure. A **Bernoulli trial** is observing one element in a Bernoulli population.

EXAMPLE 5.13

The toss of a coin results in a Bernoulli population, because each toss results in a head (success) or a tail (failure).

EXAMPLE 5.14

In a local election, the voters either favor a candidate for mayor or they do not. Thus we could view the population of voter choices as a Bernoulli population, where a success is one who favors the candidate and a failure is one who opposes the candidate.

We see that the opinion of voters on a specific issue is not unlike the simple experiment of tossing a coin. A random sample from a Bernoulli population consists of n objects in which each object is a success or a failure. The result, which is a sequence of Bernoulli trials, is called a *binomial experiment*.

Definition

A **binomial experiment** is an experiment that consists of n repeated independent Bernoulli trials in which the probability of success on each trial is π and the probability of failure on each trial is $1 - \pi$.

So we see that an experiment that consists of repeatedly drawing independently from a Bernoulli population is called a binomial experiment. As stated, the outcomes are labeled as success or failure and normally one will be interested in the number of successes out of the *n* trials.

Definition

The random variable x, which gives the number of successes in the *n* trials of a binomial experiment, is called a **binomial random variable.** The range of values of x will be

$$R_x = \{0, 1, 2, \ldots, n\}$$

The random variables x = no. of heads in 3 tosses of a coin and w = no. of boys in a family with 6 children, considered in the previous section, are both binomial random variables.

EXAMPLE 5.15

Are the following random variables binomial random variables? Explain why or why not.

(a) Forty percent of all airline pilots are over 40 years of age. A company has 15 pilots. Let the random variable x be the number of pilots in the company who are over 40 years of age.

(b) Suppose a salesperson makes sales to 20% of her customers. One day she started counting her customers until she made a sale. Let x be the number of customers until her first sale.

(c) A room contains six women and four men. Three people are selected to form a committee. Let the random variable x be the number of women on the committee.

SOLUTION

(a) Each of the 15 pilots constitutes a trial, and each trial results in one of two outcomes—success is over 40 years of age and failure is not over 40 years of age. The trials are independent because one pilot's age has no relationship to another's age. The probability of a success is $\pi = .4$ and of a failure is $(1 - \pi) = .6$, and these probabilities remain the same from trial to trial because 40% of all pilots are over 40 years of age. Thus we have met the conditions of a binomial experiment. Because x is the number of successes, we have a binomial random variable.

(b) The salesperson may never make a sale, hence the range of x is

$$R_x = \{1, 2, 3, \ldots\}$$

which clearly is not the sample range of a binomial random variable.

Had x been defined to be the number of sales out of say 50 customers, then x would have been a binomial random variable, because it would have been the number of successes out of 50 trials.

(c) The probability of a woman being selected on the first draw is 6/10. On the second draw, however, it is 5/9 if a woman was obtained on the first draw or 6/9 if a man was obtained on the first draw. Clearly the trials are dependent and the probability does not remain constant from trial to trial. The experiment does not satisfy the conditions of a binomial experiment; therefore the random variable x is not a binomial random variable.

The binomial random variable is discrete because its range of values is

$$R = \{0, 1, 2, \ldots, n\}$$

Of the examples of discrete random variables given in Section 5.1, pages 167–168, item 2—the number of students passing the psychology test; item 3—the number of students favoring beer being sold on campus; and item 5—the number of computer terminals in use are binomial random variables.

The combinatorial method at the end of Section 5.2 suggests a general formula for the binomial probabilities. Suppose there are n trials with π being the probability of success on each trial. For there to be exactly k successes there must be $n - k$ failures. Moreover, there are

$$\binom{n}{k}$$

different possibilities for the k successes. Because the probability of each success is π, the probability of a given arrangement of k successes and $n - k$ failures is

$$\pi^k (1 - \pi)^{n-k}$$

Thus, the probability of k successes in n trials is

$$p(k) = \binom{n}{k} \pi^k (1 - \pi)^{n-k}$$

For certain values of n and π these probabilities are given in the binomial tables in Appendix B (Table B2). We now discuss how to read those tables.

To aid our discussion an excerpt from the Binomial Table is reproduced here.

$n = 10$

$k \mid \pi$	.10	.20	.30	.40	.50	.60	.70	.80	.90
0	.349	.107	.028	.006	.001	.000+	.000+	.000+	.000+
1	.387	.268	.121	.040	.010	.002	.000+	.000+	.000+
2	.194	.302	.233	.121	.044	.011	.001	.000+	.000+
3	.057	.201	.267	.215	.117	.042	.009	.001	.000+
4	.011	.088	.200	.251	.205	.111	.037	.006	.000+
5	.001	.026	.103	.201	.246	.201	.103	.026	.001
6	.000+	.006	.037	.111	.205	.251	.200	.088	.011
7	.000+	.001	.009	.042	.117	.215	.267	.201	.057
8	.000+	.000+	.001	.011	.044	.121	.233	.302	.194
9	.000+	.000+	.000+	.002	.010	.040	.121	.268	.387
10	.000+	.000+	.000+	.000+	.001	.006	.028	.107	.349

(.000+ means that the probability rounded to three decimal points is 0, but is not exactly 0 because that is associated with an impossible event. The "+" means it is slightly greater than 0.)

This section of the table corresponds to $n = 10$ independent trials. Down the left column we see values for k of 0, 1, 2, 3, 4, 5, 6, 7, 8, 9, and 10. This corresponds to k successes out of the 10 trials. Along the top we see values for π of .10, .20, .30, and so on until we get to .90. (Table B2 also includes values for .01, .05, .95, and .99.) Remember that π stands for the probability of a single success in the Bernoulli population. The body of the table gives the probabilities corresponding to the particular values of k and π for $n = 10$. For example, the probability of 4 successes out of 10 trials when $\pi = .7$ is .037.

EXAMPLE 5.16

Suppose that the probability of successfully rehabilitating a convicted criminal in a penal institution is .4. If we let r represent the number successfully rehabilitated out of a random sample of 10 convicted criminals, then r is a binomial random variable with $n = 10$ independent trials and $\pi = .4$. The probability distribution for r then is given by the binomial probability distribution table.

(a) Find the probability that 6 of the 10 prisoners will be successfully rehabilitated.

(b) Find the probability that no more than 2 of the 10 are successfully rehabilitated.

SOLUTION

(a) Focusing on the .4 column, we see the various probabilities associated with all the values of k when $n = 10$. Thus the probability that 6 of the 10 prisoners will be rehabilitated is .111.

(b) No more than 2 implies that k could be either 0 or 1 or 2. Thus to find the probability we simply add those probabilities in the .4 column that correspond to the values 0, 1, and 2 in the k column, and we get

$$.006 + .040 + .121 = .167$$

We can say that there is a 16.7% chance that 2 or fewer are successfully rehabilitated.

If we plot all the probabilities for $n = 10$ and $\pi = .4$, we get a graph of this binomial distribution, which is given in Figure 5.7. Had π been only .1 we could go down the .10 column and get all the probabilities associated with the possible values of k. Figure 5.8(a) is a graph of this distribution.

Note that for small values of π the distribution is heavily concentrated on the lower values of k, which means that a fewer number of successes is more likely. Figure 5.8(b) shows, for large values of π (in this case $\pi = .9$), the binomial distribution is heavily concentrated on the larger values of k, which means that a higher number of successes is more likely. Note also that the distributions for $\pi = .1$ and $\pi = .9$ are mirror images

FIGURE 5.7

Binomial probability distribution for $n = 10$ and $\pi = .4$

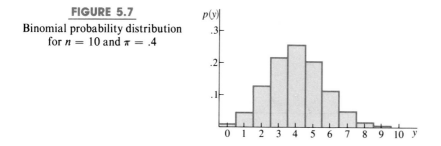

FIGURE 5.8

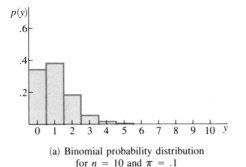

(a) Binomial probability distribution
for $n = 10$ and $\pi = .1$

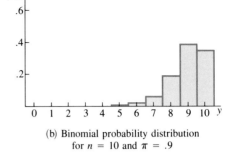

(b) Binomial probability distribution
for $n = 10$ and $\pi = .9$

FIGURE 5.9

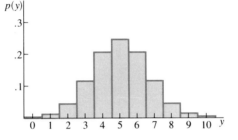

of each other, as would be the case for $\pi = .2$ and $\pi = .8$, and so on. If $\pi = .5$, the binomial distribution is symmetrical as is shown in Figure 5.9.

It is apparent that the values of n and π are very important, because they distinguish between the various binomial distributions. We saw in Section 3.1 that such values that characterize a distribution are called the *parameters*. From the parameters we get other characteristics of the distribution, such as the mean and the standard deviation.

In Section 5.3 we learned how to find the expected value of an arbitrary random variable. In the special case of the binomial distribution, those procedures give that the mean is the product of n and π; that is,

$$\text{mean} = n\pi$$

and the standard deviation is given by

$$\text{standard deviation} = \sqrt{n\pi(1 - \pi)}$$

EXAMPLE 5.17

In Example 5.16, find the mean and the standard deviation of the variable r; that is, find the mean number of prisoners successfully rehabilitated out of the 10 prisoners. Also find the standard deviation.

SOLUTION   The parameters are

$$n = 10 \quad \text{and} \quad \pi = .4$$

Therefore the mean is

$$n\pi = (10)(.4) = 4$$

and the standard deviation is

$$\sqrt{n\pi(1 - \pi)} = \sqrt{(10)(.4)(.6)} = \sqrt{(2.4)} = 1.55$$

So we would expect 4 of the 10 to be successfully rehabilitated with a standard deviation of 1.55 prisoners.

An interesting feature of the binomial distribution is what happens when n increases and π remains constant. Figure 5.10 gives the binomial distributions for $\pi = .2$ and $n = 5, 10, 20,$ and 30. Note that as n increases from 5 to 30 the distribution becomes more "bell shaped." Recall that the Empirical rule can be applied to bell-shaped distributions.

FIGURE 5.10

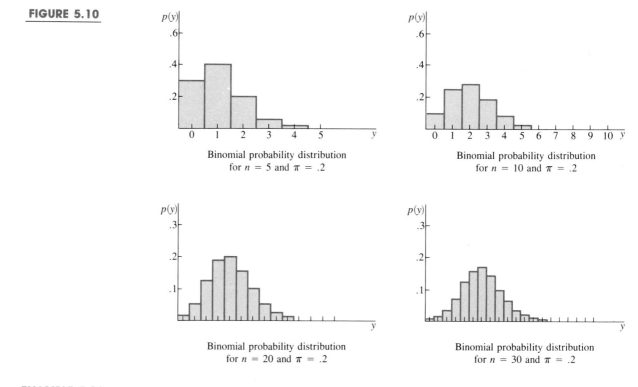

Binomial probability distribution
for $n = 5$ and $\pi = .2$

Binomial probability distribution
for $n = 10$ and $\pi = .2$

Binomial probability distribution
for $n = 20$ and $\pi = .2$

Binomial probability distribution
for $n = 30$ and $\pi = .2$

EXAMPLE 5.18

A certain drug manufacturer claims that its vaccine is 80% effective; that is, each person vaccinated stands an 80% chance of developing immunity. Suppose 100 people are vaccinated. If we let v be the number that develop immunity, then v is a binomial random variable with parameters

$$n = 100 \quad \text{and} \quad \pi = .8$$

and thus the mean is

$$n\pi = (100)(.8) = 80$$

so, we expect 80 to develop immunity. The standard deviation is

$$\sqrt{n\pi(1 - \pi)} = \sqrt{(100)(.8)(.2)} = \sqrt{16} = 4$$

So, using the Empirical rule, we can say that approximately 95% of the distribution is between 72 and 88. That is, two standard deviations below the mean will be

$$80 - 2(4) = 80 - 8 = 72$$

and two standard deviations above the mean will be

$$80 + 2(4) = 80 + 8 = 88$$

So we know that if $\pi = .8$, then out of 100 people, we expect somewhere between 72 and 88 to develop immunity. Anything outside those limits is considered unusual.

$$p(k) = \binom{n}{k} \pi^k (1 - \pi)^{n-k}$$

$$\frac{1}{10}\left(\frac{6}{10}\right) + \frac{1}{12}\left(\frac{6}{10}\right) + \frac{1}{10}\left(\frac{6}{10}\right)$$

EXERCISES 5.4

5.20 Suppose x is a binomial random variable; compute the probabilities corresponding to the following values of n, π, and k.

(a) $n = 10$, $\pi = .8$, $k = 7$

(b) $n = 15$, $\pi = .3$, $k = 6$

(c) $n = 6$, $\pi = .5$, $k = 3$

(d) $n = 13$, $\pi = .2$, $k = 5$

5.21 In each of the following, graph the binomial distribution.

(a) $n = 10$, $\pi = .8$

(b) $n = 15$, $\pi = .3$

(c) $n = 6$, $\pi = .5$

(d) $n = 4$, $\pi = .9$

5.22 For each part in Exercise 5.21, find the mean and standard deviation.

5.23 If y is a binomial random variable with n and π given as follows, calculate the unknown probability.

(a) $n = 10$, $\pi = .3$, $P(y \leq 4)$

(b) $n = 6$, $\pi = .8$, $P(y > 3)$

(c) $n = 15$, $\pi = .9$, $P(y > 13)$

(d) $n = 5$, $\pi = .2$, $P(y < 3)$

5.24 How could the probabilities, related to the binomial distribution, be obtained if $n > 20$?

5.25 How could the probabilities, related to the binomial distribution, be obtained if $\pi = .68$?

5.26 Assume that 60% of student loan applications are approved. Out of 10 applications:

(a) What is the probability that 8 or more are approved?

(b) How many are expected to be approved?

(c) What is the standard deviation of the number approved out of 10 applications?

5.27 A new TV show has a 20% chance of being successful. NBC will introduce eight new shows this season. What is the probability that fewer than three will be successful? How many are expected to succeed?

5.28 A presidential aid believes that half of the U.S. Congress favors a particular action taken by the President. If this conjecture is true, what is the probability that out of a sample of 15 members of Congress more than 8 favor the action? Of the 15, how many are expected to favor the action?

5.29 A box contains six marbles—two of which are white. Three are drawn with replacement. What is the probability that two of the three are white?

5.30 Which of the following are binomial random variables?

(a) The number of accidents per week involving alcohol in your state

(b) The number of violent crimes per month committed in your city

(c) The number of successful heart transplants out of five patients

(d) The length of a prison term for possession of marijuana

(e) The number of approved food stamp recipients out of 50 applications

5.31 Forty percent of the married couples with children agree on methods of child discipline. Out of 18 married couples, what is the probability that less than half agree on methods of child discipline?

5.32 A small town is 60% black. A selected jury consists of eight whites and four blacks. Was the jury randomly selected? Hint: Find the probability that four or fewer blacks were selected.

5.5   Normal Distribution

Often the numerical outcomes of an experiment are the result of some measurement process in which the possible values are not a discrete set of numbers but numbers on an interval of the real number line. In those cases the random variable associated with the outcomes is called a *continuous random variable*. Examples include:

1. the time to complete a task
2. the age at which infants begin to walk
3. the weight or height of 2-year-old children
4. the grade point average of college students
5. the IQ of prisoners on death row
6. the income of heavy equipment operators

For continuous random variables, a function, called a *probability density function* or *frequency curve*, is defined over the interval of real numbers that the variable could assume, with the property that probabilities are represented by areas under the curve.

Definition

A **probability density function,** $f(y)$, which describes the probability distribution for a continuous random variable y, has the following properties:

(a) $f(y) \geq 0$.

(b) The total area under the probability density curve is 1.00, which corresponds to 100%.

(c) $P(a \leq y \leq b)$ = area under the probability density curve between a and b.

EXAMPLE 5.19

Consider the case in which the variable y represents the gross annual income of a blue-collar worker selected at random from the adult males in the city of Detroit. The probability density function might be described by the density curve shown in Figure 5.11. The shaded area represents the percentage of workers with incomes between $16,000 and $28,000. It can also be thought of as the probability that a worker selected at random will have an income between $16,000 and $28,000. It appears that the shaded area represents approximately 60% of the total area, so we write it as

▲ Because probabilities are represented by areas under the curve, it does not matter whether the end points of the interval are included. In other words, we have
$P(a < y < b) = P(a \leq y \leq b)$.

$$P(16 \leq y \leq 28) \approx .60$$

FIGURE 5.11

Distribution of the incomes (in thousands of dollars) of male blue-collar workers in Detroit

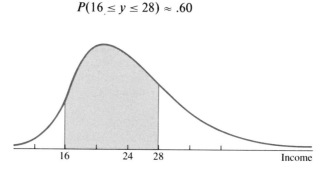

Observe that the probability density curve in Example 5.19 is centered at about 24 (i.e., 24,000), is somewhat spread out, and tails off to the right. The shape of a probability density curve indicates what values of the random variable are more likely to occur when sampled. Two important characteristics of a random variable are its expected value and its variance. The expected value of a continuous random variable, like that of a discrete random variable, is the value of the random variable that will balance the frequency curve. Both the expected value and the variance of a continuous random variable can be calculated from the probability density function, but this is beyond the scope of this book.

One of the most commonly observed random variables is one whose probability density curve is characterized by its expected value (mean) and its variance; it is called the *normal random variable*. Its probability density curve is *symmetrical* (left half is mirror image of right half) and bell shaped, as shown in Figure 5.12.

FIGURE 5.12
A normal probability density curve

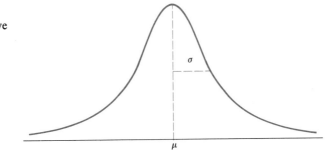

The center of the normal distribution locates the expected value or mean and is denoted by $E(y)$ or the Greek letter μ. The amount of spread is determined by the standard deviation (square root of the variance) and is denoted by the Greek letter σ. Note in Figure 5.12 that at the peak of the probability density curve, the curvature is concave downward, whereas on the tails it is concave upward. The point of transition from concave downward to concave upward is called a *point of inflection*. For the normal curve, there are two points of inflection on either side of the mean. The distance from the mean, μ, to the point of inflection is one standard deviation, σ. Clearly if the standard deviation were to be increased, the density curve would become more spread out. Also if the mean were to change, the density curve would shift so that it would be centered at the mean. Thus these two quantities change the appearance of the density curve, which is why we say they characterize the normal distribution. Figure 5.13

FIGURE 5.13
Probability density curves for three different normal distributions

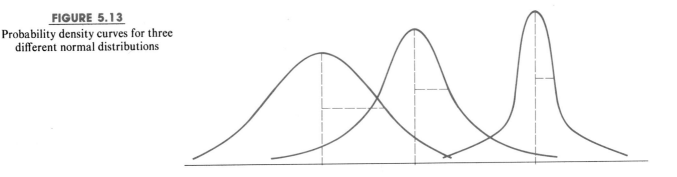

shows three separate normal curves with different means and standard deviations. We see that by changing the two parameters, μ and σ, we can describe any number of different normal distributions.

In addition to being symmetrical about its mean, the normal density curve continues infinitely in both directions. However most of the distribution lies within three standard deviations on either side of μ. In fact, the Empirical rule was derived from the normal distribution. More precisely, for a normal distribution we have that

68.26% of the distribution is between $\mu - \sigma$ and $\mu + \sigma$

95.44% of the distribution is between $\mu - 2\sigma$ and $\mu + 2\sigma$

99.74% of the distribution is between $\mu - 3\sigma$ and $\mu + 3\sigma$

EXAMPLE 5.20

Scores on the Stanford-Binet IQ test are assumed to be normally distributed with a standardized mean of 100 and standard deviation of 16. What percent of the population have IQ's between 100 and 116?

▲ It is suggested that a picture be drawn of the distribution and that the desired area be shaded.

SOLUTION   Figure 5.14 shows a normal curve centered at 100, with the desired area from 100 to 116 shaded. Observe that 116 is 16 points or exactly one standard deviation above the mean. From the Empirical rule we have that 68.26% of the distribution is one standard deviation on *either* side of the mean, so the upper half would be one-half that; hence

$$P(100 \leq IQ \leq 116) = .3413$$

FIGURE 5.14

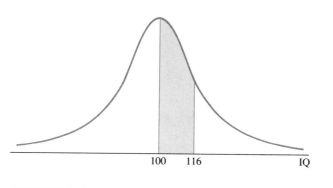

EXAMPLE 5.21

Assume that the time, t, it takes a bank teller to cash a check is normally distributed with a mean of 30 seconds and a standard deviation of 10 seconds. What percent of the customers will be served in less than 20 seconds?

SOLUTION   Figure 5.15 is a graph of the distribution, in which the desired area (below 20) is shaded. Because of the symmetry of the curve, 50% of the area is below the mean of 30. Because 20 is 10 points, or one standard deviation below 30, we know that 34.13% of the area lies between 20 and 30. Thus

$$P(t < 20) = .5000 - .3413 = .1586$$

or approximately 16% of the area lies below 20.

FIGURE 5.15

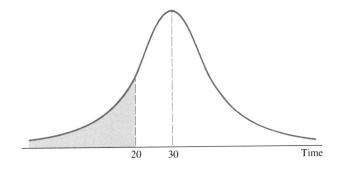

In the two preceding examples, the probability was determined by using the Empirical rule, which gives probabilities associated with one, two, or three standard deviations above or below the mean. But suppose we are not dealing with an integer number of standard deviations above or below the mean? For example, what percent of the population has an IQ (on the Stanford-Binet) between 100 and 124? The probability is illustrated in Figure 5.16 (at the top of page 192).

Clearly 124 is 24 points above the mean, which is not an exact multiple of the standard deviation of 16; thus we cannot apply the Empirical rule.

We do have that 124 is $124 - 100 = 24$ points above μ, which represents $24/16 = 1.5$ standard deviations above μ. So translated, the problem becomes: What percent of a normal distribution lies between the mean μ and 1.5 standard deviations above the mean? The *standard* normal probability distribution table in Appendix B (Table B3) is constructed so as to answer this question. Because a normal curve is symmetrical, it is necessary to give the area only on one side of the mean. Thus each entry in Table B3 corresponds to the area from the mean to a point that is z standard deviations above the mean. A portion of Table B3 is reproduced here for our discussion.

z	.00	.01	.02	.03	.04	.05	. . .
.0	.0000	.0040	.0080	.0120	.0160	.0199	. . .
.1	.0398	.0438	.0478	.0517	.0557	.0596	. . .
.2	.0793	.0832	.0871	.0910	.0948	.0987	. . .
.	.	.	.	.	.	.	
.	.	.	.	.	.	.	
1.4	.4192	.4207	.4222	.4236	.4251	.4265	. . .
1.5	.4332	.4345	.4357	.4370	.4382	.4394	. . .
.	.	.	.	.	.	.	
.	.	.	.	.	.	.	
.	.	.	.	.	.	.	

To complete the problem, simply find 1.5 down the z column, and find the associated probability of .4332. Thus 43.32% of the area lies between the mean and 1.5 standard deviations above the mean. Applying this to the problem, we can say that 43.32% have IQ's between 100 and 124.

We can also say that 43.32% have IQ's between 76 and 100, because 76 is

▲ *Remember* that the table value is the area *between* the mean and the given score.

$$\frac{76 - 100}{16} = -24/16 = -1.5$$

FIGURE 5.16

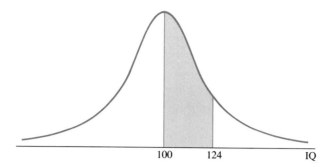

standard deviations from the mean. The minus sign simply means that it is 1.5 standard deviations *below* the mean. And because of the symmetry of the normal distribution about the mean, the area from 76 to 100 is the same as the area from 100 to 124. Therefore 2(.4332) = .8664, or 86.64% have IQ's between 76 and 124.

To solve problems of this type, one must determine the number of standard deviations that a given score is from the mean. This is accomplished by calculating the standardized *z*-score:

$$z = \frac{\text{score} - \text{mean}}{\text{std. dev.}}$$

which gives the number of standard deviations the score is from the mean. The distribution of *z* has a mean of 0 and a standard deviation of 1, and is known as the *standard* normal distribution.

Probabilities associated with a normal distribution other than the standard normal can be found by finding the probability associated with the corresponding *z*-score in the standard normal distribution table. The additional columns in Table B3 are for *z*-scores carried out to two decimal places.

EXAMPLE 5.22

What percent of the population has an IQ between 80 and 120 on the Stanford-Binet test?

SOLUTION   The shaded area shown in Figure 5.17 represents the percent in question. Looking at the right half of the distribution, we see that 120 has a *z*-score of

$$z = (120 - 100)/16 = 20/16 = 1.25$$

FIGURE 5.17

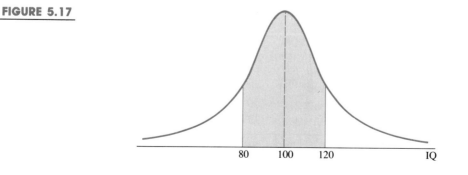

Thus 120 is 1.25 standard deviations above the mean. Going down the z column of the standard normal table to 1.2 and over to the .05 column, we find .3944. So the probability of an IQ between 100 and 120 is .3944 or 39.44%. Also we see that 80 has a z-score of

$$z = (80 - 100)/16 = -20/16 = -1.25$$

That is, 80 is 1.25 standard deviations below the mean. Because of the symmetry of the normal distribution, we also have 39.44% between 80 and 100. Therefore

$$P(80 \leq IQ \leq 120) = .3944 + .3944 = .7888$$

or 78.88% of the population has an IQ between 80 and 120.

If a tail area is desired, the table value must be subtracted from .5000.

EXAMPLE 5.23

What percent of the population has an IQ above 120 on the Stanford-Binet test?

The shaded area of Figure 5.18 gives the desired probability. As in the previous example, the area from 100 to 120 is .3944. Moreover the total area above 100 is .5. Thus

$$P(IQ > 120) = .5000 - .3944 = .1056$$

or 10.56% of the area lies above 120.

FIGURE 5.18

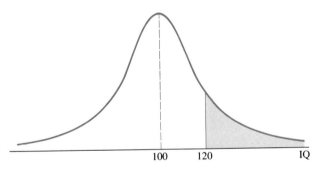

EXAMPLE 5.24

What is the probability of an IQ more than two standard deviations from the mean?

SOLUTION   Two standard deviations is $2(16) = 32$ points. For an IQ to be more than 32 points from the mean it would have to be either below 68 or above 132. The shaded area of Figure 5.19 gives the desired probability.

We must first find the area above 132 and add that to the area below 68. As stated previously, 132 is

$$z = (132 - 100)/16 = 32/16 = 2$$

FIGURE 5.19

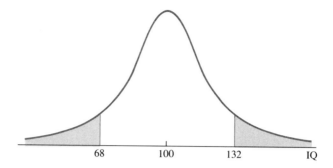

standard deviations above the mean. Looking up a z of 2, we find .4772 between 100 and 132; so $.5000 - .4772 = .0228$ is above 132. Because 68 is two standard deviations below the mean, we also have .0228 below 68. Therefore the probability of an IQ more than two standard deviations from the mean is

$$.0228 + .0228 = .0456$$

The standard normal probability table works for all probability problems involving normally distributed populations. The following steps may help.

> **To Work Probability Problems for Normally Distributed Populations**
>
> 1. Draw a graph of a normal curve, label the mean, and shade the desired area.
>
> 2. Find the number of standard deviations the given score is from the mean by finding the z-score.
>
> 3. Find the associated probability for the z-score in the standard normal probability table.
>
> 4. Relate the result to the problem at hand.

EXAMPLE 5.25

An important measurement in the textile industry is the tensile strength of a produced material. Suppose that the tensile strength of a roll of woven polypropylene is normally distributed with a mean of 89 and a standard deviation of 4. What is the probability that the tensile strength will be somewhere between 80 and 100?

SOLUTION   The probability we are seeking is represented by the shaded areas, A_1 and A_2, in Figure 5.20. We first find A_1, the area to the right of the mean, and then A_2, the area to the left of the mean. We then add the two together to find the desired probability. To find A_1 we compute the z-score for 100 as follows:

$$z = \frac{100 - 89}{4} = 11/4 = 2.75$$

FIGURE 5.20

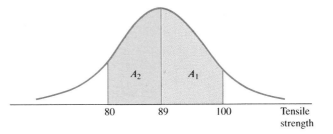

From Table B3 we find that

$$A_1 = .4970$$

The z-score for 80 is

$$z = \frac{80 - 89}{4} = -9/4 = -2.25$$

From Table B3 we find that

$$A_2 = .4878$$

So the probability that the tensile strength is between 80 and 100 is

$$A_1 + A_2 = .4970 + .4878 = .9848$$

A slightly harder problem is to find the area between two values that lie on the same side of the mean, as illustrated in Example 5.26.

EXAMPLE 5.26

Referring to Example 5.25, find the probability that the tensile strength of the woven polypropylene is between 95 and 100.

SOLUTION   The probability is depicted as A in Figure 5.21. We have just seen in the previous example that the z-score associated with 100 is

$$z = \frac{100 - 89}{4} = 11/4 = 2.75$$

which has an associated table probability of .4970. We must remember, however, that .4970 corresponds to all the area from 89 to 100. Thus to find A we must subtract the

FIGURE 5.21

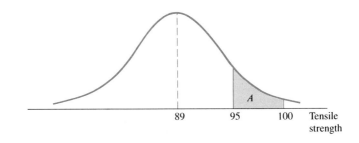

area from 89 to 95. This area can be located by finding the z-score for 95 and looking up the associated probability. Thus

$$z = \frac{95 - 89}{4} = 6/4 = 1.5$$

which has a corresponding probability of .4332. Therefore

$$A = .4970 - .4332 = .0638$$

Often instead of finding a probability associated with a certain score, we wish to find the score corresponding to a probability.

EXAMPLE 5.27

Suppose we have a normal distribution with a mean of 70 and a standard deviation of 15. Find the value of b so that 80% of the distribution lies below it. In other words, if x is a normal random variable with mean 70 and standard deviation of 15, find b so that $P(x \le b) = .80$.

SOLUTION This problem is the reverse of the previous examples. Here we are given the probability and asked to find the associated score. We know that 50% of the distribution lies below the mean of 70. Therefore b is above 70, because 80% of the distribution lies below b. In fact, b is so far above 70 that the area under the curve from 70 to b is 30% (50% + 30% = 80%) as illustrated in Figure 5.22. Looking in the z-tables we find a z-score of .84 corresponding to .2995 (the closest value to .3000). Hence b is .84 standard deviation above 70. We have that

$$b = 70 + .84(15)$$
$$= 70 + 12.6$$
$$= 82.6$$

This value of b is called the 80th percentile.

FIGURE 5.22

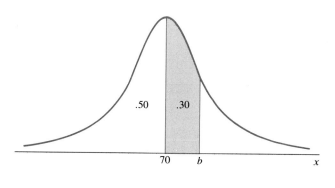

Definition

The ***p*th percentile** is the score, so that $p\%$ of the distribution lies below it.

EXAMPLE 5.28

What IQ score on the Stanford-Binet test corresponds to the 95th percentile?

SOLUTION   The 95th percentile is the point, so that 95% of the scores are below it. From Figure 5.23 we see that we are looking for the score above the mean, so that .95 of the area is below it or only .05 of the area is above it. We can also say that .4500 of the area is between the mean and the score. The z corresponding to an area closest to .4500 (.4495) is 1.64. Thus we can say that the score we are looking for is *1.64 standard deviations* above the mean, and thus the 95th percentile score is

▲ Perhaps we should have used 1.645 instead of 1.64 because .4500 is half way between .4495 and .4505.

$$b = 100 + 1.64(16) = 100 + 26.24 = 126.24$$

FIGURE 5.23

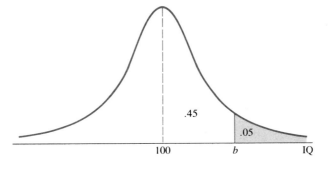

The normal distribution is very important because many physical measurements and natural phenomena are closely approximated by it. However a more fundamental reason for studying the normal distribution involves the theoretical properties of the *sample* mean, which allows us to conduct statistical inferences. This will be discussed at length in Chapter 7.

EXERCISES 5.5

5.33 Suppose z has a standard normal distribution. Find the percent of the distribution between

(a) $z = 0$ and $z = 2.0$ (b) $z = 0$ and $z = 2.6$
(c) $z = 0$ and $z = 1.36$ (d) $z = -2.0$ and $z = 0$
(e) $z = -1.42$ and $z = 0$ (f) $z = -2.82$ and $z = 0$

5.34 Suppose z has a standard normal distribution. Find

(a) $P(z < 1.64)$ (b) $P(z \geq 1.96)$
(c) $P(-1.35 \leq z \leq 1.35)$ (d) $P(1.22 \leq z \leq 2.47)$
(e) The value of z so that 5% of the area lies below it

5.35 Suppose x is a normally distributed random variable with a mean of 50 and a standard deviation of 10. Find

(a) $P(40 \leq x \leq 56)$ (b) $P(x \geq 64)$

5.36 Suppose x is a normally distributed random variable with a mean of 8 and a standard deviation of 3. Find

(a) $P(5 < x < 10)$ (b) $P(x > 9)$

5.37 Suppose x is a normally distributed random variable with a mean of 8 and a standard deviation of 3. Find b such that

$$P(x < b) = .9$$

5.38 The accompanying figure shows a possible density curve for the variable w that represents the possible weight of professional football players.

(a) Is w continuous or discrete?

(b) Can we apply the Empirical rule to find probabilities? Explain.

(c) Can the normal tables be used to find probabilities? Explain.

(d) Most of the professional players have weights between what values?

(e) Interpret the shaded area.

(f) Does the density curve seem reasonable? Explain.

FIGURE FOR EXERCISE 5.38

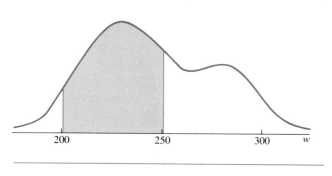

5.39 What is the probability that it will take the bank teller in Example 5.21 more than a minute to cash a check? Recall that $\mu = 30$ seconds and $\sigma = 10$ seconds.

5.40 Suppose a test of coordination for first graders is scored so that the mean for all first graders is 50 and the standard deviation is 15. If we assume further that the distribution is normal, what percent of the first graders score

(a) between 40 and 70?

(b) below 30?

(c) above 75?

5.41 Suppose the mean account in an investment firm is $8,000 and the standard deviation is $2,000. Assuming that the distribution of accounts is normal, what percent of the accounts are

(a) above $13,000?

(b) below $9,000?

(c) below $5,000?

(d) between $3,000 and $6,000?

5.42 A public health department closes the beach to bathers when the contamination level index is in its 80th percentile. From research we know that the index level is normally distributed with a mean of 160 and a standard deviation of 20.

(a) What is the probability the index level exceeds 190?

(b) What is the probability the index level is between 150 and 170?

(c) At what level is the index when the beach is closed?

5.43 Employees of a company are given a test for which the scores are distributed normally with mean 100 and variance 25. The top 5% will be awarded top positions with the company. What score is necessary to get one of the top positions?

5.6   Computer Session (optional)

Three Minitab subcommands relate to the material in this chapter.

PDF BINOMIAL

Recall from Section 5.4 that a binomial random variable gives the number of successes in n independent trials of a binomial experiment. To determine the probabilities associated with each of the possible values of the random variable, we could either work them out with a mathematical formula or look them up in the binomial tables in Appendix B2. Working them out by hand leaves a lot to be desired. To use the tables in Appendix B2, the value of π (probability of a success) must be a specific value. If neither of these alternatives seems practical, we can use Minitab to compute the probabilities in which any value of π is permissible. The command is PDF with the subcommand BINOMIAL.

```
PDF [PUT RESULTS IN C];
    BINOMIAL PROBABILITIES FOR N = K, AND P = K.
```

Upon specifying the values of N and P (here P corresponds to the probability of a success, π), a table of binomial probabilities is printed out. Putting the results in a column is optional.

To compute the probabilities associated with a binomial random variable with 20 trials and a .68 probability of success, the command is

```
PDF;
    BINOMIAL PROBABILITIES FOR N = 20, AND P = .68.
```

EXAMPLE 5.29

Suppose 68% of all requests for financial aid are approved by a university. Determine the probabilities associated with the number of approvals out of 20 requests.

SOLUTION   See the Minitab printout in Figure 5.24 at the top of page 200.

RANDOM BINOMIAL

In Section 4.6 we used the command BERNOULLI to simulate a Bernoulli experiment. For example, to simulate tossing a fair coin 20 times, we use the command

```
RANDOM 20 PUT IN C1;
    BERNOULLI WITH P = .5.
```

We can then count the number of heads in the 20 tosses. But suppose we were interested only in the number of heads in 20 tosses and wished to conduct the experiment 30 times. We could repeat the RANDOM BERNOULLI command 30 times, each time counting the number of successes, or we could use the RANDOM command with the subcommand BINOMIAL. The command is

```
RANDOM K OBSERVATIONS INTO C,C,...,C;
    BINOMIAL WITH N = K, P = K.
```

To simulate 30 experiments of tossing a fair coin 20 times, the command is

```
RANDOM 30 OBSERVATIONS PUT IN C1;
    BINOMIAL WITH N = 20, P = .5.
```

FIGURE 5.24

Minitab output for Example 5.29

```
MTB > PDF;
SUBC> BINOMIAL 20, P = .68.

     BINOMIAL WITH N =   20   P = 0.680000
         K           P( X = K)
         4             0.0000
         5             0.0001
         6             0.0005
         7             0.0019
         8             0.0066
         9             0.0188
        10             0.0440
        11             0.0849
        12             0.1354
        13             0.1770
        14             0.1881
        15             0.1599
        16             0.1062
        17             0.0531
        18             0.0188
        19             0.0042
        20             0.0004
```

EXAMPLE 5.30

An experiment consists of rolling five dice and observing how many turn up on 6. The number of sixes that turn up is a binomial random variable with $N = 5$ and $P = 1/6 = .16666\ldots$. Simulate the experiment 20 times. How many times did all five dice turn up on six? How many times did a six fail to turn up? Give a summary of the results.

SOLUTION   See the Minitab printout in Figure 5.25.

FIGURE 5.25

Minitab output for Example 5.30

```
MTB > RANDOM 20 OBSERVATIONS INTO C1;
SUBC> BINOMIAL WITH N = 5, P = .166667.

MTB > PRINT C1

C1
      0     0     2     1     1     0     0     2     0     1
      2     1     0     0     0     1     0     1     2     2

MTB > TABLE C1

  ROWS: C1

            COUNT

      0        9
      1        6
      2        5
    ALL       20
```

We see that on nine occasions none of the dice turned up on a 6. On six occasions there was one 6 showing and on five occasions there were two sixes showing. Not once did all five show sixes.

RANDOM NORMAL   RANDOM with the subcommand NORMAL simulates a random sample from a normal population with a specified mean and standard deviation. The results are stored in a column to be printed out. The command is

```
RANDOM K OBSERVATIONS INTO C,C,...,C;
  NORMAL WITH MEAN = K, SIGMA = K.
```

Suppose we wish to simulate 80 IQ scores. Assuming the mean IQ is 100 and the standard deviation is 16, we have

```
RANDOM 80 OBSERVATIONS PUT IN C1;
  NORMAL WITH MEAN = 100, SIGMA = 16.
```

EXAMPLE 5.31

Simulate 80 IQ's as just discussed, and construct a histogram of the results. Does the histogram resemble a normal distribution?

SOLUTION   See the Minitab printout in Figure 5.26.

FIGURE 5.26
Minitab output for Example 5.31

```
MTB > RANDOM 80 OBSERVATIONS IN C2;
SUBC> NORMAL WITH MEAN = 100, SIGMA = 16.


MTB > PRINT C2
C2
    113.006    109.484     97.721     90.110    114.953    127.437
     99.436    111.732     88.115     91.017     89.302     98.536
     90.875     85.375     84.950    110.233    111.657     66.234
     98.050    116.854     91.791     77.559     95.755     96.931
     69.480    111.972    104.536    118.506    106.731     62.067
    110.547     99.925    103.646    103.808     83.173     96.865
    107.066     88.035     88.340     88.764    100.586     88.728
     82.782    101.662     97.994    120.832     65.713     99.372
    114.277    104.477     91.171    132.329    116.450    118.570
    117.329    102.736    112.923    106.062     81.273    114.770
    103.728    104.148     97.128    104.033    108.585    110.435
     77.356    117.150    117.636    122.879    125.168    102.167
     78.825     89.632     95.394     83.914     96.914     78.713
     84.850    102.274
```

```
MTB > HISTOGRAM C2

Histogram of C2    N = 80

Midpoint    Count
      60        1   *
      70        3   ***
      80       10   **********
      90       13   *************
     100       25   *************************
     110       16   ****************
     120        9   *********
     130        3   ***
```

The histogram appears to approximate a normal distribution.

EXERCISES 5.6

5.44 It was reported (*USA Today,* October 23, 1984) that 12.9% of the 1983 medical school seniors went into internal medicine. If 30 seniors are completing medical school at a university, determine the probability distribution of the number who plan to go into internal medicine. What is the probability that none of the 30 go into internal medicine? How many are expected to go into internal medicine? What is the standard deviation of the distribution?

5.45 Suppose 25 universities plan to graduate 30 medical students each. Simulate the number of students that plan to go into internal medicine for the 25 universities (see Exercise 5.44). Out of all 25 universities, how many students plan to go into internal medicine?

5.46 In 1982, 39% of the business/management degrees were awarded to women, as compared with only 8.1% in 1971 (*USA Today,* October 23, 1984). Determine the probability that out of 26 business/management graduates in 1982 over half are women.

5.47 Simulate the drawing of a random sample of size 50 from a normal population that has a mean of 70 and a standard deviation of 8. Construct a histogram of the sample.

5.48 Suppose the average cost for 1 hour of consultation with a lawyer is $88.64 and the standard deviation is $5.23. Assuming that the distribution of the costs for various lawyers across the nation is normally distributed, simulate the fees of 20 different lawyers. Construct a stem and leaf plot of the generated data.

Key Concepts

☑ A *random variable* is a rule that assigns a numerical value to the outcomes of an experiment. If the range of values of the random variable is a discrete set of values, then the random variable is called *discrete*. If the range of values is a continuous set of values, then it is called a *continuous* random variable.

☑ The *probability distribution* of a random variable is a table or function that lists the range of values of the variable and the probability with which it assumes those values. This table can be used to find probabilities of events.

☑ The *expected value* of a random variable is an average value of the random variable. It is the value that will balance the distribution. The *standard deviation* is a measure of the variability of the random variable. It is found by first finding the *variance*, and then taking the square root. An interpretation of the amount of variability of a random variable can be given with either the *Empirical rule* or *Chebyshev's rule*. The Empirical rule applies to bell-shaped distributions, and Chebyshev's rule applies to any distribution.

☑ Two very important distributions are the *binomial* and *normal* distributions. A *binomial experiment* is an experiment consisting of several independent trials in which each trial results in either a success or a failure. The *binomial random variable* is a discrete variable that represents the number of successes in a binomial experiment. Probabilities associated with the binomial random variable can be found in the binomial probability tables in Appendix B.

☑ The *normal random variable* is a continuous random variable that is associated with the measurements of some variable measured on at least an interval scale. The normal probability density curve (which corresponds to probabilities) is a bell-shaped curve that is symmetrical. It is uniquely determined by specifying its mean and standard deviation. Probabilities associated with the normal distribution can be found by determining how far a score is from the mean by calculating the *z*-score, and then finding the associated probability in the standard normal tables in Appendix B.

Learning Goals

Having completed this chapter you should be able to:

1. Understand the concept of a random variable and be able to distinguish between a discrete and a continuous random variable.   *Section 5.1*

2. Give the probability distribution of a random variable.   *Section 5.2*

3. Understand the concept of expected value and variance of a random variable.   *Section 5.3*

4. List the characteristics of the binomial experiment.   *Section 5.4*

5. Find probabilities associated with the binomial distribution.   *Section 5.4*

6. List the characteristics of the normal distribution.   *Section 5.5*

7. Find probabilities associated with the normal distribution.   *Section 5.5*

To test your skills answer the following questions.

5.49 Identify each of the following random variables as either discrete or continuous:

(a) The amount of water pumped into a tank overnight

(b) The number of tickets drawn from a barrel until a lottery winner is found

(c) The cost of a week's groceries for a family of 4

(d) The number of patients treated in the emergency room over a 24-hour period

(e) The length of life of a computer chip

5.50 A device to measure one's resistance to pain has a scale that is assumed to be normally distributed with a mean of 30 and a standard deviation of 5. What proportion of those using the device score from 22 to 30? above 34?

5.51 Sketch a graph of the distribution of a binomial variable when $n = 5$ and $\pi = .3$.

5.52 Answer *true* (T) or *false* (F). The following experiments satisfy the conditions of a binomial experiment.

(a) Select 20 random voters and record whether they favor reelection of the President.

(b) Select eight tickets from a barrel and record the names that appear on the tickets.

(c) Draw three balls without replacement from an urn containing five red and seven white balls and record the color.

(d) Roll a pair of fair dice 10 times and each time observe if the total is 7.

(e) A drug to relieve pain is administered to 10 patients. After 2 hours each patient reports that he or she is better, worse, or no change.

(f) The IQ of 20 college students is recorded.

5.53 It is believed that the proportion of convicted felons having a history of juvenile delinquency is about .7. Out of 10 convicted felons

(a) what is the probability that no more than five have a history of juvenile delinquency?

(b) what is the expected number to have had a history of juvenile delinquency?

5.54 Suppose the grade point average of students at a university is normally distributed with a mean of 2.5 and a standard deviation of .5. If the top 10% in grade point average makes the dean's list, what grade point average do they need?

SUPPLEMENTAL EXERCISES FOR CHAPTER 5

5.55 Suppose x is a discrete random variable with probability distribution table as follows:

x	0	2	4	5
p(x)	1/10	2/10	3/10	4/10

(a) What value of x is most likely to come up?

(b) What is the probability that x will come up even?

(c) What is the probability that x will be less than 4?

(d) What are $E(x)$ and $V(x)$?

5.56 A typical bar exam for lawyers might include some persons on their fourth or fifth attempt at the exam. Suppose the following is the distribution of the number of attempts at the bar exam for prospective lawyers.

No. of attempts	1	2	3	4	5
Probability	.4	.3	.1	.1	.1

Find the expected number of attempts and the standard deviation. Sketch a graph of the probability distribution and locate the mean μ. Construct the interval $\mu \pm 2\sigma$ and give an interpretation using either the Empirical rule or Chebyshev's rule.

5.57 Ninety percent of heroin addicts confess to the use of marijuana prior to the use of heroin. Out of four heroin addicts, what is the probability all used marijuana prior to the use of heroin?

5.58 In a local clinic, 30% of the patients have high blood pressure (HB), 40% are overweight (OW), and 10% are both. Out of 15 patients,

(a) What is the probability that no more than 3 have HB?

(b) What is the probability that no more than 3 are OW?

(c) What is the probability that no more than 3 are both?

(d) How many are expected to have HB?

(e) How many are expected to be OW?

(f) How many are expected to be both?

5.59 Suppose 30% of the student body live off campus. Five students are selected at random. Let the random variable x represent the number of students that live off campus out of the five selected.

(a) What kind of random variable is x?

(b) Give the probability distribution table for x and graph it.

(c) How many do you expect to live off campus out of the five selected?

(d) What is the probability that no more than two live off campus?

5.60 A new drug, Nimodipine, holds considerable promise of providing relief for people suffering from migraine headaches, who did not respond to other drugs. Clinical trials have shown that 90% of the patients with severe migraines experience relief of their pain without suffering allergic reactions or side effects. Suppose 16 migraine patients try Nimodipine.

(a) What is the probability all 16 experience relief?

(b) What is the probability at least 14 experience relief?

(c) How many are expected to experience relief?

5.61 A recent article reports that 40% of all cancer patients are cured. Out of 12 patients, what are the chances that more than half are cured?

5.62 In a large metropolitan area, 70% of the families with four or more children are receiving welfare benefits. Out of eight randomly selected families with four or more children, what is the probability that all are on welfare?

5.63 Eighty percent of AIDS patients eventually die from the disease. If 6 patients are hospitalized with AIDS, what is the probability that none will survive?

5.64 It has been determined genetically that the sex of an offspring is determined by the male parent. In the horse industry the female offspring is desirable for marketability reasons. Suppose a stallion is bred to 10 mares, and assume that the sex of each offspring is equally likely to be male or female. What is the probability that the stallion will produce 8 fillies out of the 10 offspring? How many are expected to be fillies?

5.65 Suppose 70% of the students at a university live in dorms. Twenty students are selected at random. What is the probability that over half of them live in dorms? Of the 20 how many are expected to live in dorms?

5.66 A professional football quarterback has a 60% completion record this season. In his next 15 passes,

(a) What is the probability he will complete from 7 to 9 passes?

(b) How many is he expected to complete?

(c) What is the standard deviation of the number he is to complete?

5.67 Suppose a basketball player has a lifetime average of .80 at the free-throw line. He starts shooting until he misses. Let x be the number of shots until he misses. List the range of values of x. Is x a discrete or continuous random variable?

5.68 Suppose that the test scores for a college entrance exam are normally distributed with a mean of 450 and a standard deviation of 100.

(a) What percent of those taking the exam score between 350 and 550?

(b) A student scoring above 400 is automatically admitted. What percent score above 400?

(c) The upper 5% receive scholarships. What score must they make on the exam to get a scholarship?

5.69 The mortality rate for a certain disease is 30%. Of 10 patients who have the disease, what is the probability that more than half will die from the disease? Of the 10 patients how many are expected to die from the disease?

5.70 Suppose the scores on a reading ability exam for 12 year olds are normally distributed with a mean of 40 and a standard deviation of 6. What is the probability that a 12 year old will score

(a) above 60?

(b) below 45?

(c) between 50 and 70?

5.71 Suppose the scores on a National Merit Scholarship exam are normally distributed with a mean of 70 and a standard deviation of 8.

(a) What percent score between 60 and 80?

(b) What percent score between 75 and 90?

(c) What percent score below 67?

(d) What is the 80th percentile?

5.72 A physical education instructor told his class that they could earn an A for the triple jump if they could jump farther than 24 feet.

(a) If the distance jumped is normally distributed with a mean of 22 feet and a standard deviation of 3 feet, what percent of the class will earn A's?

(b) If the instructor wants 70% of his class to make a C or better, how far will they have to jump to get a C or better?

5.73 A job satisfaction index score for nurses is normally distributed with a mean of 50 and a standard deviation of 10. What is the probability that a nurse selected at random will have an index score greater than 55?

5.74 It is claimed that the distribution of scores on a college placement exam is normally distributed with a mean of 50 and a standard deviation of 10. A group of 25 students from a particular private school all scored above 55. What is the probability of scoring above 55? Do you think it is unusual that all 25 scored above 55?

5.75 The time it takes an ordinary mouse to run through a particular maze is assumed to be normally distributed with a mean of 15 seconds and a standard deviation of 3 seconds. The top 10% in speed are claimed to be super mice and will be selected for another experiment. The bottom 10% are fed to the CAT!

(a) What percent of the mice have times between 10 and 20 seconds?

(b) What time will they have to beat in order to be called super mice?

(c) What time do the mice have to beat in order to run again?

5.76 An estimated 2.3 million people are poisoned each year by dangerous chemicals and products found in the home. Sixty-four percent involve children under the age of 6. Out of the next four calls into the poison control center, what is the probability at least two are for children under the age of 6?

5.77 The scores of the STEP science test are assumed to be normally distributed with a mean of 20 and a standard deviation of 5. What percent will score

(a) above 30?

(b) below 17?

(c) between 18 and 23?

5.78 The top 25% on the STEP science test described in Exercise 5.77 receive scholarships. What must they score in order to receive a scholarship?

5.79 Ninety percent of the trees planted by a landscaping firm survive. What is the probability that 10 or more of the 14 trees just planted will survive?

5.80 Suppose you know that the amount of time your watch gains or loses (i.e., gains negatively) in seconds per day is normally distributed with a mean of 0 and standard deviation 1; that is, it is a standard normal distribution. What is the probability that

(a) it gains no more than 1.5 seconds in a day?

(b) it gains at least .5 second in a day?

(c) it loses .5 second in a day?

(d) it gains between 0.7 and 1.4 seconds in a day?

5.81 If a population of measurements can be assumed to be normally distributed, what two quantities distinguish it from another normal population?

5.82 How do you measure the relative magnitude of an arbitrary measurement from a population?

6

Describing a Distribution

The normal distribution considered in the previous chapter is a very important distribution. Many real-world applications follow normal theory; the IQ score is a good example of a quantity that follows normal theory. Yet there are real-world applications that deviate from normality. All distributions are not symmetrical, as is the normal distribution. The tails of all distributions do not drop off nicely as do the tails of the normal distribution. In this chapter some of the other possibilities will be investigated. Procedures will be given that will help classify a distribution in a major category. The classification will then suggest the path to follow for further analysis of the data.

Scholastic Aptitude Test Scores

Much concern has been expressed in recent years about the declining Scholastic Aptitude Test (SAT) scores of high school seniors. The decline has persisted from 1963 (when the average SAT math score was 502 and the average SAT verbal score was 478) through 1981. Many educators were encouraged in 1982, when both the math and the verbal SAT scores rose for the first time in 19 years. That encouragement was short lived, however, because the average SAT verbal score dropped one point in the following year. Yet both rose again in 1984 when the average SAT math was 471 and the SAT verbal was 426. Many educators are hopeful that we are on a continuous upswing.

Many reasons have been given for the presistent decline, such as the permissive attitude of our society and the state funding of education. As a result, in 1984–1985, state tax appropriations for higher education jumped an average of 16% above the funding in 1982–1983. Some state legislatures have commissioned studies of the problem, such as the one conducted by the General Assembly of the state of Georgia (The Atlanta *Journal,* September 23, 1983). Their study suggested that the decline was linked directly to the per-student funding by the states; in states where funding was high, the scores were high, and where funding was low, the scores were low. Critics of the study suggest that it is not fair to compare state averages, however, because in some states as many as 65% of the students take the exam, and in other states only 3% take it. The 3% represent the so-called cream of the crop, and one would expect their average to be high. So the critics possibly have a valid argument.

From this discussion, we see the need for thorough methods of organizing and summarizing the data so that some of the questions raised might be addressed.

The following table lists the 1983–1984 SAT averages by state, together with the change in scores from the previous year and the percent of students taking the exam. In Section 6.5 we will analyze these scores.

State	Total	Verbal	Change	Math	Change	%
Iowa	1,089	519	+1	570	−3	3
S. Dak.	1,086	520	+3	566	+6	3
N. Dak.	1,054	500	−5	554	−6	3
Kans.	1,051	502	+4	549	+9	5
Utah	1,045	503	−5	542	−3	4
Nebr.	1,041	493	−1	548	+2	6
Mont.	1,034	490	+10	544	+9	9
Wyo.	1,034	489	−3	545	+15	5
Minn.	1,020	481	−1	539	+1	7
N. Mex.	1,014	487	+3	527	+8	8
Okla.	1,009	484	−5	525	+4	5
Tenn.	1,009	486	+3	523	+4	8
Wis.	1,007	475	+2	532	−1	10
Ark.	1,003	482	0	521	+3	4
Ky.	997	479	+4	518	+5	6
Idaho	992	480	+1	512	−1	7
Miss.	992	480	−1	512	+6	3

(table continues)

(table continued)

State	Total	Verbal	Change	Math	Change	%
Ariz.	987	469	+4	509	+4	11
Ill.	981	463	+1	518	+1	14
Mo.	981	469	+3	512	+2	11
La.	981	472	+3	508	−6	5
Colo.	979	465	−4	514	−6	17
Mich.	976	461	+3	515	+4	11
W. Va.	976	466	0	510	−2	7
Ala.	970	467	+1	503	−5	6
Ohio	968	460	+2	508	+4	16
Wash.	968	463	0	505	−5	19
Nev.	931	442	+4	489	+9	17
N.H.	931	448	+4	483	+2	57
Alaska	915	443	+6	471	−9	30
Oreg.	907	435	+3	472	+3	42
Vt.	907	437	+3	470	−2	54
Conn.	904	436	+3	468	+3	69
Del.	902	433	0	469	+2	50
Calif.	897	421	0	476	−2	38
Md.	897	429	+2	468	+2	50
Mass.	896	429	+2	467	+4	66
N.Y.	894	424	0	470	+4	62
Va.	894	428	+1	466	+3	51
Maine	892	429	+1	463	+2	46
Fla.	890	423	0	467	+3	38
Pa.	887	425	0	462	+1	52
R.I.	885	424	+2	461	+2	61
N.J.	876	418	0	456	+3	65
Hawaii	869	396	+2	474	+3	47
Tex.	866	413	+1	453	0	32
Ind.	864	410	0	454	0	47
N.C.	827	395	+1	432	+1	47
Ga.	822	392	+2	430	+2	49
S.C.	803	384	+1	419	+4	49

SOURCE: The College Board.

6.1 Distributional Shapes

▲ The word *parent* is used because it is the distribution that, in a sense, produces all the scores.

In the study of statistics the word *population* can be interpreted in two ways. For instance, suppose we are studying the SAT scores of college students. Technically we think of the population as the complete collection of college students, and then we investigate the *variable* SAT score. Informally, however, we often simply think of the population of SAT scores corresponding to the college students. On the one hand the population consists of the students and on the other it consists of the SAT scores. In either case we are interested in the *distribution* of the SAT scores for college students. We will often refer to it as the *parent distribution*.

Definition

The distribution of the measurements in the original population is called the **underlying** or **parent distribution.**

To describe the parent distribution, such as the distribution of SAT scores, the statistician is concerned with three things. He or she is first concerned with the general *shape* of the distribution, second, with some measure of where the distribution is *centered* and, third, with some measure of *variability* in the distribution. First we will identify several distributional shapes.

In Chapter 5 we were introduced to the discrete and the continuous random variables. The binomial is an example of a discrete random variable, and the normal is an example of a continuous random variable. In the discrete case, the probability distribution is described by a rule that denotes what quantitative values are possible and how likely they are to occur.

In the continuous case, the distribution is described by a frequency curve that would resemble a stem and leaf plot of the entire population if it were available. The shape of the normal distribution, given by the frequency curve in Figure 5.12, is the most familiar distributional shape one will encounter. It is the standard with which other population distributions are compared. In this section we will consider some of the other possible distributional shapes that deviate from normality.

▲ A distribution is symmetrical about some value if the distribution to the left of that value is a mirror image of the distribution to the right of that value.

Recall that the normal distribution is symmetrical about its mean. Distributions that are not symmetrical are classified as *skewed.*

Definition

A distribution whose frequency curve has one tail longer than the other is called **skewed.** If the left tail is longer than the right, it is called **skewed left.** If the right tail is longer than the left, it is called **skewed right.**

A stem and leaf plot or histogram of a random sample can give some indication of the shape of the population distribution.

EXAMPLE 6.1

Following is the stem and leaf plot of the murder rates for the major cities given in Example 3.3:

```
0*  | 1  1
0t  | 3  2  2  2  3  3  2
0f  | 4  4  4  4  4  4  4  4  5  5  5  5
0s  | 6  6  7  6
0–  | 9  8  9  9  9
1*  | 1  0
1t  |
1f  | 4
1s  | 6
1–  | 9
```

This stem and leaf plot tails off to the right (toward the larger values); hence we classify it as *skewed right*. The frequency curve for the population from which this sample came from might look like Figure 6.1.

FIGURE 6.1

A skewed right frequency distribution

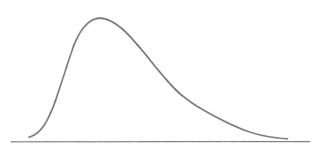

Figure 6.2 shows a *skewed left* distribution. If the distribution is not skewed but symmetrical, the length of its tails can be compared with the tails of the normal distribution.

Definition

A symmetrical distribution is called **short-tailed** if the tails of its frequency curve drop off more rapidly than the tails of a normal curve. It is called **long-tailed** if the tails of its frequency curve drop off less rapidly than the tails of a normal curve.

▲ If a distribution is not symmetrical, then it would not be classified as being either short-tailed or long-tailed.

FIGURE 6.2

A skewed left frequency distribution

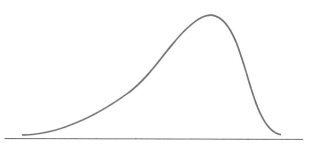

Figure 6.3 depicts a short-tailed and a long-tailed distribution, both compared with a normal frequency curve.

FIGURE 6.3

A short-tailed, long-tailed, and
normal distribution

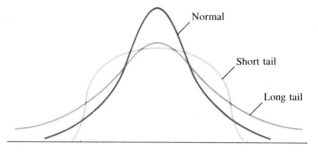

EXAMPLE 6.2

A class of 30 fifth graders completed a standardized reading test with the following
scores:

86	103	92	115	94	102	123	81	108	93
97	105	73	94	117	83	99	101	98	94
48	106	100	134	98	149	95	67	102	107

Construct a stem and leaf plot and comment on the distributional shape. (An ordered
stem and leaf plot for this data appears in Example 3.13.)

SOLUTION

Stem and leaf plot

```
 4 | 8
 5 |
 6 | 7
 7 | 3
 8 | 6  1  3
 9 | 2  4  3  7  4  9  8  4  8  5
10 | 3  2  8  5  1  6  0  2  7
11 | 5  7
12 | 3
13 | 4
14 | 9
```

The stem and leaf plot tails show several extreme observations on both ends, indicating
a *long-tailed* distribution.

We will see, later, that the long-tailed distribution is particularly important in that
it is a source of potential trouble. Some of the more commonly used statistical
procedures are misleading in the presence of a long-tailed distribution. Thus it is
important that we be able to identify the shape of our population distribution.

Another source of potential trouble is the *bimodal distribution* (see Figure 6.4).

FIGURE 6.4

A bimodal frequency distribution

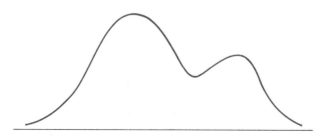

Definition

A distribution is called **bimodal** if its frequency curve has two peaks. A single peak is called a **mode**.

Drawing conclusions about bimodal populations can be dangerous, because the modes are usually the result of some extraneous variable that is confounded with the data. If the extraneous variable can be identified, then perhaps the population can be separated into two populations and then analyzed separately.

EXAMPLE 6.3

Following are the income levels (rounded to the nearest $100) of the assistant professors of a small university:

$20,500	19,000	23,000	20,000	17,500	19,500	22,000	20,600	21,000
18,000	22,500	17,000	19,500	21,500	22,000	18,500	18,700	17,600
20,400	21,800	21,400	17,300	21,300	20,900	18,900	22,500	19,200
18,200	18,400	21,100	17,800	21,400	21,700	18,800	18,300	20,700
19,400	18,800	19,800	20,900	18,500	23,100	23,500	18,000	20,000
18,900	21,600	17,500	22,300	21,600	19,200	18,100	21,000	18,200
19,100	18,500	19,600	21,200	20,100	21,500			

Construct a stem and leaf plot and classify the distributional shape.

SOLUTION   The stem and leaf plot can be simplified by dropping the last two digits of each salary and by plotting only the first three digits. The lowest salary is $17,000, so our first stem is 17. The highest salary is $23,500, so our last stem is 23. Our stem and leaf plot becomes:

```
17 | 5 0 6 3 8 5
18 | 0 5 7 9 2 4 8 3 8 5 0 9 1 2 5
19 | 0 5 5 2 4 8 2 1 6
20 | 5 0 6 4 9 7 9 0 1            × 10²
21 | 0 5 8 4 3 1 4 7 6 6 0 2 5
22 | 0 5 0 5 3
23 | 0 1 5
```

A double-stem stem and leaf might be more informative, and we will order as it is constructed.

```
17* | 0 3
17– | 5 5 6 8
18* | 0 0 1 2 2 3 4
18– | 5 5 5 7 8 8 9 9
19* | 0 1 2 2 4
19– | 5 5 6 8
20* | 0 0 1 4                    × 10²
20– | 5 6 7 9 9
21* | 0 0 1 2 3 4 4
21– | 5 5 6 6 7 8
22* | 0 0 3
22– | 5 5
23* | 0 1
23– | 5
```

Clearly the stem and leaf plot exhibits two peaks; hence we classify the distribution as bimodal. After reexamining the salaries, it was determined that the salaries clustering about $18,500 most generally belonged to women and those clustering about $21,000 belonged to men. Hence we have identified the extraneous variable, sex, that is creating the two modes.

We have just seen that the stem and leaf plot provides some guidelines in determining the general shape of a population distribution. Later in this chapter we will study other ways to determine whether the population distribution is skewed, long-tailed, short-tailed, bimodal, or possibly something else.

EXERCISES 6.1

6.1 A test to measure aggressive tendencies was given to a group of teenage boys who were members of a street gang. The test is scored from 10 to 50 with a high score indicating more aggression. Construct a double-stem stem and leaf plot of the following scores, and comment on the distributional shape.

38	27	44	39	41	26	35
45	39	28	16	37	11	36
33	46	42	37	40	19	24
29	32	34	31	30	32	43

6.2 The Tennessee Self Concept Scale was given to the group of teenage boys in Exercise 6.1. Construct a stem and leaf plot of the Self Criticism score that follows, and comment on the distributional shape.

26	19	23	27	24	33	25
29	14	30	20	25	5	18
7	28	31	37	28	3	20
25	45	29	22	41	34	22

6.3 Following are the starting salaries for 25 new Ph.D. psychologists selected randomly from universities in the East and South. Construct a stem and leaf plot, and comment on the distributional shape.

$27,900	28,300	14,400	25,200	28,900
23,700	31,000	20,100	23,100	26,600
21,200	29,400	29,100	20,900	28,700
29,200	24,600	22,000	29,000	24,800
23,300	34,000	24,200	23,300	30,400

6.4 The reaction times of 30 senior citizens (over 65 years of age) applying for driver's license renewals were measured with the following results:

93	105	66	94	64	98	109	71	86	31
101	128	97	85	96	60	94	42	64	99
84	107	77	55	98	80	90	79	90	96

Construct a stem and leaf plot, and comment on the distributional shape.

6.5 Violent crimes are offenses of murder, forcible rape, robbery, and aggravated assault. Following are the rates per 100,000 inhabitants for each of the 50 states and the District of Columbia for the year 1983. Construct a stem and leaf plot, and comment on the distributional shape.

Ala.	416.0	Ky.	322.2	N. Dak.	53.7
Alaska	613.8	La.	640.9	Ohio	397.9
Ariz.	494.2	Maine	159.6	Okla.	423.4
Ark.	297.7	Md.	807.1	Oreg.	487.8
Calif.	772.6	Mass.	576.8	Pa.	342.8
Colo.	476.4	Mich.	716.7	R.I.	355.2
Conn.	375.0	Minn.	190.9	S.C.	616.8
Del.	453.1	Miss.	280.4	S. Dak.	120.0
D.C.	1,985.4	Mo.	477.2	Tenn.	402.0
Fla.	826.7	Mont.	212.6	Tex.	512.2
Ga.	456.7	Nebr.	217.7	Utah	256.0
Hawaii	252.1	Nev.	655.2	Vt.	132.6
Idaho	238.7	N.H.	125.1	Va.	292.5
Ill.	553.0	N.J.	553.1	Wash.	371.8
Ind.	283.8	N. Mex.	686.8	W. Va.	171.8
Iowa	181.1	N.Y.	914.1	Wis.	190.9
Kans.	326.6	N.C.	409.6	Wyo.	237.2

SOURCE: *Uniform Crime Reports for the U.S., 1983,* Department of Justice, Washington, D.C., pp. 52–63.

6.6 In a hospital, a semiprivate room usually accommodates two patients but may contain as many as four. Following are the average daily rates for semiprivate rooms across the United States. Construct a stem and leaf plot, and comment on the distributional shape.

Ala.	$160.24	Del.	214.34	Ind.	182.11
Alaska	265.69	D.C.	280.41	Iowa	175.59
Ariz.	191.93	Fla.	177.24	Kans.	181.22
Ark.	143.48	Ga.	149.83	Ky.	168.15
Calif.	275.77	Hawaii	227.53	La.	149.30
Colo.	209.05	Idaho	189.34	Maine	211.26
Conn.	198.48	Ill.	237.29	Md.	183.98

Mass.	214.66	Nebr.	153.31	N.C.	141.05	R.I.	200.64	Tex.	155.56	Wash.	223.21
Mich.	255.00	Nev.	238.76	N. Dak.	170.56	S.C.	133.21	Utah	177.18	W. Va.	165.30
Minn.	183.33	N.H.	196.42	Ohio	226.76	S. Dak.	159.49	Vt.	208.07	Wis.	165.05
Miss.	109.66	N.J.	184.37	Okla.	165.23	Tenn.	138.96	Va.	163.29	Wyo.	151.83
Mo.	183.38	N. Mex.	191.44	Oreg.	219.60						
Mont.	189.32	N.Y.	223.48	Pa.	254.24						

SOURCE: *USA Today*, October 29, 1984 and the Health Insurance Association of America

Diagnostic Procedures

▲ The inference made from the collected data is only as good as the data itself, which means that it is very important that the sample be *representative* of the population.

It is difficult to classify correctly, with 100% accuracy, the exact shape of a population distribution, because usually only sample data are available. We infer a statistical shape that is consistent with the collected data. For example, as in the previous section, if a stem and leaf plot is skewed right, then we would infer that the population frequency curve is skewed right. In this section additional procedures are given to infer shape and support the conjecture made by viewing a stem and leaf plot.

Midsummary Analysis

The midsummaries, introduced in Chapter 3, can be used to help diagnose the shape of the frequency distribution.

Midsummary Analysis

1. If the distribution is near symmetric, then the midsummaries will be nearly equal.

2. If the distribution is skewed right, then the midsummaries will become progressively larger.

3. If the distribution is skewed left, then the midsummaries will become progressively smaller.

EXAMPLE 6.4

Duplicated here is the quantile summary diagram found in Example 3.13 for the reading scores of a group of fifth graders.

	30		Midsummaries
$M(15.5)$	98.5		98.5
$Q(8)$	93	106	99.5
$E(4.5)$	82	116	99
R	48	149	98.5

The closeness of the midsummaries suggests that the distribution from which the sample came is nearly symmetric.

EXAMPLE 6.5

The stem and leaf plot and quantile summary diagram found in Example 3.12 are repeated here. Find the midsummaries and comment on the distributional shape.

										26	
0	1 4 6										
1	3 6 7 8 9 9					$M(13.5)$			28.5		
2	2 4 6 8 9					$Q(7)$	18			53	
3	4 6 8					$E(4)$	13			73	
4	2 5					R	1			104	
5	3 6										
6	4										
7	3										
8	1										
9	4										
10	4										

SOLUTION   Averaging Q_1 and Q_3 we get

$$\text{Mid}Q = (53 + 18)/2 = 35.5$$

Averaging E_1 and E_7 we get

$$\text{Mid}E = (73 + 13)/2 = 43.0$$

Averaging HI and LO we get

$$\text{Mid}R = (104 + 1)/2 = 52.5$$

Thus we have

	26		Midsummaries
$M(13.5)$	28.5		28.5
$Q(7)$	18	53	35.5
$E(4)$	13	73	43.0
R	1	104	52.5

Here the midsummaries get progressively larger, suggesting that the frequency distribution is skewed right. This is obvious by investigating the stem and leaf plot; in some problems, however, it is not so obvious, and the preceding procedure can be useful in detecting skewness.

Pseudostandard Deviation Analysis

The range of a data set covers 100% of the data, the E-spread $= E_7 - E_1$, covers the middle 3/4 of the data, and the Q-spread $= Q_3 - Q_1$, covers the middle 1/2 of the data. Thus we have that the range is greater than the E-spread, which, in turn, is greater than the Q-spread. The difference in magnitude, however, can give some indication of the length of the tails of the parent distribution. If the range is considerably larger than either the E-spread or Q-spread, then it is likely that the parent distribution has long tails. Moreover for symmetric distributions, the spreads can be compared with the corresponding spreads for a normal distribution.

If a normal distribution has standard deviation σ, then the

$$q\text{-spread} = 1.35\sigma \quad \text{and} \quad e\text{-spread} = 2.3\sigma$$

Thus

$$\sigma = q\text{-spread}/1.35 = e\text{-spread}/2.3$$

Lowercase q and e are used here because the spreads refer to those of a population distribution. Capital Q and E are used to refer to the spreads of a sample. Using the sample information, these ratios are called *pseudostandard deviations*, PSD_q and PSD_e. That is,

$$PSD_q = Q\text{-spread}/1.35 \quad \text{and} \quad PSD_e = E\text{-spread}/2.3$$

▲ The numerical values, 1.35 and 2.3, used in computing the pseudostandard deviations are derived from the normal distribution.

Comparing the sample standard deviation, s, with PSD_q and PSD_e can give some insight into the length of the tails of symmetric distributions.

Length of Tails of Symmetric Distributions

1. If s, PSD_q, and PSD_e are close in value, then a close-to-normal distribution is indicated.

2. If s is somewhat larger than both PSD_q and PSD_e, then a distribution with long tails (as compared with normal) is indicated.

3. If s is somewhat smaller than both PSD_q and PSD_e, then a distribution with short tails (as compared with normal) is indicated.

EXAMPLE 6.6

The midsummary analysis of the reading scores for the fifth graders (Example 6.4) indicates that the distribution is symmetric. Comment on the length of the tails by conducting an analysis of the pseudostandard deviations.

SOLUTION   From Example 6.4, we have $Q_3 = 106$ and $Q_1 = 93$ so that

$$Q\text{-spread} = 106 - 93 = 13$$

Also $E_7 = 116$ and $E_1 = 82$ so that

$$E\text{-spread} = 116 - 82 = 34$$

Thus we have

$$PSD_q = Q\text{-spread}/1.35 = 13/1.35 = 9.63$$

and

$$PSD_e = E\text{-spread}/2.3 = 34/2.3 = 14.78$$

From the data (Example 3.13) and using a hand-held calculator, we find that

$$s = 18.92$$

We observe that s is larger than either pseudostandard deviation, which suggests that the distribution has long tails. We have numerical evidence to support the conjecture that the parent distribution is symmetric with long tails.

Normal Probability Plot

The pseudostandard deviation analysis is a useful *numerical* procedure for determining whether a sample has come from a normal population. An alternate (and perhaps more reliable) *graphical* procedure for checking normality is the normal probability plot. The probability plot is a plot of the sample data against the values we would have expected had the sample come from a normal population.

If the sample is indeed from a normal population, then the probability plot should be closely approximated by a straight line. Deviations from a straight line indicate non-normality. In fact, if the probability plot curves upward on the right and downward on the left, as in Figure 6.5, then there is evidence that the sample is from a long-tailed distribution. Data from a short-tailed distribution would exhibit the opposite pattern; that is, the probability plot would curve downward on the right and upward on the left.

FIGURE 6.5

A normal probability plot of data from a long-tailed distribution

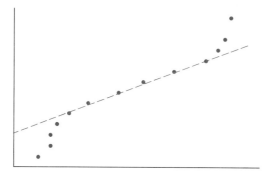

By hand, the computations for a probability plot are rather tedious and will not be presented here. On the other hand, Minitab (or almost any other statistical package) makes the job plausible. The construction of the normal probability plot will be given in the Minitab Session (Section 6.6) at the end of the chapter.

EXERCISES 6.2

6.7 A social scientist examined the records at several hospitals and found the following sample of ages of women at the birth of their first child:

30	18	35	22	23	22	36	24	23	28
19	23	25	24	33	21	24	19	33	23
19	32	21	18	36	21	25	17	21	24
39	22	23	18	22	28	18	15	25	21
23	26	38	24	20	36	27	21	28	26
22	28	33	18	17	21	15	20	16	21
23	15	20	38	16	24	42	22	24	24
20	17	26	39	22	21	28	20	29	14
25	20	19	17	21	24	26			

In Exercise 3.21 you were asked to construct a quantile summary diagram to include midsummaries and spreads. Use the midsummaries to comment on the shape of the distribution.

6.8 In Exercise 6.7, calculate s and the pseudostandard deviations. Use the results to comment on the length of the tails of the distribution.

6.9 In Exercise 6.1, a test to measure aggressive tendencies was given to a group of teenage boys who were members of a street gang. Their scores were

38	27	44	39	41	26	35
45	39	28	16	37	11	36
33	46	42	37	40	19	24
29	32	34	31	30	32	43

Construct a quantile summary diagram to include the midsummaries. From the midsummaries, comment on the distributional shape and compare your answer with the answer obtained in Exercise 6.1.

6.10 In Exercise 6.9, calculate s and the pseudostandard deviations. Use the results to comment on the length of the tails of the distribution.

6.11 In Exercise 6.2, the Tennessee Self Concept Scale was given to the group of teenage boys. The scores were as follows:

26	19	23	27	24	33	25
29	14	30	20	25	5	18
7	28	31	37	28	3	20
25	45	29	22	41	34	22

Construct a quantile summary diagram to include the midsummaries. From the midsummaries, comment on the distributional shape and compare your answer with the answer obtained in Exercise 6.1.

6.12 In Exercise 6.11, calculate s and the pseudostandard deviations. Use the results to comment on the length of the tails of the distribution.

6.13 In Exercise 6.3, the starting salaries for 25 new Ph.D. psychologists were recorded as follows:

$27,900	28,300	14,400	25,200	28,900
23,700	31,000	20,100	23,100	26,600
21,200	29,400	29,100	20,900	28,700
29,200	24,600	22,000	29,000	24,800
23,300	34,000	24,200	23,300	30,400

Construct a quantile summary diagram to include the midsummaries. From the midsummaries, comment on the distributional shape and compare your answer with the answer obtained in Exercise 6.3.

6.14 In Exercise 6.13, calculate s and the pseudostandard deviations. Use the results to comment on the length of the tails of the distribution.

6.15 In Exercise 6.4, the reaction times of 30 senior citizens applying for driver's license renewals were recorded as follows:

93	105	66	94	64	98	109	71	86	31
101	128	97	85	96	60	94	42	64	99
84	107	77	55	98	80	90	79	90	96

Construct a quantile summary diagram to include the midsummaries. From the midsummaries, comment on the distributional shape and compare your answer with the answer obtained in Exercise 6.4.

6.16 In Exercise 6.15, calculate s and the pseudostandard deviations. Use the results to comment on the length of the tails of the distribution.

6.17 In Exercise 6.6, the average daily rates for semiprivate hospital rooms throughout the United States were as follows:

Ala.	$160.24	Ky.	168.15	N. Dak.	170.56
Alaska	265.69	La.	149.30	Ohio	226.76
Ariz.	191.93	Maine	211.26	Okla.	165.23
Ark.	143.48	Md.	183.98	Oreg.	219.60
Calif.	275.77	Mass.	214.66	Pa.	254.24
Colo.	209.05	Mich.	255.00	R.I.	200.64
Conn.	198.48	Minn.	183.33	S.C.	133.21
Del.	214.34	Miss.	109.66	S. Dak.	159.49
D.C.	280.41	Mo.	183.38	Tenn.	138.96
Fla.	177.24	Mont.	189.32	Tex.	155.56
Ga.	149.83	Nebr.	153.31	Utah	177.18
Hawaii	227.53	Nev.	238.76	Vt.	208.07
Idaho	189.34	N.H.	196.42	Va.	163.29
Ill.	237.29	N.J.	184.37	Wash.	223.21
Ind.	182.11	N. Mex.	191.44	W. Va.	165.30
Iowa	175.59	N.Y.	223.48	Wis.	165.05
Kans.	181.22	N.C.	141.05	Wyo.	151.83

SOURCE: *USA Today*, October 29, 1984, and the Health Insurance Association of America.

Construct a quantile summary diagram to include the midsummaries. From the midsummaries, comment on the distributional shape and compare your answer with the answer obtained in Exercise 6.6.

6.18 In Exercise 6.17, calculate s and the pseudostandard deviations. Use the results to comment on the length of the tails of the distribution.

The Box Plot

The stem and leaf plot provides information about the general shape of a frequency distribution. We now introduce the *box plot*, which also provides this information, and also provides additional detailed information about the tails of the distribution. It pinpoints certain *outliers* in the data. As described in Chapter 3, their presence tends to inflate s and distort $\bar{y}$. Therefore it is necessary that we have a method of detecting these outliers so that, if necessary, summary measures other than $\bar{y}$ and s can be investigated.

Theory has shown that $\bar{y}$ and s perform favorably when the parent distribution is normal in shape. Furthermore

---

Winsor's Principle

The variation of frequency in the center of the distribution usually will be closely approximated by that of a normal distribution.

Thus if a frequency distribution deviates from normality, it usually occurs *in the tails*. The box plot will be used to identify this deviation from normality. A scaling factor of 1.5 times the Q-spread is found so that we may determine whether any observations fall beyond a region that would be expected of a normal distribution. Any observations outside this region are classified as outliers.

Construction of the Box Plot

The construction of the box plot is made easy by using the information in the stem and leaf plot and the quantile summary diagram. This information is used to construct a *box plot summary diagram.*

The ingredients for the box plot summary diagram are

1. The *scaling factor*, *SF*, which is $1.5 \times Q$-spread.
2. The *inner fences, f*, on the lower and upper sides.
   The lower inner fence is $(Q_1 - SF)$. The upper inner fence is $(Q_3 + SF)$.
3. The *outer fences, F*, on the lower and upper sides.
   The lower outer fence is $(Q_1 - 2SF)$. The upper outer fence is $(Q_3 + 2SF)$.
4. The *adjacent values, a*, which are the observations from the stem and leaf plot that are closest but inside the inner fences.

▲ We would expect that most observations from a normal distribution would fall inside the inner fences.

Any observations located between the inner and outer fences are classified as *mild outliers,* and any observations located beyond the outer fences are classified as *extreme outliers*. All this is summarized in the box plot summary diagram:

	SF		
a	lower	upper	
f	lower	upper	
mild outliers	#	#	mild outliers
F	lower	upper	
extreme outliers	#	#	extreme outliers

EXAMPLE 6.7

An ordered stem and leaf plot and *quarter* summary diagram for the data in Example 6.4 are repeated here. Construct a box plot summary diagram.

4	8										
5											
6	7										
7	3										
8	1	3	6								
9	2	3	4	4	4	5	7	8	8	9	
10	0	1	2	2	3	5	6	7	8		
11	5	7									
12	3										
13	4										
14	9										

$$
\begin{array}{c|cc}
 & \multicolumn{2}{c}{30} \\
M(15.5) & \multicolumn{2}{c}{98.5} \\
Q(8) & 93 & 106 \\
\end{array} \quad Q\text{-spread} = 13
$$

SOLUTION   The scaling factor is found to be

$$SF = 1.5 \times Q\text{-spread} = 1.5 \times 13 = 19.5$$

Subtracting this from Q_1, we find the lower inner fence to be

$$f = Q_1 - SF = 93 - 19.5 = 73.5$$

subtracting another scaling factor, we find the lower outer fence to be

$$F = f - SF = 73.5 - 19.5 = 54$$

Adding SF to Q_3, we find the upper inner fence to be

$$f = Q_3 + SF = 106 + 19.5 = 125.5$$

adding another scaling factor, we find the upper outer fence to be

$$F = f + SF = 125.5 + 19.5 = 145$$

This information is placed in the summary diagram and we have thus far

$$
\begin{array}{c|cc}
 & \multicolumn{2}{c}{19.5} \\
a & & \\
f & 73.5 & 125.5 \\
 & & \\
F & 54 & 145 \\
\end{array}
$$

The adjacent value on the lower side is found from the stem and leaf plot to be the closest observation at least as large as 73.5. This would be 81. The adjacent value on the upper side is found from the stem and leaf plot to be the closest observation no larger than 125.5. This would be 123. Placing this in the diagram, we now have

▲ Notice that the numbers in the left-hand column decrease as you go down, and the numbers in the right-hand column increase as you go down.

$$
\begin{array}{c|cc}
 & \multicolumn{2}{c}{19.5} \\
a & 81 & 123 \\
f & 73.5 & 125.5 \\
 & & \\
F & 54 & 145 \\
\end{array}
$$

Next we determine whether there are any outliers. Any observation between the inner and outer fences is classified as a mild outlier, and any observation beyond the outer fence is classified as an extreme outlier. On the lower side we see from the stem and leaf plot that there are two observations (67 and 73) between the inner and outer fences and one observation (48) beyond the outer fence. On the upper side we see from the stem and leaf plot that there is one observation (134) between the fences and one observation (149) beyond the outer fence. Placing this in the diagram, we complete the box plot summary diagram.

		19.5		
a	81	123		
f	73.5	125.5		
67, 73	two	one	134	
F	54	145		
48	one	one	149	

From the box plot summary diagram, we are now ready to draw a *box plot*. The box of the box plot is a rectangle that has ends at Q_1 and Q_3 with an interior line drawn at M. The adjacent values are marked with an $\times$ and joined to the box with lines. Mild outliers are drawn as open circles, and extreme outliers are drawn as solid circles.

EXAMPLE 6.8

Draw a box plot from the information in Example 6.7.

SOLUTION   The information needed to draw the box plot is the box plot summary diagram and the quarter summary diagram. They are repeated here:

Box plot summary diagram

		19.5		
a	81	123		
f	73.5	125.5		
67, 73	two	one	134	
F	54	145		
48	one	one	149	

Quartile summary diagram

		30	
$M(15.5)$		98.5	
$Q(8)$	93		106

A scale will be needed to cover the entire range of the data from 48 to 149. The box is drawn from $Q_1 = 93$ to $Q_3 = 106$ with a vertical line drawn at $M = 98.5$. Thus far we have

Placing an × at the two adjacent values (81 and 123) and connecting to the box, we have

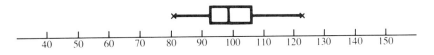

Drawing open circles for mild outliers and closed circles for extreme outliers, we have the completed box plot:

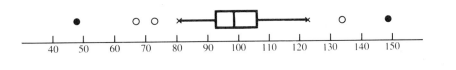

Interpretation of the Box Plot

The box contains the middle 50% of the data with the median dividing the lower 25% from the upper 25%. The position of the median line gives some indication about the shape of the middle of the distribution. If the median line is somewhat in the middle of the box, the middle 50% of the distribution is symmetric. If the median line is toward one end of the box, the middle 50% of the distribution is skewed in the opposite direction.

The length of the whiskers (tails) of the box plot give some indication about symmetry or skewness of the rest of the data. If the whiskers are about the same length, the distribution without outliers is symmetric. If one whisker is longer than the other, the distribution is skewed in that direction. Outliers beyond a long whisker are even stronger evidence that the distribution is skewed in that direction.

For a sample drawn from a normal population, one would expect to see about 7 mild outliers out of every 1,000 observations and no more than 2 extreme outliers out of every 1 million observations. If more outliers than this are observed in a sample, then there is evidence that the distribution is not normal. Certainly no extreme outliers are expected from a sample of 30 observations drawn from a normal population. Thus the box plot in Example 6.8 suggests that the parent distribution is symmetric and has tails somewhat longer than the normal distribution. This reaffirms the information obtained from the analysis of the midsummaries and the pseudostandard deviations.

EXERCISES 6.3

For Exercises 6.19 through 6.22, construct a box plot and interpret the results.

6.19 Exercise 6.1 Section 6.1. (See Exercises 6.9 and 6.10 of Section 6.2)

6.20 Exercise 6.2 Section 6.1. (See Exercises 6.11 and 6.12 of Section 6.2)

6.21 Exercise 6.3 Section 6.1. (See Exercises 6.13 and 6.14 of Section 6.2)

6.22 Exercise 6.4 Section 6.1. (See Exercises 6.15 and 6.16 of Section 6.2)

6.23 Following is a list of murder rates (no. per 100,000) for 30 cities selected from the South:

12	10	10	13	12	12	14	7	16	18
8	29	12	14	33	10	6	18	11	25
8	16	14	11	10	20	14	11	12	13

(a) Construct a stem and leaf plot.

(b) Construct a quantile summary diagram to include midsummaries and spreads.

(c) From the midsummaries comment on the distributional shape.

(d) Construct a box plot summary diagram and draw a box plot.

(e) Calculate PSD_q and PSD_e and compare with s; then comment on the shape of the distribution.

6.4 Choosing a Parameter

Using the procedures outlined in the previous sections of this chapter, we can gain valuable insight about the shape of a population distribution. To further describe the population distribution we should also look at the numerical quantities associated with it. We recall from Chapter 3 that these descriptive measures are called parameters.
Examples of parameters are:

1. μ—the population mean
2. μ_T—a trimmed population mean
3. θ—the population median
4. σ—the population standard deviation
5. π—the population proportion
6. ρ—the population correlation coefficient
7. $\mu_1 - \mu_2$—the difference of two population means
8. $\theta_1 - \theta_2$—the difference of two population medians
9. $\pi_1 - \pi_2$—the difference of two population proportions

There are many population parameters that could be studied. We will limit ourselves to those just listed, however, because they describe the characteristics of the various populations that will be studied in this text. Presently the task is to decide which parameter best describes the population characteristic of interest. As suggested at the beginning of Section 6.1, we are usually concerned with the center of the distribution and the amount of variability.

Measures of Center

Of the parameters listed, μ, μ_T, and θ are measures of the center of a single measurement distribution. To decide which one best describes the center of a distribution it is convenient to consider two cases, symmetric and asymmetric distributions.

Symmetric Distributions. The choice of a parameter that measures the *center* of a symmetric distribution is quite easy, because all natural measures of center, including the mean, median, and trimmed means, coincide. Quite simply, we think of the population mean, μ, as the measure of center associated with a symmetric distribution.

EXAMPLE 6.9

A group of 25 college students applying for graduate school takes the Miller Personality Test for admission purposes. Their scores are:

21	18	20	25	23
19	30	24	29	14
25	22	35	26	23
16	33	25	22	17
34	22	31	27	25

Construct a stem and leaf plot and a box plot. Comment on the distributional shape and determine what parameter best describes a "typical" score.

SOLUTION   Following is a stem and leaf plot and box plot of the data.

Ordered stem and leaf plot

1*	4
1–	6 7 8 9
2*	0 1 2 2 2 3 3 4
2–	5 5 5 5 6 7 9
3*	0 1 3 4
3–	5

Quantile summary diagram

		25		Midsummaries	Spreads
$M(13)$		24		24	
$Q(7)$	21		27	24	6
$E(4)$	18		31	24.5	13
R	14		35	24.5	21

$\bar{y} = 24.24$ $s = 5.555$
$\text{PSD}_q = 4.44$
$\text{PSD}_e = 5.65$

Box plot summary diagram

	9	
a	14	35
f	12	36
	none	none
F	3	45
	none	none

Box plot

All midsummaries are very close, strongly suggesting that the data are coming from a symmetrical population. Also because PSD_q, PSD_e and s are close, we conclude that the population is possibly normal. For normal and symmetrical populations, the mean, μ, is an excellent representation of the center of the distribution and a typical score.

Skewed Distributions.   For skewed distributions the choice of a parameter that measures the center is not so straightforward. The choice depends on the problem setting. For example, an insurance company concerned with losses incurred by its clients is more interested in the "mean payout" than it is in the "median payout," because premiums are determined by the expected payouts. In fact, it is conceivable that the insurance company will be interested in estimating a quantity even beyond the mean payout.

More generally, however, in disciplines such as economics, education, and the sciences, the population median is the preferred measure of the center of a skewed distribution.

EXAMPLE 6.10

A psychological test for measuring racial prejudice is given to a random sample of 25 high school students. From the recorded data, construct a stem and leaf plot and box plot. Analyze the midsummaries and comment on the shape of the parent distribution. What parameter best describes a "typical" score?

59	54	41	51	87
42	65	42	44	46
74	41	58	83	58
47	62	48	48	45
72	79	52	61	48

SOLUTION   The ordered stem and leaf plot and quantile summary diagram are as follows:

```
4*  | 1 1 2 2 4
4-  | 5 6 7 8 8 8
5*  | 1 2 4
5-  | 8 8 9
6*  | 1 2
6-  | 5
7*  | 2 4
7-  | 9
8*  | 3
8-  | 7
```

	25		**Midsummaries**	**Spreads**
$M(13)$		52	52	
$Q(7)$	46	62	54	16
$E(4)$	42	74	58	32
R	41	87	64	46
			$\bar{y} = 56.28$	$s = 13.66$

The box plot summary diagram and box plot are as follows:

	24		
a	41	83	
f	22	86	
	none	one	87
F	-2	110	
	none	none	

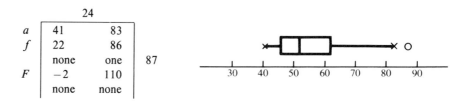

The midsummaries and box plot indicate that the parent distribution is skewed right. Because the distribution is skewed, the median is a more representative measure of the center of the distribution and a typical score.

Measures of Variability

The population standard deviation, σ, is the most common measure of variability in a measurement population. Because it is a measure of variation about the *mean*, it should be used when μ is the parameter used to measure the center of the distribution, that is, in symmetric distributions.

For skewed distributions, the population median is the preferred measure of the center of the distribution. Recall that the population median, θ, divides the population frequency distribution in half, and the population quartiles, θ_1 and θ_3, further subdivide it into quarters. (See Figure 3.4.) The distance between the first and third quartiles, called the q-spread $= \theta_3 - \theta_1$, is the preferred measure of dispersion in skewed populations.

▲ Note that small q is used here to denote the q-spread because we are referring to the population q-spread. The sample Q-spread $= Q_3 - Q_1$, where Q_1 and Q_3 are the first and third sample quartiles.

Proportions

The parameter of interest in a Bernoulli population is the proportion of successes, π.

Measures of Association

If two variables are related, the population correlation coefficient, ρ, measures the degree of association.

Comparing Two Parameters

Often we wish to compare the characteristics of two populations, such as comparing the effectiveness of two diet plans. Generally such comparisons reduce to a comparison of parameters associated with the two populations. Depending upon the shape of the distribution, this could be a comparison of population means, population medians, or population trimmed means. In the case of Bernoulli populations, the comparison would be a comparison of the proportion of successes in the two populations. Problems such as these will be addressed in Chapter 10.

EXERCISES 6.4

In Exercises 6.24 through 6.28, identify each of the parameters from the list of parameters at the beginning of this section.

6.24 The mean IQ is 100 as measured by the Stanford-Binet IQ test.

6.25 Eighty percent of the adult population drive cars.

6.26 The difference in median salaries of men and women in the teaching profession is $3,000.

6.27 A score of 20 on a test divides the upper 50% from the lower 50%.

6.28 The leading cause of death among 15- to 19-year-olds is car accidents, accounting for 45% of teenage deaths.

6.29 An animal behaviorist is studying the time it takes an animal to perform a certain behavioral task. The variable measured is the time from the beginning of the experiment until each individual animal has performed. The researcher is interested in the mean time of performance. Some animals are slow to respond and, in fact, some may never respond properly, making the calculation of the mean impossible. What parameter would be a meaningful measure of the "typical" time of performance?

6.30 In studying the response time to a drug or poison, some subjects will react very quickly, whereas others will not react for a long period of time. What would be a meaningful measure of the average response time?

Using the Descriptive Tools

In this chapter we have learned how to classify a distribution shape and how to choose a descriptive measure of the center of a distribution and a measure of the variability in the distribution. In this section we will put together all of the techniques learned in the previous sections and give a thorough analysis of a data set.

EXAMPLE 6.11

A kilowatt-hour of electricity is the amount of energy required to burn ten 100-watt light bulbs for 1 hour. Following are the rates per kilowatt-hour for each of the 50 states and the District of Columbia. Construct a stem and leaf plot and a box plot. Comment on the distributional shape. Choose an appropriate measure of central tendency.

Ala.	6.34	Ky.	5.63	N. Dak.	5.61
Alaska	8.38	La.	6.25	Ohio	7.34
Ariz.	7.35	Maine	6.98	Okla.	6.11
Ark.	6.68	Md.	6.84	Oreg.	3.88
Calif.	6.74	Mass.	8.44	Pa.	7.36
Colo.	6.14	Mich.	6.71	R.I.	9.02
Conn.	9.12	Minn.	6.21	S.C.	6.00
Del.	9.17	Miss.	6.08	S. Dak.	6.07
D.C.	6.31	Mo.	6.01	Tenn.	4.80
Fla.	7.26	Mont.	4.21	Tex.	7.17
Ga.	5.94	Nebr.	5.91	Utah	6.88
Hawaii	11.29	Nev.	6.02	Vt.	6.33
Idaho	3.58	N.H.	8.97	Va.	6.57
Ill.	8.47	N.J.	10.01	Wash.	3.38
Ind.	6.11	N. Mex.	7.34	W. Va.	5.50
Iowa	6.72	N.Y.	10.40	Wis.	6.70
Kans.	7.10	N.C.	6.19	Wyo.	5.20

SOURCE: *USA Today*, October 18, 1984, and the Energy Information Administration.

SOLUTION   A complete analysis follows:

Stem and leaf plot

```
 3 | 38, 58, 88
 4 | 21, 80
 5 | 20, 50, 61, 63, 91, 94
 6 | 00, 01, 02, 07, 08, 11, 11, 14, 19, 21, 25, 31, 33, 34, 57, 68, 70, 71, 72, 74, 84, 88, 98
 7 | 10, 17, 26, 34, 34, 35, 36
 8 | 38, 44, 47, 97
 9 | 02, 12, 17                 × 10⁻²
10 | 01, 40
11 | 29
```

Quantile
summary diagram

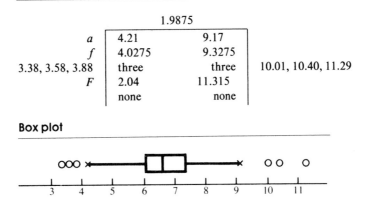

		51		Midsummaries	Spreads	PSD
$M(26)$		6.57		6.57		
$Q(13.5)$	6.015		7.34	6.6775	1.325	.98
$E(7)$	5.50		8.97	7.235	3.47	1.51
R	3.38		11.29	7.335	7.91	
				$\bar{y} = 6.76$		$s = 1.61$

Box plot
summary diagram

		1.9875	
a	4.21		9.17
f	4.0275		9.3275
3.38, 3.58, 3.88	three		three	10.01, 10.40, 11.29
F	2.04		11.315
	none		none

Box plot

(box plot diagram showing values from 3 to 11 with outliers OOO at left near 3–4, box from about 6 to 7.5, and outliers O O O near 10–11)

The midsummaries are increasing slightly, suggesting that the distribution is slightly skewed right, if not symmetrical. The pseudostandard deviations, in comparison with s, rule out normality. The box plot strongly suggests a long-tailed distribution. Even though the midsummaries are increasing, the box plot evidence is stronger, and therefore we should classify this distribution as a long-tailed distribution. As such, a trimmed mean would be a meaningful measure of the center. If indeed the population distribution is symmetric, then the trimmed mean coincides with the population mean, in which case the problem becomes one of estimating the population mean. In Chapter 8 we will look at methods of estimating the mean of a long-tailed distribution.

EXAMPLE 6.12

Analyze the SAT data given at the beginning of this chapter.

SOLUTION   As with any data set, first construct an ordered stem and leaf plot. (After experimenting with first a single stem and then a double stem, it was decided that a five-stem stem and leaf plot would better illustrate the data.)

8*	03
8t	22, 27
8f	
8s	64, 66, 69, 76
8–	85, 87, 90, 92, 94, 94, 96, 97, 97
9*	02, 04, 07, 07, 15
9t	31, 31
9f	
9s	68, 68, 70, 76, 76, 79
9–	81, 81, 81, 87, 92, 92, 97
10*	03, 07, 09, 09, 14
10t	20, 34, 34
10f	41, 45, 51, 54
10s	
10–	86, 89

From this stem and leaf plot we would classify the distributional shape as being bimodal. In Section 6.1 it was pointed out that bimodal data indicate that some extraneous variable is acting on the data. If possible, we would like to identify that variable. Returning to the data set, it appears that the higher SAT scores belong to states where the percent of student population that took the test varied from 3% to 19% (median = 7%). The lower SAT scores come from states where the percent of the student population that took the test varied from 30% to 69% (median = 49%). Thus the extraneous variable creating the modes is the percent of the student population that took the test. The presence of this extraneous variable indicates that we have two populations, and thus the data should be analyzed as two data sets.

We now construct stem and leaf plots and quantile summary diagrams for the two data sets.

Percent of population taking SAT greater than or equal to 30%

8*	03
8t	22, 27
8f	
8s	64, 66, 69, 76
8–	85, 87, 90, 92, 94, 94, 96, 97, 97
9*	02, 04, 07, 07, 15
9t	31

Percent of population taking SAT less than or equal to 19%

9t	31
9f	
9s	68, 68, 70, 76, 76, 79
9–	81, 81, 81, 87, 92, 92, 97
10*	03, 07, 09, 09, 14
10t	20, 34, 34
10f	41, 45, 51, 54
10s	
10–	86, 89

	22		Midsummaries	Spreads
$M(11.5)$	893		893	
$Q(6)$	869	902	885.5	33
$E(3.5)$	845.5	907	876.25	61.5
R	803	931	867	128

	28		Midsummaries	Spreads
$M(14.5)$	1000		1000	
$Q(7.5)$	980	1034	1007	54
$E(4)$	970	1051	1010.5	81
R	931	1089	1010	158

From those with lower SAT scores we detect a downward trend in the midsummaries, which means the distribution is skewed left, indicating that the median is the better measure of the center of the distribution. From those with higher SAT scores we detect an upward trend in the midsummaries, which means the distribution is skewed right, again indicating that the median is the appropriate measure of the center of the distribution. In comparison, the lower group had a median SAT score of 893, whereas the upper group had a median SAT score of 1,000. To better compare the data sets we present side-by-side box plots (box plots graphed with a common scale).

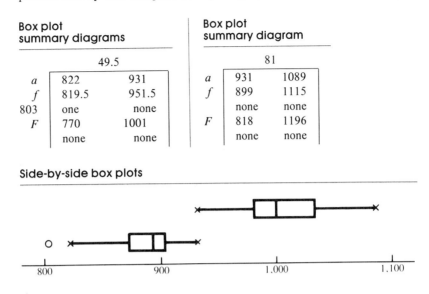

Box plot summary diagrams

	49.5	
a	822	931
f	819.5	951.5
803	one	none
F	770	1001
	none	none

Box plot summary diagram

	81	
a	931	1089
f	899	1115
	none	none
F	818	1196
	none	none

Side-by-side box plots

The box plots clearly demonstrate that there are two distinct groups. In states where, for example, less than 20% take the SAT test, we can expect scores around 1,000, and in states where more than 30% take the test, we can expect scores around 893, over 100 points lower. A scatterplot of the original data set (SAT scores vs percent taking test) also shows a definite relationship between the SAT scores and the percent taking the test. A measure of the association between the two variables with the correlation coefficient will show that the relationship is strong.

EXAMPLE 6.13

A number of faculty members at a particular state university (University I) are concerned with their declining salaries, as compared with the salaries at other universities. One specific complaint is that the faculty at University I are paid, on average, lower salaries than faculty at other state universities of comparable size and type. To investigate further, data were collected on 50 faculty members from University I and on 50 faculty members from University II, another state university of comparable size and type. The following data represent the 9-month salary (in $100) and the rank (1 = assistant professor, 2 = associate professor, 3 = full professor) for each faculty member in the two samples.

University I

278 (3)	264 (3)	277 (3)	286 (3)	284 (3)	243 (2)	232 (2)	203 (1)	274 (3)	267 (3)
280 (3)	274 (3)	275 (3)	292 (3)	183 (1)	193 (1)	276 (3)	188 (1)	267 (3)	282 (3)
239 (2)	229 (2)	196 (1)	200 (1)	290 (3)	229 (2)	239 (2)	210 (1)	302 (3)	242 (2)
271 (3)	310 (3)	257 (2)	199 (1)	241 (2)	276 (3)	199 (1)	270 (3)	230 (2)	263 (3)
246 (2)	195 (1)	302 (3)	232 (2)	285 (3)	293 (3)	224 (2)	269 (3)	234 (2)	243 (2)

208 (1)	230 (2)	301 (3)	212 (1)	297 (3)	253 (2)	249 (2)	261 (2)	242 (2)	229 (1)
256 (2)	247 (2)	210 (1)	246 (2)	287 (3)	314 (3)	229 (1)	225 (1)	225 (1)	305 (3)
237 (2)	305 (3)	257 (2)	237 (1)	211 (1)	298 (3)	247 (2)	238 (2)	307 (3)	300 (3)
246 (2)	250 (2)	247 (2)	221 (1)	216 (1)	255 (2)	298 (3)	222 (1)	232 (1)	310 (3)
312 (3)	263 (2)	238 (2)	306 (3)	235 (1)	219 (1)	250 (1)	252 (2)	219 (1)	243 (2)

Perform an analysis of this data by comparing the salaries at University I with those at University II.

SOLUTION From the data, we construct back-to-back stem and leaf plots (two stem and leaf plots with a common stem) and quantile summary diagrams.

```
            University I              University II
                   83 │ 18 │
                99653 │ 19 │
                   30 │ 20 │ 8
                    0 │ 21 │ 012699
                  994 │ 22 │ 125599
               994220 │ 23 │ 0257788
                63321 │ 24 │ 23667779
                    7 │ 25 │ 0023567
                97743 │ 26 │ 13
            876654410 │ 27 │
                65420 │ 28 │ 7
                  320 │ 29 │ 788
                   22 │ 30 │ 015567
                    0 │ 31 │ 024
```

	50		Midsummaries
$M(25.5)$		260	260
$Q(13)$	229	277	253
$E(7)$	199	286	242.5
R	183	310	246.5

	50		Midsummaries
$M(25.5)$		247	247
$Q(13)$	229	287	258
$E(7)$	219	305	262
R	208	314	261

The back-to-back stem and leaf plots of the salaries at the two universities do not reveal any significant differences. The quantile summary diagrams reveal that the median salary at University I is $26,000 and the median salary at University II is $24,700. On the surface it appears that University I does not have a case.

Taking a closer look, however, we see that the stem and leaf plot for University I appears to be trimodal and the stem and leaf for University II appears to be bimodal, which indicates an extraneous variable confounded with the data. Investigating further, we decide to look at the salaries at the various ranks. Table 6.1 gives back-to-back stem and leaf plots and summaries at the three ranks.

TABLE 6.1
Summary of results at each rank

University I		Stem		University II	

Assistant Professor / Stem / Assistant Professor

University I		Stem		University II	
Assistant Professor		**Stem**		**Assistant Professor**	
Number	10	3	18*	Number	17
Mean	$19,660	8	18–	Mean	$22,353
		3	19*		
Standard		9965	19–	Standard	
Deviation	$759	30	20*	Deviation	$1,110
			20–	8		
Median	$19,750	0	21*	012	Median	$22,200
			21–	699		
			22*	12		
			22–	5599		
			23*	2		
			23–	57		

HI: 250

Associate Professor / Stem / Associate Professor

University I		Stem		University II	
Associate Professor		**Stem**		**Associate Professor**	
Number	15	4	22*		Number	20
Mean	$23,733	99	22–		Mean	$24,785
		4220	23*	0		
		99	23–	788		
Standard		3321	24*	23	Standard	
Deviation	$847	6	24–	667779	Deviation	$843
			25*	023		
Median	$23,900	7	25–	567	Median	$24,700
			26*	13		

Full Professor / Stem / Full Professor

University I		Stem		University II	
Full Professor		**Stem**		**Full Professor**	
Number	25	43	26*		Number	13
Mean	$28,028	977	26–		Mean	$30,308
		4410	27*			
		87665	27–			
Standard		420	28*		Standard	
Deviation	$1,239	65	28–	7	Deviation	$730
		320	29*			
Median	$27,700		29–	788	Median	$30,500
		22	30*	01		
			30–	5567		
		0	31*	024		

We observe that the median salary at University I is below the median salary at University II at all three ranks. We now understand University I's concern. On the whole, their median salary is greater than the median salary at University II; at each rank, however, they are below University II. Finally side-by-side box plots (Figure 6.6) at each rank demonstrate the seriousness of the salary differences at the various ranks. Can you explain how this can happen?

FIGURE 6.6

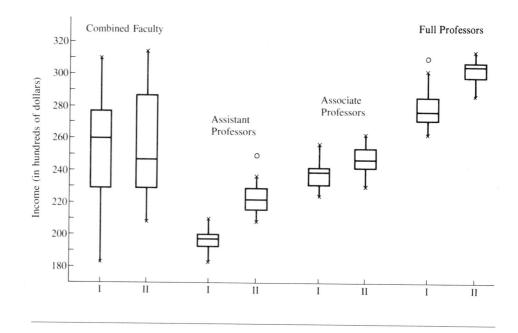

EXERCISES 6.5

6.31 The number of women in U.S. state legislatures is as follows:

Ala.	9	Nebr.	6
Alaska	8	Nev.	6
Ariz.	19	N.H.	121
Ark.	7	N.J.	11
Calif.	14	N. Mex.	9
Colo.	25	N.Y.	23
Conn.	43	N.C.	23
Del.	10	N. Dak.	17
Fla.	28	Ohio	11
Ga.	19	Okla.	12
Hawaii	17	Oreg.	20
Idaho	16	Pa.	10
Ill.	28	R.I.	22
Ind.	18	S.C.	12
Iowa	17	S. Dak.	14
Kans.	23	Tenn.	9

Ky.	10	Tex.	13
La.	5	Utah	9
Maine	41	Vt.	32
Md.	35	Va.	11
Mass.	26	Wash.	28
Mich.	16	W. Va.	18
Miss.	5	Wis.	26
Mo.	25	Wyo.	22
Mont.	19		

SOURCE: *USA Today*, October 25, 1984, and Center for the American Woman and Politics.

Construct a box plot of the data, comment on the shape of the distribution of women legislators across the states, and give a measure of the center of the data.

6.32 The median price of a single-family home in the United States was $73,600 in the third quarter of 1984, according to a National Association of Realtors survey. Following are the median prices in 37 markets across the United States:

Albany, N.Y.	$ 52,400
Anaheim, Calif.	134,900
Atlanta	64,600
Baltimore	65,200
Birmingham, Ala.	66,600
Boston	102,000
Chicago	77,500
Cincinnati	59,600
Cleveland	65,600
Columbus, Ohio	60,400
Dallas-Fort Worth	83,400
Denver	85,000
Detroit	48,000
Fort Lauderdale, Fla.	76,000
Houston	79,600
Indianapolis	54,200
Kansas City, Mo.	57,600
Los Angeles	115,300
Louisville, Ky.	49,500
Memphis, Tenn.	65,200
Miami	84,400
Milwaukee	69,600
Minneapolis-St. Paul	75,100
Nashville, Tenn.	64,000
New York metro area	106,900
Oklahoma City	63,600
Philadelphia	59,500
Providence, R.I.	61,400
Rochester, N.Y.	62,100
St. Louis	64,400
Salt Lake City	67,600
San Antonio, Texas	71,600
San Diego	102,900
San Francisco	132,600
San Jose, Calif.	121,600
Tampa, Fla.	60,800
Washington	92,900

SOURCE: *USA Today*, November 12, 1984, and National Association of Realtors; California Association of Realtors.

Construct a box plot of the data and interpret the results.

6.33 The average teachers' salaries across the states for 1973–1974 and for 1983–1984 are as follows:

State	Average salary 1973–1974	1983–1984
Ala.	$ 9,226	$18,000
Alaska	15,667	36,564
Ariz.	10,414	21,605
Ark.	7,820	16,929
Calif.	13,113	26,403
Colo.	10,131	22,895
Conn.	11,030	22,624
Del.	11,304	20,925
D.C.	N/A	27,659
Fla.	10,018	19,545
Ga.	9,392	18,505
Hawaii	11,112	24,357
Idaho	8,383	18,640
Ill.	11,871	23,345
Ind.	10,508	21,587
Iowa	9,863	20,140
Kans.	8,894	19,598
Ky.	8,295	19,780
La.	9,166	19,100
Maine	9,238	17,328
Md.	11,741	24,095
Mass.	11,121	22,500
Mich.	12,545	28,877
Minn.	11,122	24,480
Miss.	7,604	15,895
Mo.	9,530	19,300
Mont.	$ 9,429	$20,657
Nebr.	9,174	18,785
Nev.	11,549	23,000
N.H.	9,613	17,376
N.J.	11,920	23,044
N. Mex.	9,100	20,760
N.Y.	13,371	26,750
N.C.	10,223	18,014
N. Dak.	8,493	20,363
Ohio	10,107	21,421
Okla.	8,238	18,490
Oreg.	10,180	22,833
Pa.	10,921	22,800
R.I.	11,407	24,641
S.C.	8,654	17,500
S. Dak.	8,150	16,480
Tenn.	8,840	17,900
Tex.	8,920	20,100
Utah	9,146	20,256
Vt.	8,932	17,931
Va.	9,919	19,867
Wash.	11,295	24,780
W. Va.	8,467	17,482
Wis.	10,830	23,000
Wyo.	9,668	24,500
Average	10,778	22,019

SOURCE: *USA Today*, November 13, 1984, and Educational Research Service; National Education Association.

Clearly the average salary for 1973–1974 is lower than for 1983–1984, but other than that, are the distributions similar? How variable are the salaries? What are the shapes of the two distributions? To investigate these questions, construct a quantile summary diagram to include midsummaries and spreads, calculate the standard deviations and pseudostandard deviations and compare them, and finally construct side-by-side box plots to shed light on the length of the tails of the two distributions. From this information compare the shapes of the two distributions.

6.34 The Student Government Association at a large university is interested in investigating the amount spent on a date by a typical male university student. They conducted a poll of 50 males to find the following amounts spent on a typical weekend date:

$22	17	5	25	20	10	8	5	40	15
0	10	15	25	0	18	15	25	12	30
0	15	25	35	30	18	22	35	20	10
0	12	23	15	10	0	5	10	50	20
15	25	15	55	25	20	17	13	30	25

Study the distribution of the amounts spent by constructing a quantile summary diagram and box plot.

6.35 Construct a quantile summary diagram and box plot of the salaries of the executives for the major corporations given in Appendix A. Comment on the distributional shape and give a measure of the center of the data.

Computer Session (optional)

Concerned with the decline in Scholastic Aptitude Test (SAT) scores by high school seniors, a researcher randomly selected 50 high school students who had taken the SAT test during their senior year. Some of the students took a special course to prepare them for the SAT test. Those who took the course are coded as a 1, and those who did not are coded 0 (see Table 6.2). The students are further classified according to sex (0 = male, 1 = female), and parental education (1 = no high school diploma; 2 = high school diploma, no college; 3 = college graduate). The grade point average (GPA), and the verbal, math, and combined SAT scores are also given for each student.

Some questions that might be addressed are:

1. Is there evidence that the special course helped prepare students for the SAT test?
2. Do males and females differ in ability on the SAT test?
3. Does the parents' education level give any indication of what the student might score?

Minitab commands that relate to the material in this chapter and that might be used to summarize and analyze this SAT data are:

1. COPY
2. STEM AND LEAF
3. BOX PLOT
4. NSCORES

First we must enter the data into the computer. This is accomplished with the READ command as is illustrated:

```
READ SEX, ED, CS, GPA, SATM, SATV INTO C1-C6
0,2,0,2.4,479,453
0,1,1,1.2,490,442
1,1,0,2.2,397,602

    . . .
    . . .

0,2,0,2.4,475,338

ADD C5,C6 PUT TOTAL SAT INTO C7

NAME C1='SEX', C2='ED', C3='COURSE', C4='GPA'
NAME C5='SATM', C6='SATV', C7='TSAT'
```

Note that the SAT total was not read in with the READ command, but rather was calculated from C5 and C6 and then stored in C7.

Now that the data are stored in the computer, we are in a position to analyze the data. First we will see if there is a difference in SAT scores and GPA scores for males and females. Recall that male is coded 0 and female is coded 1. The COPY command can be used to separate the SAT scores for males and females.

TABLE 6.2

SAT scores for 50 high school seniors

Student ID	Sex	Parental education	Preparation course	GPA	SAT scores		
					Math	Verbal	Total
1	0	2	0	2.4	479	453	932
2	0	1	1	1.2	490	442	932
3	1	1	0	2.2	397	602	999
4	1	3	1	3.8	599	517	1116
5	0	2	1	3.4	559	575	1134
6	1	2	0	2.2	539	396	935
7	0	3	0	3.6	730	525	1255
8	0	3	0	4.0	623	674	1297
9	1	2	1	2.8	521	528	1049
10	0	3	1	2.8	570	509	1079
11	0	2	1	3.4	530	454	984
12	1	1	0	2.9	552	478	1030
13	1	2	0	2.8	476	554	1030
14	1	1	0	2.0	520	430	950
15	1	1	0	1.5	439	451	890
16	0	1	0	2.4	521	663	1184
17	1	3	0	3.1	665	567	1232
18	0	2	0	2.6	523	441	964
19	1	2	0	3.1	495	412	907
20	0	3	0	3.2	648	384	1032
21	1	1	0	2.0	375	470	845
22	1	2	1	2.4	541	453	994
23	0	2	1	2.9	484	703	1187
24	1	2	1	3.1	602	595	1197
25	1	1	1	2.6	526	550	1076
26	1	1	0	0.7	478	229	707
27	1	1	1	0.7	410	389	799
28	1	3	1	3.4	669	559	1228
29	1	2	0	2.8	488	447	935
30	0	2	0	3.2	589	465	1054
31	0	2	0	1.7	426	473	899
32	0	1	0	1.4	458	421	879
33	1	3	1	3.1	597	601	1198
34	1	2	0	1.7	351	427	778
35	1	2	0	2.7	484	483	967
36	1	1	0	1.9	475	515	990
37	1	3	1	3.5	629	582	1211
38	0	1	0	1.3	501	476	977
39	0	2	0	2.0	456	392	848
40	0	3	1	3.0	527	552	1079
41	1	2	0	2.3	491	343	834
42	1	2	0	2.2	505	408	913
43	0	2	0	3.1	547	440	987
44	0	3	0	2.5	596	484	1080
45	0	3	0	2.7	539	564	1103
46	1	1	0	2.1	359	386	745
47	1	2	0	2.2	559	526	1085
48	1	3	0	4.0	658	482	1140
49	0	3	1	4.0	616	593	1209
50	0	2	0	2.4	475	338	813

COPY

The COPY command has two subcommands, USE and OMIT. In our example we will copy the GPA and TSAT scores corresponding to the males in C8 and C9 and corresponding to the females in C10 and C11. The commands with their subcommands are now given.

```
COPY C4, C7 INTO C8, C9;
   USE 'SEX' = 0.
NAME C8='GPA.M', C9='TSAT.M'
COPY C4, C7 INTO C10, C11;
   USE 'SEX' = 1.
NAME C10='GPA.F', C11='TSAT.F'
```

The OMIT subcommand is the opposite of USE. It indicates what not to copy. There are other options with the COPY command that are described in the Minitab handbook.

To compare the GPA scores for males and females we will use the STEM-AND-LEAF command described in Chapter 2, the DESCRIBE command given in Chapter 3, and the BOXPLOT command.

BOXPLOT

The form of this command is

```
BOXPLOT FOR C2 (LEVELS IN C1)
```

which produces a box plot of the data in C2.

In Figure 6.7, we will compare the GPA scores for males, which have been stored in GPA.M, and for females, which have been stored in GPA.F.

FIGURE 6.7

Minitab output

```
MTB > DESCRIBE 'GPA.M', 'GPA.F'

              N      MEAN    MEDIAN    TRMEAN    STDEV    SEMEAN
GPA.M        22     2.691     2.750     2.700    0.807     0.172
GPA.F        28     2.493     2.500     2.504    0.799     0.151

             MIN       MAX        Q1        Q3
GPA.M      1.200     4.000     2.300     3.250
GPA.F      0.700     4.000     2.025     3.100


MTB > STEM 'GPA.M', 'GPA.F'

STEM-AND-LEAF OF GPA.M      N  = 22

Leaf Unit = 0.10

    3     1 234
    4     1 7
    8     2 0444
   (5)    2 56789
    9     3 012244
    3     3 6
    2     4 00
```

```
STEM-AND-LEAF OF GPA.F        N = 28

Leaf Unit = 0.10

     2    0 77
     2    1
     5    1 579
    14    2 001222234
    14    2 678889
     8    3 11114
     3    3 58
     1    4 0


MTB > BOXPLOT 'GPA.M'


                              -----------------
    *  ---------------------I     +      I----------------
                              -----------------
    ------+---------+---------+---------+---------+---------+GPA.M
        1.50      2.00      2.50      3.00      3.50      4.00


MTB > BOXPLOT 'GPA.F'


                           ---------------
    ------------------I      +       I-------------
                           ---------------
    ----+---------+---------+---------+---------+---------+--GPA.F
      0.70      1.40      2.10      2.80      3.50      4.20
```

Similar commands can be given to compare the SAT scores for males and females (see Figure 6.8). Recall that with the COPY command, we stored the total SAT scores for males in TSAT.M and for females in TSAT.F.

FIGURE 6.8
Minitab output

```
MTB > DESCRIBE 'TSAT.M' 'TSAT.F'

               N      MEAN    MEDIAN    TRMEAN     STDEV    SEMEAN
TSAT.M        22    1041.3    1043.0    1039.9     133.5      28.5
TSAT.F        28     992.1     992.0     993.9     150.2      28.4

              MIN       MAX        Q1        Q3
TSAT.M      813.0    1297.0     932.0    1146.5
TSAT.F      707.0    1232.0     894.3    1108.3


MTB > STEM 'TSAT.M', 'TSAT.F'
```

```
STEM-AND-LEAF OF TSAT.M     N  = 22

Leaf Unit = 10

       2     8 14
       4     8 79
       6     9 33
      10     9 6788
      11    10 3
      11    10 5778
       7    11 03
       5    11 88
       3    12 0
       2    12 59


STEM-AND-LEAF OF TSAT.F     N  = 28

Leaf Unit = 10

       2     7 04
       4     7 79
       6     8 34
       7     8 9
      11     9 0133
      (5)    9 56999
      12    10 334
       9    10 78
       7    11 14
       5    11 99
       3    12 123
```

MTB > BOXPLOT 'TSAT.M'

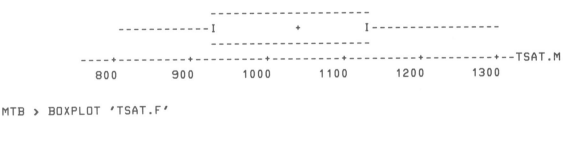

MTB > BOXPLOT 'TSAT.F'

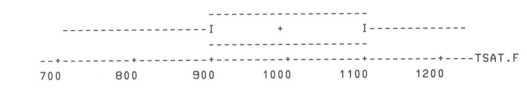

The normal probability plot that was discussed in Section 6.2 can be constructed by using the PLOT command (discussed in Section 2.6) and the NSCORES command.

After entering the sample data in C1, the probability plot is formed by the commands:

```
NSCORES OF DATA IN C1 PUT IN C2
PLOT C1 VS C2
```

For example, Figure 6.9 gives a probability plot of the GPA scores for males that have been stored in GPA.M.

FIGURE 6.9

Minitab output

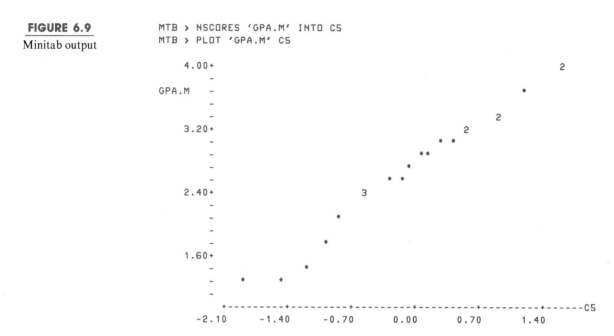

```
MTB > NSCORES 'GPA.M' INTO C5
MTB > PLOT 'GPA.M' C5
```

With the exception of the one point on the left, the probability plot appears in a straight line, which indicates normality.

EXERCISES 6.6

Exercises 6.36 through 6.39 refer to the SAT scores for the 50 high school seniors.

6.36 With the COPY command and the USE subcommand store the total SAT scores for those who took the preparatory course in TSAT.C and those who did not take the course in TSAT.NC. Compare the two groups with the DESCRIBE, STEM-AND-LEAF, and BOXPLOT commands.

6.37 With the COPY command and the USE subcommand, separate the GPA scores into three groups corresponding to the parents' education. Store the scores in GPA.E1, GPA.E2, and GPA.E3 and then compare the three groups with the DESCRIBE, STEM-AND-LEAF, and BOXPLOT commands.

6.38 Check the normality of the SAT MATH scores by constructing a normal probability plot.

6.39 Construct a probability plot of the GPA scores for females. Do the points of the plot fall in a straight line? Comment on the shape of the distribution.

6.40 Use the RANDOM command to generate a sample of size 50 from a normal population with a mean of 100 and a standard deviation of 20. Make a normal probability plot of the data. Do the points of the plot fall in a straight line?

Key Concepts

☑ This chapter is aimed at describing a population distribution via the collected data. The description involves three things: the shape of the distribution, measures of location, and measures of variability. *Distributional shapes* are classified as skewed, symmetric short tailed, symmetric long tailed, normal, and multimodal.

☑ A *skewed* distribution is a distribution with one tail significantly longer than the other. It is said to be skewed in the direction of the long tail. A *normal* distribution is described in Chapter 5. A *symmetric short-tailed* distribution has tails that drop off more rapidly than the tails of a normal curve. A *symmetric long-tailed* distribution has tails longer than the tails of a normal curve. A *multimodal* distribution has more than one peak. Typically the *bimodal* distribution with its two peaks is more commonly observed in real data. The occurrence of more than one mode in a data set is an indication of an extraneous variable that is confounded with the data. If possible, the factor should be identified.

☑ The *midsummaries* can be used to detect skewness. The lengths of the tails of symmetric distributions can be diagnosed using the *pseudostandard deviations*.

☑ The *box plot* is a graphical tool to picture the data and detect possible outliers in the data. The techniques discussed in this chapter are used collectively to describe a data set and get as much information from it as possible.

Learning Goals

Having completed this chapter you should be able to:

1. Identify the various population distributional shapes.   *Section 6.1*
2. Use the midsummaries to diagnose shape.   *Section 6.2*
3. Use the pseudostandard deviations to diagnose shape.   *Section 6.2*
4. Construct a box plot.   *Section 6.3*
5. Interpret the results of a box plot.   *Section 6.3*
6. Choose a parameter that measures the population characteristic of interest. *Section 6.4*
7. Use the descriptive tools to describe the distribution.   *Section 6.5*

Review Questions

To test your skills answer the following questions.

6.41 A psychologist designs an experiment to study the effect of electroshock on the time required to complete a difficult task. Following are the number of trials necessary to complete the task:

6	11	14	3	9
15	4	7	12	8
14	16	6	10	15
3	9	18	31	26
21	25	20	48	23

(a) Construct a stem and leaf plot.

(b) Construct a quantile summary diagram to include midsummaries and spreads.

(c) From the midsummaries comment on the distributional shape.

(d) Calculate $\bar{y}$ and s using your calculator.

(e) Calculate PSD_q and PSD_e and compare with s, and comment on the shape of the distribution.

(f) Construct a box plot summary diagram, draw a box plot, and comment on the distribution.

6.42 The following data represent the education level attained by employees in three different industries located in a medium-sized community:

Industry

A	B	C
8	8	12
8	9	12
1	9	8
12	8	9
14	7	12
9	6	9
12	12	18
14	16	12
12	2	14

Does it appear that the educational levels for the three industries differ? Does there appear to be a large variation in education levels? Construct side-by-side box plots and compare the distributions.

SUPPLEMENTAL EXERCISES FOR CHAPTER 6

6.43 From the following data construct a stem and leaf plot and quantile summary diagram to include the midsummaries and spreads. Use the midsummaries to evaluate the skewness of the distribution.

11.7	12.8	17.2	14.5	17.3	15.9	12.3	13.6
11.5	19.0	15.8	14.7	11.7	19.6	16.9	13.8
13.5	13.9	17.6	13.7	12.6	13.3	15.0	14.6
15.7	13.7	18.0	13.4	13.5	12.8	20.9	14.4
16.4	18.5	16.8	12.1	19.5	14.9	14.8	21.6
13.9							

6.44 Use the information in Exercise 6.43 to calculate the pseudostandard deviations and construct a box plot. Use the pseudostandard deviations to comment on the length of the tails of the distribution. From the box plot and other information, make a general statement about the shape of the distribution. What parameter best describes a "typical" score?

6.45 Following are the fertility rates for all states including the District of Columbia, which were first considered in Exercise 3.68 of the Chapter 3 supplementary exercises.

Ala.	1.9	D.C.	1.5
Alaska	2.3	Fla.	1.7
Ariz.	2.1	Ga.	1.9
Ark.	2.0	Hawaii	2.1
Calif.	1.9	Idaho	2.5
Colo.	1.8	Ill.	1.9
Conn.	1.5	Ind.	1.8
Del.	1.8	Iowa	2.0

Kans.	2.0	N. Dak.	2.1
Ky.	1.9	Ohio	1.8
La.	2.2	Okla.	2.0
Maine	1.7	Oreg.	1.8
Md.	1.6	Pa.	1.6
Mass.	1.5	R.I.	1.5
Mich.	1.8	S.C.	1.8
Minn.	1.9	S. Dak.	2.4
Miss.	2.2	Tenn.	1.7
Mo.	1.9	Tex.	2.1
Mont.	2.1	Utah	3.2
Nebr.	2.0	Vt.	1.7
Nev.	1.8	Va.	1.6
N.H.	1.7	Wash.	1.8
N.J.	1.6	W. Va.	1.8
N. Mex.	2.2	Wis.	1.9
N.Y.	1.6	Wyo.	2.4
N.C.	1.6		

SOURCE: Population Reference Bureau

Construct a stem and leaf plot and quantile summary diagram to include the midsummaries and spreads. Use the midsummaries to evaluate the skewness of the distribution.

6.46 Construct a box plot of the data in Exercise 6.45. Calculate PSD_q and PSD_e and compare with s, and comment on the length of the tails of the distribution. Make a general statement about the shape of the distribution and identify a parameter that would best measure the center of the distribution.

6.47 Following are the average annual sales commissions (1983) paid lottery agents in the 18 states with lotteries. These data were first considered in Exercise 3.69 of the Chapter 3 supplementary exercises.

State	No. of lottery agents	Average sales commission
Ariz.	2,100	$ 2,228
Colo.	2,500	4,880
Conn.	3,000	3,200
D.C.	850	3,986
Del.	324	5,247
Ill.	7,710	3,320
Maine	1,629	490
Md.	1,239	20,399
Mass.	3,600	5,279
Mich.	7,293	4,758
N.H.	1,150	574
N.J.	4,000	12,000
N.Y.	12,449	4,281
Ohio	5,400	3,704
Pa.	7,700	8,013
R.I.	883	7,667
Vt.	750	307
Wash.	4,711	1,804

SOURCE: *Gaming and Wagering Business* and *USA Today*, January 15, 1985.

Using the average sales commission data, construct a stem and leaf plot and quantile summary diagram to include the midsummaries and spreads. Use the midsummaries to evaluate the skewness of the distribution.

6.48 Construct a box plot of the data in Exercise 6.47. Calculate PSD_q and PSD_e and compare with s, and comment on the length of the tails of the distribution. Make a general statement about the shape of the distribution and identify a parameter that would best measure the center of the distribution.

6.49 Following are the annual food expenditures for a random sample of 40 single households in the state of Ohio first considered in Exercise 3.70 of the Chapter 3 supplementary exercises.

$2,845	3,170	2,352	4,978	3,820	2,475	3,160
5,780	2,175	2,648	2,872	4,250	3,970	2,534
6,870	2,734	2,847	4,670	5,176	3,640	2,765
1,180	3,679	3,320	7,580	2,416	3,743	2,830
3,127	3,249	2,648	1,976	2,784	3,869	2,086
5,587	3,420	2,645	8,147	4,367		

Construct a stem and leaf plot and quantile summary diagram to include the midsummaries and spreads. Use the midsummaries to evaluate the skewness of the distribution.

6.50 Construct a box plot of the data in Exercise 6.49. Calculate PSD_q and PSD_e and compare with s, and comment on the length of the tails of the distribution. Make a general statement about the shape of the distribution and identify a parameter that would best measure the center of the distribution.

6.51 Following are the fees for a coronary bypass for a random sample of 20 physicians first considered in Exercise 3.73 of the Chapter 3 supplementary exercises.

$3,820	3,540	2,800	4,260	3,920
3,400	4,370	3,890	4,420	3,860
3,970	4,250	5,200	3,270	3,950
4,180	4,470	3,920	4,840	4,650

Construct a stem and leaf plot and quantile summary diagram to include the midsummaries and spreads. Use the midsummaries to evaluate the skewness of the distribution.

6.52 Construct a box plot of the data in Exercise 6.51. Calculate PSD_q and PSD_e and compare with s, and comment on the length of the tails of the distribution. Make a general statement about the shape of the distribution and identify a parameter that would best measure the center of the distribution.

7

Sampling Distributions and the Central Limit Theorem

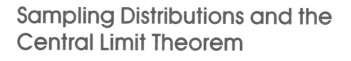

CONTENTS

The computed value of a statistic is dependent upon the sample that is observed. Thus for each different sample there is possibly a different value for the statistic. In other words, prior to sampling, a statistic is a random variable whose distribution corresponds to the different values that the statistic could assume in repeated sampling.

The distributions, called sampling distributions, of selected statistics are presented in this chapter. You will see that many are approximated by the normal distribution. Also, through examples, you will see how sampling distributions can be used to evaluate the generalizations about a population from a sample.

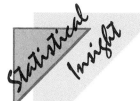

Pollsters Analyze Their Results

By Timothy Kenny

The nation's political pollsters were busy Wednesday poring over their results from Tuesday's election—trying to decide what they did right and wrong, and debating which of the polls was really the "closest" to the outcome.

Gallup Poll correctly predicted Reagan's preliminary winning margin—18 points—with 59 percent of the vote going to President Reagan.

"We predicted it right on the nose," said Gallup's Ann Osborne. "Everybody was pretty close, but nobody hit it right on the mark as we did."

One difference between Gallup and other pollsters was its equal allocation of undecided voters—4 percent in its Nov. 2–3 poll. Each side was simply given half the undecideds.

Other polls, including USA TODAY's, simply reported the undecided votes, without trying to divide them between the candidates.

Some pollsters feel you can't compare the two types of polls.

"I feel throughout this fall there's been an undue concentration on the margin and not enough on the actual numbers," said Laurily Epstein, polling consultant for NBC.

The network predicted Reagan would win 58 percent of the vote and Walter Mondale 34 percent, with 8 percent undecided. Reagan took 59 percent of the vote while Mondale won 41 percent.

Said CBS pollster Warren Mitofsky, who also projected 58 percent for Reagan with an 18 percent gap. "I'm satisfied with what we got. I'm glad we got near the right mark.

"I don't pretend we can do better than that," he added.

USA TODAY pollster Gordon Black, who predicted 60 percent for Reagan with a 25 percent gap and 5 percent undecided, agreed with Epstein.

"The media focuses on the gap," said Black. "If you look at the numbers, four polls predicted within 1 percent the correct Reagan vote."

Those four: NBC, Gallup, USA TODAY and CBS, which said Reagan would win 58 percent of the vote with a 21 percent gap.

Black said variances in the margin between the candidates "was produced in the Mondale numbers."

Humphrey Taylor, president of Louis Harris & Associates, Inc., which predicted Reagan would win 56 percent of the vote with a 12 percent gap, said, "We would have liked to have been closer on the margin. I'm pleased we called the House races right and we were almost the only people to say the turnout would not go up."

Why the polls differed

Allocating "undecided" voters was one reason polls differed from each other:

Poll	Reagan	Mondale	Undecided	Margin
USA TODAY	60	35	5	25
NBC	58	34	8	24
Time	54	30	16	24
CBS/N.Y. Times	58	37	5	21
Gallup	59	41	0	18
Actual result	59	41	—	18
Newsweek	57	40	3	17
ABC/Wash. Post	54	40	6	14
NPR/Harris	56	44	0	12
Roper	52.5	42.5	5	10

SOURCE: *USA Today*, November 7, 1984. Copyright, 1984 *USA Today*. Reprinted with permission.

The preceding article points out that, of nine polls, only the Gallup poll correctly predicted the actual results of the 1984 presidential election. Although this is indeed the case, can we say the other polls were wrong? For instance, the *Newsweek* poll predicted 57% for Reagan, 40% for Mondale, and 3% undecided. Suppose the 3% undecided vote were divided equally among the two candidates, as Gallup did, then the *Newsweek* figures would read

Poll	Reagan	Mondale
Newsweek	58.5%	41.5%

In this case, the *Newsweek* poll is 1/2 of 1% off the actual results. Is this prediction incorrect?

We must realize that this prediction was obtained from a sample of probably less than 2,000 voters. Yet the prediction was within 1/2% of the results from millions of voters. We might say that the prediction was quite accurate. Whenever an estimate is given from a sample there will be a certain amount of error, called the margin of error (not the same as the margin given in the article). In national polls of this type the margin of error usually is from 3% to 5%. Even though the *Time* magazine poll appears to differ substantially from the actual results, we can split the undecided vote and their prediction will be

Poll	Reagan	Mondale
Time	62%	38%

which is within 3% of the actual results.

Following are the results of the same polls with the undecided vote split equally (a common practice) among the two candidates:

Poll	Reagan	Mondale
USA Today	62.5%	37.5%
NBC	62%	38%
Time	62%	38%
CBS/*NY Times*	60.5%	39.5%
Gallup	59%	41%
Actual results	59%	41%
Newsweek	58.5%	41.5%
ABC/*Wash. Post*	57%	43%
NPR/Harris	56%	44%
Roper	55%	45%

All polls except the Roper and *USA Today* are within 3% of the actual results. Note that four of the polls overestimated the percent of voters favoring Reagan and four underestimated the percent favoring Reagan. This is certainly plausible, because the margin of error can be either positive or negative. It is possible that the margin of error in the Roper and *USA Today* polls was ±4%, in which case their predictions were within the margin. Also there might have been reason to divide the undecided in a way other than equally among the two candidates. After all, *USA Today* was within 1% of the actual results for Reagan.

The point of this discussion is that estimates vary from sample to sample, and they do not have to be the same to be valid estimates. The purpose of a poll of this type is to predict the next President of the United States, and *all* of these polls correctly predicted Ronald Reagan by a wide margin.

7.1 Introduction

In Chapter 6 we saw that describing a population distribution involves determining the shape of the distribution, specifying a measure of the center of the distribution, and giving a measure of the variability. The parameters that measure the center and the variability pertain to the entire population, and thus are usually unknown. To study the population further, a sample must be selected and used to estimate the unknown parameters. Because a *statistic* is a numerical quantity calculated from the observations in a sample, we must decide which statistic best estimates the unknown parameter.

Examples of statistics include:

1. $\bar{y}$—the sample mean
2. $\bar{y}_T$—a trimmed sample mean
3. M—the sample median
4. s—the sample standard deviation
5. p—the sample proportion
6. r—the sample correlation coefficient
7. $\bar{y}_1 - \bar{y}_2$—the difference of two sample means
8. $M_1 - M_2$—the difference of two sample medians
9. $p_1 - p_2$—the difference of two sample proportions

Although the value of a population parameter is usually unknown, and may never be known, the value of a statistic is known because it can be calculated from the collected sample. We must realize, however, that the value assumed by the statistic is dependent on the observed sample. The value of the statistic we happen to observe is the value yielded by the sample we collected. A large number of possible samples can be chosen from a population, and each sample will yield its own value of the statistic. Thus there is a distribution of potential values that the statistic can assume, and it is called the *sampling distribution* of that statistic.

Definition

The **sampling distribution of a sample statistic** is the probability distribution associated with the various values that the statistic could assume in repeated sampling.

In an earlier chapter we were concerned with the percent of registered voters that favored an incumbent mayor seeking reelection. The percent of *all* registered voters in favor of the mayor is the parameter π, which we wish to study. The statistic, p, which we will use to estimate the unknown value of π, is the percent of voters in a sample that are in favor of the mayor. Suppose that in a sample of 1,000 registered voters we find that 380 favor the mayor seeking reelection. Then we would say that the statistic realized the value

$$p = 380/1000 = .38$$

Using this as our estimate of π, we are led to believe that approximately 38% of all registered voters favor the mayor seeking reelection.

Had we selected another sample of 1,000 registered voters, it is almost certain that there would *not* have been exactly 380 in favor of the mayor as before. The value of the statistic p will vary from sample to sample. Then we might ask, "Is it reasonable to use the value of $p = .38$ to estimate the unknown value of π?" The answer is yes, if the *sampling variability* of p is not too erratic. It is desirable that a statistic be *stable* from sample to sample. If the statistic assumed widely varying values from sample to sample, then the conclusions we drew about the population parameter would be less than reliable because we did not know which of the different values of the statistic to take as being close to the parameter. Thus it is through the sampling variability that we can determine how precise an estimate of a parameter will be. The amount of variability associated with a statistic is measured by its *standard error.*

Definition

The **standard error of a statistic** is the standard deviation of its sampling distribution.

▲ The Empirical rule states that 95% of a bell-shaped distribution is within two standard deviations of its mean. The standard deviation of the sampling distribution is the standard error, and thus we would expect the estimate to be within two standard errors of the mean.

Suppose, for example, that the standard error of the statistic p used to estimate the percent of registered voters in favor of the mayor is .015. From the Empirical rule we have that the estimated proportion of .38 should be within $2(.015) = .03$ of the true proportion of voters in favor of the mayor. That is, the true proportion should be somewhere between .35 and .41. We call .03, which is 2 times the standard error, the *margin of error* of the estimate. In the sections to follow we will learn how to compute the standard error (and hence the margin of error), and will see that it is dependent upon the sample size. Typically as the sample size increases the margin of error becomes smaller, yielding a more stable estimate.

EXERCISES 7.1

In Exercises 7.1 through 7.5, identify each of the statistics from the list of statistics given at the beginning of this section.

7.1 The median salary of 50 workers in a plant is $320 per week.

7.2 In a 1984 *USA Today* poll of 750 lawyers, 93% said that one needs a lawyer to help draw up a will.

7.3 The median monthly income for households in western states was $245 higher than those in southern states in 1983.

7.4 From a sample of size 2,000, it was found that 60% of the teenagers killed in auto accidents were drinking before their accidents.

7.5 The average weight loss for patients on a certain diet plan was 4.6 pounds.

7.6 Consider a population that is described by the following probability distribution:

y	1	2	3	4	5
p(y)	.1	.2	.3	.3	.1

(a) Find the mean and variance of y.

Suppose a sample of size 2 is chosen from this population. Assuming that the observations are independent, verify that all possible samples of size 2 and their probabilities are as listed:

Sample	Prob.	Sample	Prob.	Sample	Prob.
(1, 1)	.01	(2, 1)	.02	(3, 1)	.03
(1, 2)	.02	(2, 2)	.04	(3, 2)	.06
(1, 3)	.03	(2, 3)	.06	(3, 3)	.09
(1, 4)	.03	(2, 4)	.06	(3, 4)	.09
(1, 5)	.01	(2, 5)	.02	(3, 5)	.03

Sample	Prob.	Sample	Prob.
(4, 1)	.03	(5, 1)	.01
(4, 2)	.06	(5, 2)	.02
(4, 3)	.09	(5, 3)	.03
(4, 4)	.09	(5, 4)	.03
(4, 5)	.03	(5, 5)	.01

(b) Find the sampling distribution of the sample mean, $\bar{y}$.

(c) What is the mean of $\bar{y}$?

(d) What is the standard error of $\bar{y}$?

(e) What is the probability that $\bar{y}$ exceeds 4.5?

7.7 Consider a population that is described by the following probability distribution:

y	1	2	6
$p(y)$	1/3	1/3	1/3

(a) Find the mean and variance of y.

(b) Find the sampling distribution of $\bar{y}$ that is calculated from a random sample of size 3. (Hint: List the possible samples of size 3 as in Exercise 7.6.)

(c) Find the sampling distribution of M, the sample median, which is found from a sample of size 3.

(d) Which statistic, $\bar{y}$ or M, do you think would be the better measure of the center of this distribution? Explain.

7.2 The Sampling Distribution of $\bar{y}$ and the Central Limit Theorem

In the previous section we observed that the value of a statistic will vary from sample to sample. This sampling variability is described by the sampling distribution of the statistic. In this section we will study the sampling distribution of the sample mean $\bar{y}$. In Section 7.3 we will consider the sampling distribution of the sample proportion, p.

Consider, for the moment, that we have a population that is normally distributed with a mean of 50 and a standard deviation of 15. The distribution curve is given in Figure 7.1.

FIGURE 7.1

Population distribution with mean 50 and standard deviation 15

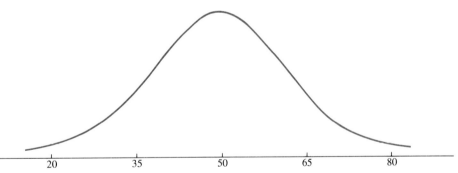

If we sample from this population, most observations will be between 20 and 80, because 20 is two standard deviations below the mean of 50 and 80 is two standard deviations above the mean of 50. In Figure 7.2 we see a stem and leaf plot of a sample of 100 observations selected at random from the population, with the frequency distribution curve superimposed over the stem and leaf plot.

We see that most observations do indeed fall between 20 and 80, with only a few falling outside those limits. Moreover the observations tend to follow the frequency distribution of the population as would be expected.

The average of this sample is 50.5, which is very close to 50, the mean of the population. Certainly another sample of 100 observations will not yield the same stem and leaf plot nor the same average, but one can expect to see something similar. In fact,

FIGURE 7.2

Population distribution with mean 50 and standard deviation 15 superimposed over a stem and leaf plot of 100 observations

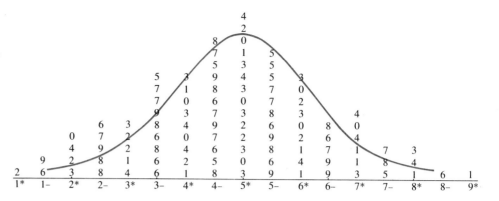

the next sample taken yielded an average of 49.8, which again is close to 50. Other samples will yield still different sample averages, and thus we have a distribution of potential values that $\bar{y}$ could assume in repeated sampling. Then we might ask:

1. What is the general shape of the distribution of potential values $\bar{y}$ can assume?
2. Are the potential values of $\bar{y}$ centered about a certain quantity?
3. How much variability is associated with the potential values of $\bar{y}$?

These questions are answered with a description of the sampling distribution of $\bar{y}$.

Prior to sampling, the statistic can be thought of as a random variable, because different samples can lead to different values of the statistic. Its sampling distribution is then the probability distribution of that random variable. Having the probability distribution of a random variable, we are able to find the mean, variance, and probabilities associated with the random variable. Consequently if we know the sampling distribution of $\bar{y}$, we can find its mean, variance, and the probabilities associated with the various values that it can assume. That is, we can answer the three questions posed earlier.

▲ Remember that all random variables have a probability distribution.

Let us take a closer look at the sample mean, $\bar{y}$, and the population described in Figure 7.1. Recall that the first random sample of size 100 had an average of 50.5, and a second random sample of 100 had an average of 49.8. What we have are two *realizations* of the random variable $\bar{y}$. Table 7.1 gives 50 realizations of the random variable $\bar{y}$ obtained from 50 random samples each of size 100.

TABLE 7.1

Fifty realizations of the random variable $\bar{y}$

50.5	49.8	50.3	49.7	50.2	50.0	48.9	49.5	51.4	50.9
52.2	51.6	50.1	50.2	51.6	46.4	51.4	48.7	52.3	49.2
45.6	50.6	51.9	53.3	54.7	44.3	48.5	51.1	50.5	49.8
47.9	50.1	50.3	49.6	50.4	49.1	48.4	55.1	50.0	49. 3
52.6	48.3	49.9	50.0	50.8	51.4	49.5	50.7	49.7	50.6

A histogram of these 50 $\bar{y}$'s will give the general appearance of the theoretical sampling distribution of $\bar{y}$. Figure 7.3 is such a histogram superimposed over the frequency distribution curve of the parent population.

Notice that the histogram of values of $\bar{y}$ tends to mound up around 50, the mean of the parent population. Also observe that the standard deviation of the parent population is 15, but the sampling distribution of $\bar{y}$ doesn't tend to be spread out nearly that much. In fact, approximately 95% of the sample averages are between 47 and 53.

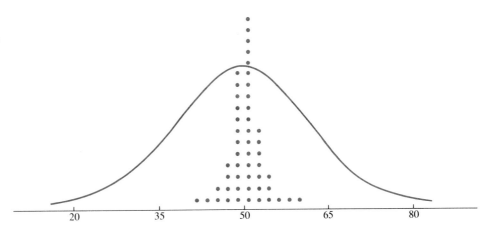

If we think about it, the 100 observations in each sample are somewhere between 20 and 80, so each average should be around 50, the mean of the parent population. Of course, some samples will give averages that are somewhat removed from 50, but most should cluster around 50.

The amount of variability associated with the sampling distribution is *measured* by its standard error. Working backward we see that, if 95% of the $\bar{y}$'s range is from 47 to 53, then the standard error will be about 1.5. (If the standard deviation is 1.5, then two standard deviations on either side of 50 will range from

$$50 - 2(1.5) = 47 \quad \text{to} \quad 50 + 2(1.5) = 53)$$

We will see later that the standard error is given by

$$\sigma/\sqrt{n} = 15/\sqrt{100} = 1.5$$

We are now in a position to answer the three questions about the sampling distribution of $\bar{y}$ when sampling from a normally distributed population with mean 50 and standard deviation of 15.

1. The general shape appears to be bell shaped; in fact it is normally distributed.
2. The distribution of potential values tends to be clustered around 50, the mean of the parent population.

▲ We use the term standard error for the standard deviation of the sampling distribution so as not to confuse it with the standard deviation of the parent population.

3. The standard error appears to be

$$\sigma/\sqrt{n} = 15/\sqrt{100} = 1.5$$

where σ is the standard deviation of the parent population.

The preceding observations are based on a specific example. In particular, the parent population is normally distributed centered at 50, with a standard deviation of 15 and a sample size of 100. However the results can be generalized to arbitrary populations (i.e., not necessarily bell shaped) with any mean μ and any standard deviation σ.

The Central Limit Theorem

One of the most important theorems in statistics, called the Central Limit Theorem (CLT), verifies all that we have observed about the sampling distribution of $\bar{y}$ when sampling from arbitrary populations.

The graphs in Figure 7.4 provide a visual representation of the Central Limit Theorem as it applies to the sampling distribution of $\bar{y}$. The graphs in Figure 7.4(a) represent the distributions of four different parent populations. Figure 7.4(b) shows the sampling distributions of $\bar{y}$ when only two observations are taken. Figure 7.4(c) shows the

FIGURE 7.4

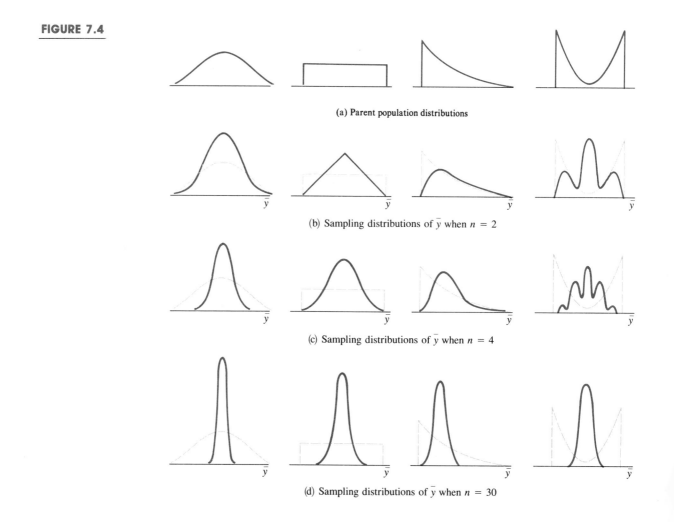

(a) Parent population distributions

(b) Sampling distributions of $\bar{y}$ when $n = 2$

(c) Sampling distributions of $\bar{y}$ when $n = 4$

(d) Sampling distributions of $\bar{y}$ when $n = 30$

sampling distributions of $\bar{y}$ based on a random sample of size 4; and finally, Figure 7.4(d) shows the sampling distributions of $\bar{y}$ based on a random sample of size 30. At each step the sampling distribution depends on the parent population distribution, but as the sample size increases, the sampling distribution of $\bar{y}$ becomes more mound shaped, less variable, and less dependent on the parent distribution. When the sample size is large, all the sampling distributions of $\bar{y}$ look similar even though the parent populations are very different. For *n larger than 30*, the sampling distribution of $\bar{y}$ appears to be approximately normally distributed and centered at the mean of the parent population as is stated in the Central Limit Theorem.

Note that in Figure 7.4, the first population is a normal population as considered in the earlier discussion. An important point is that the sampling distribution of $\bar{y}$ is exactly normal for *all* sample sizes when sampling from a normal population, but is only approximately normal when sampling from a nonnormal population. The approximation becomes better as the sample size becomes larger. The more the parent population deviates from normality, the larger the sample size must be in order to get a reasonable approximation. In any case, a sample size in excess of 30 seems sufficient.

▲ In the future we will use the sample size of 30 as the dividing line between large samples and small samples.

In the remainder of this section we will give some applications of the Central Limit Theorem.

EXAMPLE 7.1

A social worker would like to know something about the income level of welfare recipients in her state. The mean income level of all welfare recipients in the state is the unknown population parameter, μ, which she would like to estimate. She decides to take a sample of 100 welfare recipients and calculate the sample mean, $\bar{y}$, and use it as an estimate of μ. The sample average turns out to be \$8,380 and she then estimates the mean income level of all welfare recipients to be \$8,380. If the standard deviation of the income level of all welfare recipients is \$2,000—that is, $\sigma = 2,000$—how close will her estimate be to the true mean income level of all welfare recipients?

SOLUTION   Because $\sigma = 2,000$ and $n = 100$ the standard error of $\bar{y}$ is

▲ We will often write the standard error of $\bar{y}$ as $SE(\bar{y})$.

$$SE(\bar{y}) = \sigma/\sqrt{n} = 2,000/\sqrt{100} = 200$$

Thus we can be reasonably certain that the observed $\bar{y}$ will be within \$400 ($2 \times SE = 2 \times 200 = 400$) of the population mean. That is, the estimate of \$8,380 should be within \$400 of its target.

Knowing the sampling distribution of the sample mean, $\bar{y}$, we are able to evaluate a $\bar{y}$ once it is calculated from a sample.

EXAMPLE 7.2

Suppose census data indicate that the distribution of annual income for schoolteachers in the United States has a mean of \$18,000 and a standard deviation of \$4,000. This means that the majority of schoolteachers earn somewhere between \$10,000 and \$26,000 (two standard deviations below and above the mean). Now suppose we randomly select 64 schoolteachers from a certain state, and find that their average income is \$16,500. Can anything be said about the income level of schoolteachers in that state?

SOLUTION   Suppose the salaries in this state are comparable with the salaries across the United States. Then we have, from the Central Limit Theorem, that the sampling distribution of $\bar{y}$ (obtained from samples of size 64 from this state) is approximately normally distributed with a

$$\text{mean} = 18,000 \quad \text{and} \quad \text{standard error} = 4,000/\sqrt{64} = 500$$

Thus 95% of all potential $\bar{y}$'s should fall between 17,000 and 19,000 (within two standard errors of 18,000). We observed a $\bar{y} = 16,500$, a full three standard errors below the mean. This is unusual if the mean salary in the state is $18,000. We are led to believe that the mean of the distribution of teachers' salaries in this state is not $18,000, but is somewhat below the national average.

Probability problems involving $\bar{y}$ can be solved by computing the standardized version of $\bar{y}$.

Standardized Version of $\bar{y}$

For a sufficiently large sample size,

$$z = \frac{\bar{y} - \mu}{\sigma/\sqrt{n}}$$

has an approximate *standard* normal distribution.

EXAMPLE 7.3

The average length of stay in a certain AA clinic is 17 days, and the standard deviation of the length of stay is 3 days. From a random sample of 36 patients find

(a) the probability that their average stay is more than 18.5 days.

(b) the probability that their average stay is between 16 and 19 days.

SOLUTION

(a) We have from the Central Limit Theorem that $\bar{y}$ is approximately normally distributed with a mean $= 17$ and a standard error $= \sigma/\sqrt{n} = 3/\sqrt{36} = .5$.

Then to find the probability that $\bar{y} > 18.5$, we find the z-score associated with 18.5 as follows

$$z = \frac{18.5 - 17.0}{.5}$$

$$= 1.5/.5$$

$$= 3.0$$

▲ We use the standard normal probability tables because $\bar{y}$ is approximately normally distributed, and z is the standardized version of it.

Looking in the standard normal probability table, we find that .4987 of the area lies from the mean to three standard deviations above the mean. Therefore the probability of observing a $\bar{y} > 18.5$ will be

$$.5000 - .4987 = .0013$$

which is extremely small.

(b) To find the probability that $\bar{y}$ lies somewhere between 16 and 19, we must work two problems—(1) find the area from 16 to 17, and (2) the area from 17 to 19, and then add the two together.

The z-score associated with 16 is

$$z = \frac{16.0 - 17.0}{.5}$$
$$= -1.0/.5$$
$$= -2.0$$

From the normal probability tables we find an area of .4772 associated with a z-score of 2. The z-score associated with 19 is

$$z = \frac{19.0 - 17.0}{.5}$$
$$= 2.0/.5$$
$$= 4.0$$

From the normal probability tables we find an area of .49997 associated with a z-score of 4. Consequently the desired probability is

$$.4772 + .49997 = .97717$$

which is very large, indicating that it is very likely that their average length of stay is somewhere between 16 and 19 days. In fact, with approximately 95% probability, it is somewhere between 16 and 18 days (two standard errors on either side of 17).

The sampling distribution also allows us to determine which of the potential values of the statistic are reasonable and which are unlikely. Thus when a value is obtained for a particular statistic, knowing the sampling distribution of that statistic allows us to determine immediately if that value is in agreement with our ideas about the population.

EXAMPLE 7.4

Returning to the welfare example given earlier in this section, suppose a conjecture has been made that the mean gross income of all welfare recipients is $9,000; that is, assume $\mu = \$9,000$. The 100 individuals selected at random had an average gross income of $8,380 (i.e., $\bar{y} = \$8,380$). Does that seem reasonable?

SOLUTION   If we also know that the standard deviation of the parent population is $\sigma = \$2,000$, the Central Limit Theorem says that $\bar{y}$ is approximately normally distributed with

$$\text{mean} = 9,000$$

and

$$\text{standard error} = \sigma/\sqrt{n}$$
$$= 2{,}000/\sqrt{100}$$
$$= 2{,}000/10$$
$$= 200$$

So we know from the Central Limit Theorem and the Empirical rule that almost all possible $\bar{y}$'s should be between

$$9{,}000 - 3(200) \quad \text{and} \quad 9{,}000 + 3(200)$$

that is between

$$\$8{,}400 \quad \text{and} \quad \$9{,}600$$

We obtained a $\bar{y}$ of \$8,380. Assuming that the 100 individuals selected at random are truly representative of the population, then the original conjecture that μ is \$9,000 must be wrong.

On the other hand, if $\bar{y}$ had been, say \$8,900, the conclusion is not so obvious. The test of hypothesis procedures in Chapter 9 will tell us what to do in this situation.

$$z = \frac{200 - 220}{5} = -4$$

EXERCISES 7.2

7.8 A population has mean 500 and standard deviation 100. A sample of size 200 is randomly selected from the population. Describe the sampling distribution of the sample mean $\bar{y}$.

7.9 A population has mean μ and we wish to estimate it. So we take a sample and calculate the sample mean $\bar{y}$, and use it to estimate μ. What is the accuracy in that estimate; that is, in general, how close will $\bar{y}$ be to μ?

7.10 A random sample of size n is selected from a population with a mean 25 and a standard deviation of 8. For each of the following values of n, give the mean and standard error of the sampling distribution of $\bar{y}$.

(a) $n = 10$ (b) $n = 16$
(c) $n = 30$ (d) $n = 100$

In which of these will the sampling distribution be adequately approximated by a normal distribution?

7.11 A random sample of size 100 is selected from a population with mean μ and standard deviation σ. For each of the following values of μ and σ, find the mean and standard error of the sampling distribution of $\bar{y}$.

(a) $\mu = 10, \sigma = 2$ (b) $\mu = 10, \sigma = 4$
(c) $\mu = 50, \sigma = 12$ (d) $\mu = 50, \sigma = 24$

7.12 A random sample of size 100 is selected from a population with mean 200 and standard deviation 50. Approximate the probability that

(a) $\bar{y}$ exceeds 220
(b) $\bar{y}$ is less than 193
(c) $\bar{y}$ is between 203.4 and 209.6

7.13 Suppose that, over a period of time, we took a large number of samples each of size 100 from a population and calculated the average, $\bar{y}$, of each sample. If the $\bar{y}$'s concentrated around 8,000 and varied from 7,600 to 8,400, give an approximate value of the standard error of $\bar{y}$.

7.14 Suppose a population has a mean $\mu = 8{,}000$ and a standard deviation $\sigma = 2{,}000$. Is it reasonable to expect a $\bar{y}$ of 8,380 from a sample of size 100?

7.15 A random sample of size 200 is selected from a population with mean 70 and standard deviation 15.

(a) Would you expect to see a $\bar{y}$ in excess of 75?
(b) What is the largest $\bar{y}$ you normally would observe?

7.16 The average number of days spent in the hospital for patients assigned to a particular surgical ward is 9 days and the standard deviation is 4 days. What is the probability that a random sample of 30 patients will have an average stay in excess of 9.5 days?

7.17 The average length of a field goal in the NFL is 38.2 yards, and the standard deviation is 6.4 yards. Suppose a typical kicker kicks 41 times in one season. What is the probability he averages less than 37 yards?

7.18 It was reported that for the 1984 Christmas holidays, the average price per gallon of self-service no-lead gasoline in the southeastern states was $1.18 and the standard deviation was 6 cents. A check of 36 randomly selected stations revealed an average price of $1.21. Did this average exceed $1.18 purely by chance or is there statistical evidence that the $1.18 is low?

7.19 The scores on a standardized test have a national mean of 70 and a standard deviation of 10. Is it unusual for a sample of 50

subjects to have a sample mean score above 75? To answer this question find the probability that 50 subjects will have a sample mean score above 75.

7.20 A normally distributed population has $\mu = 50$ and $\sigma = 10$. What is the probability that

(a) a single score, selected at random, will be between 48 and 54?

(b) the mean of a sample of 25 will have a value between 48 and 54?

7.21 A particular standardized psychological exam has a mean of 70 and a standard deviation of 5.

(a) Assuming that the distribution is normal, what is the probability one will score somewhere between 68 and 74?

(b) What is the probability that 36 subjects will have an average somewhere between 68 and 74?

7.3   Sampling Distribution of the Sample Proportion p

Recall that a Bernoulli population is one in which each element is either a success or failure. Moreover we generally are interested in the proportion of successes, denoted by π. If we wish to estimate π from a sample, it seems reasonable to calculate the sample proportion p. We now study the sampling distribution of this statistic, p.

In Section 7.1, we saw that a sample of 1,000 registered voters revealed that 380 favored the mayor seeking reelection. Thus we obtained a sample proportion of

$$p = 380/1000 = .38 = 38\%$$

We also observed that a second sample of 1,000 most likely would not result in exactly 380 positive responses. So just as $\bar{y}$ varies from sample to sample, so does the sample proportion p. This variability is described by its sampling distribution. As with $\bar{y}$, prior to sampling, p is a random variable and thus has a probability distribution. We call it a sampling distribution because it is arrived at by repeated sampling.

The three questions we posed with regard to the sampling distribution of $\bar{y}$ may also be asked of the sampling distribution of p; namely,

1. What is the general shape of the distribution of potential values that p can assume?
2. Are the potential values that p can assume centered about a certain quantity?
3. How much variability is there associated with the potential values that p can assume?

The following version of the Central Limit Theorem applies to the sample proportion.

> **Central Limit Theorem Applied to p**
>
> If the sample size, n, is sufficiently large, the sampling distribution of p will
>
> 1. be approximately normally distributed
>
> 2. be centered at π, the true proportion of successes in the Bernoulli population
>
> 3. have a standard error of $\sqrt{\pi(1-\pi)/n}$
>
> In practice, the sample size should be large enough so that $n\pi \geq 5$ and $n(1-\pi) \geq 5$.

The following example illustrates the Central Limit Theorem as it applies to p.

EXAMPLE 7.5

Suppose we have a Bernoulli population with $\pi = .4$. A computer selects a random sample of size 1,000 and calculates p. The value obtained is $p = .38$. This process is repeated 100 times, and consequently results in 100 realizations of p. A histogram of these 100 p's is shown in Figure 7.5. Observe that the histogram appears to be approximately normally distributed, centered around .4, the true proportion in the population, and exhibits very little variability. In other words, most of the potential values of the sample proportion p are very close to π.

FIGURE 7.5

A histogram of 100 realizations of the sample proportion p

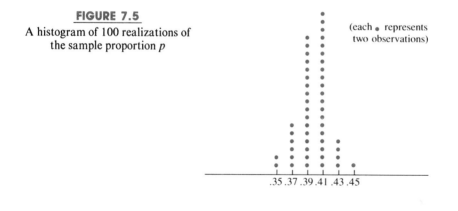

(each • represents two observations)

EXAMPLE 7.6

A Bernoulli population has parameter $\pi = .7$. If the sample proportion p is calculated from a sample of size 1,600, how is p distributed? What is its mean? What is its standard error?

SOLUTION   Because 1,600 is a large sample size, the CLT says that the sampling distribution of p is approximately normal with a mean of .7 (the same as π) and a standard error of

$$SE(p) = \sqrt{\pi(1 - \pi)/n}$$
$$= \sqrt{(.7)(.3)/1600}$$
$$= \sqrt{.21}/40$$
$$= .0115$$

In the preceding example, the standard error was $\sqrt{\pi(1 - \pi)/n} = .0115$. What would the standard error be if π had been some other value? Consider the following table of values of the standard error for various values of π when $n = 1,600$.

π	$SE(p) = \sqrt{\pi(1 - \pi)/n}$
.2	.01
.3	.0115
.4	.0122
.5	.0125
.6	.0122
.7	.0115
.8	.01

We see that the standard error increases until we get to $\pi = .5$, and then it decreases. The maximum standard error occurs when $\pi = .5$. This is true for any sample size. Also note the symmetry in the table; the standard error is the same for $\pi = .2$ and .8, the same for $\pi = .3$ and .7, and the same for $\pi = .4$ and .6.

EXAMPLE 7.7

A Bernoulli population has an unknown parameter π. A random sample of size $n = 400$ gives a sample proportion $p = .62$. What value do you think π is?

SOLUTION   Because a sample of size 400 is large, the CLT says that the distribution of p tends to mound up around π. Moreover the standard error is maximum when $\pi = .5$, so we know that the standard error is no more than

$$\sqrt{(.5)(.5)/400} = .5/20 = .025$$

Thus 95% of all p should lie within $2(.025) = .05$ of the true value of π. So we are led to believe that our value of $p = .62$ should be very close to π. In fact, we can estimate π to be $.62 \pm .05$.

EXAMPLE 7.8

It has been reported that 46% of all U.S. homes have more than one television set. Out of a random sample of 500 homes, what is the probability that less than half have more than one television set?

SOLUTION   The p calculated from the sample of 500 homes will have a sampling distribution that is approximately normal with a mean of .46 and a standard error of

$$\sqrt{(.46)(.54)/500} = .0223$$

The probability that less than half of the homes have more than one TV is written as

$$P(p < .5)$$

To find this probability we must find the z-score corresponding to .5 as follows:

$$z = \frac{.5 - .46}{\sqrt{(.46)(.54)/500}} = \frac{.04}{.0223} = 1.79$$

From Table B3 we find the associated probability to be .4633 so that

$$P(p < .5) = .5 + .4633 = .9633$$

It is highly probable that fewer than half will have more than one TV set.

EXAMPLE 7.9

Suppose it has been reported that 40% of the adult population in the United States believe that abortion is acceptable under any circumstances. In a random sample of 1,000 adults, 450 said abortion is acceptable under any circumstance. Does this cast doubt on the prior claim?

SOLUTION   If the true proportion of adults who believe that abortion is acceptable is 40%, then the CLT says that the sampling distribution of the random variable p is approximately normal with mean .4 and standard error $\sqrt{(.4)(.6)/1000} = .0155$. So we would expect approximately 95% of the potential p's to be between

$$.4 - 2(.0155) \quad \text{and} \quad .4 + 2(.0155)$$

which reduces to

$$.369 \quad \text{and} \quad .431$$

However the p we obtained from our sample was

$$p = 450/1000 = .45$$

which is more than two standard errors above .4. Consequently we have sufficient evidence to question the validity of the claim that $\pi = .40$.

This section and the previous section described the sampling distributions of two often used statistics. This will be useful information as we proceed to statistical inferences about the two population parameters, μ and π. In order to conduct a statistical inference we must first determine an appropriate sample size. This topic will be investigated in the next section.

EXERCISES 7.3

7.22 A sample of 400 observations is randomly selected from a Bernoulli population with parameter $\pi = .8$. Describe the sampling distribution of the sample proportion p.

7.23 A random sample of size n is selected from a Bernoulli population with 70% successes. For each of the following values of n, give the mean and standard error of the sampling distribution of p.

(a) $n = 100$ (b) $n = 400$

(c) $n = 1000$ (d) $n = 1600$

7.24 A random sample of 1,600 is selected from a Bernoulli population. For each of the following values of π, find the mean and standard error of the sampling distribution of p.

(a) $\pi = .1$ (b) $\pi = .3$

(c) $\pi = .5$ (d) $\pi = .7$

(e) $\pi = .9$ (f) $\pi = .575$

7.25 A random sample of $n = 400$ is selected from a Bernoulli population with 80% successes. Approximate the probability that the sample proportion, p,

(a) exceeds 78%

(b) is less than 76%

(c) is between .76 and .83

7.26 A random sample of size $n = 100$ is selected from a Bernoulli population with $\pi = .3$.

(a) Would you expect to see a p in excess of .4?

(b) What is the largest p you normally would expect to see from the sample?

7.27 A random sample of 1,000 adults revealed that 23% of the residents in western states dine out using a credit card. How close would the 23% be to the true proportion of westerners who dine out using a credit card?

7.28 A drug company claims that DMSO will reduce arthritis pain in 80% of all cases. What is the probability that 75 or more out of 100 sufferers of arthritic pain will experience relief, if the drug company's claim is true?

7.29 It is reported that 70% of the population does not have adequate hospitalization coverage. In a large hospital 200 patients were admitted in one week. What is the probability that more than 125 do not have adequate coverage?

7.30 The poverty line for a family of 4 was $10,178 in 1980. Fifty-three percent of all families in Tunica, Mississippi, are below that poverty line. Is it unusual that a sample of 100 families in Tunica will have more than 60 families below the poverty line?

7.31 Forty-one percent of all applicants passed the California bar exam in 1983. Out of a group of 100 applicants, what is the probability that over half passed?

7.32 A survey of 40,000 working women by *McCall's* magazine revealed that 80% are married. In a random sample of 1,000 working women, what is the probability that fewer than 750 are married?

7.33 President Reagan got 59% of the popular vote in the November 1984 election for a landslide victory. A random sample of 400 voters is selected. What is the probability that more than 240 voted for Reagan?

7.34 A presidential aide stated that over half of the nation agreed with the invasion of Grenada. However a Roper poll of 2,000 Americans found only 34% supporting the invasion. Did the 34% simply happen by chance or is there statistical evidence that the aide is in error?

7.35 One third of all ex-convicts return to jail within three years. One state is trying a new rehabilitation system for its prisoners. After releasing 200 prisoners who had participated in the new system, only 48 have returned within three years. Is there statistical evidence that the new system is working to reduce the proportion of repeat offenders?

The Relation between Sample Size and Sampling Distribution

A question often asked when one begins to sample a population is,

How large of a sample do I need?

The answer depends on the accuracy desired when estimating unknown quantities. We will see that the more accuracy desired, the larger the sample size will have to be.

Recall that a statistic such as $\bar{y}$ or p varies from sample to sample, and thus will not consistently give the same value. So we ask, how closely will it approximate the true parameter value when the sample is taken? The answer lies in the sampling distribution and its standard error.

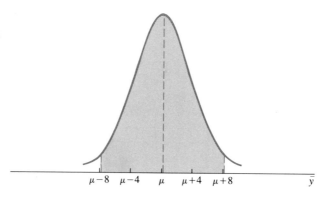

For example, when the sample size is large, the sampling distribution of $\bar{y}$ is approximately normal about μ (the population mean) with standard error $\sigma/\sqrt{n}$. Figure 7.6 gives the general shape of this sampling distribution when $n = 100$ and $\sigma = 40$. The shaded area illustrates that most $\bar{y}$'s (95% of them) fall within

$$2(\sigma/\sqrt{n}) = 2(4) = 8 \text{ units of } \mu$$

What would happen to this sampling distribution if n were increased to 400? It is still approximately normally distributed around μ, but the standard error becomes

$$\sigma/\sqrt{n} = 40/\sqrt{400} = 40/20 = 2$$

So now 95% of the $\bar{y}$'s computed from samples of size 400 will fall within

$$2(\sigma/\sqrt{n}) = 2(2) = 4 \text{ units of } \mu$$

Thus the sampling distribution is more tightly clustered about μ, the true mean of the parent population, when the sample size is increased to 400. Figure 7.7 shows the two sampling distributions of $\bar{y}$—one based on a sample of size 100 and the other on a sample of size 400.

We see that with a sample of size 400 we tend to get a more accurate estimate of μ than if n were only 100. The accuracy improves as the sample size increases. This is clearly illustrated by the fact that n appears in the denominator of the standard error $\sigma/\sqrt{n}$. We see that as n increases, the standard error becomes smaller, hence the greater accuracy.

▲ Although accuracy improves with a larger sample, it may be that the gain in accuracy is not worth the extra cost associated with taking a larger sample.

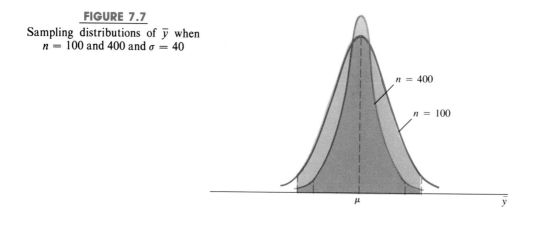

The same is true of the sampling distribution of p, because again it is approximately normally distributed about π and the standard error is

$$\sqrt{\pi(1 - \pi)/n}$$

Again n appears in the denominator. Thus as n increases, the standard error decreases. Figure 7.8 gives two sampling distributions of p, one based on a sample of size 400 and the other based on size 1,600. Clearly p is more closely distributed about π when the sample size is 1,600.

FIGURE 7.8

Sampling distributions of p when $n = 400$ and 1,600

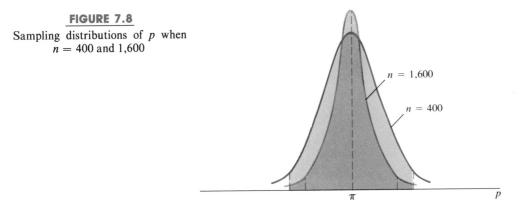

EXAMPLE 7.10

A Bernoulli population has unknown parameter π. What is the maximum standard error of p if the sample size is $n = 100$? What if $n = 400$?

SOLUTION   The standard error of p is given by

$$SE(p) = \sqrt{\pi(1 - \pi)/n}$$

This quantity is maximized when $\pi = .5$. Hence the maximum standard error when $n = 100$ is

$$\max SE(p) = \sqrt{(.5)(.5)/100} = .5/10 = .05$$

and when $n = 400$ is

$$\max SE(p) = \sqrt{(.5)(.5)/400} = .5/20 = .025$$

We see that the sample size must be increased fourfold in order that the standard error be halved.

EXAMPLE 7.11

The Stanford-Binet IQ test has a standard deviation of $\sigma = 16$. How large is the standard error of $\bar{y}$ if we have a sample of size $n = 100$? If $n = 400$?

SOLUTION   The standard error of $\bar{y}$ is given by

$$SE(\bar{y}) = \sigma/\sqrt{n}$$

Because $\sigma = 16$ and $n = 100$, the standard error becomes

$$\text{SE}(\bar{y}) = \sigma/\sqrt{n} = 16/\sqrt{100} = 1.6$$

If $n = 400$, it becomes

$$\text{SE}(\bar{y}) = 16/\sqrt{400} = .8$$

In summary, we can say that the sampling distributions of both $\bar{y}$ and p tend to cluster more tightly about μ and π respectively, when the sample size is increased. Thus estimates are more accurate when the sample size is increased. In Chapter 8 we will develop the formulas that are used to determine the sample size for a specified amount of accuracy in specific estimation problems.

EXERCISES 7.4

7.36 Determine the maximum standard error of p for the following sample sizes.
(a) 25   (b) 100   (c) 200
(d) 500   (e) 1,000   (f) 2,000

7.37 Suppose a population of measurements has a standard deviation of 20. Determine the standard error of $\bar{y}$ for the following sample sizes.
(a) 25   (b) 50   (c) 100
(d) 200   (e) 500   (f) 1,000

7.38 A new drug is proposed as a cure for lung cancer. A sample of 100 patients is to be tested. What is the maximum standard error of the proportion of successfully treated patients? What would be the maximum standard error of the estimate if the sample size is 900?

7.39 A standardized science test has a national mean of 250 and a standard deviation of 50. The test is to be given to a group of prospective science majors. What is the standard error of the estimate, $\bar{y}$, calculated from a sample of size 64? from a sample of size 400?

7.5   Sampling Distribution of the Difference of Two Statistics

In the chapters to follow we will see the importance of being able to compare the difference in two parameters. For example, we might wish to investigate the difference in the mean retirement ages of men and women. Clearly we would have to investigate the difference in two sample means—the mean retirement age for a random sample of men and the mean retirement age for a random sample of women. Specifically to compare the means of two populations, we need to compare the means of two independent random samples taken from the two populations. To assess how close the difference in the two sample means will be to the difference in the two population means, we need to investigate the sampling distribution of

$$\bar{y}_1 - \bar{y}_2$$

where $\bar{y}_1$ is the mean of a random sample from population 1 and $\bar{y}_2$ is the mean of an independent random sample from population 2.

Suppose population 1 has mean μ_1 and standard deviation σ_1 and population 2 has mean μ_2 and standard deviation σ_2. Further suppose that two independent random samples of size n_1 and n_2 are taken respectively from the two populations. If n_1 and n_2 are sufficiently large ($n_1 > 30$ and $n_2 > 30$), we have from the Central Limit Theorem that the sampling distributions of $\bar{y}_1$ and $\bar{y}_2$ are both approximately normally distributed. Because the samples are assumed to be independent, it can be shown that the difference $\bar{y}_1 - \bar{y}_2$ is approximately normally distributed. Further if the mean of the sampling distribution of $\bar{y}_1$ is μ_1 and the mean of the sampling distribution of $\bar{y}_2$ is μ_2, it follows that the mean of the sampling distribution of $\bar{y}_1 - \bar{y}_2$ is $\mu_1 - \mu_2$.

The standard errors of $\bar{y}_1$ and $\bar{y}_2$ are $\sigma_1/\sqrt{n_1}$ and $\sigma_2/\sqrt{n_2}$, respectively; again assuming independence of the two samples, it can be shown that the standard error of $\bar{y}_1 - \bar{y}_2$ is

$$SE(\bar{y}_1 - \bar{y}_2) = \sqrt{\frac{\sigma_1^2}{n_1} + \frac{\sigma_2^2}{n_2}}$$

These results are summarized as follows.

Sampling Distribution of $\bar{y}_1 - \bar{y}_2$

Two independent random samples of sizes n_1 and n_2 are selected from two populations that have means μ_1 and μ_2 and standard deviations σ_1 and σ_2, respectively. If n_1 and n_2 are each larger than 30, the sampling distribution of the difference in the sample means, $\bar{y}_1 - \bar{y}_2$, is

1. approximately normally distributed

2. centered at $\mu_1 - \mu_2$

3. with a standard error of

$$\sqrt{\frac{\sigma_1^2}{n_1} + \frac{\sigma_2^2}{n_2}}$$

EXAMPLE 7.12

A manufacturer of a new gasoline additive claims that his product will increase gas mileage by 3 miles per gallon. Suppose μ_1 denotes the mean miles per gallon for a particular type automobile with the additive, and μ_2 denotes the mean miles per gallon without the additive. Suppose also that $\sigma_1 = \sigma_2 = 8$ miles per gallon. If samples of size 50 and 54 are taken with and without the additive and the average miles per gallon computed, describe the sampling distribution of $\bar{y}_1 - \bar{y}_2$.

SOLUTION The claim is that the additive will increase the mileage by 3 miles per gallon; therefore $\mu_1 - \mu_2 = 3$. Furthermore we have

$$\sqrt{\frac{\sigma_1^2}{n_1} + \frac{\sigma_2^2}{n_2}} = \sqrt{\frac{64}{50} + \frac{64}{54}} = 1.57$$

Now we have that the sampling distribution of $\bar{y}_1 - \bar{y}_2$ is approximately normally distributed with a mean of 3 and a standard error of 1.57.

We also wish to investigate the sampling distribution of the difference in two sample proportions. Assume Bernoulli population 1 has proportion π_1 of successes and Bernoulli population 2 has proportion π_2 of successes. Two independent random samples of size n_1 and n_2 from the two populations will yield two sample proportions, p_1 and p_2.

The sampling distribution of the difference in sample proportions is summarized as follows.

Sampling Distribution of $p_1 - p_2$

If n_1 and n_2 are sufficiently large, then the sampling distribution of $p_1 - p_2$ is

1. approximately normally distributed

2. centered at $\pi_1 - \pi_2$

3. with a standard error of

$$SE(p_1 - p_2) = \sqrt{\frac{\pi_1(1 - \pi_1)}{n_1} + \frac{\pi_2(1 - \pi_2)}{n_2}}$$

▲ Recall that a sample size n is sufficiently large to estimate a population proportion π if $n\pi \geqslant 5$ and $n(1 - \pi) \geqslant 5$.

EXAMPLE 7.13

Simmons Market Research Bureau claims that 36% of all adult men and 26% of women read magazines in the bathroom (*USA Today,* October 9, 1985). If these figures are accurate, describe the sampling distribution of the difference in sample proportions from samples of size 200 and 180 of men and women, respectively.

SOLUTION   The claim is that π_1 is .36 and π_2 is .26, so that $\pi_1 - \pi_2 = .10$. Also we have

$$SE(p_1 - p_2) = \sqrt{\frac{\pi_1(1 - \pi_1)}{n_1} + \frac{\pi_2(1 - \pi_2)}{n_2}}$$

$$= \sqrt{\frac{(.36)(.64)}{200} + \frac{(.26)(.74)}{180}}$$

$$= .047$$

So the sampling distribution of $p_1 - p_2$ is approximately normally distributed with a mean of .10 and a standard error of .047.

Computer Session (optional)

The computer can be used to approximate the sampling distribution of a statistic. The idea is to have the computer draw a sample from a given population and then calculate the statistic. Repeating the procedure, we can generate several realizations of the statistic that can be graphed to give an approximation to the sampling distribution. For example, we can use RANDOM with the subcommand NORMAL to generate observations from a normal population. The values generated can then be averaged to give a single $\bar{y}$. Repeating this, we can generate as many $\bar{y}$'s as we desire. A histogram of the generated $\bar{y}$'s will give an approximation of the sampling distribution of $\bar{y}$ taken from a normal population. The subcommand UNIFORM can be used to generate data from a uniform population and thus show that the sampling distribution of $\bar{y}$ is approximately normal when sampling from a nonnormal population.

EXAMPLE 7.14

Perform a simulation study to illustrate the Central Limit Theorem. Generate a sample of size 12 from a normal population with mean 50 and standard deviation 10, and then calculate the sample mean. Repeat the process 100 times and then make a histogram of the 100 computed $\bar{y}$'s and draw a box plot. Does the distribution of the $\bar{y}$'s appear to be normally distributed?

SOLUTION   The command to generate the normal data is

```
RANDOM 100 OBSERVATIONS INTO C1-C12
   NORMAL MU = 50, SIGMA = 10
```

Note that each of the 100 samples of size 12 appears in the rows of the worksheet. Thus to find the average of the sample we will average across the rows with the command RMEANS. The MINITAB session shown in Figure 7.9 gives the solution.

FIGURE 7.9

Minitab output for Example 7.14

```
MTB > RANDOM 100 OBSERVATIONS INTO C1-C10;
SUBC> NORMAL MU =50, SIGMA = 10.

MTB > RMEANS C1-C10 PUT IN C14
MTB > NOTE RMEAN CALCULATES THE ROW MEANS
MTB > NOTE C14 CONTAINS THE 100 AVERAGES

MTB > HISTOGRAM C14

Histogram of C14   N = 100

Midpoint    Count
      42        1   *
      44        8   ********
      46       12   ************
      48       18   ******************
      50       25   *************************
      52       24   ************************
      54        9   *********
      56        2   **
      58        1   *
```

(Figure 7.9 cont.)

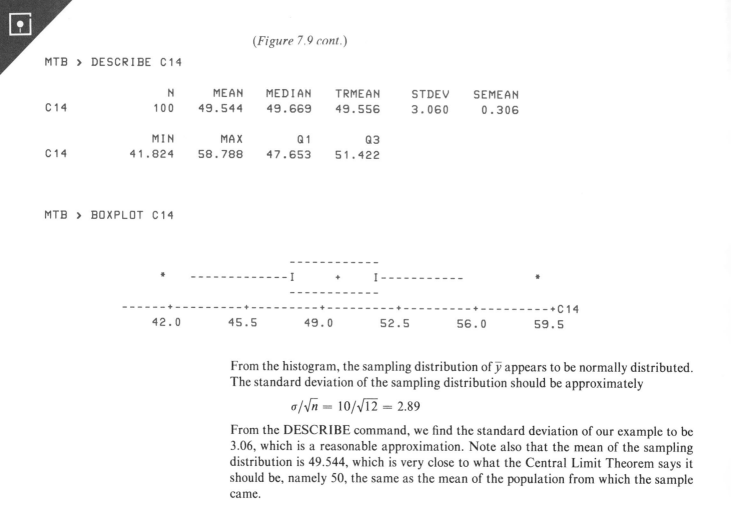

```
MTB > DESCRIBE C14

               N      MEAN    MEDIAN    TRMEAN    STDEV    SEMEAN
C14          100    49.544    49.669    49.556    3.060    0.306

             MIN       MAX        Q1        Q3
C14       41.824    58.788    47.653    51.422


MTB > BOXPLOT C14


                             -----------
              *  ------------I    +    I-----------           *
                             -----------
        ------+---------+---------+---------+---------+---------+C14
            42.0      45.5      49.0      52.5      56.0      59.5
```

From the histogram, the sampling distribution of $\bar{y}$ appears to be normally distributed. The standard deviation of the sampling distribution should be approximately

$$\sigma/\sqrt{n} = 10/\sqrt{12} = 2.89$$

From the **DESCRIBE** command, we find the standard deviation of our example to be 3.06, which is a reasonable approximation. Note also that the mean of the sampling distribution is 49.544, which is very close to what the Central Limit Theorem says it should be, namely 50, the same as the mean of the population from which the sample came.

EXERCISES 7.6

7.40 Perform a simulation study to illustrate the Central Limit Theorem. Generate a sample of size 12 from a uniform population, and then calculate the sample mean. Repeat the process 100 times, and then make a histogram and box plot of the 100 $\bar{y}$'s. Does the distribution of the $\bar{y}$'s appear to be normally distributed?

7.41 Repeat Exercise 7.40, except this time generate data from a Laplace distribution with the subcommand LAPLACE.

7.7 Summary and Review

Key Concepts

☑ The *sampling distribution* of a statistic describes the variability associated with the statistic in repeated sampling. It is desirable that the sampling distribution of a statistic remain stable from sample to sample.

☑ As with any distribution, describing the sampling distribution of a statistic involves three things. They are:

1. Describe the general shape of the distribution.
2. Give a measure of where the distribution is centered.
3. Give a measure of the amount of variability in the distribution.

The variability in the sampling distribution of a statistic is measured by the *standard error* of the statistic, which is the standard deviation of the sampling distribution.

☑ The *Central Limit Theorem,* a very important theorem in statistics, describes the sampling distribution of the sample mean, $\bar{y}$. It says that, regardless of the shape of the parent distribution, the sampling distribution of $\bar{y}$ is approximately normal for sufficiently large sample size. It also says that the mean of the sampling distribution is the same as the mean of the parent distribution, and the standard error is $\sigma/\sqrt{n}$.

☑ Also the Central Limit Theorem applies to the sampling distribution of the sample proportion, p. For sufficiently large sample size, its sampling distribution is approximately normally distributed with a mean of π and a standard error of $\sqrt{\pi(1-\pi)/n}$.

☑ The *accuracy* of an estimate is directly related to the sample size. To increase accuracy, you increase the sample size.

☑ The sampling distribution of the difference of two sample means or the difference of two sample proportions is also approximately normally distributed for sufficiently large sample sizes.

Learning Goals

Having completed this chapter you should be able to:

1. Understand the concept of a sampling distribution.   *Section 7.1*
2. Describe the standard error of a statistic.   *Section 7.1*
3. State and understand the Central Limit Theorem.   *Section 7.2*
4. Describe the sampling distribution of $\bar{y}$.   *Section 7.2*
5. Describe the sampling distribution of p.   *Section 7.3*
6. Discuss the relationship between sample size and sampling distribution. *Section 7.4*
7. Describe the sampling distribution of the difference in two sample means or two sample proportions.   *Section 7.5*

To test your skills answer the following questions.

7.42 A population has mean μ and standard deviation σ. A sample of size n is randomly selected from the population. Describe the sampling distribution of the sample mean $\bar{y}$ by stating the Central Limit Theorem.

7.43 Suppose the mean of a population is 65 and the standard deviation is 15. For samples of size 100, give the mean and standard error of the sampling distribution of $\bar{y}$. Will the sampling distribution be adequately approximated by a normal distribution? Why?

7.44 The average time to install new brakes on an automobile is 56.7 minutes, and the standard deviation is 9.3 minutes. What is the probability that a random sample of 36 installations had an average exceeding 1 hour?

7.45 A sample of n observations is randomly selected from a Bernoulli population with parameter π. Describe the sampling distribution of the sample proportion p by stating the Central Limit Theorem.

7.46 The cure rate for colon/rectal cancer is 70% if detected early. Out of 40 patients, what is the probability that more than 30 are cured?

7.47 What is the maximum standard error of the sample proportion when the sample size is 200?

7.48 Two independent random samples are taken from two populations with means 100 and 150, and standard deviations of 20 and 30, respectively. If the sample sizes are 40 and 45, what is the standard error of the sampling distribution of the difference in the sample means?

SUPPLEMENTARY EXERCISES FOR CHAPTER 7

7.49 In 1968 Nixon won 43.4% of the popular vote and Humphrey won 42.7% of the popular vote. Are these figures parameters or statistics?

7.50 A Gallup poll in September 1968 showed that 43% favored Nixon and 28% favored Humphrey. Are these figures parameters or statistics?

7.51 One-half of all Americans are involved in alcohol-related accidents in their lifetime (Source: U.S. Department of Transportation). Is the one-half (50%) a parameter or a statistic?

7.52 The death rate due to automobile accidents in 1983 was 18.3 per 100,000. Is 18.3 a parameter or a statistic?

7.53 According to a 1984 survey of 6,000 service stations, the average price per gallon of gasoline was \$1.22, which was 3 cents lower than in 1983. Is the \$1.22 a parameter or a statistic?

7.54 According to 1983 figures, 12.6% of U.S. households earn more than \$50,000. Is it unusual that a random sample of 200 households will have more than 30 with incomes in excess of \$50,000?

7.55 Suppose a standardized test has a mean of 70 and a standard deviation of 15. Is it unusual that a random sample of 40 will average more than 75?

7.56 Consider a population that is described by the following probability distribution:

y	0	2	5	7
$p(y)$	.2	.3	.4	.1

(a) Find the mean and variance of y.

(b) Find the sampling distribution of $\bar{y}$ that is calculated from a random sample of size 2.

(c) Find the sampling distribution of M, the sample median, which is found from a sample of size 2.

(d) Which statistic, $\bar{y}$ or M, do you think would be the better measure of the center of this distribution?

7.57 Suppose a bimodal population has a mean of 20 and a standard deviation of 4. A sample of size 100 is to be randomly selected from the population, and the sample mean is to be computed. Prior to selection of the sample, how will the potential values of the sample mean be distributed? Give the mean and standard deviation of the sampling distribution of the sample mean.

7.58 Annual incomes for intracity social workers are assumed to be normally distributed with a mean of \$18,500 and a standard deviation of \$2,400.

(a) What percent of the workers receive an income greater than \$20,000?

(b) What is the probability that 36 workers will have an average income in excess of \$20,000?

7.59 A national science test for tenth graders has a mean of 75 and a standard deviation of 20. A sample of 100 tenth grade students from the New York City Public School system had an average of 72 on the test. Obtain bounds that include the mean score of almost all New York City tenth graders.

7.60 Suppose the starting salary for new Ph.D. psychologists is normally distributed with a mean of $28,000 and a standard deviation of $4,000.

(a) What percent have starting salaries above $33,000?

(b) Suppose a certain university has 16 Ph.D. graduates, and their average starting salary is $33,000. Is this unusual?

7.61 An insurance company's records show that the mean payout for all automobile claims is $1,800 and the standard deviation is $400. Suppose 90 claims are filed in 1 week. What is the probability they average more than $1,900?

7.62 A standardized personality test has a mean of 25 and a standard deviation of 5. Suppose 75 college applicants take the test. What is the probability their average is between 24 and 26.5?

7.63 The Bayley Scale of Infant Development, which has a mean of 100 and a standard deviation of 16, was given to a group of 34 infants who were exposed to a new educational procedure developed at a university. Their average was 104.6. Is there evidence that this group of children scored unusually high on the test?

7.64 An index score is given to new nurses after having been on the job for 2 months. From past records the mean index score is 45 and the standard deviation is 8 points. What is the probability that a group of 35 new nurses will have an average index score less than 43?

7.65 An airline company has historical data that suggest that the mean number of passengers on its flights is 212 and the standard deviation is 42 passengers. What is the probability that the next 50 flights will average less than 200 passengers?

7.66 In the eastern states, 12% of those dining out pay by credit card. Out of a random sample of 160 people from eastern states, what is the probability that fewer than 15 will use a credit card to pay for their meal?

7.67 We wish to estimate the proportion of students who have smoked marijuana on at least three different occasions. What will be the maximum standard error of our estimate if we randomly select 200 students? What if we randomly select 500 students?

7.68 The owner of a local ski resort would like to estimate the mean daily amount of money spent at the resort by the guests. Assuming that $15.00 is a reasonable estimate of the standard deviation of the amounts spent by guests, what is the standard error of the estimated mean daily amount of money spent if the owner randomly selects 50 guests?

7.69 The state highway patrol would like to estimate the average speed of motorists on a certain section of interstate. Assume the standard deviation of the speeds of the motorists is 12 mph. What is the standard error of their estimate if they randomly select 150 motorists?

7.70 A zoologist is interested in estimating the life span of the white-tailed deer. Assuming the standard deviation in the life span is 2.5 years, what is the standard error of the estimate if the zoologist randomly selects 40 white-tailed deer?

7.71 To illustrate the variability associated with a statistic, conduct the following project: From the classified section of your local Sunday newspaper, randomly select 30 houses that are for sale. Calculate the average price of the 30 houses. From the same newspaper randomly select another 30 houses and again calculate the average. Do this several times and then compare the averages you found. Remember that the average cost of a house in your area for a given week is not variable—it is a fixed quantity, μ. The $\bar{y}$'s you calculate will vary, however; each one is a separate estimate of μ. The variability associated with the $\bar{y}$'s is the idea of the sampling variability of $\bar{y}$.

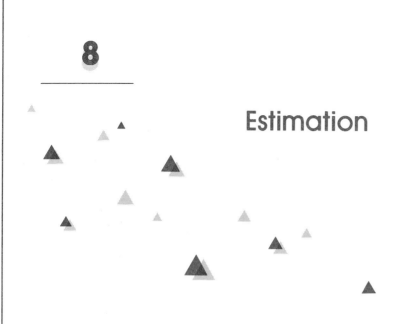

8

Estimation

In the preceding chapters we learned how to examine a data set intelligently and how to describe some of its more important characteristics. Now we wish to use that information to go one step beyond and describe the characteristics of a population. This is known as the area of statistical inference. Using the information in a sample, we will infer generalizations to the population. Inferential statistics usually falls into two categories—estimation and hypothesis testing. In this chapter we will examine the area of estimation.

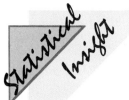

The Cost of a New Home in the United States

A new home is the most expensive item most of us will ever buy. In fact, many may never be able to afford one. The cost of a new home depends on several factors, including labor, material, land, and the real culprit, financing. In 1949, financing accounted for only 5% of the total cost of a new home. Today it accounts for close to 20% of the total cost. Also the price of a new home can vary significantly, depending on the region of the country. For example, the Census Bureau reported that in 1984 the average cost of a new home in the West was $109,400, in the Midwest was $107,800, in the Northeast was $106,200, and in the South was $86,100. The difference between the West and South was around $23,000.

Instead of reporting the average cost of a new home, the National Association of Realtors reports the median cost of a new home. Their figures differ considerably from the Census Bureau's figures. They reported that in 1984 the median cost of a new home in the West was $97,100, in the Midwest was $56,700, in the Northeast was $82,400, and in the South was $73,300 (*USA Today*, September 28, 1984).

So we see that not only do the costs vary across regions, but the choice of statistic used to estimate the costs varies. Which is the better measure of the cost of a new home—the mean or the median? The answer depends on the distributional shape of the population of housing costs. In this chapter we will take a closer look at these issues.

Table 8.1 lists the median prices of single-family homes in 37 markets across the United States, as reported by the National Association of Realtors. In Exercise 8.44 and Exercise 8.75 you are asked to analyze these data further.

TABLE 8.1

Albany, N.Y.	$ 52,400
Anaheim–Santa Ana, Calif.	134,900
Atlanta, Ga.	64,600
Baltimore, Md.	65,200
Birmingham, Ala.	66,600
Boston, Mass.	102,000
Chicago, Ill.	77,500
Cincinnati, Ohio	59,600
Cleveland, Ohio	65,600
Columbus, Ohio	60,400
Dallas–Fort Worth, Tex.	83,400
Denver, Colo.	85,000
Detroit, Mich.	48,000
Fort Lauderdale, Fla.	76,000
Houston, Tex.	79,600
Indianapolis, Ind.	54,200
Kansas City, Mo.	57,600
Los Angeles, Calif.	115,300
Louisville, Ky.	49,500
Memphis, Tenn.	65,200
Miami, Fla.	84,400
Milwaukee, Wis.	69,600
Minneapolis–St. Paul, Minn.	75,100
Nashville, Tenn.	64,000
New York metro area	106,900
Oklahoma City, Okla.	63,600
Philadelphia, Pa.	59,500
Providence, R.I.	61,400
Rochester, N.Y.	62,100
St. Louis, Mo.	64,400
Salt Lake City, Utah	67,600
San Antonio, Tex.	71,600
San Diego, Calif.	102,900
San Francisco, Calif.	132,600
San Jose, Calif.	121,600
Tampa, Fla.	60,800
Washington, D.C.	92,900

SOURCE: National Association of Realtors.

8.1　Properties of Point Estimators

The objective of estimation is to compute a value of a statistic that will be reasonably close to the unknown value of a population parameter. Deciding which sample statistic to use in a given situation is an important decision. Obviously we want to choose the one that will, in general, give the best estimate of the parameter. The statistic that we decide to use to estimate the parameter is called a *point estimator* of the parameter. The estimator is the rule that tells how to compute the estimate.

Definition

A **point estimator** of a parameter is a statistic whose value should be a close approximation to the true value of the parameter. The actual numerical value that the point estimator assumes from the collected data (the sample) is called the **point estimate.**

EXAMPLE 8.1

Following is a list of starting salaries (in dollars) of 50 computer science majors selected at random from a large midwestern university.

$19,400	19,720	17,600	19,500	20,520
20,230	18,100	20,400	13,200	19,950
20,200	22,330	22,410	20,530	24,200
22,670	20,300	16,950	19,760	21,460
17,950	21,500	26,250	19,510	17,980
20,350	19,280	18,300	23,100	16,930
20,890	22,840	17,760	16,950	21,250
19,970	18,960	20,680	15,480	19,940
21,760	18,640	14,100	21,320	18,130
18,600	17,720	23,550	21,380	19,900

Following is a list of possible estimators of a typical salary, together with the resulting estimate:

1. If the estimator is $\bar{y}$, then the estimate is $19,808.
2. If the estimator is M, then the estimate is $19,945.
3. If the estimator is $\bar{y}_{T.10}$, then the estimate is $19,845.
4. If the estimator is the midrange, then the estimate is $19,725.

In the preceding example we must decide which estimator is the most reliable in estimating the average salary. That is, which estimator will give consistently closer values to the unknown parameter? By studying the sampling distribution of an estimator, we can investigate the various values it can assume and possibly determine whether its potential values are "close" to the value of the parameter.

The concept of "closeness" is characterized by the following three criteria for an estimator:

1. Unbiased estimator
2. Efficient estimator
3. Consistent estimator

Unbiased Estimator   An estimator, being a statistic, has a sampling distribution and therefore has a mean and a standard deviation. If the mean of the sampling distribution of the estimator is equal to the parameter of interest in the parent population, we say that estimator is an *unbiased* estimator of the parameter.

> **Definition**
>
> If the sampling distribution of a statistic has a mean equal to the parameter being estimated, the statistic is called an **unbiased estimator** of the parameter. If the mean of the sampling distribution is not equal to the parameter, the statistic is said to be **biased.**

In many cases the sampling distribution is mound shaped (like the normal distribution), and hence if the estimator is unbiased, the potential values of the estimator tend to cluster about the value of the unknown parameter. Clearly this is desirable.

EXAMPLE 8.2

Suppose that a parent distribution has unknown mean, μ, and a standard deviation, σ. A random sample $y_1, y_2, y_3, \ldots, y_n$ is selected from the population. The Central Limit Theorem says that if n is large, the sampling distribution of $\bar{y}$, the sample mean, is approximately normally distributed with a mean μ and a standard deviation of $\sigma/\sqrt{n}$.

If $\bar{y}$ is the chosen estimator of μ, then it is an unbiased estimator because its sampling distribution has mean μ.

Figure 8.1 illustrates the sampling distribution of $\bar{y}$, and we see that the potential values of $\bar{y}$ tend to cluster about μ, the mean of the parent distribution.

If, in addition, the parent distribution is continuous and symmetric, then the sample median and the trimmed means are also unbiased estimators of μ.

FIGURE 8.1
Sampling distribution of $\bar{y}$

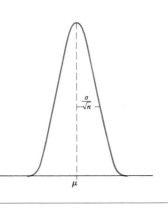

EXAMPLE 8.3

Consider a Bernoulli population with unknown parameter π. Let p be the proportion of successes in a sample of size n. The Central Limit Theorem says that the sampling distribution of p is approximately normal with a mean of π and a standard deviation of

$$\sqrt{\pi(1 - \pi)/n}$$

Because the sampling distribution has mean π, we can say that p is an unbiased estimator of π.

In repeated sampling, an unbiased estimator will average out to be equal to the parameter in question. Using it will not result in a systematic overestimate or underestimate of the unknown parameter as would be the case with a biased estimator.

EXAMPLE 8.4

▲ This is the definition form of s^2. Recall that there is also a computational form that is usually easier to use.

Suppose that we wish to estimate a population variance σ^2. A reasonable estimator of σ^2 is the sample variance, which is given by the formula

$$s^2 = \frac{\Sigma(y_i - \bar{y})^2}{n - 1}$$

It is beyond the scope of this book, but it can be shown that the sampling distribution of s^2 has mean σ^2. Thus s^2, as defined, is an unbiased estimator of σ^2.

Suppose the sample variance had been defined as

$$V = \frac{\Sigma(y_i - \bar{y})^2}{n}$$

with a denominator of n instead of $n - 1$. It can be shown that the sampling distribution of V has a mean of

$$[(n - 1)/n]\sigma^2$$

Because V does not have mean σ^2, it is a biased estimator of σ^2 and will tend to underestimate it. For example, if $n = 10$, then V will, on the average, estimate 9/10 of σ^2.

Efficient Estimators

If an estimator is unbiased, there is no guarantee that its value from a particular sample will be close to the parameter that we wish to estimate. For example, the estimator A in Figure 8.2 is an unbiased estimator of μ but has a high probability of considerable deviation from μ. This is because the variability in its sampling distribution is large.

On the other hand, unbiased estimator B in Figure 8.2 has a sampling distribution whose variability is small. Consequently the potential values (from the various samples) that B could assume are tightly clustered about μ; hence there is a high probability of being close to μ. We say that estimator B is *more efficient* than estimator A.

Definition

If two estimators are unbiased, we say that one is **more efficient** than the other if its standard error is smaller for the same sample size.

FIGURE 8.2

Sampling distribution of two
unbiased estimators of μ

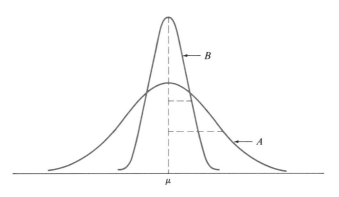

▲ If estimator A is less efficient than
estimator B, then the sample size for
A can be increased to the point that
it is just as efficient as B. However, B
is more efficient because we can gain
the same efficiency with fewer
observations.

Recall that the standard error is a measure of the variability associated with the sampling distribution of the estimator. So restricting ourselves to unbiased estimators, we can say that the smaller the standard error, the more tightly clustered will be the potential values of the estimator about the unknown parameter. If the standard error of an unbiased estimator were zero, then every value obtained for the estimator would be the same and that value would coincide with the value of the parameter. That is, we could estimate the unknown parameter with 100% accuracy. However there are no estimators (except in trivial cases) with standard error zero. What we strive for is an unbiased estimator whose standard error is smaller than the standard error of any other unbiased estimator.

EXAMPLE 8.5

If the parent distribution is *normal* with mean μ and standard deviation, σ, then the sample median and the sample mean are both unbiased estimators of μ. Which is better?

SOLUTION   We know that the standard error of the sample mean is

$$SE(\bar{y}) = \sigma/\sqrt{n}$$

▲ Don't forget that in this example we are assuming that the parent distribution is normal.

Also it can be shown that the standard error of the sample median is approximately 1.2533 times as great as the standard error of the sample mean. That is,

$$SE(M) = 1.2533\sigma/\sqrt{n}$$

Because $SE(\bar{y})$ is less than $SE(M)$, the sample mean, $\bar{y}$, is more efficient than the sample median. Consequently the potential values of $\bar{y}$ are more tightly clustered about μ than the potential values of the sample median. Further under the normality condition, it can be shown that $\bar{y}$ is more efficient than any other unbiased estimator of μ.

Using the more efficient estimator means that, when we obtain a value of the estimator (the estimate), there is a greater chance that it is closer to the unknown parameter than had we used a less efficient estimator.

If the normality assumption in Example 8.5 is dropped, it is possible that the sample median might be more efficient than the sample mean. We can't say for sure because it depends on the distribution of the parent population.

The Laplace distribution (a discussion of its characteristics can be found in a more advanced book) is a symmetric distribution with long tails. In this case, it can be shown that for large samples, the sample median has standard error given by

$$\text{SE}(M) = .7071\sigma/\sqrt{n}$$

Because $\text{SE}(\bar{y}) = \sigma/\sqrt{n}$, we have that, in the case of this long-tailed distribution, the median is more efficient than the mean.

In many applications we will not know whether we are sampling from a normal or Laplace or some other type distribution. So we cannot say for sure which is the better estimator. In those cases we need a *robust estimator*—one that works well in a wide variety of population distributions. One such robust estimator of the center of a distribution is the trimmed mean, $\bar{y}_T$. A long-tailed distribution tends occasionally to produce outliers that have an adverse effect on $\bar{y}$. Trimming eliminates the outliers, and hence $\bar{y}_T$ is not adversely affected by them.

▲ Remember that the pseudostandard deviations and the box plot can be used to give some indication of the lengths of the tails of the parent distribution.

Consistent Estimators

The final desirable property that an estimator can have, which we will consider, is that of *consistency*.

Definition

An estimator is **consistent** if the probability that it is not close to the unknown parameter approaches zero as the sample size increases.

More specifically, an unbiased estimator whose standard error approaches zero as the sample size increases is said to be consistent. Thus a consistent estimator becomes more reliable as the sample size is increased.

EXAMPLE 8.6

The standard error of the unbiased estimator $\bar{y}$ is $\sigma/\sqrt{n}$, which clearly approaches zero as n increases (n is in the denominator). Thus $\bar{y}$ is a consistent estimator of μ. If the parent distribution is normal, then the sample median is also consistent because its standard error is $1.2533\sigma/\sqrt{n}$, which approaches zero as n increases.

Any one of these properties alone does not produce the best estimator but, using them collectively, we can usually come up with a good estimator of the parameter in question.

The underlying parent distribution must be considered when deciding what estimator to use in a particular situation. For example, in a highly skewed or otherwise far from normal distribution, it is possible for a biased estimator to have a higher concentration of probability about the parameter than an efficient unbiased estimator.

In the next section we will look at some typical situations and see how an estimator is chosen.

▲ Remember that the midsummaries and the box plot are useful tools in observing skewness and deviations from normality.

EXERCISES 8.1

8.1 Why is p an unbiased estimator of π?

8.2 Show that p is a consistent estimator of π.

8.3 Suppose the parent population is *not* normal and has mean μ.

(a) Does $\bar{y}$ have to be an unbiased estimator of μ?

(b) Is there an estimator possibly more efficient than $\bar{y}$ in estimating μ?

8.4 Sketch a picture of the sampling distribution of a biased estimator of μ.

8.5 Should $\bar{y}_T$ always be used to estimate μ in a nonnormal population?

8.6 Assume the bull's-eye in the targets shown in the accompanying figure represents a parameter that is to be estimated. Each shot represents a value of the statistic that is to be used to estimate the parameter. To illustrate the concepts of unbiasedness and efficiency, classify the four targets as having high or low bias and high or low efficiency.

8.7 The manufacturer of a computerized airline ticket reservation system has contracted to deliver a system that will check out reservations in a mean time of 40 seconds. Identify the parameter of interest. Give two possible statistics that might be used to estimate this parameter. From the following sample of 20 reservation times (in seconds), calculate the values of your two statistics. Which do you prefer?

| 40 | 45 | 29 | 58 | 48 | 39 | 30 | 50 | 33 | 42 |
| 63 | 31 | 27 | 47 | 58 | 32 | 51 | 34 | 39 | 51 |

8.8 The time (in minutes) it takes a subway to travel from the airport to downtown was recorded on 30 different trips. Identify a parameter of interest. From the following sample calculate the values of two statistics that might be used to estimate the parameter. Which do you prefer?

34.4	41.6	39.9	40.1	40.4	39.7
37.6	39.2	40.1	38.8	41.2	37.9
40.3	42.5	38.9	39.5	39.9	40.4
42.5	41.1	39.4	37.7	40.2	40.1
39.3	39.6	40.0	41.3	40.4	38.3

FIGURE FOR EXERCISE 8.6

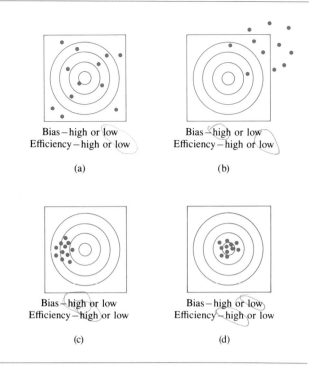

Bias—high or low
Efficiency—high or low

(a)

Bias—high or low
Efficiency—high or low

(b)

Bias—high or low
Efficiency—high or low

(c)

Bias—high or low
Efficiency—high or low

(d)

8.2 Estimators of Some Population Parameters

For a population whose distribution is continuous, we are concerned with measures of the center of the distribution (mean and median) and measures of dispersion (standard deviation). For Bernoulli populations we are concerned with the proportion of successes. In this section we will learn how to estimate these parameters.

Population Mean

In Section 6.4 it was pointed out that all natural measures of the center of a symmetric distribution coincide. Thus the problem of estimating the center of a symmetric distribution becomes one of estimating the point of symmetry. For simplicity, we will estimate the population mean, μ.

There are many different point estimators that might be used to estimate μ. Among the possible choices are the sample mean, the sample median, the midrange, or any of the possible trimmed means. The appropriate choice depends on the underlying parent distribution.

Normal Parent Distribution. If the parent distribution is normal, there is no better estimator of μ than the sample mean, $\bar{y}$. This follows from the fact that

1. $\bar{y}$ is an unbiased estimator of μ, regardless of the parent distribution (Example 8.2).

2. Under the normality condition, $\bar{y}$ is more efficient than any other unbiased estimator of μ (Example 8.5).

3. $\bar{y}$ is a consistent estimator of μ (Example 8.6).

Thus $\bar{y}$ is the appropriate choice for estimating μ when we are convinced that the parent distribution is normal or near normal.

▲ From a sample it is impossible to determine if the parent distribution is *exactly* normal. With our descriptive tools however, we can decide if the sample supports the idea that the parent distribution is normal. Possibly we could say that the distribution is close or near to being normal.

EXAMPLE 8.7

A psychologist has developed an IQ test that should not discriminate against any ethnic group. To obtain a norm for the test, she gives the test to a random sample of 100 citizens in her community. Following is a stem and leaf plot and a box plot of the sample data.

Ordered stem and leaf plot

7*	2
7:	5 7 7
8*	1 2 3 4 4
8:	6 6 7 8 8 8 8 9 9
9*	0 1 1 2 2 3 3 3 4 4
9:	5 5 6 6 6 6 6 6 7 7 7 7 8 8 9
10*	0 0 0 0 0 0 1 1 1 1 1 2 2 2 3 3 4 4
10:	5 5 6 6 6 7 7 8 8 8 8 9 9 9
11*	0 0 1 1 2 3 4 4 4
11:	5 5 6 6 6 7 7 8
12*	2 2 3 3 4
12:	5 5
13*	0

Quantile summary diagram

		100	Midsummaries	Spreads
$M(50.5)$		101	101	
$Q(25.5)$	93	109.5	101.25	16.5
$E(13)$	88	116	102	28.0
R	72	130	101	58.0

$\bar{y} = 101.35$

$s = 12.3$

$PSD_q = 12.2$

$PSD_e = 12.2$

Box plot summary diagram

		24.75
a	72	130
f	68.25	134.25
	none	none
F	43.5	159.0
	none	none

Box plot

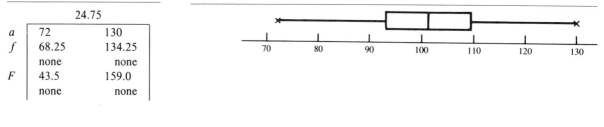

Comment on the shape of the underlying parent distribution and estimate the mean score on the IQ test.

SOLUTION   From the stem and leaf and box plot and the fact that s, PSD_q, and PSD_e are all very close in value, we are led to believe that the data are coming from a population that is near normal in shape, hence

$$\bar{y} = 101.35$$

will be our estimate of the mean score.

Symmetric Long-tailed Distribution.   If the parent distribution is symmetric and has long tails (as compared with the normal distribution), it has been shown that $\bar{y}$ can be a terrible estimator of μ because it is very sensitive to outliers. And with a long-tailed distribution, it is likely that a few outliers will appear in any sample. As pointed out in Section 8.1, the trimmed mean is a robust estimator that is not adversely affected by outliers. How much to trim is a matter of debate, but generally a 10% or 20% trimmed mean is suggested. Trimming should be symmetric and enough to remove outliers.

EXAMPLE 8.8

Following are the educational levels, in years, of a sample of 40 auto workers in a plant in Detroit.

22	16	11	11	8	12	21	8
12	12	17	14	12	5	10	8
6	12	8	12	11	10	9	14
13	12	10	10	12	1	11	13
14	10	12	12	12	12	9	12

Comment on the shape of the underlying parent distribution of educational levels, and estimate the mean of the distribution.

SOLUTION   Following is a stem and leaf and box plot of the data.

Ordered stem and leaf plot

```
0*  | 1
0t  |
0f  | 5
0s  | 6
0-  | 8  8  8  8  9  9
1*  | 0  0  0  0  0  1  1  1  1
1t  | 2  2  2  2  2  2  2  2  2  2  2  2  2  3  3
1f  | 4  4  4
1s  | 6  7
1-  |
2*  | 1
2t  | 2
```

Quantile summary diagram

	40		Midsummaries	Spreads
$M(20.5)$	12		12	
$Q(10.5)$	10	12	11	2
$E(5.5)$	8	14	12	6
R	1	22	11.5	21

$$\bar{y} = 11.4 \qquad s = 3.699$$
$$\bar{y}_{T.10} = 11.25 \qquad PSD_q = 1.48$$
$$PSD_e = 2.61$$

**Box plot
summary diagram**

	3		
a	8	14	
f	7	15	
5, 6	two	two	16, 17
F	4	18	
1	one	two	21, 22

Box plot

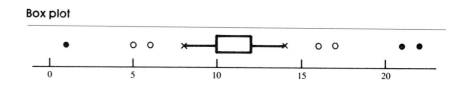

Because of the presence of outliers and the fact that s is considerably larger than both PSD_q and PSD_e, it appears that the parent distribution has tails somewhat longer than a normal distribution. Consequently a trimmed mean is our choice of an estimator of μ. A 10% trimming will remove the outliers, so our estimate of μ is

$$\bar{y}_{T.10} = 11.25 \text{ years}$$

Other Symmetric Distributions. It is difficult to determine the exact shape of the parent distribution. Recent research has shown however, that under the assumption that the parent distribution is symmetric, $\bar{y}$ and the trimmed means are reasonably good estimators of μ. Generally speaking, the longer the tails of the distribution the more one should trim. In the extreme case of no tails (Uniform Distribution), the best estimator of μ is the midrange; however $\bar{y}$ performs reasonably well in this case.

EXAMPLE 8.9

An engineer for NASA wishes to investigate the heat resistance of a particular type of electronic component to be used in a space probe. He measures the temperature at which 30 such components fail and records the results in the following stem and leaf plot and box plot.

**Ordered
stem and leaf plot**

2*	14	22	23	25	27						
2–	51	53	55	67	73	80	80	85	87	97	99
3*	09	10	13	17	19	25	47	48	49		
3–	85	87	89	90	90						

**Quantile
summary diagram**

	30		Midsummaries	Spreads
$M(15.5)$	298		298	
$Q(8)$	255	347	301	92
$E(4.5)$	226	386	306	160
R	214	390	302	176

$$\bar{y} = 300.53 \qquad s = 54.919$$
$$\bar{y}_{T.10} = 299.5 \qquad PSD_q = 68.148$$
$$PSD_e = 69.565$$

**Box plot
summary diagram**

	138	
a	214	390
f	117	485
	none	none
F	–21	623
	none	none

▲ The inner fences of the box plot were labeled to illustrate that the tails are very short.

Box plot

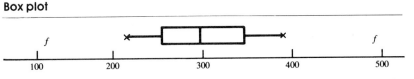

Comment on the shape of the underlying parent distribution and estimate the mean temperature of failure of this particular type component.

SOLUTION   Because s is considerably smaller than both PSD_q and PSD_e, and from the appearance of the box plot, it is reasonable to assume that the parent distribution is symmetric with short tails. We observe that the midrange, $\bar{y}$, and $\bar{y}_{T.10}$ are all very close, and any would be a reasonable estimate of the mean temperature of failure. We choose to use $\bar{y} = 300.53$ as our estimate.

It has been shown that the performance of the estimator $\bar{y}$ is not nearly as adversely affected by a short-tailed distribution as it is by a long-tailed distribution. Although the midrange is more efficient in the uniform distribution (a no-tail distribution), it is reasonable to use $\bar{y}$ as the estimator of the center of a distribution whose tails are shorter than normal. In fact, we recommend using $\bar{y}$ to estimate the mean of any symmetric distribution whose tails are not excessively long. But when a box plot or other descriptive tools indicate long tails, it is suggested that we use a trimmed mean to estimate μ. It should be pointed out that there are numerous other robust estimators of the point of symmetry of a symmetrical distribution; however it is beyond the scope of this book to consider the different possibilities.

Estimator of the Population Mean

1. The sample mean, $\bar{y}$, is the recommended estimator of the population mean when the parent distribution is normal, near normal, or symmetric with tails that are *not* excessively long.

2. A trimmed sample mean, $\bar{y}_T$, is the recommended estimator of the population mean when the parent distribution is symmetric with long tails.

For nonsymmetrical parent distributions, we may consider parameters other than the population mean.

Population Median

If the parent distribution is skewed, it may be that the median is a better indicator of the center of the distribution. In such cases, the sample median should be used to estimate the population median.

EXAMPLE 8.10

Following is a list of 25 scores on the first test in a statistics class taught at a small midwestern university. Give an estimate of the typical score.

61	38	64	70	81
11	61	92	77	47
98	76	83	58	97
78	14	80	41	65
90	81	98	72	78

SOLUTION   Following is a stem and leaf plot and a box plot of the data.

Ordered stem and leaf plot

1	1 4
2	
3	8
4	1 7
5	8
6	1 1 4 5
7	0 2 6 7 8 8
8	0 1 1 3
9	0 2 7 8 8

Quantile summary diagram

		25		Midsummaries	Spreads
$M(13)$		76		76	
$Q(7)$	61		81	71	20
$E(4)$	41		92	66.5	51
R	11		98	43.5	87
				$\bar{y} = 68.44$	$s = 23.475$

Box plot summary diagram

	30	
a	38	98
f	31	111
11, 14	two	none
F	1	141

Box plot

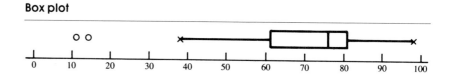

From the stem and leaf plot and box plot and the fact that the midsummaries get progressively smaller, we are led to believe that the distribution is skewed left. Therefore the median of 76 is our point estimate of a typical score. It is interesting to note that $\bar{y} = 68.44$ is somewhat below the median and will tend to misrepresent the center of the data.

Population Proportion

To estimate π, the proportion of successes in a Bernoulli population, our intuition suggests using p, the proportion of successes in the sample. Our intuition is correct for

1. p is an unbiased estimator of π (Example 8.3).
2. p is a consistent estimator of π because it is unbiased, and

$$\text{SE}(p) = \sqrt{\pi(1 - \pi)/n}$$

approaches zero as n gets larger.

3. p is an efficient estimator of π because it has a variance smaller than the variance of any other unbiased estimator of π.

EXAMPLE 8.11

What percent of the population feel they have to wait too long when they visit their family physician? A random sample of 400 patients revealed that 280 felt they had to sit in the waiting room too long. Give an estimate of π, the proportion of the population that believe they have to wait too long.

SOLUTION   Because 280 of the 400 indicated that they had excessive waiting periods, our point estimate of π is

$$p = 280/400 = .70$$

or 70%.

Population Variance

The population variance, σ^2, is a measure of the dispersion in the population. The most widely used estimator of the population variance is the sample variance.

Estimator of the Population Variance

The sample variance, s^2, is an unbiased, consistent estimator of the population variance, σ^2.

Although s^2 is not the most efficient estimator, it is somewhat efficient when compared with other estimators.

The standard deviation, s, which is the square root of the sample variance, is generally used as an estimator of the population standard deviation even though it is biased. For large samples the bias becomes negligible, and thus s seems to be a reasonable estimator of σ.

Because the computation of s involves squared deviations, it is affected by outliers even more than $\bar{y}$. The Q-spread is a robust estimator of spread that is not as affected by outliers. Hence in the presence of outliers in a long-tailed distribution, the Q-spread seems to be a reasonable estimator of the population variability.

▲ The Q-spread is just an estimate of population variability and not specifically the population standard deviation.

The Empirical rule can be used to give a quick estimate of the standard deviation. Recall that for distributions that are somewhat normal in shape, the Empirical rule states that most (95%) of the distribution lies within two standard deviations of the mean. Thus the range should cover roughly four standard deviations. Consequently dividing the range by 4 should give an approximate estimate of the standard deviation. In summary, we have

---

An Estimator of the Standard Deviation Based on the Range

Estimated standard deviation = Range/4

EXAMPLE 8.12

▲ Almost all of a distribution is covered by six standard deviations so it is not uncommon to use Range/6 as the estimate of the standard deviation. Here the estimate would be $600/6 = 100$, which is the standard deviation of SAT scores.

The minimum SAT verbal score is 200 and the maximum is 800. Thus the range is

$$R = 800 - 200 = 600$$

Dividing by 4 we have that an estimate of the standard deviation is

$$R/4 = 600/4 = 150$$

Remember that Range/4 is a rough estimate of the standard deviation and should be used only when accuracy is not an important issue. It should not serve as a replacement for s.

EXERCISES 8.2

8.9 What statistic would you use to estimate the center of a distribution if a box plot of the sample suggested that the parent population was

(a) approximately normal in shape?

(b) has tails significantly longer than the normal distribution?

(c) has tails significantly shorter than the normal distribution?

(d) is highly skewed left?

(e) is highly skewed right?

8.10 To study the distance a professional golfer can drive a ball, the following distances in yards were recorded by 20 pros.

259	270	248	262	271
255	261	242	251	238
273	271	265	268	251
273	265	241	239	254

Construct a stem and leaf plot and box plot. Comment on the general shape of the underlying parent distribution, and estimate the center of the distribution.

8.11 A survey of freshmen was conducted to find out how many hours per week they studied for their courses. The sample of 50 freshmen revealed the following scores:

30	23	32	40	33	25	15	29	37	30
28	42	28	19	14	35	20	35	44	5
32	25	17	39	40	10	41	38	12	30
20	19	28	43	30	8	41	35	38	45
10	39	15	20	40	36	25	16	22	18

Construct a stem and leaf plot and box plot. Comment on the general shape of the underlying parent distribution, and estimate the center of the distribution.

8.12 A standardized exam for math competency has a national mean of 70 and a standard deviation of 12. The exam is given to 31 entering freshmen at a small community college with the following results:

61	25	73	68	76	82	90	42	75	34
53	70	92	75	100	61	46	65	90	77
115	73	85	68	75	87	70	69	65	50
70									

Construct a stem and leaf plot and box plot. Comment on the general shape of the underlying parent distribution, and estimate the center of the distribution.

8.13 A sociologist studying ethnic problems randomly selects 50 Puerto Rican families in Miami. She finds the following weekly family incomes (in dollars):

$ 50	180	75	90	205
280	190	200	70	215
180	155	235	80	100
110	250	260	400	125
145	75	80	160	220
245	80	125	175	180
225	250	90	120	165
170	235	210	170	140
135	255	270	160	180
250	290	150	275	150

Construct a stem and leaf plot and box plot. Comment on the general shape of the underlying parent distribution, and estimate the center of the distribution.

8.14 The sociologist in Exercise 8.13 has a random sample of 225 names from the registered voter roll in Miami and finds that 45 are Puerto Rican. Estimate the percent of registered voters in Miami who are Puerto Rican.

8.15 A survey of 400 randomly selected homes in a large community reveals that a television set is turned on an average of

6.2 hours per day. Does this mean all sets are on 6.2 hours per day? Does it mean that the average number of hours is 6.2 for all large communities? What about small or rural communities?

8.16 An American Council on Education survey of 182,370 freshmen students at 345 schools reported that 68% enter college to make more money. Along those lines the report gave the following distribution of incomes of the parents of the students attending universities:

Income	Percent
Up to $9,999	6%
$10,000 to 19,999	13%
$20,000 to 29,999	18%
$30,000 to 39,999	20%
$40,000 and up	43%

Construct a relative frequency histogram of the data, and comment on the distributional shape. What measure of central tendency seems appropriate?

8.17 The report in Exercise 8.16 also gave the following distribution of incomes of the parents of the students attending predominantly black schools:

Income	Percent
Up to $9,999	25%
$10,000 to 19,999	28%
$20,000 to 29,999	19%
$30,000 to 39,999	13%
$40,000 and up	15%

Construct a relative frequency histogram of the data, and comment on the distributional shape. What measure of central tendency seems appropriate?

8.18 From the data on the starting salaries of Computer Science majors in Example 8.1, construct a stem and leaf plot and a box plot. Analyze the midsummaries, and comment on the shape of the parent distribution. Of the measures of central tendency listed in the example, which one seems appropriate?

8.19 Suppose that in Example 8.1 we wish to investigate the amount of variability in salaries of the Computer Science majors. What statistic from the sample should we calculate to estimate this variability? Give two possible choices together with the estimate obtained from the data. Of those that you list, which do you think will be the most reliable estimator of the variability?

Confidence Interval for the Mean of a Symmetric Distribution

Point estimators are subject to sample variability. The estimate obtained clearly depends on the sample that was selected. Instead of trying to come up with a single number as the estimate of an unknown parameter, perhaps it would be better to come up with an interval of values that contains the parameter with high probability.

In practice an estimate of an unknown parameter is usually given in the form of the estimate ± a margin of error. For example, the Bureau of Labor Statistics may estimate the number of unemployed in a certain area as 1,500 ± 300, being rather confident that the actual number is somewhere between 1,200 and 1,800. If a degree of confidence can be attached to this interval, we have a *confidence interval* for the unknown parameter.

Three things are needed in order to develop a confidence interval:

1. A good point estimator of the parameter
2. The sampling distribution (or approximate sampling distribution) of the point estimator
3. The desired confidence level, which is usually stated as a percent

Knowledge of the sampling distribution of the estimator allows us to make a probability statement that involves the estimator and the parameter with the desired confidence level.

We will now develop a confidence interval for the population mean μ. In Section 8.2

it was decided that if we wished to estimate the mean of a symmetric population, we should use either $\bar{y}$ or $\bar{y}_T$ as the estimator. If the parent distribution is symmetric with long tails, we should use $\bar{y}_T$ to estimate μ. In all other symmetrical cases, including the normal distribution, we should use $\bar{y}$ to estimate μ.

In Section 8.4 we will construct a confidence interval for μ based on $\bar{y}_T$ and in this section we will construct a confidence interval for μ based on $\bar{y}$.

Large-Sample Confidence Interval Based on $\bar{y}$

Suppose that a symmetrically distributed population has unknown mean μ and known standard deviation σ, and we desire a 95% confidence interval for μ. Assuming that we have a large sample, the Central Limit Theorem says that, regardless of the parent distribution, the sampling distribution of $\bar{y}$ is approximately normal with mean μ and a standard error

$$\text{SE}(\bar{y}) = \sigma/\sqrt{n}$$

The normal probability tables show that a normal variable will be within 1.96 standard deviations of its mean with 95% probability. Because $\bar{y}$ is approximately normally distributed and its standard error is $\sigma/\sqrt{n}$, it should be within $1.96\sigma/\sqrt{n}$ of its mean, μ, with 95% probability. The value

$$\text{ME} = 1.96\sigma/\sqrt{n}$$

is the *margin of error* in using $\bar{y}$ to estimate μ with 95% probability. Adding and subtracting the margin of error to the estimator, we have that a 95% confidence interval estimate of μ is

$$\bar{y} \pm \text{ME} = \bar{y} \pm 1.96\sigma/\sqrt{n}$$

Suppose we desire a confidence level other than 95%; for example, suppose we desire 98% confidence or perhaps only 80% confidence? The estimate will remain $\bar{y}$, but what is the margin of error?

We see from Figure 8.3 that $z_{\alpha/2}$ denotes the value of a standard normal variable, which is chosen so that $\alpha/2$ of the probability lies above it; or, $(1-\alpha)100\%$ of the probability lies between $-z_{\alpha/2}$ and $+z_{\alpha/2}$.

FIGURE 8.3
Definition of $z_{\alpha/2}$

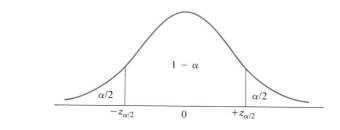

The following table gives a few values of $z_{\alpha/2}$ that are obtained from the standard normal probability tables.

$1-\alpha$	.80	.85	.90	.95	.98	.99
$z_{\alpha/2}$	1.28	1.44	1.645	1.96	2.33	2.58

EXAMPLE 8.13

Substituting $z_{\alpha/2}$ for 1.96, we have that the margin of error in using $\bar{y}$ to estimate μ with $(1 - \alpha)100\%$ confidence is

$$\text{ME} = z_{\alpha/2}\sigma/\sqrt{n}$$

The form of the large-sample $(1 - \alpha)100\%$ confidence interval for μ based on $\bar{y}$ becomes

$$\bar{y} \pm z_{\alpha/2}\sigma/\sqrt{n}$$

A new test is designed to measure the rehabilitation potential of mentally ill patients. The test was given to 40 patients who had an average score of 68. Assuming that the test scores have a standard deviation of 10 points, find a 98% confidence interval for μ, the mean score on the new test.

SOLUTION   Because a 98% confidence interval is desired, we choose $\alpha = .02$, hence

$$z_{\alpha/2} = z_{.01} = 2.33$$

so that the confidence interval given by

$$\bar{y} \pm z_{\alpha/2}\sigma/\sqrt{n}$$

becomes

$$\bar{y} \pm 2.33\sigma/\sqrt{n}$$

Substituting in the numerical values, we have

$$68 \pm (2.33)(10)/\sqrt{40} = 68 \pm 3.684$$

which is 64.316 to 71.684.

We are 98% confident that this interval contains the mean score of the new test.

Observe in the preceding example that in order to calculate the confidence interval, it is necessary that the population standard deviation, σ, be given.

If σ is unknown, which is usually the case, we would not be able to compute the margin of error and hence the limits of the confidence interval. When n is large, however, as is assumed here, the population standard deviation, σ, is closely approximated by the sample standard deviation, s. Consequently using s as an estimate of σ, we have

$$\text{estSE}(\bar{y}) = s/\sqrt{n}$$

and

$$\text{ME} = z_{\alpha/2}s/\sqrt{n}$$

We have the following general statement:

A large-sample $(1 - \alpha)100\%$ confidence interval for μ based on $\bar{y}$ is given by the limits

$$\bar{y} \pm z_{\alpha/2}s/\sqrt{n}$$

(A large sample is where $n > 30$.)

EXAMPLE 8.14

Using the data in Example 8.7, find a 95% confidence interval for the mean score on the IQ test.

SOLUTION   In Example 8.7 it was decided that the best estimator of the mean score is $\bar{y}$, so the confidence interval will be based on $\bar{y}$. Because the desired confidence level is 95%, we have

$$z_{\alpha/2} = z_{.025} = 1.96$$

Returning to the data, we find that

$$\bar{y} = 101.35 \quad \text{and} \quad s = 12.3$$

Thus we have that

$$\text{estSE}(\bar{y}) = s/\sqrt{n} = 12.3/\sqrt{100} = 1.23$$

so that

$$\text{ME} = (1.96)(1.23) = 2.41$$

The confidence interval

$$\bar{y} \pm 1.96s/\sqrt{n}$$

becomes

$$101.35 \pm 2.41$$

or

$$98.94 \text{ to } 103.76$$

Interpreting Confidence Intervals

It is important to realize that the probability attached to an interval is only relevant prior to sampling. After sampling and calculating the endpoints of the interval, the probability statements no longer apply. That is, in Example 8.14 it is meaningless to say that μ lies between 98.94 and 103.76 with 95% probability. Either μ is between the two values or it is not—there is no probability associated with it.

What is meant by a 95% confidence interval is that if 100 different intervals are obtained from 100 different samples, then it is likely that 95 of those intervals will contain μ and 5 will not. This does not mean that in practice we calculate 100 different intervals—we calculate only one. But once it is found, we know there is a very good chance that the interval does indeed contain μ. Unfortunately there is always the chance that it does not, but this is rare.

Another important point about confidence intervals is that once an interval is obtained, no preferential treatment is given to any value in the interval. True, $\bar{y}$ is the center of the interval and is the point estimate of μ, but for confidence intervals it is just one of the values in the interval.

Validity and Precision of Confidence Intervals

The usefulness of a confidence interval is judged by its validity and precision. Validity is measured by the confidence level, which is the probability that the interval contains the true value of the parameter. Precision is measured by the length of the interval. In general, the length is twice the margin of error, because the form of the interval is:

estimator ± the margin of error. Consequently the smaller the margin of error, the more precise the interval.

As stated, the confidence level is a measure of the validity. Ideally we would like to have a 100% confidence interval, that is, one that would guarantee that the parameter would be somewhere in the interval. However that would be at the expense of precision—there would be NO precision because the margin of error would be infinitely large.

Thus the validity is fixed at a reasonable level, say 95%, and then, given the assumptions, we search for the most precise interval. For example, assuming that the parent distribution is normal, the most precise confidence interval for μ at any validity level (i.e., the one giving the smallest margin of error) is the one based on $\bar{y}$ given above.

Given another parent distribution, say one that is highly skewed or one with long tails, the confidence interval based on $\bar{y}$ may not be the most precise. It may be that the interval should be based on the median or a trimmed mean rather than the sample mean, $\bar{y}$.

In general, the two characteristics, validity and precision, compete with each other. Recall that the margin of error is

$$\text{ME} = z_{\alpha/2}\sigma/\sqrt{n}$$

Clearly for a fixed sample size, n, the only way to reduce the margin of error (increase precision) is to decrease the value of $z_{\alpha/2}$, which in effect reduces validity. On the other hand, to raise validity, say from 95% confidence to 99% confidence, the $z_{\alpha/2}$ value increases from 1.96 to 2.58, which increases the margin of error, which decreases precision.

It appears that it is not possible to increase precision while maintaining a fixed validity level. Because the sample size, n, appears in the denominator, however, it is possible to decrease the margin of error, and hence increase precision, by increasing n. In fact, we can have as much precision as we desire if we can afford the increased sample size. Remember that large samples are usually costly.

Steps to Follow in Determining Sample Size

1. Specify the desired confidence level.

2. Specify the desired precision by giving a bound for the margin of error.

3. Set the bound equal to the margin of error and solve the equation for n.

EXAMPLE 8.15

The precision of the 98% confidence interval obtained in Example 8.13 is quantified by its margin of error, which was $\text{ME} = 3.684$. Suppose we wish to maintain the same validity, namely 98%, but wish to increase the precision so that the margin of error is only 2.5 points on the test. How large should the sample size be?

SOLUTION   Because the population standard deviation is known, the equation for the margin of error is

$$\text{ME} = z_{\alpha/2}\sigma/\sqrt{n}$$

For a 98% confidence interval the equation becomes

$$\text{ME} = 2.33\sigma/\sqrt{n}$$

Substituting $\text{ME} = 2.5$ and $\sigma = 10$, we can solve for n as follows:

$$2.5 = (2.33)(10)/\sqrt{n}$$
$$\sqrt{n} = (2.33)(10)/2.5$$
$$\sqrt{n} = 9.32$$

Hence

$$n = (9.32)^2 = 86.86$$

Thus we would need a sample of size 87.

Sample Size Determination for Estimating μ

In general, to estimate μ with a $(1 - \alpha)100\%$ confidence interval so that the bound on the margin of error is B, we solve the following equation for n:

$$B = \frac{z_{\alpha/2}\sigma}{\sqrt{n}}$$

The solution is

$$n = \frac{z_{\alpha/2}^2\sigma^2}{B^2}$$

In the previous example we were able to determine the sample size because the value of σ was known. As has been pointed out, however, the value of σ is usually not available, and must be estimated. Normally the sample standard deviation, s, is used to estimate σ. Here we are trying to determine the sample size, in which case, s is not obtainable. We must find some other means of estimating σ. It may be that a previous study will give a sample standard deviation that can be used; or, as was pointed out in Section 8.2, the range/4 is a reasonable estimate of σ.

EXAMPLE 8.16

A study was undertaken to estimate the number of credit hours delivered by a typical University of North Carolina professor. It is assumed that no professor will deliver more than 400 credit hours. Find the sample size necessary to estimate the mean number of credit hours delivered by all professors in the UNC system to within 20 hours with a 90% confidence interval.

SOLUTION   Because a 90% confidence interval is desired, we have

$$z_{\alpha/2} = z_{.05} = 1.645$$

and the margin of error is

$$\text{ME} = 1.645\sigma/\sqrt{n}$$

It is given that no professor will deliver more than 400 credit hours; therefore the range is $400 - 0 = 400$. An estimate of σ is

$$\text{Range}/4 = 400/4 = 100$$

Substituting 20 hours for the margin of error, we have

$$20 = (1.645)(100)/\sqrt{n}$$

We now solve for n. Or using the formula developed previously, we have

$$n = \frac{(1.645)^2(100)^2}{(20)^2}$$
$$= 67.65$$

Thus we need a sample of 68 professors.

The results of this section are summarized as follows:

Large-Sample Confidence Interval for μ Based on $\bar{y}$

APPLICATION: Symmetric distributions whose tails are not excessively long

ASSUMPTION: $n > 30$

ESTIMATOR: $\bar{y}$, the sample mean

STANDARD ERROR OF THE ESTIMATOR: $\text{SE}(\bar{y}) = \sigma/\sqrt{n}$, $\text{estSE}(\bar{y}) = s/\sqrt{n}$

A $(1 - \alpha)100\%$ confidence interval for μ is given by the limits

$$\bar{y} \pm z_{\alpha/2}s/\sqrt{n}$$

SAMPLE SIZE: $n = \dfrac{z_{\alpha/2}^2\sigma^2}{B^2}$

EXAMPLE 8.17

For a random sample of 68 professors, the average number of credit hours delivered was 265 and the standard deviation was 80 hours. Find a 90% confidence interval for the mean number of credit hours delivered.

SOLUTION   Assuming a symmetrical distribution in which the tails are not excessively long, the confidence interval will be based on $\bar{y}$.

From the collected data we have $\bar{y} = 265$ credit hours and $s = 80$ hours. The estimated standard error of $\bar{y}$ is

$$\text{estSE}(\bar{y}) = s/\sqrt{n} = 80/\sqrt{68} = 9.70$$

hours.

Thus a 90% confidence interval for the mean number of credit hours delivered is

$$265 \pm 1.645(9.70)$$

or

$$265 \pm 15.96$$

which gives the limits

$$249.04 \text{ to } 280.96$$

To display a confidence interval graphically, we mark off the endpoints on a line graph and put a circle at the location of the center of the interval. For Example 8.17 we have the graph shown in Figure 8.4. Clearly we can observe the length of the interval and thus evaluate the precision of the interval. Later we will see that we can also use this graphical technique to compare confidence intervals.

FIGURE 8.4

Displaying a confidence interval

240   260   280   300

EXERCISES 8.3

8.20 A random sample of size n is selected from a population with unknown mean μ and standard deviation $\sigma = 20$. Calculate a 95% confidence interval for μ based on $\bar{y}$ for each of the following samples:

(a) $n = 30$, $\bar{y} = 94.3$ (b) $n = 45$, $\bar{y} = 96.4$
(c) $n = 100$, $\bar{y} = 95.6$ (d) $n = 200$, $\bar{y} = 95.8$

8.21 A random sample of 60 observations selected from a population with unknown mean μ and standard deviation $\sigma = 50$ yielded $\bar{y} = 465.8$.

(a) Find a 95% confidence interval for μ based on $\bar{y}$.

(b) Find a 99% confidence interval for μ based on $\bar{y}$.

8.22 Find the following confidence intervals for μ based on $\bar{y}$ from the given information.

(a) $n = 36$, $\bar{y} = 72.4$, $s = 11.2$, confidence level = .95

(b) $n = 64$, $\bar{y} = 128.3$, $s = 32.4$, confidence level = .98

(c) $n = 100$, $\bar{y} = 465$, $s = 112$, confidence level = .99

8.23 A city planner randomly samples 100 apartments in the inner-city area in order to estimate the mean living area per apartment. Find a 95% confidence interval based on $\bar{y}$ for the mean living area if the sample yielded an average of 1,325 square feet and a standard deviation of 42.4 square feet.

8.24 To determine the diameter of Venus, an astronomer makes 36 measurements of the diameter and finds $\bar{y} = 7848$ miles and $s = 310$ miles. Find a 95% confidence interval estimate based on $\bar{y}$ of the diameter of Venus.

8.25 Determine the sample size so that the margin of error in determining the diameter of Venus in Exercise 8.24 is only 50 miles.

8.26 You wish to estimate the mean number of ounces of coffee dispensed from a certain coffee machine. A sample of 40 cups yielded $\bar{y} = 5.2$ ounces and $s = .24$ ounce. Find a 99% confidence interval estimate based on $\bar{y}$ of the mean number of ounces dispensed per cup.

8.27 A consumer researcher sampled 100 electric razors, all of the same make and model. The sample mean life (hours of operation before failure) was 560 hours, and the standard deviation was 35 hours. Construct a 90% confidence interval estimate based on $\bar{y}$ of the true mean life span.

8.28 To study the birth weights of infants whose mothers smoke, a physician records the weights of 100 newborns whose mothers smoke. Find a 98% confidence interval based on $\bar{y}$ for the mean birth weight of children of smoking mothers if $\bar{y}$ was found to be 6.1 pounds and $s = 2.1$ pounds.

8.29 Determine the sample size so that the margin of error in determining the mean birth weight in Exercise 8.28 is only .3 pound.

8.30 Construct a 99% confidence interval based on $\bar{y}$ for the mean number of hours freshmen study from the data given in Exercise 8.11.

8.31 A banker would like to estimate the average amount of the loans made to farmers in his community for the past growing season. He randomly selects 40 accounts and finds that the average is $78,460, and the standard deviation is $22,260. Assuming the distribution of loan amounts is near normal in shape, find a 90% confidence interval for the mean loan amount.

8.4 Robust Confidence Intervals for the Center of a Distribution

The goal of interval estimation is to come up with an interval estimate that gives good results in a large variety of cases. It has been pointed out that the precision of the confidence interval based on $\bar{y}$ is significantly reduced in the presence of long tails. In Section 8.2 it was suggested that a trimmed mean (a robust estimator) be used as the estimator of the population mean when the parent distribution is symmetrical with long tails. We will now construct a confidence interval for μ based on a trimmed mean.

▲ Here a robust estimator would be one that is not affected by strange behavior in the tails of the parent distribution.

Large-Sample Confidence Interval Based on $\bar{y}_T$

We will assume that we have a large sample from a symmetric population, which has mean μ and standard deviation σ. For large samples, the sampling distribution of $\bar{y}_T$ is approximately normal with mean μ and a standard error that we will estimate presently.

Recall that the estimated standard error of $\bar{y}$ is $s/\sqrt{n}$. To estimate the standard error of $\bar{y}_T$ it is tempting simply to calculate the ordinary standard deviation of the trimmed sample and divide by $\sqrt{k}$, where k is the size of the trimmed sample. It has been shown, however, that a better estimator of the standard error of $\bar{y}_T$ is obtained by calculating the standard deviation of the trimmed sample, call it s_T, based on the so-called Winsorized sample.

Definition

The **Winsorized sample** is obtained by replacing the trimmed values with the values that were next in line for trimming.

EXAMPLE 8.18

The education levels of the auto workers in Example 8.8 exhibited a long-tailed distribution. From the data, which are repeated here, construct the Winsorized sample based on a 10% trimming.

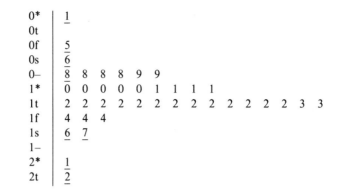

```
0*  | 1
0t  |
0f  | 5
0s  | 6
0–  | 8  8  8  8  9  9
1*  | 0  0  0  0  0  1  1  1  1
1t  | 2  2  2  2  2  2  2  2  2  2  2  2  2  3  3
1f  | 4  4  4
1s  | 6  7
1–  |
2*  | 1
2t  | 2
```

SOLUTION   The sample size is 40 so a 10% trimming will trim four observations from each end. They are underlined in the preceding stem and leaf plot. As stated, the Winsorized sample is obtained by replacing those values with the values next in line for trimming. That will be 8 on the low end and 14 on the upper end. We now present a stem and leaf plot of the Winsorized sample.

```
0–  | 8  8  8  8  8  8  8  9  9
1*  | 0  0  0  0  0  1  1  1  1
1t  | 2  2  2  2  2  2  2  2  2  2  2  2  2  3  3
1f  | 4  4  4  4  4  4  4
```

Definition

The **standard deviation of a trimmed sample** based on the Winsorized sample is given by

$$s_T = s_W \sqrt{(n - 1)/(k - 1)}$$

where s_W is the ordinary sample standard deviation of the Winsorized sample.

EXAMPLE 8.19

Calculate s_T for the Winsorized sample in Example 8.18.

SOLUTION   To find s_T we first find the ordinary sample standard deviation s_W of the Winsorized sample.

From the data in the stem and leaf plot we find

$$s_W = 2.015$$

Thus

$$
\begin{aligned}
s_T &= s_W \sqrt{(n - 1)/(k - 1)} \\
&= 2.015\sqrt{39/31} \\
&= (2.015)(1.122) \\
&= 2.26
\end{aligned}
$$

We now give the estimated standard error of the trimmed mean $\bar{y}_T$.

Definition

The **estimated standard error of the trimmed mean,** $\bar{y}_T$, is given by

$$\text{estSE}(\bar{y}_T) = s_T/\sqrt{k}$$

where s_T is the standard deviation of the trimmed sample and k is the size of the trimmed sample.

As stated earlier, for a sufficiently large sample the sampling distribution of $\bar{y}_T$ is approximately normal. Therefore $\bar{y}_T$ should not deviate more than 1.96 standard errors from μ with 95% probability. That is, the estimate of μ is $\bar{y}_T$ with a margin of error of

$$\pm 1.96 s_T/\sqrt{k}$$

Replacing 1.96 with $z_{\alpha/2}$, we have

Large-Sample Confidence Interval for μ Based on $\bar{y}_T$

APPLICATION: Symmetric distribution with long tails

ASSUMPTION: $n > 30$

ESTIMATOR: $\bar{y}_T$, the trimmed sample mean

ESTIMATED STANDARD ERROR OF THE ESTIMATOR: $\text{estSE}(\bar{y}_T) = s_T/\sqrt{k}$

A $(1 - \alpha)100\%$ confidence interval for μ is given by the limits

$$\bar{y}_T \pm z_{\alpha/2} s_T/\sqrt{k}$$

where s_T is the standard deviation of the trimmed sample based on the Winsorized sample and k is the size of the trimmed sample.

EXAMPLE 8.20

▲ Recall that $\bar{y}_{T.10}$ represents the 10% trimmed mean.

Calculate a 98% confidence interval based on $\bar{y}_{T.10}$ for the mean education level of the auto workers described in Example 8.8. Show that it is more precise than an interval based on $\bar{y}$.

SOLUTION   From Example 8.8 we have

$$\bar{y}_{T.10} = 11.25$$

and from Example 8.19 we have

$$s_T = 2.26$$

The 98% confidence interval for μ takes the form

$$\bar{y}_{T.10} \pm 2.33 s_T/\sqrt{k}$$

which is

$$11.25 \pm (2.33)(2.26)/\sqrt{32}$$

or

$$11.25 \pm .931$$

So the 98% confidence interval for μ based on $\bar{y}_{T.10}$ is from 10.319 to 12.181.

A 98% confidence interval for μ based on $\bar{y}$ takes the form

$$\bar{y} \pm 2.33s/\sqrt{n}$$

and thus, using the data in Example 8.8, we have

$$11.4 \pm (2.33)(3.699)/\sqrt{40}$$

which becomes

$$11.4 \pm 1.363$$

or

$$10.037 \text{ to } 12.763$$

▲ Remember that validity is the confidence level (here 98%) and the precision is measured by the length of the interval.

Figure 8.5 compares the two intervals for μ.

Clearly the interval based on $\bar{y}_T$ is more precise and equally valid.

FIGURE 8.5

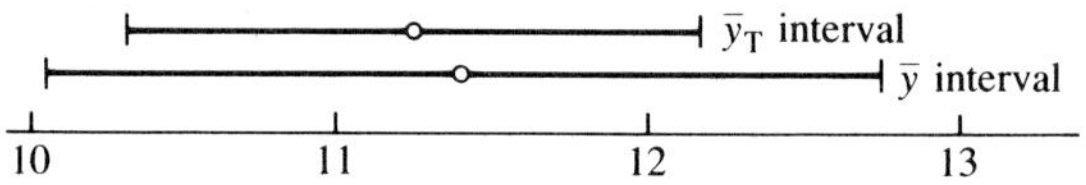

Focus on Problem Solving

Many of the hand-held calculators simplify the calculations of trimmed means and standard deviations. Any calculator that has statistical functions such as a standard deviation will allow one to enter data and also to remove data. One could enter all the sample data, calculate the mean and standard deviation, and then remove the observations that are to be trimmed. Recalculating the sample mean would then be the trimmed mean. One could then enter the additional observations for the Winsorized sample and calculate the standard deviation, which would be the standard deviation based on the Winsorized sample. With a little practice, the calculations become quite easy.

Large-Sample Confidence Interval for the Population Median

In Section 8.2 it was suggested that if the parent distribution is significantly skewed, then the median may be a better indicator of the center of the distribution than is the mean. We will now construct a large sample confidence interval for the population median, θ.

The interval we will give is based on the binomial distribution and is commonly referred to as the confidence interval obtained from the sign test (refer to a text on nonparametric statistics). Assuming that we have a large sample, we can use the Central Limit Theorem and approximate the binomial distribution with the normal distribution.

Up to now, a sample size of 30 has been used to distinguish large samples from small samples. In constructing a confidence interval for the population median, however, a sample size of 20 is used to distinguish between large and small samples. This is because the binomial distribution is reasonably approximated by a normal curve when $n > 20$.

To construct the interval we must first order the sample in an ordered stem and leaf plot. Then the $(1 - \alpha)100\%$ confidence interval will be formed by two of the sample observations, C_1 and C_2. C_1 is located by counting in from the low end of the ordered stem and leaf plot, and C_2 is located by counting in from the high end of the ordered stem and leaf plot (the same way we find Q_1 and Q_3). The distance we count in from each end, call it Location of C, is approximated by the normal distribution and is given by

$$\text{Location of } C = [n - z_{\alpha/2}\sqrt{n}]/2$$

Normally this Location of C will not be a whole number, in which case we round up to the next whole number.

EXAMPLE 8.21

Find a 90% confidence interval for the population median, θ, from the test scores in Example 8.10.

SOLUTION For a 90% confidence interval we have

$$z_{\alpha/2} = z_{.05} = 1.645$$

From the ordered stem and leaf plot obtained in Example 8.10 (which is repeated here) we find that $n = 25$. Substituting these values in the formula for Location of C, we find

Ordered stem and leaf plot							Location of C
1	1	4					$= [n - z_{\alpha/2}\sqrt{n}]/2$
2							$= [25 - 1.645\sqrt{25}]/2$
3	8						$= (25 - 8.2)/2$
4	1	7					$= 16.8/2$
5	8						$= 8.4$
6	1	1	4	5			≈ 9
7	0	2	6	7	8	8	
8	0	1	1	3			
9	0	2	7	8	8		

Thus $C_1 = 64$ (the ninth observation from the low end) and $C_2 = 80$ (the ninth observation from the high end). The 90% confidence interval for the population median is given by the interval from 64 to 80.

Although this confidence interval is suggested when the parent distribution is skewed, it is appropriate for estimating the population median regardless of the shape of the parent distribution.

8.32 For the following sample sizes determine the location in an ordered stem and leaf plot of the two observations that would form a 95% confidence interval for the population median, θ.

(a) $n = 25$ (b) $n = 50$

(c) $n = 100$ (d) $n = 200$

8.33 For a sample size of 100 determine the location in an ordered stem and leaf plot of the two observations that would form a confidence interval for the population median, θ, with the following confidence:

(a) 90%   (b) 98%

(c) 95%   (d) 99%

8.34 Construct a 99% confidence interval estimate based on $\bar{y}_{T.10}$ for the mean math competency score for the freshmen at the small community college in Exercise 8.12.

8.35 Construct a 90% confidence interval estimate for the median weekly incomes for the Puerto Rican families of Miami in Exercise 8.13.

8.36 A college coach is interested in estimating the mean amount spent on recruiting a high school football player. He tabulates the amounts spent on recruiting his last 35 players in the following stem and leaf plot.

0	00, 00, 00, 00, 00
1	00, 20
2	10, 50
3	00, 25, 40, 50, 50, 70, 90
4	00, 25, 50, 50, 75, 90, 95
5	00, 20, 30, 50, 50, 65, 75, 80
HI	1000, 1050, 1500, 2000

Find a 90% confidence interval based on $\bar{y}_{T.10}$ for the mean amount spent per player.

8.37 A new pain reliever is to be tested on patients in a hospital. The drug is administered to a random sample of 50 patients, and the time to relief is recorded. From the following data construct a stem and leaf plot and decide whether M or $\bar{y}_T$ is the best estimator of a typical time to relief. Construct a 95% confidence interval for the center of the distribution. The following times are recorded in minutes.

4.1	3.8	4.7	3.9	4.6
5.4	4.5	3.6	3.9	5.1
4.9	5.6	6.4	5.3	4.8
3.5	5.2	3.6	5.7	6.8
6.4	8.3	3.8	4.7	5.4
6.3	7.4	4.6	3.8	5.3
3.9	5.2	5.9	6.7	8.4
3.1	4.5	7.1	4.8	5.9
4.0	3.4	5.7	8.8	3.3
4.9	5.0	3.6	3.9	4.8

8.38 From the following ordered stem and leaf plots, construct a 95% confidence interval for the population median, θ.

(a)
0	5
1	2 8
2	1 2 3 5 7 8
3	1 3 4 5 5 6 9
4	2 5 8 8 9
5	1 4
6	3
7	4

(b)
0	0 0 1 5
1	2 4 5 5 8 9 9
2	1 2 3 5
3	3 4 6
4	2 5
5	1 4
6	3
7	4
8	1

(c)
0	0 0 0 1 5 7 9
1	2 5 8 9
2	1 3 5
3	3 4 6
4	2 5 7
5	1
6	3
7	4
8	1

8.39 The time to repair the body of an automobile involved in a collision can vary from a reasonably short time to a rather long time when special parts must be ordered. It is reasonable to expect that the distribution of times for repairs will be skewed positively. From the following recorded times (in hours) for repairing 22 automobiles, construct a 98% confidence interval for the median time of a repair:

10.3	4.1	8.6	2.1	3.7	5.6	11.4	5.7
3.8	4.5	2.7	22.6	5.5	9.6	4.8	3.9
10.4	4.8	5.9	6.7	7.6	12.0		

8.40 A university is thinking of underwriting its own health insurance. In order to get an idea of the size of a typical claim, a random sample of 30 claims submitted under the old plan is chosen for analysis. From the following claim amounts construct a 90% confidence interval based on $\bar{y}_{T.10}$ for the mean claim amount:

$126	201	910	540	350	425
55	460	310	456	1,122	350
575	375	560	1,025	550	94
280	375	460	650	385	20
450	720	275	85	428	360

8.41 A realtor is studying the prices of single-family homes in a small city in a southern state. From the following prices of homes, construct a 99% confidence interval based on $\bar{y}_{T.10}$ for the mean cost of a home (amounts are in hundreds of dollars):

$825	955	810	610	832	543
840	722	880	645	316	875
700	980	815	1,280	857	743
815	689	875	445	842	1,100
868	758	847	665	844	1,350
860	780	875	880	714	847
839	885	1,000	890	855	841

8.42 The daily receipts in a small hardware store for 36 working days are recorded as follows. Find a 95% confidence interval based on $\bar{y}_{T.10}$ for the mean daily receipt.

$ 98.50	224.65	231.78	35.60	156.80	124.70
210.50	375.60	217.44	114.00	251.70	195.00
155.70	133.50	287.50	151.40	423.50	241.98
222.37	65.40	172.30	234.85	115.70	168.90
205.11	214.70	216.20	196.30	257.80	195.70
233.70	528.50	227.82	224.31	27.18	229.50

8.43 Top executives are in demand by the major corporations. From the following random sample of advertised salaries, construct a 95% confidence interval for the median salary offered general managers in 1984:

$ 95,000	85,000	65,500	98,000	150,000
75,000	60,000	78,000	100,000	75,500
90,000	85,000	150,000	120,000	55,000
70,000	85,500	120,000	85,000	99,900
110,000	150,000	95,000	60,000	185,000
75,000				

8.44 The median prices of single-family homes in 37 markets across the United States as reported by the National Association of Realtors are listed as follows. From the data construct a 98% confidence interval for the median price of a single-family home in the United States.

Albany, N.Y.	$ 52,400
Anaheim–Santa Ana, Calif.	134,900
Atlanta, Ga.	64,600
Baltimore, Md.	65,200
Birmingham, Ala.	66,600
Boston, Mass.	102,000
Chicago, Ill.	77,500
Cincinnati, Ohio	59,600
Cleveland, Ohio	65,600
Columbus, Ohio	60,400
Dallas–Fort Worth, Tex.	83,400
Denver, Colo.	85,000
Detroit, Mich.	48,000
Fort Lauderdale, Fla.	76,000
Houston, Tex.	79,600
Indianapolis, Ind.	54,200
Kansas City, Mo.	57,600
Los Angeles, Calif.	115,300
Louisville, Ky.	49,500
Memphis, Tenn.	65,200
Miami, Fla.	84,400
Milwaukee, Wis.	69,600
Minneapolis–St. Paul, Minn.	75,100
Nashville, Tenn.	64,000
New York metro area	106,900
Oklahoma City, Okla.	63,600
Philadelphia, Pa.	59,500
Providence, R.I.	61,400
Rochester, N.Y.	62,100
St. Louis, Mo.	64,400
Salt Lake City, Utah	67,600
San Antonio, Tex.	71,600
San Diego, Calif.	102,900
San Francisco, Calif.	132,600
San Jose, Calif.	121,600
Tampa, Fla.	60,800
Washington, D.C.	92,900

SOURCE: National Association of Realtors.

8.5   Confidence Interval for a Population Proportion

Consider a Bernoulli population that consists of a collection of successes and failures. Let π be the proportion of successes. For example, the population could be the voters who either favor an issue (a success) or oppose it (a failure). The value of π represents the proportion of voters that favors the issue. In this section we will learn how to estimate π with a confidence interval.

In most situations when we are estimating a proportion, the sample size will be large. For example, public opinion polls usually involve a large number of people. The

▲ Recall that the approximation is satisfactory if $n\pi \geq 5$ and $n(1 - \pi) > 5$.

Central Limit Theorem applies to sample proportions and says that for large n the sample proportion, p, has a sampling distribution that is approximately normal, with a mean π and standard error, $\text{SE}(p) = \sqrt{\pi(1 - \pi)/n}$.

Because the sampling distribution is approximately normal, we have that the margin of error in using p to estimate π is

$$\text{ME} = z_{\alpha/2}\sqrt{\pi(1 - \pi)/n}$$

Moreover, because n is large, the standard error can be estimated with

$$\text{estSE}(p) = \sqrt{p(1 - p)/n}$$

(replacing π with p) and thus the margin of error is estimated to be

$$\text{estME} = z_{\alpha/2}\sqrt{p(1 - p)/n}$$

Adding and subtracting the estimated margin of error to the estimator p, we have the following confidence interval for π:

$$p \pm z_{\alpha/2}\sqrt{p(1 - p)/n}$$

EXAMPLE 8.22

To investigate the proportion of smokers who believe that smoking causes cancer, a researcher randomly samples 500 smokers from all walks of life and finds that 350 believe smoking can cause cancer. Obtain a 99% confidence interval for the true proportion of smokers who believe that it can cause cancer.

SOLUTION The form of a 99% confidence interval for π, the true population proportion, is

$$p \pm 2.58\sqrt{p(1 - p)/n}$$

The sample reveals that

$$p = 350/500 = .7 \quad \text{and thus} \quad (1 - p) = .3$$

so the interval for π becomes

$$.7 \pm 2.58\sqrt{(.7)(.3)/500}$$

which gives

$$.7 \pm .0529$$

The interval is from .6471 to .7529 and we are 99% confident that this interval does contain the proportion of smokers that believes smoking causes cancer.

Using the formula for the margin of error, we can determine the sample size necessary to guarantee a given margin of error for a given validity level.

EXAMPLE 8.23

The estimated margin of error in Example 8.22 is .0529. Find the sample size necessary to reduce that margin to .03.

SOLUTION   To retain the same validity the margin of error is

$$ME = 2.58 \sqrt{\pi(1 - \pi)/n}$$

which must also equal .03. Thus we have the formula

$$.03 = 2.58 \sqrt{\pi(1 - \pi)/n}$$

and we must solve for n.

To solve the equation we need a value for π. However π is unknown (we are trying to find the sample size to estimate it). In a situation such as this we have two choices. First we can use $\pi = .7$, which is the estimate obtained with the sample above; that is, use a value obtained in a previous experiment or pilot study. Or, if we have no idea of the value of π, we can use $\pi = .5$, which tends to be a conservative approach.

▲ Recall that the standard error and hence the margin of error of p is maximized when $\pi = .5$.

In this example, because we have the estimate of $\pi = .7$, we will use it. The equation becomes

$$.03 = 2.58 \sqrt{(.7)(.3)/n}$$

So

$$\sqrt{n} = \frac{(2.58) \sqrt{(.7)(.3)}}{.03}$$

$$\sqrt{n} = 39.41$$

and

$$n = (39.41)^2$$
$$= 1,553.1481$$

Thus we need a sample of size 1,554 to estimate π with a 99% confidence interval with a margin of error of no more than .03.

Sample Size Determination for Estimating π

In general, to estimate π with a $(1 - \alpha)100\%$ confidence interval so that the bound on the margin of error is B, we solve the following equation for n:

$$B = \frac{z_{\alpha/2} \sqrt{\pi(1 - \pi)}}{\sqrt{n}}$$

The solution is

$$n = \frac{z_{\alpha/2}^2 \pi(1 - \pi)}{B^2}$$

The results of this section are summarized as follows:

Large-Sample Confidence Interval for π

APPLICATION: Bernoulli populations

ESTIMATOR: p, the sample proportion

STANDARD ERROR OF THE ESTIMATOR: $SE(p) = \sqrt{\pi(1 - \pi)/n}$, estSE$(p)$ $= \sqrt{p(1 - p)/n}$

A $(1 - \alpha)100\%$ confidence interval for π is given by the limits

$$p \pm z_{\alpha/2}\sqrt{p(1 - p)/n}$$

SAMPLE SIZE: $n = \dfrac{z_{\alpha/2}^2\pi(1 - \pi)}{B^2}$

EXERCISES 8.5

8.45 Of 900 people treated with a new drug, 180 showed an allergic reaction. Estimate with a 90% confidence interval the proportion of the population that would show an allergic reaction. How large a sample is necessary to ensure that the margin of error is no more than .03?

8.46 The state of Massachusetts matched bank statements with welfare recipients and found that of 1,000 welfare recipients 200 had accounts ranging from $20,000 to $50,000, 50 had accounts ranging from $50,000 to $100,000, and 10 had accounts in excess of $100,000. The state of Massachusetts says that if a person has a bank account in excess of $20,000, he or she should not receive welfare benefits. Estimate with a 99% confidence interval the percent of welfare recipients who should not receive benefits. How large a sample is necessary to ensure that the margin of error is no more than .03?

8.47 To learn whether financial aid is a determining factor in the choice of a college, The College Board conducted a survey of 1,183 students. They determined that 722 high school graduates entered their first choice of college, regardless of financial offers (*USA Today,* October 30, 1984). Find a 99% confidence interval for the proportion that did not use financial aid as a determining factor in the choice of their college.

8.48 What is the margin of error in the interval estimate of Exercise 8.47? What sample size will reduce the margin of error to no more than .03?

8.49 A report by the NCAA showed that 635 of 2,048 athletes at 11 schools has used anti-inflammatory painkillers at least once in the past 12 months (Charlotte *Observer,* January 11, 1985). Obtain a 90% confidence interval for the proportion of all athletes that use anti-inflammatories.

8.50 In 1983, a record 64,877 applicants took bar exams in the United States. The accompanying table gives the number of takers and the number who passed in three states. Construct separate 95% confidence intervals for the pass rates in the three states. Graphically display all three confidence intervals on the same line axis.

State	Test takers	No. passed
California	12,499	5,112
Iowa	379	344
Ohio	2,054	1,660

8.51 A Roper poll of 2,000 American adults (*USA Today,* January 15, 1985) showed that 1,440 felt that chemical dumps are among the most serious environmental problems. Estimate with a 98% confidence interval the proportion of the population that considers chemical dumps among the most serious environmental problems.

8.52 A 1984 survey of 17,000 seniors in approximately 140 public and private high schools nationwide (by the University of Michigan's Institute for Social Research) showed that 10,540 had tried an illegal drug. Construct a 99% confidence interval for the proportion of all high school seniors who have tried an illegal drug.

8.53 *Newsweek* magazine interviewed 757 adults by telephone on January 3 and 4, 1985 (*Newsweek,* January 14, 1985), and found that 159 felt that abortions should be legal under all circumstances. Construct a 90% confidence interval for the proportion of all adults that feels this way. What is the maximum margin of error for the survey? How large a sample will yield a maximum margin of error of no more than .03?

8.54 It is believed that 30% of all adults are overweight. How large a sample is necessary to estimate the true proportion of adults that are overweight with a 95% confidence interval so that the margin of error of the estimate is no more than 2 percentage points?

8.55 A pollster believes that approximately 60% of the registered voters will be in favor of a particular issue. How many people will he need to interview so that the margin of error of his 98% confidence interval estimate will be no more than 3%?

8.56 Suppose one wishes to estimate the percent of American citizens who receive some sort of government aid. How large a sample will be needed in order that the margin of error of a 95% confidence interval estimate will be no more than 2%?

Small-Sample Confidence Interval for the Mean of a Normal Distribution—Student's *t* Distribution

The confidence intervals for μ, constructed in the previous sections, relied on the fact that for large samples, the sampling distributions of $\bar{y}$ and $\bar{y}_T$ are approximately normal regardless of the parent distribution. We realize that the larger n is the better the approximation; but also, the closer the parent distribution resembles a normal distribution the better the approximation. In fact, if the parent distribution is normal, then the distribution of $\bar{y}$ is exactly normal regardless of the sample size. We shall use this fact in the development of the small-sample confidence interval for μ.

To develop the small-sample confidence interval for μ we need a basic understanding of a new sampling distribution, called *Student's t distribution*. This distribution was developed by a British chemist, W. S. Gosset, in 1908. Gosset worked for a brewery company that did not allow its chemists to publish their research; however Gosset did publish his results under the pseudonym "Student." Thus the distribution became known as Student's *t* distribution, using the letter *t* to distinguish it from *z*, the standard normal.

Like the standard normal distribution, Student's *t* distribution is bell shaped and symmetrical about zero. Its standard deviation is not 1, however, but some value larger than 1, which means that its spread is more than the spread of the standard normal. The actual magnitude of the standard deviation depends on the sample size, and for $n > 3$ is given by

$$\sqrt{(n-1)/(n-3)}$$

▲ This is not an equation that you need to worry about. It is simply here to illustrate that as n gets larger, the standard deviation of t gets closer to 1.

Clearly as n gets larger this standard deviation gets closer to 1, in fact, the *t* distribution approaches the standard normal distribution as n becomes infinitely large.

But if n is small, we see from Figure 8.6 that the *t* distribution has thicker and longer tails than does the standard normal distribution. The general shape of the *t* distribution depends on a quantity called its *degrees of freedom*. For the application here, the degrees of freedom are $n - 1$, and hence the dependence on the sample size.

FIGURE 8.6

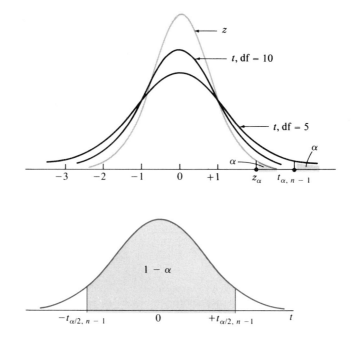

FIGURE 8.7

Referring to Figure 8.6, we see the general shape of a few t curves in comparison to the standard normal curve. Just as z_α is the value of the standard normal variable such that α of the area lies above it, $t_{\alpha, n-1}$ denotes the value of the t variable with $n-1$ degrees of freedom such that α of the area lies above it. Because of the symmetry about zero, the limits that include the middle $(1 - \alpha)100\%$ (see Figure 8.7) will be

$$-t_{\alpha/2, n-1} \quad \text{to} \quad +t_{\alpha/2, n-1}$$

Because the t has longer tails than the standard normal, the limits $\pm t_{\alpha/2, n-1}$ that include the middle $(1 - \alpha)100\%$ will naturally be larger than the limits, $\pm z_{\alpha/2}$ found in the normal tables.

Table B4 gives the t values for different values of α and different values of the degrees of freedom (abbreviated df). For example, for df $= 10$; $t_{.025, 10} = 2.228$.

Because the t is symmetric about 0, the middle 95% will be between -2.228 and $+2.228$. Recalling that the middle 95% of a standard normal is between -1.96 and $+1.96$, we see that the t is indeed spread out more than the standard normal.

EXAMPLE 8.24

Using Table B4, find the upper and lower 5% points of the t distribution with 15 degrees of freedom.

SOLUTION   With 15 degrees of freedom, the upper 5% point is found to be

$$t_{.05, 15} = 1.753$$

Because the distribution is symmetrical about 0, the lower 5% point is

$$-t_{.05, 15} = -1.753$$

We then have that the middle 90% of the distribution lies between -1.753 and $+1.753$.

Returning to the development of the confidence interval for μ, we will assume that we have a random sample from a *normal* population with mean μ and standard deviation σ. As stated, for large or small samples, the distribution of the sample mean $\bar{y}$ is exactly normally distributed with mean μ and standard deviation $\sigma/\sqrt{n}$ if the parent distribution is normal.

Computing the standardized version of $\bar{y}$, that is, the z-score for $\bar{y}$, we have that

$$z = \frac{\bar{y} - \mu}{\sigma/\sqrt{n}}$$

has a standard normal distribution.

▲ It is when we have a large sample that we don't need the normality assumption.

Usually σ is unknown and must be estimated with the sample standard deviation, s. In replacing σ with s, the distribution becomes a t distribution. That is,

$$t = \frac{\bar{y} - \mu}{s/\sqrt{n}}$$

is distributed as a t distribution with $n - 1$ degrees of freedom. For a large sample size the substitution of s for σ does not have an appreciable effect on the distribution because the t approaches the standard normal when n becomes large.

For small samples, however, the distribution of t does depend on the sample size through its degrees of freedom. This means that in the construction of a small-sample confidence interval for μ, the value $z_{\alpha/2}$ found in the large-sample interval will have to be replaced with a $t_{\alpha/2,\, n-1}$ value found in Table B4. Consequently the small-sample $(1 - \alpha)100\%$ confidence interval for μ becomes

▲ This is the same interval as the large-sample interval for μ except that σ has been replaced with s and the $z_{\alpha/2}$ value has been changed to the $t_{\alpha/2,\, n-1}$ value.

$$\bar{y} \pm t_{\alpha/2,\, n-1}s/\sqrt{n}$$

Summarizing, we have

Small-Sample Confidence Interval for μ Based on $\bar{y}$

APPLICATION: Normal populations

ASSUMPTIONS: A small sample from a normal population

ESTIMATOR: $\bar{y}$, the sample mean

ESTIMATED STANDARD ERROR OF THE ESTIMATOR: $\text{estSE}(\bar{y}) = s/\sqrt{n}$

A $(1 - \alpha)100\%$ confidence interval for μ is given by the limits

$$\bar{y} \pm t_{\alpha/2,\, n-1}s/\sqrt{n}$$

EXAMPLE 8.25

Following are the Miller Personality Test scores for 25 college students who have applied for admission to graduate school. Refer to Example 6.9 for a descriptive analysis of the data.

21	18	20	25	23	19	30	24	29	14	25	22	35
26	23	16	33	25	22	17	34	22	31	27	25	

Find a 99% confidence interval for the mean score of college students on the Miller Personality Test.

SOLUTION In Example 6.9, it was determined that the scores are from a distribution that appears to be normal in shape. Because the sample size is small ($n = 25$), we will find a t-interval for μ based on $\bar{y}$; it takes the form

$$\bar{y} \pm t_{\alpha/2,\,n-1} s/\sqrt{n}$$

Because a 99% interval is suggested, we find

$$t_{\alpha/2,\,n-1} = t_{.005,\,24} = 2.797$$

Also, in Example 6.9 we found that

$$\bar{y} = 24.24 \quad \text{and} \quad s = 5.555$$

Thus the 99% confidence interval for μ becomes

$$24.24 \pm (2.797)(5.555)/\sqrt{25}$$

which is

$$24.24 \pm 3.1075$$

Thus we are 99% confident that the interval from

$$21.1325 \quad \text{to} \quad 27.3475$$

does indeed contain the true mean Miller Personality Test Score.

EXERCISES 8.6

8.57 For the following sample sizes and values of α, find $t_{\alpha/2,\,n-1}$.

(a) $n = 14, \quad \alpha = .05$ (b) $n = 26, \quad \alpha = .05$

(c) $n = 10, \quad \alpha = .10$ (d) $n = 20, \quad \alpha = .10$

(e) $n = 22, \quad \alpha = .01$ (f) $n = 41, \quad \alpha = .01$

8.58 From the following summary data, construct 95% confidence intervals for the population mean. (Assume the data are randomly selected from normal populations.)

(a) $n = 15, \ \Sigma y = 810, \ \Sigma y^2 = 63{,}780$

(b) $n = 24, \ \Sigma y = 244.8, \ \Sigma y^2 = 4{,}876.46$

8.59 From the following ordered stem and leaf plots, construct 99% confidence intervals for μ based on $\bar{y}$. (Assume the data are randomly selected from normal populations.)

(a)
1*	0 3
1–	5 7 8 8
2*	0 2 3 4 4
2–	5 6 6 7 8 9
3*	0 1 3 4
3–	6 7 9
4*	0 1
4–	5

(b)
1*	0
1–	5 7
2*	0 2 4
2–	5 6 9
3*	0 1 2 3
3–	5 6 7 7 8
4*	0 1 2 4
4–	5 7 9
5*	3 4
5–	7

(c)

1*	0	2	3		
1–	5	7	8	8	9
2*	0	2	3	4	4	4
2–	5	6	7	9	
3*	0	1	3		
3–	5	6			
4*	0				

8.60 A sample of 22 public high school students spent an average of 6.5 hours studying per week, and the standard deviation was 2.3 hours. Assuming the distribution of hours spent studying is near normal, find a 95% confidence interval for the mean number of hours spent studying by public high school students.

8.61 A new drug to anesthetize patients was tested on 10 patients who were to undergo surgery. Their recovery times, in hours, were

2.6	3.0	2.8	3.1	3.5
2.9	3.1	2.7	2.9	3.3

Assuming recovery time is normally distributed, construct a 90% confidence interval for the mean recovery time for the new drug.

8.62 On August 20, 1984, a skull and almost complete skeleton of a 12-year-old male *Homo erectus* who died some 1.6 million years ago was found on the west side of Lake Turkana in northern Kenya. It is believed that the 5 foot 6 inch youth would have matured to 6 feet had he lived (*USA Today,* October 19, 1984). This is contrary to the assumption that early humans were smaller than modern man. *Homo erectus* has been found in Java, China, and various African sites. Suppose we were able to determine the adult height of five *Homo erectus* skeletons to be 64, 70, 73, 69, and 70 inches. Assume that the parent distribution is near normal. Find a 95% confidence interval for the mean adult height of *Homo erectus*. Does the interval contain 6 feet?

8.63 In order to estimate the mean number of visitors per day to the John F. Kennedy Library, the number of visitors was recorded on 12 randomly selected days throughout the year. Assuming the parent distribution is near normal, estimate with a 90% confidence interval the mean number of visitors per day.

198,286	249,821	294,653	255,728
267,475	231,759	275,641	191,374
228,391	307,613	250,834	259,540

8.64 An air pollution index for 15 randomly selected days during the summer months was recorded for a major western city. Assuming the parent distribution is near normal, estimate the mean air pollution index with a 98% confidence interval.

57.6	61.2	59.4	65.6	58.3
44.7	63.2	48.8	55.7	59.0
64.3	43.2	59.7	71.2	45.6

8.7 Small-Sample Confidence Interval for the Center of a Nonnormal Distribution (optional)

If the parent distribution does not depart drastically from normality, then the small-sample confidence interval for μ, obtained in Section 8.6, is appropriate. The normality assumption is reasonable in a large number of applications, because the normal distribution approximates many real-life situations. In this section we address the problem that occurs when the normality assumption is not appropriate.

As described by the Winsor principle, departures from normality occur in the tails of the distribution. Usually they are classified as either short tailed, long-tailed, or skewed. We have seen that traditional statistical techniques deteriorate drastically in the presence of the long-tailed distribution. With the added assumption of symmetry, however, we observed in Section 8.4 (large-sample case) that the long-tailed distribution could be dealt with by using $\bar{y}_T$ as the estimate of μ. Recent research indicates that the same is possibly true for small-sample theory. However, until more concrete procedures are available for estimating the mean of a nonnormal population with a small sample, we shall limit our study to that of estimating the population median.

Population Median

There are many situations where the population median is the parameter to estimate. For example, if the parent distribution is skewed, then the median is a better indicator of the center of the distribution than is the mean. Also in Section 8.1 we saw that in the case of at least one symmetric long-tailed distribution (the Laplace) the sample median is a more efficient estimator of the point of symmetry than is the sample mean.

To obtain a small-sample ($n \leq 20$) confidence interval for the population median, we need first to order the sample (ordered stem and leaf plot). Then the confidence interval will be formed by two of the sample observations, C_1 and C_2. C_1 is located by counting from the low end of the ordered stem and leaf plot, and C_2 is located by counting from the high end of the ordered stem and leaf plot (the same way we found C_1 and C_2 in Section 8.4). Unlike Section 8.4, the distance we count from each end (Location of C) is found from the binomial tables with $p = \frac{1}{2}$ or from Table 8.2. We will use Table 8.2. For example, if we desire a 95% confidence interval for the population median and the sample size is, say $n = 15$, then Table 8.2 gives the

$$\text{Location of } C = 4$$

Thus C_1 is the fourth observation from the low end of the ordered stem and leaf plot, and C_2 is the fourth observation from the high end of the ordered stem and leaf plot.

EXAMPLE 8.26

A study was undertaken to investigate the number of days to recovery for patients who suffer from a certain type of viral disease, after being treated with a new drug. From the data recorded in the following stem and leaf plot, obtain a 95% confidence interval for the "average" number of days to recovery.

Recovery time in days

0–	5 6 7 8 8 9
1*	1 2 2 3
1–	7 8
2*	
2–	8
3*	2
3–	7

From the stem and leaf plot, it appears that the data are skewed right. Therefore we choose to find a small-sample confidence interval for the median number of days to recovery. From Table 8.2 with 15 observations and 95% confidence we find

$$\text{Location of } C = 4$$

Thus C_1, the fourth value from the low side, is 8, and C_2, the fourth value from the high side, is 18. Therefore the 95% confidence interval for the median number of days is given by the limits

$$(8, 18)$$

▲ The exact confidence level of 96.48% is found using the binomial probability tables. You do not have to be able to compute the exact confidence level.

We should point out that the confidence level given in Table 8.2 is only approximate, because of the discrete nature of the binomial distribution. For example, the exact confidence level for the interval obtained in Example 8.26 is 96.48%, which is reasonably close to 95%. It is as close to 95% as we can get, because we must count to the fourth observation—we cannot count in a fraction of an observation.

TABLE 8.2
Location of C

n	90% C.I.	95% C.I.	99% C.I.
4	1		
5	1		
6	2	1	
7	2	1	
8	3	2	1
9	3	2	1
10	3	2	1
11	3	2	1
12	4	3	2
13	4	3	2
14	5	4	3
15	5	4	3
16	5	4	3
17	6	5	4
18	6	5	4
19	6	5	4
20	7	6	5

EXERCISES 8.7

8.65 For the following sample sizes, determine the location of the two observations in an ordered stem and leaf plot that would form a 95% confidence interval for the population median, θ.

(a) $n = 10$ (b) $n = 12$
(c) $n = 15$ (d) $n = 19$

8.66 For a sample size of 18, determine the location of the two observations in an ordered stem and leaf plot that would form a confidence interval for the population median, θ, with the following confidence:

(a) 90%   (b) 95%   (c) 99%

8.67 From the following ordered stem and leaf plots, construct 95% confidence intervals for the population median.

```
(a)  4 | 0  3  5
     5 | 0  2  3  5  5  7  9
     6 | 1  2  5
     7 | 4
     8 | 1
     9 | 2
```

```
(b)  4 | 0
     5 | 0  2  5  9
     6 | 1  2  4  5  5  8
     7 | 4  7  9
     8 | 1
     9 | 2  5
    10 | 1
```

```
(c)  5 | 0  2  9
     6 | 1  4  5  8
     7 | 4  9
     8 | 1
     9 | 2
    10 | 1  4
```

8.68 The female alcoholic generally begins drinking at a later age than does the male alcoholic. Following are the ages at which 14 female alcoholics began drinking:

18	16	24	14	19	22	28
16	19	21	22	35	20	15

Find a 99% confidence interval for the median age.

8.69 Shoplifting and employee theft cost retailers over $2 billion in 1983. Following are the values of merchandise found in the possession of shoplifters apprehended in a department store over a busy weekend:

$38	20	12	100	22	45
5	75	150	20	25	19

Construct a 90% confidence interval for the median value of shoplifted merchandise.

8.70 In 1984 the average medical malpractice award approached $1 million ($954,858), almost tripling the 1979 average of $367,319 (Jury Verdict Research, Inc.; *USA Today*, January 17, 1985). Considering that some malpractice awards are extremely large, it might be more appropriate to report the median award. From the following data construct a 90% confidence interval for the median malpractice award (data recorded in thousands of dollars):

760	380	125	250	2,800	450
100	150	2,000	180	650	275
850	1,700	1,500	3,000	390	

8.71 In 1983 Houston, Texas, had the highest motor vehicle traffic death rate among cities over 1 million people. The accompanying table gives the reported rates for 6 major cities:

City	Rate per 100,000
Houston	20.3
Los Angeles	13.8
Detroit	8.7
Chicago	8.1
New York	7.2
Philadelphia	7.1

SOURCE: National Safety Council; *USA Today*, November 27, 1984.

Use the rates to construct a 90% confidence interval for the median traffic death rate.

8.72 Had one invested $10,000 in coins in 1980, by the end of 1984 his or her investment would have grown to $23,602 (*Fact* magazine, *USA Today,* October 29, 1984). Over this 5-year period, coins were the best investment, with over 27% per year return on the investment. Investing in the New York Stock Exchange would have yielded an average rate of return of 10.7% per year. Suppose that an investor is investigating 12 possible investments. He or she calculates the return on each investment for the past year with the following results:

12.6	9.8	13.2	11.6	12.1	8.7
15.6	10.4	18.4	11.2	9.4	10.5

Construct a 90% confidence interval for the median return on investment.

8.73 Lewiston, Idaho, has the least expensive electricity in the United States. A 3-month bill for 1,500 kilowatt hours during December 1983, January and February 1984 was only $31.14 (National Association of Regulatory Utility Commissioners). Average home electric bills in the winter can vary as much as 600%, depending on where you live. Following are the costs of some 3-month bills (1,500 kilowatt hours) at 15 randomly selected cities on the West Coast:

$ 78.26	184.28	124.36	35.28	144.70
197.75	120.60	110.18	131.49	148.14
94.50	129.34	101.26	135.70	110.90

Construct a 99% confidence interval for the median 3-month bill for cities on the West Coast.

Computer Session (optional)

▲ A computer routine for constructing a confidence interval for the population median is not given since the procedure is easily accomplished by hand.

There are two Minitab commands that relate to the construction of confidence intervals. Both commands are used to construct confidence intervals based on $\bar{y}$ for the population mean. ZINTERVAL is used when the population standard deviation is known, and TINTERVAL is used when the population standard deviation is not known and the parent distribution is assumed to be normally distributed.

ZINTERVAL

The form of the command is

```
ZINTERVAL K PERCENT, SIGMA = K, FOR C,C,...,C
```

The K percent is the desired confidence level specified as a decimal. SIGMA is the value of the known population standard deviation.

EXAMPLE 8.27

Are bounced-check fees too high? The California Supreme Court thinks so. The court unanimously ordered a lower court to decide on the fairness of bounced-check fees in California (*USA Today,* July 19, 1985). The court said that California banks collect over $200 million each year from the bounced-check fee when the actual cost of processing a bounced check is about $1.00. Following are the fees charged (in dollars) at some randomly selected banks across the United States.

$20	15	20	25	20
10	30	10	35	10
5	20	35	25	20
15	30	25	20	5
30	20	30	15	25

Construct a 90% confidence interval for the mean bounced-check fee. Use the ZINTERVAL command with 5 as the value of σ.

SOLUTION The solution appears in Figure 8.8. First, the data are stored in the computer with the SET command. They are then printed back out to check for errors. Before construction of the confidence interval, a stem and leaf plot and a box plot are constructed to check out the lengths of the tails of the distribution to be certain that we do not have a long-tailed or skewed distribution.

FIGURE 8.8

Minitab output for Example 8.27

```
MTB > SET DATA IN C1
DATA> 20 15 20 25 20 10 30 10 35 10 5 20 35 25 20 15 30
DATA> 25 20 5 30 20 30 15 25
DATA> END

MTB > PRINT C1

C1
    20    15    20    25    20    10    30    10    35    10
     5    20    35    25    20    15    30    25    20     5
    30    20    30    15    25
```

```
MTB > STEM C1

Stem-and-leaf of C1          N  = 25
Leaf Unit = 1.0

      2     0 55
      5     1 000
      8     1 555
    (7)     2 0000000
     10     2 5555
      6     3 0000
      2     3 55


MTB > BOXPLOT C1


                                -------------------
             ----------------I        +         I----------------
                                -------------------
          ----+---------+---------+---------+---------+---------+--C1
           6.0       12.0      18.0      24.0      30.0      36.0


MTB > ZINTERVAL .90, SIGMA = 5.0, C1

THE ASSUMED SIGMA =5.00


                  N     MEAN    STDEV  SE MEAN   90.0 PERCENT C.I.
C1               25    20.60     8.58    1.00  (   18.95,    22.25)
```

TINTERVAL

The form of the command is

```
TINTERVAL K PERCENT, FOR C,C,...,C
```

EXAMPLE 8.28

Repeat the confidence interval in Example 8.27, except assume σ is unknown and use the TINTERVAL command.

SOLUTION   Normally we would not do a t-interval and a z-interval on the same data set; only one is appropriate. However to illustrate the command we will construct the t-interval using the same data. Assuming it has been entered in C1 and we have checked the lengths of the tails, the command and output are presented in Figure 8.9. We see that the t-interval is somewhat longer than the z-interval. There are two reasons for this: (1) the sample standard deviation is considerably larger than the assumed value of σ of 5.0 and, (2) generally t-intervals are longer than z-intervals because the t distribution is more spread out than the normal distribution.

FIGURE 8.9

Minitab output for Example 8.28

```
MTB > TINTERVAL .90, C1

          N      MEAN    STDEV   SE MEAN    90.0 PERCENT C.I.
C1       25     20.60     8.58      1.72   (   17.66,    23.54)
```

EXERCISES 8.8

8.74 Using the RANDOM command and the NORMAL sub-command, randomly generate 10 samples of 50 observations each from a normal population with mean 70 and standard deviation of 12. Construct 10 90% confidence intervals for μ (one for each of the 10 samples), using $\sigma = 12$ and the ZINTERVAL command.

(a) How many of the 10 intervals contain μ, which is specified to be 70?

(b) How many intervals would you expect to contain μ?

(c) If the confidence level had been 99% instead of 90%, would the interval be longer or shorter?

(d) Would you expect more of the intervals to contain μ if the confidence level is 99%?

8.75 Enter the housing cost data at the beginning of the chapter (Statistical Insight problem) into the computer.

(a) Construct a 98% confidence interval for the population mean using the TINTERVAL command.

(b) Construct a stem and leaf plot and a box plot of the data.

(c) Comment on the shape of the distribution.

(d) Considering the intervals found in part (a) and in Exercise 8.44, which do you think best describes the center of the data?

8.76 From the Executive Salary Data in Appendix A, enter the salaries of the executives, and compute a 95% confidence interval for the mean salary.

8.77 Is the normality assumption required for the construction of the t-interval in Exercise 8.76? Explain.

Summary and Review

☑ In Chapter 6 we learned to classify the different distributional shapes. Once the shape of the distribution is determined, we can decide which parameter best describes the characteristic of interest. For example, is it better to examine the population mean or the population median? Once the parameter is decided upon, we are in a position to determine what statistic best estimates the parameter. In making this decision we must be familiar with the desirable properties of an *estimator* (the statistic used to estimate the parameter). The three properties studied are *unbiasedness, efficiency,* and *consistency.*

☑ For symmetric distributions, the mean, the median, and all the trimmed means coincide; therefore to estimate a typical score we simply estimate the mean. The estimator used to estimate the mean is determined by the type of symmetric distribution we have observed.

☑ If the distribution is normal or, in the large-sample case, symmetrical with tails that are not excessively long, the confidence interval for the mean should be based on $\bar{y}$. For large samples it takes the form

$$\bar{y} \pm z_{\alpha/2} s/\sqrt{n}$$

For a normal parent distribution and a small sample size it takes the form

$$\bar{y} \pm t_{\alpha/2,\, n-1} s/\sqrt{n}$$

☑ If the distribution is symmetric with long tails, then the confidence interval for the mean should be based on $\bar{y}_T$. For large samples it takes the form

$$\bar{y}_T \pm z_{\alpha/2} s_T/\sqrt{k}$$

☑ If the distribution is skewed, the best measure of center is the median and therefore the confidence interval should be for the population median. The confidence interval is formed by two sample observations, (C_1, C_2), where C_1 is found by counting in from the low end of an ordered stem and leaf plot, and C_2 is found by counting in from the high end of the ordered stem and leaf plot. The distance we count is given by the Location of C. For large samples,

$$\text{Location of } C = [n - z_{\alpha/2} \sqrt{n}]/2$$

For small samples the Location of C is found in Table 8.2.

☑ Finally the large-sample confidence interval for a population proportion is given by

$$p \pm z_{\alpha/2} \sqrt{p(1 - p)/n}$$

Having completed this chapter you should be able to:

1. List the desirable characteristics of an estimator. *Section 8.1*

2. Determine what statistic best estimates the parameter of concern. *Section 8.2*

3. Construct a large-sample confidence interval for the population mean based on $\bar{y}$. *Section 8.3*

4. Construct a large-sample confidence interval for the population mean based on $\bar{y}_T$.
   Section 8.4

5. Construct a large-sample confidence interval for the population median.
   Section 8.4

6. Construct a confidence interval for a population proportion. *Section 8.5*

7. Construct a small-sample confidence interval for the population mean based on $\bar{y}$.
   Section 8.6

8. Construct a small-sample confidence interval for the population median.
   Section 8.7

9. Interpret a confidence interval. *Section 8.3*

10. Understand the concepts of validity and precision of confidence intervals.
    Section 8.3

11. Determine sample size. *Sections 8.3 and 8.5*

Review Questions

To test your skills answer the following questions.

8.78 Suppose $\alpha = .05$ and $n = 16$. Find

(a) z_α (b) $z_{\alpha/2}$

(c) $t_{\alpha, n-1}$ (d) $t_{\alpha/2, n-1}$

8.79 We wish to estimate with a 95% confidence interval the proportion of TV viewers who would desire a 1-hour national news program. How large a sample is necessary so our precision of estimation will be within 3 percentage points?

8.80 We wish to estimate with a confidence interval the center of a population distribution using the random sample presented here in a stem and leaf plot:

```
0 | 1 4 6
1 | 6 8 9 3 9 7
2 | 8 6 9 4 2
3 | 4 8 6
4 | 5 2
5 | 3 6
6 | 4
7 | 3
8 | 1
9 | 4
```

(a) It appears that the distribution is

   (i)   symmetric with long tails
   (ii) symmetric with a long tail
   (iii) bimodal
   (iv) skewed right
   (v)   skewed left

(b) Will you base your confidence interval on

   (i) $\bar{y}$
   (ii) $\bar{y}_T$
   (iii) Median
   (iv) p

(c) Construct a 99% confidence interval for the center of the distribution based on your answer to part (b).

8.81 Suppose you wish to estimate the mean income of new Ph.D. psychologists. A random sample of 16 new Ph.D. psychologists had an average income of $32,300 and a standard deviation of $4,000.

(a) Obtain a 98% confidence interval for the mean income of new Ph.D. psychologists.

(b) What assumption did you make about the distributional shape?

(c) How might precision be increased and still maintain the same validity?

8.82

(a) What is meant by an unbiased estimator?

(b) When is estimator A more efficient than estimator B?

(c) Suppose the parent population is normal with mean μ and standard deviation σ. A random sample $y_1, y_2, \ldots, y_n$ is selected. What statistic would you use to estimate μ? Explain why.

8.83 In studying the percent of convicted felons who have a history of juvenile delinquency:

(a) How large a sample will be needed to estimate the percent to within .03 with a 99% confidence interval?

(b) A sample of 1,600 convicted felons revealed that 1,120 had a history of juvenile delinquency. Find a 99% confidence interval for the true percent. Interpret the results.

8.84 To estimate the self-concept of a group of college administrators, a psychologist administers the Tennessee Self Concept Scale exam to 16 administrators, and finds that their average self-concept score is 23 and the standard deviation is 3. Assume the distribution of scores is close to normal.

(a) Find a 95% confidence interval for the mean self-concept score.

(b) How might precision be increased and still maintain the same validity?

8.85 In a study to determine whether alcoholism has a genetic basis, several genetic markers were observed on a group of 50 Caucasian alcoholics. For 10 of the 50 the antigen B15 was present. Estimate with a 90% confidence interval the proportion having this antigen in the population of Caucasian alcoholics.

8.86 Suppose one wishes to estimate the mean monthly expenditures of women students on campus. Suppose the standard deviation of the monthly expenditures is $27.

(a) How large a sample is needed so that the margin of error in a 99% confidence interval estimate is no more than $10?

(b) Suppose a random sample of 45 women was selected, and the average monthly expenditure was found to be $128. Find a 99% confidence interval based on $\bar{y}$ for the mean monthly expenditure of all women students.

8.87 A random sample of 64 homes in a small community yielded an average of 160 gallons of heating oil consumed over a given period of time. If it is known that the amount of heating oil consumed is close to normally distributed with a standard deviation of 32 gallons, find a 95% confidence interval based on $\bar{y}$ for the mean number of gallons of heating oil consumed over this period by all residents of the community.

8.88 From a random sample of 2,400 college students it is found that 960 believe that the penalties for the use of marijuana should be reduced.

(a) Find a 99% confidence interval for the proportion of all college students who favor the proposition.

(b) How large a sample is needed so that the margin of error of the estimate is no more than 2%?

8.89 We wish to give a personality test to a certain group of people, and we know that their scores will range from 0 to 20. How large a sample is needed to estimate the mean test score with a 95% confidence interval if the margin of error is no more than .5?

8.90 The average violent crime rate (number per 100,000 population) in 25 randomly selected areas of the South was 486, and the standard deviation was 94. Assuming the distribution is normal, find a 99% confidence interval for the mean violent crime rate in the South.

8.91 The distribution of monthly salaries of unskilled construction workers is assumed to be long tailed. A random sample of 64 workers yielded the following information:

$$\bar{y}_{T.10} = \$1,530.00$$
$$s_T = \$285.00$$

Find a 98% confidence interval based on $\bar{y}_{T.10}$ for the mean monthly salary for all unskilled construction workers.

8.92 How large a sample is needed to estimate the average hospital costs to within $30 with a 90% confidence interval if the population is near normal with a standard deviation of $120.

8.93 In Exercise 8.92, a random sample of 45 receipts yielded an average of $250. Find a 90% confidence interval based on $\bar{y}$ for the mean cost.

8.94 The registrar at a university would like to estimate the percent of students registered for the spring semester that plan to attend summer school.

(a) If the registrar wishes to be 98% confident of obtaining a sample percent within 3% of the actual percent that plan to attend summer school, what sample size is needed?

(b) The registrar selected a random sample of 500 students and found that 94 plan to attend summer school. Find a 98% confidence interval for the actual percent of students that plan to attend summer school.

8.95 A reading teacher would like to estimate the mean reading speed of students in the fifth grade. A sample of 16 students had an average reading speed of 285 words per minute and a standard deviation of 48. In order to construct a confidence interval for the mean reading speed based on $\bar{y}$, what assumption must be made about the parent population? Assuming the assumption is correct, find a 95% confidence interval for the mean reading speed.

8.96 The ages of nine randomly selected people attending a movie rated PG were

19   18   26   17   22   16   25   20   17

Find a 99% confidence interval based on $\bar{y}$ for the mean age of people attending the movie.

8.97 A sample of 32 commuters was asked how far they travel to work each day. Following are the data measured in miles:

2.4	4.6	3.1	.2	3.8	5.9	4.2	5.7
27.8	6.4	3.8	6.4	5.8	3.9	.1	4.6
3.6	2.7	.1	5.2	6.7	7.8	3.4	4.9
5.8	4.7	31.0	4.7	2.8	3.9	.3	5.5

Construct a 95% confidence interval for the median miles traveled.

8.98 A pollster believes approximately 40% of the registered voters will be in favor of a particular issue; however, he wishes to estimate the true proportion with a 98% confidence interval.

(a) How many people will he need to interview so that the error of estimation is no more than 4%?

(b) The pollster found 260 in favor of the issue. What is the 98% confidence interval for the true proportion?

8.99 A company wishes to estimate the proportion of accounts that are paid on time.

(a) How large of a sample is needed to estimate the true proportion to within 3% with a 95% confidence interval?

(b) Construct a 95% confidence interval if 300 out of 400 paid their accounts on time.

8.100 A sample of 35 hotels in Miami yielded an average daily rate of $62.40 and a standard deviation of $13.50. Find a 90% confidence interval based on $\bar{y}$ for the mean daily rate of hotels in Miami.

8.101 In the early 1980s, samples of Tylenol were found to have been tampered with and contained traces of poison. The manufacturers of Tylenol designed a new safety bottle and launched a mass advertising campaign to convince their customers that it was safe to use their product. After the advertising campaign, 320 out of a random sample of 400 users of Tylenol said they would continue to use the product. Find a 98% confidence interval for the proportion of all Tylenol users who would continue to use the product.

8.102 A random sample of 100 cups of coffee from a coffee vending machine contained an average of 6.7 ounces in a 7-ounce cup. The standard deviation was .2 ounce. Find a 96% confidence interval based on $\bar{y}$ for the mean amount of coffee served by the machine.

8.103 In determining the flow characteristics of oil through a valve, the inlet oil temperature is measured in degrees Fahrenheit. A sample of 12 readings yielded

93	99	97	99	94	91
93	90	89	92	90	93

Construct a 95% confidence interval based on $\bar{y}$ for the mean temperature. What assumption is required for the interval to be 95% valid?

8.104 A laboratory tested 15 batteries manufactured by a company and found the following lives in hours:

19	18	26	17	22
16	25	20	17	18
17	19	20	18	19

Construct a 95% confidence interval for the median life of all batteries produced by the company.

8.105 A psychologist wishes to examine the amount of learning exhibited by schizophrenics after taking a specified dose of a tranquilizer. One hour after taking the drug, 19 patients were given a standard exam. Their scores were

36	29	30	32	37	15	34	23	47	32
39	24	10	25	34	13	52	30	33	

(a) Decide if a confidence interval should be based on $\bar{y}$ or M.

(b) Construct a 90% confidence interval based on the answer to part (a).

8.106 The number of women in U.S. state legislatures is as follows:

Ala.	9	Nebr.	6
Alaska	8	Nev.	6
Ariz.	19	N.H.	121
Ark.	7	N.J.	11
Calif.	14	N. Mex.	9
Colo.	25	N.Y.	23
Conn.	43	N.C.	23
Del.	10	N. Dak.	17
Fla.	28	Ohio	11
Ga.	19	Okla.	12
Hawaii	17	Oreg.	20
Idaho	16	Pa.	10
Ill.	28	R.I.	22
Ind.	18	S.C.	12
Iowa	17	S. Dak.	14
Kans.	23	Tenn.	9
Ky.	10	Tex.	13
La.	5	Utah	9
Maine	41	Vt.	32
Md.	35	Va.	11
Mass.	26	Wash.	28
Mich.	16	W. Va.	18
Miss.	5	Wis.	26
Mo.	25	Wyo.	22
Mont.	19		

Based on your observation about the shape of the distribution given in Exercise 6.31, construct a 99% confidence interval for the center of the data.

8.107 In order to estimate the income made by a typical tobacco farmer, a random sample of 50 tobacco farmers was selected. Following are the incomes recorded for the 50 farmers:

$ 6,280	9,690	7,858	8,820	6,500
7,468	8,719	6,790	8,650	9,400
7,843	12,170	9,760	9,280	14,897
5,438	9,980	7,654	10,190	7,823
9,840	5,790	6,874	10,690	9,450
11,657	6,470	19,357	6,794	7,865
8,747	9,347	8,785	7,589	12,768
8,658	24,860	9,793	6,680	8,749
9,845	7,895	14,678	8,980	5,897
9,879	8,370	8,530	10,250	8,450

Comment on the distributional shape of the incomes of tobacco farmers and decide what parameter best describes a "typical" income. Find a 99% confidence interval for that parameter.

8.108 An instrument used to measure anxiety has a standardized mean of 50 and a standard deviation of 15. The instrument will be used to estimate the average anxiety level of emotionally disturbed children. How large a sample is necessary so that the standard error of the estimate is no more than 2 points?

8.109 Have the students in your class determine the amount they spent in the community during the last weekend (make sure that it is a normal weekend, which would tend to be representative of all weekends). Using your collected data, construct a 95% confidence interval for the mean amount that students spend in the community on a weekend.

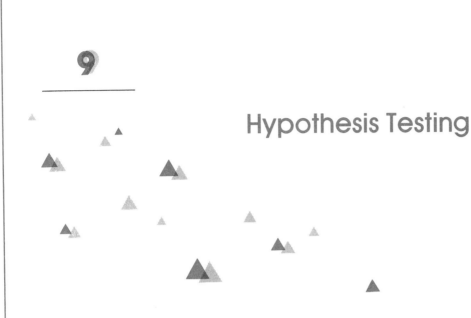

9

Hypothesis Testing

Many of our statistical questions can be answered by estimation procedures; however others may require a verification that can only be accomplished with a test of hypothesis.

Hypothesis testing is a means by which decisions can be made. For example, a new medication will be marketed if it is more than 70% effective. This statement can be formulated in a hypothesis and tested with data from a random sample. If the data support the hypothesis, then the decision will be made to market the product. In this chapter we will investigate several different problems that can be answered with a test of hypothesis.

49 mpg:
Honda, Chevy, Nissan lead 1985 EPA mileage chart

By Wayne Beissert

Japanese-made cars again dominate the Environmental Protection Agency's annual mileage ratings, taking the top three spots and six of the top 10.

Topping the list for the second year in a row is the Honda Civic Coupe HF, which the EPA says gets 49 miles per gallon in city driving and 54 mpg on the highway.

The Honda is followed by the Chevrolet Sprint (made by Suzuki) at 47/53 mpg and the Nissan Sentra diesel at 45/50 mpg.

The top USA-produced cars are the Ford Escort and Lincoln-Mercury Lynx diesel models, rated at 43/52 mpg.

Foreign-made cars also dominate the tail end of the ratings, taking the lowest seven positions.

The worst gas guzzlers are the British-made Rolls-Royce Camargues and Corniche/Continentals at 8/11 mpg. Also on the bottom of the list are the British-made Jaguar XJ-S at 13/17 mpg and the German-made Mercedes-Benz 500SEL at 14/16 mpg.

The biggest gas guzzlers among domestic cars are the

Chevrolet Camaro/Pontiac Firebird at 15/24 mpg and the Chevrolet Caprice Wagon/ Pontiac Parisienne Wagon at 15/22 mpg.

Overall, though, the EPA says fleet average gas mileage for domestic and foreign cars is improving, climbing to 26.8 mpg for the 1985 model year from 26.6 mpg in 1984 and 15.8 mpg in 1975.

The annual report also warns that "use of leaded fuel in cars designed for unleaded fuel" increases exhaust emissions and results in poor fuel economy.

SOURCE: *USA Today*, September 24, 1984.

How reliable is the EPA mileage chart? If one were to buy a Honda Civic, can he or she expect to get 49 miles per gallon in city driving as the preceding article suggests?

One obvious way to answer this question is to drive the car and check the mileage. Just as the EPA cautions, however, a motorist's actual mileage can vary, depending on such factors as the amount of traffic, the weather, car maintenance, and so on. Therefore one should check the mileage on several different occasions under a variety of conditions.

Suppose that the mileage was checked on 35 different occasions with the following miles per gallon recorded:

48.3	49.8	39.6	43.5	46.8	49.4	52.6
45.3	49.7	45.3	40.7	45.6	48.2	50.3
48.2	40.9	43.5	54.2	50.0	51.2	50.8
45.5	49.2	40.5	45.3	50.2	44.8	47.3
48.7	49.2	47.4	51.4	49.6	50.3	48.5

Is there evidence in these data to suggest that the average miles per gallon for the Honda Civic will be less than 49 mpg? In this chapter we will learn how to answer this question.

Hypothesis testing is an area of statistical inference in which one evaluates a conjecture about some characteristic of the parent population based upon the information contained in a random sample. Usually the conjecture concerns one of the unknown parameters of the population.

Consider the following conjecture that the mean family income in a certain rural county is $15,000 per year. This can be written as $\mu = 15,000$, where μ denotes the true mean family income of all residents of the rural county. The truth of the conjecture is to be determined with a *test of significance*.

Definition

The statement being tested in a test of significance is called the **null hypothesis.** Generally speaking, the null hypothesis is the hypothesis of "no difference" and is denoted as H_0.

Clearly if we believed that the null hypothesis were true, then there would be no need to test it. But if we doubt its truth, then data are collected and analyzed to assess the strength of evidence against the null hypothesis. If there is *strong* evidence that casts doubt as to the truth of the null hypothesis, it will be rejected in favor of the *alternative hypothesis.*

Definition

The **alternative hypothesis** is what is believed to be the truth if the null hypothesis is false. Typically the researcher believes the alternative, denoted as H_a, to be true, and thus it is also called the **research hypothesis.**

A test of significance, then, is a problem of deciding between the null and the alternative hypotheses on the basis of the information contained in a random sample. The goal will be to reject H_0 in favor of H_a, because the alternative is the hypothesis that the researcher believes to be true. If we are successful in rejecting H_0, we then declare the results to be "significant."

Suppose that, in the preceding example, the researcher believes that the mean family income of the county residents is greater than $15,000. Formulated, this says $\mu > 15,000$; and because this is what the researcher wishes to support, it is a statement of the alternative hypothesis.

The null hypothesis is

$$H_0: \mu = 15,000$$

and the alternative hypothesis is

$$H_a: \mu > 15,000$$

EXAMPLE 9.1

A conjecture is made that the mean starting salary for Computer Science graduates is $25,000 per year. Assuming that you believe it is more than $25,000, formulate appropriate null and alternative hypotheses to evaluate the claim.

SOLUTION The parameter about which the conjecture is made is the *mean* starting salary μ. The statement that "you believe it is more than \$25,000" is formulated as $\mu > 25,000$. Because this is the statement you wish to support, it is placed in the alternative (research) hypothesis. Consequently we have

$$H_0: \mu = 25,000$$

and

$$H_a: \mu > 25,000$$

EXAMPLE 9.2

A spokesperson for the school administration states that 70% of all high school students have tried marijuana. The student body government feels that this is unfair, and decides to set up a null and an alternative hypothesis to evaluate the claim by the administration. How should they set up their hypotheses?

SOLUTION Let π represent the true proportion of students that have tried marijuana. The student body government is doing the testing, and they wish to show that π is something less than 70% (the research hypothesis). Thus

$$H_0: \pi = .7$$

and

$$H_a: \pi < .7$$

The strength of evidence against the null hypothesis is reflected in a probability statement involving the observed outcome of the sample. The probability is computed under the assumption that the null hypothesis is true, and thus a large probability will support the truth of the null hypothesis. If the probability is *small*, however, there is evidence that the null hypothesis is false and should be rejected. How small is an important question that will be addressed later, but for now remember that we want to reject the null hypothesis in favor of the alternative only when there is *significant evidence* against it. Later we will compute the previously referenced probability and discuss what is meant by significant evidence.

The testing of the null hypothesis is like a trial by jury. The null hypothesis says "the defendant is not guilty," and the jury must decide if the presented evidence (observed sample) is significant enough to warrant conviction (or rejection) of the defendant. Even though we hope that the jury reaches the correct decision in all cases, we are aware that an innocent person can be convicted or a guilty person freed. Similarly we would like a test procedure that will always make the correct decision with regard to the hypothesis being tested. However, a null hypothesis can be rejected when it is true or not rejected when it is false.

Definitions

A **Type I error** is rejecting a null hypothesis that is true. The probability of committing a Type I error is denoted as α.

A **Type II error** is failing to reject a null hypothesis that is false. The probability of committing a Type II error is denoted as β.

There are four possible situations that might result in a test of hypothesis:

Null hypothesis	Decision	
	Reject H_0	Fail to reject H_0
True	Type I error	Correct decision
False	Correct decision	Type II error

In the earlier example of the mean family income of rural county residents, a Type I error is committed if we decide that the mean family income exceeds $15,000 when in reality it does not. A Type II error is committed if the mean family income does exceed $15,000, but we fail to detect it. The seriousness of the two errors depends upon the circumstances involved. Here there is not enough information to determine which is the more serious error. In other examples, however, it is quite clear which error is more serious.

Suppose the decision is whether we should accept or reject a parachute for a sky dive. The null hypothesis might be

H_0: The parachute will open

and the alternative will then be

H_a: The parachute will not open

A Type I error is committed if we reject H_0 when it is true; that is, we say the parachute will not open when in fact it will. On the other hand, a Type II error is committed if we accept H_0 when it is false; that is, we say the parachute will open when in fact it will not. Clearly in this case, we do not want to make a Type II error! We want the chance of the Type II error to be extremely small. In other situations, the opposite might be true, and the Type I error will be the more serious error.

EXAMPLE 9.3

The standard medication for a certain disease is effective in 60% of all cases. A drug company believes that its new drug is more effective than the old treatment. Formulate null and alternative hypotheses to test if there is statistical evidence to support the new drug. Interpret the Type I and Type II errors and determine which is the more serious.

SOLUTION   Let π denote the cure rate for the new drug. The drug company feels that the new drug is more effective than the standard medication; therefore the research hypothesis will be that $\pi > .60$. Thus

$H_0: \pi = .60$

and

$H_a: \pi > .60$

Rejecting the null hypothesis will amount to a claim that the new drug is more effective than the standard medication. Falsely rejecting it, or a Type I error, will imply that it is more effective when it is not. A Type II error is committed with the conclusion that the new drug is not more effective when, in fact, it is more effective. Here the Type I error is more serious because marketing an ineffective drug is worse than failing to publicize a potentially effective drug.

It would be ideal if the probabilities α and β of the Type I and Type II errors are zero or very near zero. This would mean that we would rarely make an error. For a fixed sample size, however, it is impossible to design a test so that both α and β are arbitrarily small. In fact, in attempting to decrease one, the other increases. Consequently α is fixed at a value called the *level of significance* (typical values might be .01, .05, .10), and then we attempt to find a test procedure that minimizes β.

▲ We will never set the level of significance at a value greater than .10 because we want the chance of a Type I error to be no more than 10%.

Definition

The **level of significance** is α, the probability of making a Type I error.

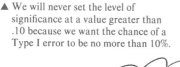

Recall that validity and precision were terms used to evaluate confidence intervals. In a similar manner, this terminology is used to evaluate a test of significance procedure. The level of significance is a measure of validity, and the precision of the test is associated with the probability of a Type II error. A thorough treatment of these concepts can be found in a more advanced statistics book.

Developing the Test Procedure

Having formulated the null and alternative hypotheses, one must develop a test procedure to assess the evidence against H_0. If the evidence is due to factors other than chance, then H_0 should be rejected in favor of the alternative, H_a.

An important step in developing the test procedure is to determine an appropriate *test statistic*; that is, choose a statistic from the sample that seems appropriate for testing the null hypothesis.

Definition

A **test statistic** is a statistic, calculated from the sample data, which is used to test the hypothesis.

The choice of test statistic depends upon the parameter being tested and the underlying parent distribution. Recall from Chapter 8 that if the underlying parent distribution is normal, then the statistic $\bar{y}$ is the best estimator of μ. Similarly using it as a test statistic, under the normality assumption, results in a test procedure that is best according to the validity and precision requirements.

If the parent distribution is not normal, then $\bar{y}$ may not be the best test statistic. For example, if the parent distribution is symmetrical with long tails, it seems reasonable that a better test statistic will be one of the trimmed sample means. In general, to test a specific parameter of a certain underlying distribution, the test statistic will be the point estimator of the parameter that was found in the previous chapter. Just as in confidence interval estimation, one must use the exploratory data analysis techniques to help choose an appropriate test statistic.

Once a test statistic is decided upon, its sampling distribution, under the assumption that H_0 is true, must be determined. With the sampling distribution it is possible to determine whether the observed data are consistent with the null hypothesis. Equivalently it is possible to determine what values of the test statistic seem reasonable for rejection of the null hypothesis. These values are collectively known as the *rejection region*.

Definition

The **rejection region** consists of those values of the test statistic that would lead to the rejection of the null hypothesis.

Having determined the rejection region, the value of the test statistic is calculated from the collected sample. If its computed value falls in the rejection region, the null hypothesis is rejected.

In summary, there are five basic steps that are followed when testing hypotheses.

Procedure Steps for Testing Hypotheses

1. Formulate the null and alternative hypotheses.

2. Decide upon an appropriate test statistic.

3. Determine the sampling distribution of the test statistic under the assumption that the null hypothesis is true.

4. Specify the rejection region and determine whether the test statistic falls in the rejection region. If so, reject the null hypothesis.

5. Interpret the results.

For an interpretation of the results, it is suggested that one gives a discussion of the results so that a nonstatistician will understand.

EXERCISES 9.1

9.1 In each of the following, formulate the null and alternative hypotheses involving the population mean, μ.

(a) Suppose you wish to show that μ exceeds 50.

(b) Suppose you wish to show that μ is less than 50.

(c) Suppose you wish to find evidence against the claim that μ is 50 when you believe it is greater than 50.

(d) Suppose you wish to find evidence against the claim that μ is 50 when you believe it is less than 50.

(e) Suppose you wish to test the claim that μ is 50.

9.2 In each of the following, formulate the null and alternative hypotheses involving the Bernoulli proportion, π.

(a) Suppose you wish to show that π exceeds 70%.

(b) Suppose you wish to show that π is less than 70%.

(c) Suppose you wish to find evidence against the claim that π is 70% when you believe it is greater than 70%.

(d) Suppose you wish to find evidence against the claim that π is 70% when you believe it is less than 70%.

(e) Suppose you wish to test the claim that π is 70%.

9.3 Suppose a testing procedure leads to the rejection of the null hypothesis. Is it possible that a Type I error was committed? Is it possible that a Type II error was committed? Explain.

9.4 If the null hypothesis is rejected, does this mean that the alternative is true? Explain.

9.5 If the null hypothesis is not rejected, does this mean that it is true? Explain.

9.6 Does $\alpha + \beta = 1$? Why or why not?

9.7 Suppose that the null hypothesis says "the patient will recover." Interpret the Type I and Type II errors. Which is more serious?

9.8 Suppose that the null hypothesis says "the fire is out." Interpret the Type I and Type II errors. Which is more serious?

✪ 9.9 A sociologist is interested in the percent of blacks in the inner city that are unemployed. He has pleaded his case with the local industries to help employ more blacks. He feels that more than 30% of the blacks in the inner city are unemployed. How would he set up the null and alternative hypotheses to support his point?

✪ 9.10 Interpret the Type I and Type II errors for Exercise 9.9.

9.11 A psychologist suspects that more than 10% of the adult population is illiterate. She takes a random sample of 1,600 adults and gives them Wechsler's Adult Intelligence Scale. State the null and alternative hypotheses to evaluate the psychologist's claim.

9.12 A claim is made that 60% of the adult population feels that there is too much violence on TV. Assuming that you feel this figure is too high, set up the null and alternative hypotheses to evaluate the claim.

Large-Sample Test of a Population Mean

Let us return to the example of the mean family income of the residents of the rural county given in Section 9.1. As stated, we wish to test

$$H_0: \mu = 15{,}000 \quad \text{vs.} \quad H_a: \mu > 15{,}000$$

Additionally let us assume that we have a large sample from a parent distribution that is symmetrical with tails that are not unusually long. In confidence interval estimation, a distribution of this type suggested that $\bar{y}$ is the proper estimator of the population mean. Similarly $\bar{y}$ is the choice of a test statistic to test the null hypothesis. We will reject H_0 in favor of H_a if $\bar{y}$ is significantly greater than 15,000.

▲ $\bar{y}$ is significantly greater than 15,000 if it falls in the rejection region.

Right-Tailed Test

This is called a *right-tailed test* (sometimes referred to as an upper-tailed test) because the rejection region consists of those values of $\bar{y}$ that are significantly greater than 15,000. Figure 9.1 depicts the rejection region on the right tail of the sampling distribution of the test statistic, $\bar{y}$.

The sampling distribution is drawn as a normal curve centered at 15,000, because, with a large sample, the Central Limit theorem states that $\bar{y}$ is approximately normally distributed with a mean the same as the mean of the parent population. Assuming H_0 is

FIGURE 9.1

Sampling distribution of $\bar{y}$ with a right-tailed rejection region

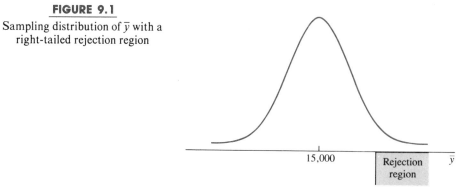

true, we have that the mean of the parent population is 15,000 and thus the sampling distribution is centered at 15,000.

To assess the evidence against H_0 and determine if $\bar{y}$ falls in the rejection region, we will calculate the probability of observing a $\bar{y}$ at least as large as the one actually observed, which we will call $\bar{y}_{obs}$.

Thus we will find

$$P(\bar{y} > \bar{y}_{obs})$$

under the assumption that the null hypothesis is true.

Figure 9.2 illustrates this probability, which is called the *p-value*.

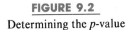
FIGURE 9.2

Determining the *p*-value

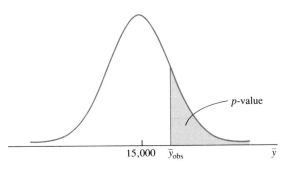

Definition

The *p-value* is the probability (computed when H_0 is assumed to be true) of observing a value of the test statistic at least as extreme as that given by the observed data.

▲ Finding the *p*-value is the same type problem as the probability problems in Chapter 7.

Knowing that the sampling distribution of $\bar{y}$ is normal, one can find the *p*-value in the normal probability table by determining the distance between $\bar{y}_{obs}$ and the hypothesized value of μ which, in this case, is 15,000.

Recall that the *z*-score gives the number of standard errors that a particular score is from the mean. That is,

$$z = \frac{\bar{y} - 15,000}{\sigma/\sqrt{n}}$$

is a measure of how close $\bar{y}$ is to 15,000. (Recall that if the population standard deviation, σ, is unknown, then the sample standard deviation, s, may be substituted in its place as long as the sample size is sufficiently large.)

Because the *p*-value is the upper-tail probability, it is found in the standard normal table by finding the probability associated with z_{obs} (the observed value of z) and subtracting it from .5. Thus

$$p\text{-value} = P(\bar{y} > \bar{y}_{obs})$$
$$= P(z > z_{obs})$$
$$= .5 - \text{table value}$$

Because the *p*-value is calculated assuming the null hypothesis is true, a small value indicates that the sample values are inconsistent with H_0. The smaller the *p*-value the greater the evidence is that the null hypothesis is false.

Suppose the researcher selects a random sample of 100 county residents and finds their average family income to be $16,200, and the standard deviation to be $4,000. Is this enough evidence to suggest that $\mu > 15,000$? The strength of evidence against the null hypothesis is reflected in the *p*-value (see Figure 9.3), which is given by

$$p\text{-value} = P(\bar{y} > 16,200)$$

where the probability is calculated under the assumption that μ is 15,000.

FIGURE 9.3

Sampling distribution of $\bar{y}$ and
$p\text{-value} = P(\bar{y} > 16,200)$

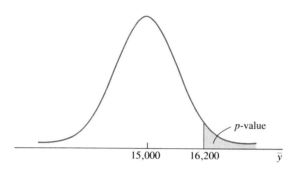

If $\mu = 15,000$, then we have

$$z_{obs} = \frac{16,200 - 15,000}{4,000/\sqrt{100}} = 1,200/400 = 3$$

Already we can conclude that the null hypothesis is false, because a *z*-score of 3 means that 16,200 is three standard errors above 15,000, which is highly unlikely. This suggests that $\mu > 15,000$. We will continue, however, and find the *p*-value. Looking up a *z*-score of 3 in the normal table, we find that .4987 of the area lies between 0 and 3, so the

$$p\text{-value} = .5 - .4987 = .0013$$

which is very small, suggesting that the null hypothesis is false.

In calculating the *p*-value, one must find the *z*-score corresponding to the observed value of $\bar{y}$. That is, calculate

$$z_{obs} = \frac{\bar{y}_{obs} - \mu_0}{\sigma/\sqrt{n}} \qquad \begin{array}{l}(\mu_0 \text{ denotes the value of } \mu \\ \text{that is being tested})\end{array}$$

and then find the *p*-value associated with z_{obs}. Thus

$$z = \frac{\bar{y} - \mu_0}{\sigma/\sqrt{n}}$$

is called the *standardized test statistic* and the *p*-value is given by

$$p\text{-value} = P(z > z_{obs})$$

Just how small of a *p*-value will lead to rejection of H_0? We saw in Figure 9.1 that the rejection region for a right-tailed test is located on the upper tail of the sampling distribution of the test statistic, $\bar{y}$. The point that divides the rejection region from the acceptance region (the complement of the rejection region) is called the *critical point*, CP. It is located on the upper tail as illustrated in Figure 9.4, so that the area above the rejection region is the level of significance α.

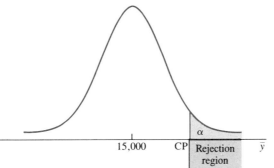

If $\bar{y}_{obs}$ falls in the rejection region, that is, if $\bar{y}_{obs} > CP$, then the p-value will be less than α and we will reject H_0 at the α level of significance.

This is illustrated more clearly in Figure 9.5, which shows both probabilities, α and the p-value. In Figure 9.5(a) we see that $\bar{y}_{obs} > CP$, consequently the p-value $< \alpha$. In Figure 9.5(b) $\bar{y}_{obs} < CP$, consequently the p-value $> \alpha$.

FIGURE 9.5

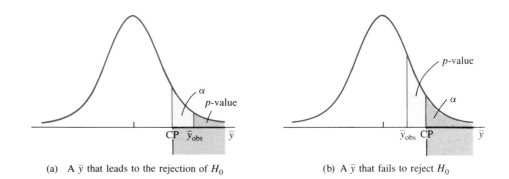

(a)   A $\bar{y}$ that leads to the rejection of H_0

(b)   A $\bar{y}$ that fails to reject H_0

In summary, if the p-value $< \alpha$, we will reject the null hypothesis and declare the results significant at the α level of significance. Most of the computer statistical packages (BMDP, SPSS, SAS, Minitab) print out the p-value, so it is very easy to determine if the results are significant.

In the example, $\bar{y}_{obs} = 16,200$ led to a p-value of .0013. Thus we could reject H_0 ($\mu = 15,000$) at any level of significance greater than .0013. This is very strong evidence that the null hypothesis is false.

A conclusion for this example could be that, on the basis of a random sample of size 100, there is strong evidence that the mean family income of all residents of the county is greater than $15,000.

Suppose that in the previous example the null hypothesis had been stated as

$$H_0: \mu \leq 15,000$$

instead of

$$H_0: \mu = 15,000$$

Recall that the p-value is calculated under the assumption that the null hypothesis is true. When H_0 states that $\mu = 15,000$, the procedure is straightforward, because we can use the specified value of μ of 15,000 in the z-score and proceed to find the p-value.

If H_0 states that $\mu \le 15{,}000$, however, which value of μ do we use in calculating the z-score? Should we, perhaps, choose 14,000 in the calculation of the z-score or just any value of μ that is less than 15,000? In Figure 9.6 we see two sampling distributions of $\bar{y}$, one assuming that $\mu = 15{,}000$ and the other assuming that μ is some value $m < 15{,}000$. Because the p-value is the tail probability beyond the observed test statistic $\bar{y}_{obs}$, it is clear that the p-value is greatest when we assume that $\mu = 15{,}000$. Thus when testing the more general null hypothesis that

▲ Recall that μ_0 simply denotes a value of μ that is being tested.

$$\mu \le \mu_0$$

we will compute the p-value under the assumption that

$$\mu = \mu_0$$

so as to obtain its maximum value. In this sense, testing

▲ The important inequality is in the statement of the alternative hypothesis H_a, because this determines the direction of the test.

$$H_0: \mu \le \mu_0 \quad \text{vs.} \quad H_a: \mu > \mu_0$$

is equivalent to testing

$$H_0: \mu = \mu_0 \quad \text{vs.} \quad H_a: \mu > \mu_0$$

FIGURE 9.6

The p-value assuming that $\mu = 15{,}000$ or a value $m < 15{,}000$

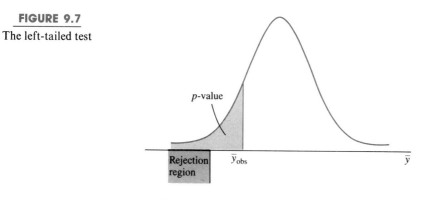

Unless stated otherwise, we will give the null hypothesis in the more general form.

Left-Tailed Test

For a *left-tailed test* the null and alternative hypotheses read as follows:

$$H_0: \mu \ge \mu_0 \quad \text{vs.} \quad H_a: \mu < \mu_0$$

Here the null hypothesis is rejected in favor of H_a if there is evidence that $\mu < \mu_0$. The test procedures are basically the same as the right-tailed test, except that the rejection

FIGURE 9.7

The left-tailed test

region is located on the *left tail* (lower tail) of the sampling distribution and the *p*-value is found by

$$p\text{-value} = P(\bar{y} < \bar{y}_{obs})$$
$$= P(z < z_{obs})$$

 ▲ The inequality on H_a points in the direction of the test.

Figure 9.7 illustrates the left-tailed test.

Two-Tailed Test

Finally a third possibility is the *two-tailed test*. Here the null hypothesis is stated as

$$H_0: \mu = \mu_0$$

and the alternative is stated as

$$H_a: \mu \neq \mu_0$$

In light of the alternative, we could conceivably reject the null hypothesis, in favor of the alternative, if $\bar{y}$ is either significantly greater than μ_0 *or* significantly less than μ_0. That is, the rejection region lies on both tails. Clearly $\bar{y}_{obs}$ cannot fall on both tails of the sampling distribution. Let us suppose that it falls on the upper tail. To find the *p*-value, we calculate

$$P(\bar{y} > \bar{y}_{obs})$$

as was done in the right-tailed test. But because this is a two-tailed test, one must account for both tails by doubling the probability. So if

$$\bar{y}_{obs} > \mu_0$$

calculate the *p*-value as in the right-tailed test and double it.
    In a similar fashion if

$$\bar{y}_{obs} < \mu_0$$

then calculate the *p*-value as in the left-tailed test and double it.
    Figure 9.8 illustrates the rejection region and calculation of the *p*-value for the two-tailed test.
    The preceding information will be reiterated through the following example.

FIGURE 9.8
The two-tailed test

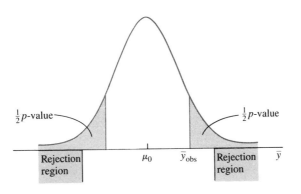

EXAMPLE 9.4

The scores for a college placement exam in mathematics are assumed to be normally distributed with a mean of 70. The exam is given to a random sample of 50 high school seniors who have been admitted to college. Their average and standard deviation on the exam were 73 and 16.8, respectively. If this is truly a random sample, is there evidence that would suggest that the mean of 70 should be raised?

SOLUTION   Let μ denote the true population mean of the placement exam. We then wish to see whether there is evidence that $\mu > 70$. This will become the research hypothesis. The null and alternative hypotheses become

$$H_0\colon \mu = 70 \quad \text{vs.} \quad H_a\colon \mu > 70$$

Because the parent distribution is assumed to be normal, the best test statistic is the sample mean $\bar{y}$. The p-value is the probability of observing a value of $\bar{y}$ at least as large as 73 if the true population mean is indeed 70. That is,

$$p\text{-value} = P(\bar{y} > 73)$$

The standardized form of the test statistic is

$$z = \frac{\bar{y} - 70}{\sigma/\sqrt{n}}$$

The observed value is

$$z_{\text{obs}} = \frac{73 - 70}{16.8/\sqrt{50}} = 1.26$$

Thus

$$
\begin{aligned}
p\text{-value} &= P(\bar{y} > 73) \\
&= P(z > 1.26) \\
&= .5 - .3962 \\
&= .1038
\end{aligned}
$$

A p-value this large exceeds any reasonable level of significance α and thus we fail to reject H_0. In summary, we declare the results of the random sample of 50 high school seniors insignificant and conclude that there is insufficient evidence to say that the mean of the college placement exam should be higher than 70.

▲ We fail to reject H_0 because the p-value $> .10$.

With a sample size of 50, as in the previous example, the Central Limit Theorem says that regardless of the shape of the parent distribution the sampling distribution of $\bar{y}$ is approximately normally distributed. Consequently the assumption that the parent distribution is normal is not necessary. On the other hand, knowing that it is normal guarantees that $\bar{y}$ is the *best* test statistic for testing the population mean. As suggested earlier, $\bar{y}$ will be the test statistic for testing a population mean in all cases where the parent distribution is symmetric with tails that are not unusually long.

The large-sample test based on $\bar{y}$ is summarized as follows:

Large-Sample Test of μ Based on $\bar{y}$

APPLICATION: Symmetric distributions whose tails are not unusually long

ASSUMPTION: $n > 30$

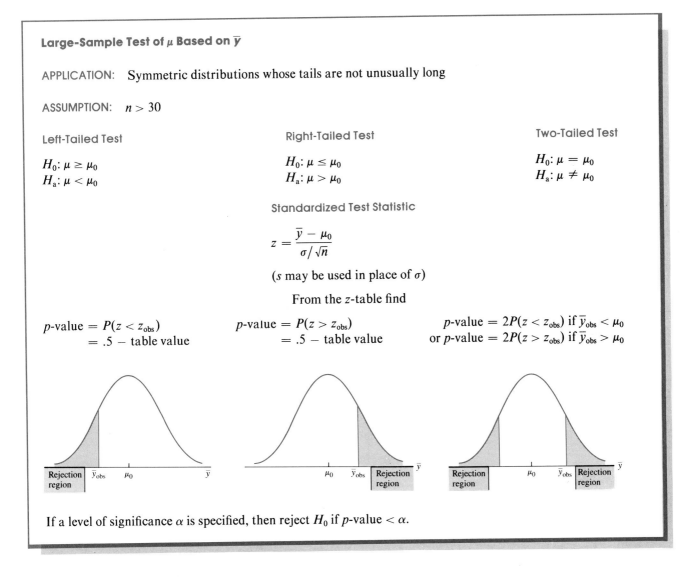

Left-Tailed Test

$H_0: \mu \geq \mu_0$
$H_a: \mu < \mu_0$

Right-Tailed Test

$H_0: \mu \leq \mu_0$
$H_a: \mu > \mu_0$

Two-Tailed Test

$H_0: \mu = \mu_0$
$H_a: \mu \neq \mu_0$

Standardized Test Statistic

$$z = \frac{\bar{y} - \mu_0}{\sigma / \sqrt{n}}$$

(s may be used in place of σ)

From the z-table find

$p\text{-value} = P(z < z_{obs})$
$\qquad = .5 - \text{table value}$

$p\text{-value} = P(z > z_{obs})$
$\qquad = .5 - \text{table value}$

$p\text{-value} = 2P(z < z_{obs})$ if $\bar{y}_{obs} < \mu_0$
or $p\text{-value} = 2P(z > z_{obs})$ if $\bar{y}_{obs} > \mu_0$

If a level of significance α is specified, then reject H_0 if $p\text{-value} < \alpha$.

EXAMPLE 9.5

To justify raising rates, an insurance company claims that the mean medical expense for all middle-class families is at least $700 per year. In a survey of 100 randomly selected middle-class families, it was found that their mean medical expense for the year was $670 and the standard deviation was $140. Assuming that the tails of the distribution of medical expenses are not unusually long, is there evidence that the insurance company is misinformed?

SOLUTION   The insurance company claims that the mean medical expense is at least $700; we doubt that, however, and believe that it is something less than $700. This will become the research hypothesis. Denoting the mean medical expense for all middle-class families as μ, we can state the null and alternative hypotheses as follows:

$$H_0: \mu \geq 700 \quad \text{vs.} \quad H_a: \mu < 700$$

Because the underlying distribution of medical expenses does not have excessively long tails, the choice for a test statistic is $\bar{y}$. Thus the standardized test statistic is

$$z = \frac{\bar{y} - 700}{s/\sqrt{n}}$$

The observed value of z becomes

$$z_{obs} = \frac{670 - 700}{140/\sqrt{100}} = \frac{-30}{14} = -2.14$$

Thus the p-value becomes

$$p\text{-value} = P(z < -2.14) = .5 - .4838 = .0162$$

which suggests that the null hypothesis should be rejected. There is significant evidence to indicate that the insurance company's claim is inaccurate; hence we might suggest that the insurance company check its source.

In Example 9.5 the p-value was .0162, and we concluded that there was significant evidence to reject H_0, whereas in Example 9.4 the p-value was .1038, and we concluded that the results were insignificant and failed to reject H_0. In practice, we can set a level of significance α prior to testing and then reject H_0 if the p-value $< \alpha$.

An alternate procedure would be to use the following rule of thumb.

Criterion for Rejection of H_0

1. If p-value $> .10$ then fail to reject H_0 and declare the results *insignificant*.

2. If $.05 < p$-value $\leq .10$ you may reject H_0, but the results are only *mildly significant*.

3. If $.01 < p$-value $\leq .05$ then reject H_0 and declare the results *significant*.

4. If p-value $\leq .01$ then reject H_0 and declare the results *highly significant*.

EXERCISES 9.2

9.13 Draw a normal curve and shade in the rejection region as in Figure 9.1 for the following test of hypothesis problems. Identify each as being right-, left-, or two-tailed test:

(a) $H_0: \mu \geq 15$
$H_a: \mu < 15$

(b) $H_0: \mu = 120$
$H_a: \mu \neq 120$

(c) $H_0: \mu \leq 650$
$H_a: \mu > 650$

(d) $H_0: \mu \geq 6000$
$H_a: \mu < 6000$

9.14 Draw a normal curve and shade in the rejection region and the p-value area as in Figure 9.7 for the following test of hypothesis problems. Calculate the p-value.

(a) $H_0: \mu \geq 15$
$H_a: \mu < 15$
$n = 48, \bar{y} = 14.2, s = 4.1$

(b) $H_0: \mu = 120$

$H_a: \mu \neq 120$

$n = 100, \bar{y} = 124.6, s = 16.3$

$z = \dfrac{124.6 - 120}{16.3/\sqrt{100}} = 2.82$

[LOOK IN TABLE]

P-VALUE $= 2[.5 - .4976] \Rightarrow .0048$

(c) $H_0: \mu \leq 650$

$H_a: \mu > 650$

$n = 250, \bar{y} = 694, s = 235.7$

9.15 From the following summary data, test the hypotheses. Be sure to calculate the *p*-value and interpret the results.

(a) $H_0: \mu \geq 2.5$

$H_a: \mu < 2.5$

$n = 150, \bar{y} = 2.33, s = 1.27$

(b) $H_0: \mu = 50$

$H_a: \mu \neq 50$

$n = 70, \bar{y} = 51.6, s = 5.9$

(c) $H_0: \mu \leq 100$

$H_a: \mu > 100$

$n = 200, \bar{y} = 101.3, s = 2.45$

9.16 A sample of 70 observations yielded $\bar{y} = 128.6$ and $s = 12.9$. Use this information to test the following hypotheses, with $\bar{y}$ being the test statistic:

$H_0: \mu \leq 125$

$H_a: \mu > 125$

Be sure to calculate the *p*-value and interpret the results.

9.17 A random sample of 40 people who received food stamps showed their average age was 39.2 years and the standard deviation was 5.2 years. Test

$H_0: \mu \geq 40$

$H_a: \mu < 40$

9.18 Suppose the mean entrance exam score for incoming freshmen at a large university is 550 and the standard deviation is 120. A sample of 100 from this year's freshman class had an average score of 580. Is this unusual? Explain with a test of hypothesis procedure.

9.19 A survey of 100 parents of first and second grade children revealed that the number of hours per week their children watch television had an average of 25.8 hours and a standard deviation of 4.4 hours. Determine whether there is statistical evidence to conclude that μ exceeds 25 hours.

9.20 A car manufacturer claims that its cars use, on the average, no more than 5.5 gallons of gas for each 100 miles. A consumer group tests 40 of the cars and finds an average consumption of 5.65 gallons per 100 miles and a standard deviation of 1.52 gallon. Do these results cast doubt on the claim made by the car manufacturer?

9.21 The average daily wage in a particular industry is $25 and the standard deviation is $5. If a company in this industry employing 36 workers pays them an average daily wage of $22, can we assume that they are paying inferior wages?

9.22 An instructor gives his class an examination which, as he knows from years of experience, yields $\mu = 78$ and $\sigma = 17$. His present class of 35 scores an average of 81.4. Is he correct in assuming that this is a superior class?

9.23 A manufacturer of TV sets claims that the average life of its picture tubes is at least 10 years. We sample 100 of the picture tubes and obtain an average of 9.6 years and a standard deviation of 2.6 years. Is there evidence to suggest that the manufacturer is in error?

9.24 A class of 50 eighth graders has completed a standardized reading test on which their scores had a mean of 107.5 and a standard deviation of 10.5. The national mean score on the test is 100. Do we have sufficient statistical evidence to proclaim that this class is superior in reading ability?

9.3   A Robust Large-Sample Test of a Population Mean

We will now consider the case where the underlying parent population is assumed to be symmetric with long tails. Under this assumption, it has been shown that a test of μ based on a trimmed mean performs better than the test based on $\bar{y}$ presented in the last section. Thus for long-tailed distributions, we now present the test of μ based on $\bar{y}_T$.

As before, the null and alternative hypotheses can take one of three forms:

Left-Tailed—$H_0: \mu \geq \mu_0$ vs. $H_a: \mu < \mu_0$

Right-Tailed—$H_0: \mu \leq \mu_0$ vs. $H_a: \mu > \mu_0$

Two-Tailed—$H_0: \mu = \mu_0$ vs. $H_a: \mu \neq \mu_0$

where μ_0 represents the value of μ we wish to test.

Section 8.4 states that for a sufficiently large sample, the sampling distribution of $\bar{y}_T$ is approximately normal with mean μ and an estimated standard error of

$$\text{estSE}(\bar{y}_T) = s_T/\sqrt{k}$$

where s_T is the standard deviation of the trimmed sample based on the Winsorized sample and k is the size of the trimmed sample.

Therefore the standardized test statistic will be

$$z = \frac{\bar{y}_T - \mu_0}{s_T/\sqrt{k}}$$

and the p-value is found, as before, by finding the observed value of z, z_{obs}, by inserting the observed values of $\bar{y}_T$ and s_T in the preceding formula for z.

The complete test is summarized as follows:

Large-Sample Test of μ Based on $\bar{y}_T$

APPLICATION: Symmetric distribution with long tails

ASSUMPTION: $n > 30$

Left-Tailed Test

$H_0: \mu \geq \mu_0$
$H_a: \mu < \mu_0$

Right-Tailed Test

$H_0: \mu \leq \mu_0$
$H_a: \mu > \mu_0$

Two-Tailed Test

$H_0: \mu = \mu_0$
$H_a: \mu \neq \mu_0$

Standardized Test Statistic

$$z = \frac{\bar{y}_T - \mu_0}{s_T/\sqrt{k}}$$

where s_T is the standard deviation of the trimmed sample based on the Winsorized sample and k is the size of the trimmed sample.

From the z-table find

p-value $= P(z < z_{\text{obs}})$
$= .5 - $ table value

p-value $= P(z > z_{\text{obs}})$
$= .5 - $ table value

p-value $= 2P(z < z_{\text{obs}})$ if $\bar{y}_{T,\text{obs}} < \mu_0$
or p-value $= 2P(z > z_{\text{obs}})$ if $\bar{y}_{T,\text{obs}} > \mu_0$

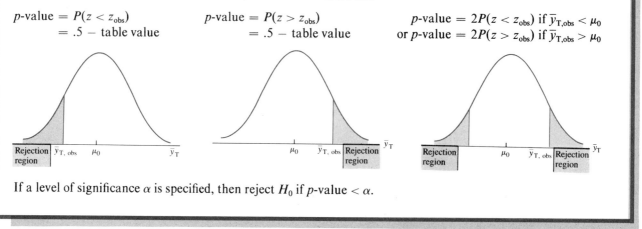

If a level of significance α is specified, then reject H_0 if p-value $< \alpha$.

An important point is that the test procedure just described is the same as the test based on $\bar{y}$ presented earlier, except that the standardized test statistic is based on $\bar{y}_T$ instead of $\bar{y}$. This similarity will be true in many of the procedures to follow.

EXAMPLE 9.6

A recruiter for a university computer science department suggests that the mean starting salary for master's level computer science majors is at least $25,000. Following is a sample of 65 starting salaries reported to the nearest $100:

$26,000	25,100	25,300	24,000	22,900
23,400	27,000	27,500	22,800	21,500
14,600	25,500	24,700	23,000	24,000
26,500	23,700	35,000	25,100	25,000
24,900	22,000	24,500	20,000	15,500
24,400	23,500	20,000	28,000	23,900
25,000	23,000	26,400	26,800	25,200
26,500	36,000	28,500	25,000	22,500
21,800	18,000	26,000	24,000	26,800
27,000	28,500	21,500	20,000	24,100
25,500	24,800	23,500	22,000	27,500
28,000	25,500	24,500	24,900	21,000
22,700	24,000	25,900	27,000	24,200

Is there evidence to deny the claim made?

SOLUTION   Denoting the mean starting salary as μ, the null and alternative hypotheses become

$$H_0: \mu \geq 25,000 \quad \text{vs.} \quad H_a: \mu < 25,000$$

which is a left-tailed test. An ordered stem and leaf plot and box plot of the data indicate that the data are symmetric with long tails.

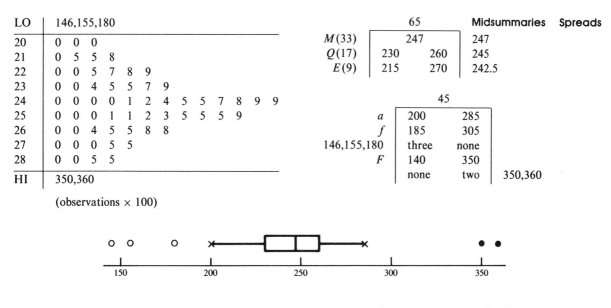

(observations $\times$ 100)

Therefore the test statistic will be $\bar{y}_T$; of which the standardized form is

$$z = \frac{\bar{y}_T - 25,000}{s_T / \sqrt{k}}$$

To find the p-value we must find the observed value of z, which means we need $\bar{y}_T$ and s_T. A 10% trimming would trim off seven observations, which seems unnecessary; therefore we will trim only 5% or 4 observations from each end. A stem and leaf plot of the Winsorized sample becomes

20	0 0 0 0 0 0
21	0 5 5 8
22	0 0 5 7 8 9
23	0 0 4 5 5 7 9
24	0 0 0 0 1 2 4 5 5 7 8 9 9
25	0 0 0 1 1 2 3 5 5 5 5 9
26	0 0 4 5 5 8 8
27	0 0 0 5 5
28	0 0 0 0 0 0

(observations $\times$ 100)

Using a calculator, we find

$$\bar{y}_{T.05} = 24,505.26$$

and

$$s_T = s_W \sqrt{(n-1)/(k-1)} = 2,301.288 \sqrt{64/56} = 2,460.18$$

Thus the observed value of z becomes

$$z_{obs} = \frac{24,505.26 - 25,000}{2,460.18/\sqrt{57}} = \frac{-494.74}{325.859} = -1.52$$

and so the p-value is

$$p\text{-value} = P(z < -1.52) = .5 - .4357 = .0643$$

This p-value is between 5% and 10%, hence we can reject H_0 and declare the results mildly significant. Our conclusion is: On the basis of a random sample of 65 starting salaries for new computer science graduates, there is some statistical evidence that the mean starting salary for all CS graduates is less than \$25,000. There is some indication that the college recruiter's claim is in error.

EXERCISES 9.3

9.25 A sample of 120 observations yielded $\bar{y}_{T.10} = 19.4$ and $s_T = 4.3$. Use this information to test the following hypotheses, with $\bar{y}_T$ being the test statistic:

$H_0: \mu \geq 20$

$H_a: \mu < 20$

Be sure to calculate the p-value and interpret the results.

9.26 A sample of 50 observations yielded $\bar{y}_{T.10} = 57.8$ and

$s_T = 6.4$. Use this information to test the following hypotheses, with $\bar{y}_T$ being the test statistic:

$H_0: \mu \geq 60$

$H_a: \mu < 60$

Be sure to calculate the p-value and interpret the results.

9.27 A sample of 80 observations yielded $\bar{y}_{T.20} = 2.6$ and $s_T = .84$. Use this information to test the following hypotheses, with $\bar{y}_T$

being the test statistic:

$H_0: \mu \leq 2.5$

$H_a: \mu > 2.5$

Be sure to calculate the *p*-value and interpret the results.

9.28 A sample of 40 observations yielded $\bar{y}_{T.10} = 655.4$ and $s_T = 241.7$. Use this information to test the following hypotheses, with $\bar{y}_T$ being the test statistic:

$H_0: \mu = 700$

$H_a: \mu \neq 700$

Be sure to calculate the *p*-value and interpret the results.

9.29 A sample of 100 students from a large high school took a college entrance exam that has a mean of 450. After an examination of the data it appeared that a trimmed mean might better represent the data than the ordinary sample mean. The 10% trimmed mean and trimmed standard deviation of the sample turned out to be 474.6 and 127.3, respectively. Determine if their average is unusually high.

9.30 A psychology test is given to 36 students in a beginning psychology class. From the following data, is there evidence to suggest that the mean of the test for all beginning psychology students is anything other than 50? Their scores were:

45	36	48	55	32	43
35	57	62	42	58	46
21	42	58	84	30	6
45	37	49	54	59	7
33	41	47	91	50	44
41	36	48	56	52	39

9.31 The reading scores of the 30 fifth grade students in Example 6.2 were shown to be symmetric with long tails (see Examples 6.4, 6.6, and 6.8). Test the hypothesis that the mean reading score is 95.

9.32 The mean cost of an evening meal at a typical restaurant in a small city is claimed to be $8.50. Suppose a random sample of 49 tickets appeared symmetrical with long tails and had a 10% trimmed mean of $9.23 and a trimmed standard deviation of $1.69. Is there statisticial evidence to refute the preceding claim about the mean cost of an evening meal?

Small-Sample Test of a Population Mean

The procedure for the small-sample test of μ is basically the same as that of the large-sample test. One difference, however, is that the Central Limit Theorem does not apply to small samples. But, as in confidence interval estimation, if it can be assumed that the underlying parent population is normally distributed, then Student's *t* distribution may be used when the sample size is small.

Small-Sample Test of the Population Mean Based on $\bar{y}$

As before, the evidence against H_0 is measured by the distance the test statistic, $\bar{y}$, is from the value of μ being tested. This can be obtained by calculating the observed value of the test statistic given by

$$t = \frac{\bar{y} - \mu_0}{s/\sqrt{n}}$$

where μ_0 represents the value of μ being tested.

With the assumption that the underlying parent population is normally distributed, we have that the test statistic, *t*, is distributed as Student's *t* is with $n - 1$ degrees of freedom (see Section 8.6).

Therefore the only modification needed is, when calculating the *p*-value, one must determine the distance between $\bar{y}_{obs}$ and μ_0 by finding the observed *t*-score as follows:

▲ You will not be able to calculate the exact *p*-value as you did in the previous sections because the *t*-table is not as extensive as the *z*-table. However, you will be able to find bounds for the *p*-value as is described in the summary and illustrated in Example 9.8.

$$t_{obs} = \frac{\bar{y}_{obs} - \mu_0}{s_{obs}/\sqrt{n}}$$

Then find the *p*-value associated with t_{obs} in the *t*-table.

The complete test is summarized as follows:

Small-Sample Test of μ Based on $\overline{y}$

APPLICATION: $n \leq 30$

ASSUMPTION: Normally distributed population

Left-Tailed Test	Right-Tailed Test	Two-Tailed Test
$H_0: \mu \geq \mu_0$	$H_0: \mu \leq \mu_0$	$H_0: \mu = \mu_0$
$H_a: \mu < \mu_0$	$H_a: \mu > \mu_0$	$H_a: \mu \neq \mu_0$

Standardized Test Statistic

$$t = \frac{\overline{y} - \mu_0}{s/\sqrt{n}}$$

From the t-table, with $n - 1$ degrees of freedom, find the p-value most closely associated with t_{obs}. If t_{obs} falls between two table values, give the two associated probabilities as bounds for the p-value. The p-value for the two-tailed test is calculated as in a one-tailed test and then doubled to account for both tails.

If a level of significance α is specified, then reject H_0 if p-value $< \alpha$.

EXAMPLE 9.7

A standardized psychology exam has a mean of 70. A research psychologist wished to see whether a particular drug had an effect on performance on the exam. He administered the exam to 18 volunteers who had taken the drug, and obtained the following scores:

68	71	75	65	61	70	70	64	71
73	62	78	70	69	76	67	69	72

Is there evidence to suggest that taking the drug reduces one's score on the exam?

SOLUTION   Let μ denote the population mean on the exam after taking the drug. The researcher hypothesis is that the drug reduces one's score on the exam; hence we have

$$H_0: \mu = 70 \quad \text{vs.} \quad H_a: \mu < 70$$

A five-stem stem and leaf plot of the data is

```
6*  | 1
6t  | 2
6f  | 4  5
6s  | 7
6–  | 8  9  9           ȳ = 69.5
7*  | 0  0  0  1  1
7t  | 2  3              s = 4.58
7f  | 5
7s  | 6
7–  | 8
```

The distribution appears normal even though the tails look a little long. A box plot would reveal no outliers. We will assume normality and use $\bar{y}$ as the test statistic. Because we have a small sample, the standardized form of the test statistic becomes

$$t = \frac{\bar{y} - 70}{s/\sqrt{n}}$$

The observed value is

$$t_{obs} = \frac{69.5 - 70}{4.58/\sqrt{18}} = -.463$$

Thus we have

$$p\text{-value} = P(t < -.463)$$

Going to the t-table with 17 degrees of freedom, we see that a tail probability of .10 (10%) corresponds to a t-score of 1.333, and the value of t_{obs} is even smaller than that. Consequently we can say that the

$$p\text{-value} > .10$$

which is insignificant, and we do not have evidence to reject the null hypothesis. In conclusion, we can say that there is insufficient evidence to indicate that the drug reduced the mean score on the exam.

EXAMPLE 9.8

Psychologists wish to investigate the learning ability of schizophrenics after having taken a specified dose of a tranquilizer. Thirteen patients were given the drug and 1 hour later were given a standardized exam. Their scores were:

15, 20, 30, 27, 24, 22, 22, 17, 21, 25, 23, 27, 25

Is there evidence that the mean score exceeds 20 after taking the tranquilizer?

SOLUTION   Let μ denote the mean score on the standardized exam by all schizophrenics who could be administered the drug. The null and alternative hypotheses are stated as follows:

$$H_0: \mu = 20 \quad \text{vs.} \quad H_a: \mu > 20$$

A stem and leaf plot of the data

1f	5
1s	7
1−	
2*	0 1
2t	2 2 3
2f	4 5 5
2s	7 7
2−	
3*	0

appears to be long tailed; however a box plot would reveal no outliers. We will assume we have met the normality assumption.

From the data we find

$$\bar{y} = 22.923 \quad \text{and} \quad s = 4.132$$

Therefore, the observed value of t becomes

$$t_{obs} = \frac{22.923 - 20}{4.132/\sqrt{13}} = 2.55$$

From the t-table with 12 degrees of freedom we see that a t-score of 2.179 has an upper-tailed probability of .025 and a t-score of 2.681 has an upper-tailed probability of .01. Because our t-score of 2.55 falls between 2.179 and 2.681 we have

$$.01 < p\text{-value} < .025$$

which is significant. Thus there is significant evidence to indicate that the mean score exceeds 20 after the tranquilizer is taken.

EXERCISES 9.4

9.33 The average power consumption of 25 randomly selected families in a community for a given period is 125.6 kilowatt hours and the standard deviation is 20.3 kw hours. Assuming that kilowatt usage is normally distributed, is there evidence that the mean usage for the whole community exceeds 120 kw hours?

9.34 The average length of time required to complete a certain aptitude test is believed to be 80 minutes. A sample of 25 students yielded an average of 86.5 minutes and a standard deviation of 15.4 minutes. Assuming normality of the parent distribution, is there evidence to reject the preceding claim? *SMALL $n < 30$*

9.35 The Chamber of Commerce of a particular city claims that the mean carbon dioxide level of air pollution is no more than 4.9 ppm. A random sample of 16 readings resulted in $\bar{y} = 5.6$ ppm and $s = 2.1$ ppm. Assuming that the carbon dioxide level is normally distributed, is there sufficient evidence against the Chamber of Commerce's claim?

9.36 Past experience indicates that the scores of the first hour exam in history are normally distributed with a mean of 72. This semester's class of 16 students had an average score of 76 and a standard deviation of 14.4. Is this class "better than usual"?

9.37 To test the claim that the average home in a certain town is within 5.5 miles of the nearest fire department, an insurance company measured the distance from 25 randomly selected homes to the nearest fire department and found $\bar{y} = 5.8$ miles and $s = 2.4$ miles. What did the insurance company find out? What assumption did you make about the underlying parent distribution?

9.38 An instructor gives his class an examination that he knows from experience has a mean of 78. His present class of 20 students had an average score of 82 and a standard deviation of 7. Is he correct in assuming that this is a superior class? What assumption did you make about the underlying parent distribution?

9.39 A random sample of 25 domestic cats had an average life span of 9.6 years and a standard deviation of 1.2 years. Is there evidence against the claim that the mean life expectancy of all domestic cats is at least 10 years? What assumption did you make about the distribution of the life expectancy of domestic cats?

9.40 A researcher claims that the Self-Criticism score on the Tennessee Self Concept Scale is at least 30 for all gifted high school students in the G-T program at their school. A sample of 20 students had the following scores:

29.8	30.4	27.5	29.8	31.0
30.2	29.5	29.0	27.0	35.8
25.4	34.2	30.7	31.5	28.2
28.9	24.2	32.4	29.2	28.5

Is there evidence to refute the researcher's claim?

9.41 A study was undertaken to determine the number of trials necessary to master a task under the influence of a drug. A sample of nine people resulted in the following:

Person	1	2	3	4	5	6	7	8	9
No. of trials	6	8	14	9	10	11	7	5	12

Assuming that the underlying distribution is normal, is there evidence against the claim that the mean number of trials is 8?

9.42 To determine the flow characteristics of oil through a valve, the inlet oil temperature is measured in degrees Fahrenheit. A sample of nine readings yields:

93, 99, 97, 99, 90, 96, 93, 90, 89

Assuming the parent distribution of oil temperatures is normal, is there evidence that the mean inlet oil temperature is significantly less than 98 degrees?

9.43 Test the hypothesis that the average statistics student will score significantly below 72 on Test 1 if 20 randomly selected students score

82	62	78	56	92	76	70	60	54	69
71	73	37	64	85	100	75	32	67	84

Test of a Population Median

We have noted in confidence interval estimation that if the parent population distribution is highly skewed, then the population median is a better indicator of the center of the distribution than is the population mean. In this section a procedure commonly known as the *sign test* is given for testing the population median. Recall that the population median is denoted by the symbol θ. Not unlike the population mean, the null and alternative hypotheses can take one of three forms:

Left-Tailed—$H_0: \theta \geq \theta_0$ vs. $H_a: \theta < \theta_0$

Right-Tailed—$H_0: \theta \leq \theta_0$ vs. $H_a: \theta > \theta_0$

Two-Tailed—$H_0: \theta = \theta_0$ vs. $H_a: \theta \neq \theta_0$

where θ_0 represents the value of θ we wish to test.

Small-Sample Test of the Population Median

Suppose that one is interested in investigating the study habits of students in a beginning college math class. Suppose, also, that one suspects that the number of hours of study outside class each week is skewed right. Therefore one is concerned with the median number of hours of study outside class each week. Further one suspects that the median number of hours is greater than 5 hours and decides to test

$$H_0: \theta = 5 \quad \text{vs.} \quad H_a: \theta > 5$$

If the median is 5 hours, one would expect approximately half the sample observations below 5 and half above 5. On the other hand, if the population median is greater than 5 (the research hypothesis claim), one would expect an abundance of observations greater than 5. Thus it seems reasonable to reject the null hypothesis if there is a large number of observations greater than 5. Had the alternative been

$$H_a: \theta < 5$$

that is, a left-tailed test, we would reject the null if there were a small number of observations greater than 5.

An appropriate test statistic for the present situation is

$$T = \text{no. of observations} > 5$$

Clearly the range of values for T is

$$\{0, 1, 2, \ldots, n\}$$

where n is the sample size. Moreover if the null hypothesis is true, then the proportion of observations greater than 5 should be 50%. As a random variable, T is distributed as a binomial random variable with parameters n and $\pi = .5$.

Suppose that out of a sample of 15 students, we find that 9 study more than 5 hours per week; that is, the observed value of T is $T_{obs} = 9$. Is this enough evidence to reject H_0 and conclude that the median exceeds 5 hours per week? The p-value is the probability of observing a value of the test statistic at least as large as we did when the null hypothesis is true. Namely

$$p\text{-value} = P(T \geq T_{obs}) = P(T \geq 9)$$

where T is binomial with $n = 15$ and $\pi = .5$.

Going to the binomial tables, we have

$$\begin{aligned} p\text{-value} &= P(T \geq 9) \\ &= 1 - P(T \leq 8) \\ &= 1 - .696 \\ &= .304 \end{aligned}$$

Clearly this p-value is insignificant, and therefore there is insufficient evidence to indicate that the true median is greater than 5 hours per week. We see from the binomial tables that it would have taken 12 or more observations greater than 5 to reject H_0 with a p-value of $.014 + .003 = .017$.

One problem that may occur when this test is conducted is that some observations may be equal to the value of the median being tested. When this occurs those observations that are "tied" with the median value are discarded, and the sample size n is reduced by that amount. For example, in the preceding example if 2 of the 15 observations were equal to 5, then the sample size would have been reduced to 13. The complete small-sample ($n \leq 20$) test of the population median is summarized at the top of page 349.

EXAMPLE 9.9

A psychologist measures the threshold reaction time (in seconds) for persons subjected to emotional stress and obtains the following measurements:

14.3, 13.7, 15.4, 14.7, 12.4, 13.1, 9.2, 14.2, 14.4, 15.8, 11.3, 15.0

Is there evidence that the median threshold reaction time is less than 15 seconds?

SOLUTION   A stem and leaf plot of the data

```
 9 | 2
10 |
11 | 3
12 | 4
13 | 1  7
14 | 2  3  4  7
15 | 0  4  8          (× 10⁻¹)
```

indicates that the distribution is skewed left, and therefore the median, θ, is the appropriate measure to investigate. We wish to obtain evidence that θ is less than 15

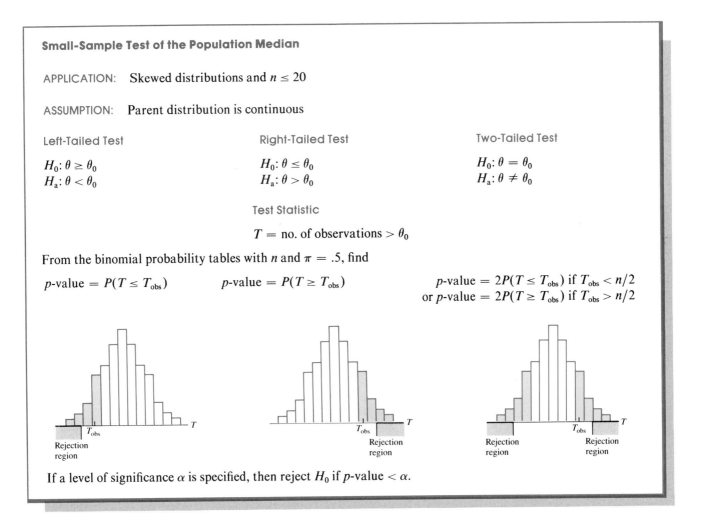

Small-Sample Test of the Population Median

APPLICATION: Skewed distributions and $n \leq 20$

ASSUMPTION: Parent distribution is continuous

Left-Tailed Test

$H_0: \theta \geq \theta_0$
$H_a: \theta < \theta_0$

Right-Tailed Test

$H_0: \theta \leq \theta_0$
$H_a: \theta > \theta_0$

Two-Tailed Test

$H_0: \theta = \theta_0$
$H_a: \theta \neq \theta_0$

Test Statistic

$T = \text{no. of observations} > \theta_0$

From the binomial probability tables with n and $\pi = .5$, find

$p\text{-value} = P(T \leq T_{obs})$

$p\text{-value} = P(T \geq T_{obs})$

$p\text{-value} = 2P(T \leq T_{obs})$ if $T_{obs} < n/2$
or $p\text{-value} = 2P(T \geq T_{obs})$ if $T_{obs} > n/2$

If a level of significance α is specified, then reject H_0 if $p\text{-value} < \alpha$.

(the research hypothesis). Stating the null and the alternative hypotheses, we have

$$H_0: \theta = 15 \quad \text{vs.} \quad H_a: \theta < 15$$

From the ordered stem and leaf plot we find that the test statistic, $T = $ no. of observations > 15, has an observed value of

$$T_{obs} = 2$$

One observation is 15, the value being tested, hence we must reduce the sample size by one. Going to the binomial tables with $n = 11$ and $\pi = .5$ we find

$$p\text{-value} = P(T \leq 2) = .032$$

We have sufficient evidence to reject the null hypothesis and conclude that the median threshold reaction time is less than 15 seconds.

Large-Sample Test of the Population Median

For large samples, we can no longer use the binomial tables, so we use a normal approximation of the binomial distribution. Recall from Chapter 5 that the mean of the binomial distribution is given by $n\pi$ and the standard deviation by $\sqrt{n\pi(1-\pi)}$. Therefore the standardized form of the test statistic T becomes

$$z = \frac{T - n\pi}{\sqrt{n\pi(1-\pi)}}$$

It can be shown that the sampling distribution of z is approximately a standard normal distribution. Therefore the p-value can be found in the normal table. For the median test, π is always equal to .5. Thus the test statistic, z, simplifies to

$$z = \frac{T - n/2}{\sqrt{n}/2} = \frac{2T - n}{\sqrt{n}}$$

In summary we have the following:

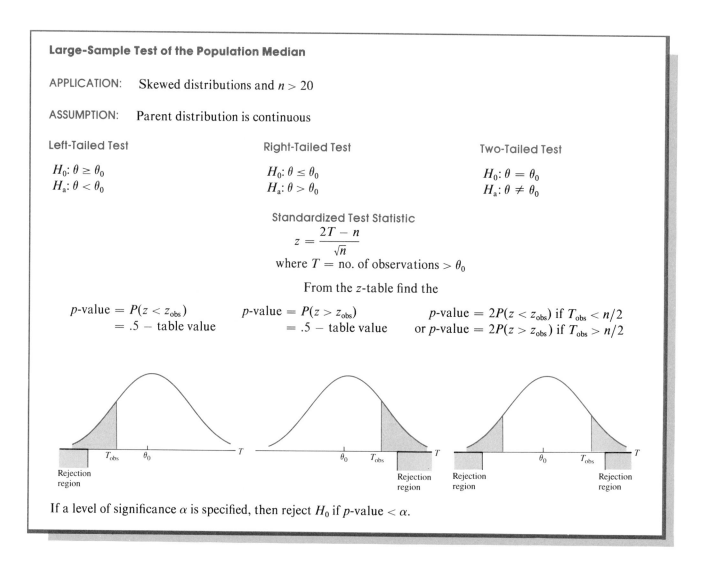

Large-Sample Test of the Population Median

APPLICATION: Skewed distributions and $n > 20$

ASSUMPTION: Parent distribution is continuous

Left-Tailed Test	Right-Tailed Test	Two-Tailed Test
$H_0: \theta \geq \theta_0$	$H_0: \theta \leq \theta_0$	$H_0: \theta = \theta_0$
$H_a: \theta < \theta_0$	$H_a: \theta > \theta_0$	$H_a: \theta \neq \theta_0$

Standardized Test Statistic

$$z = \frac{2T - n}{\sqrt{n}}$$

where T = no. of observations $> \theta_0$

From the z-table find the

p-value $= P(z < z_{obs})$ p-value $= P(z > z_{obs})$ p-value $= 2P(z < z_{obs})$ if $T_{obs} < n/2$
$\quad = .5$ − table value $\quad = .5$ − table value   or p-value $= 2P(z > z_{obs})$ if $T_{obs} > n/2$

If a level of significance α is specified, then reject H_0 if p-value $< \alpha$.

EXAMPLE 9.10

An experimental lethal drug is injected into mice, and the survival time (in seconds) is recorded. The researcher suspects that the median survival time is less than 45 seconds. From the following data, see if there is evidence to support her theory.

| 32 | 37 | 44 | 41 | 65 | 27 | 35 | 29 | 54 | 42 | 38 | 57 | 40 | 24 | 39 |
| 43 | 82 | 52 | 74 | 25 | 34 | 25 | 47 | 34 | 24 | 38 | 91 | 30 | 35 | 31 |

SOLUTION   A stem and leaf plot and box plot of the data points out that the data is skewed right.

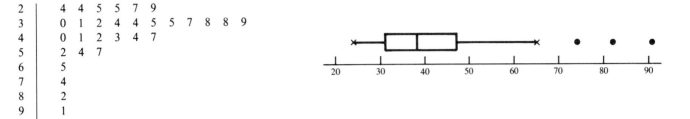

2	4 4 5 5 7 9
3	0 1 2 4 4 5 5 7 8 8 9
4	0 1 2 3 4 7
5	2 4 7
6	5
7	4
8	2
9	1

Thus the investigator will concentrate on the population median. The null and alternative hypotheses become

$$H_0: \theta = 45 \quad \text{vs.} \quad H_a: \theta < 45$$

We have a large sample, and consequently the test statistic is

$$z = \frac{2T - n}{\sqrt{n}}$$

The number of observations greater than 45 is 8. So the observed value of z becomes

$$z_{obs} = \frac{(2)(8) - 30}{\sqrt{30}} = -2.56$$

From the normal table we have

$$
\begin{aligned}
p\text{-value} &= P(z \leq -2.56) \\
&= .5 - .4948 \\
&= .0052
\end{aligned}
$$

which is highly significant. Our researcher has statistical evidence that the median survival time is less than 45 seconds.

The test procedures for the population median described in this section are intended for skewed distributions. They are appropriate, however, for testing the median of any shape continuous distribution. If the median best describes the population characteristic of concern, then these procedures (large or small sample) should be used. In fact, if the sample size is small and you cannot meet the normality requirement for the t-test described in Section 9.4, then this test is recommended.

9.44 Test the following hypotheses using the sample information. Be sure to find the p-value and interpret the results.

(a) $H_0: \theta \geq 50$
  $H_a: \theta < 50$
  $n = 14$; $T =$ no. of observations $> 50 = 5$

(b) $H_0: \theta \leq 3.5$
  $H_a: \theta > 3.5$
  $n = 8$; $T =$ no. of observations $> 3.5 = 7$

(c) $H_0: \theta = 100$
  $H_a: \theta \neq 100$
  $n = 18$; $T =$ no. of observations $> 100 = 4$

9.45 Test the following hypotheses using the sample information. Be sure to find the p-value and interpret the results.

(a) $H_0: \theta \geq 400$
  $H_a: \theta < 400$
  $n = 50$; $T =$ no. of observations $> 400 = 14$

(b) $H_0: \theta \leq 12.5$
  $H_a: \theta > 12.5$
  $n = 100$; $T =$ no. of observations $> 12.5 = 56$

(c) $H_0: \theta = 75$
  $H_a: \theta \neq 75$
  $n = 200$; $T =$ no. of observations $> 75 = 78$

9.46 From the following data, test the null hypothesis that the population median is no more than 50.

84	62	55	36	44	97	73	83	4	90
17	60	79	75	57	59	56	48	66	

9.47 Test the null hypothesis that the population median is 50, using the following sample.

97	89	25	81	11	83	16	96
44	32	98	19	68	33	25	54
74	82	17	49	33	22	62	20
92	80	62	48	71	32	59	54

9.48 According to test theory, the median mental age of the 16 girls listed below should be 100. Do their ages differ significantly from 100?

Mental ages			
87	89	93	93
93	95	95	99
99	102	108	108
113	113	114	114

9.49 The following data are the results of an attempt to assess the predictive validity of Klopfer's Prognostic Rating Scale (PRS) with subjects who received behavior modification psychotherapy. Scores have been ordered.

2.2	4.1	4.2	5.0	5.4
5.8	6.3	6.6	6.8	6.9
7.1	7.4	7.4	7.7	8.2
8.7	9.4	9.5	11.7	11.9

Test to see if there is statistical evidence to suggest that the population median exceeds 6.

9.50 Following are the ages of a sample of 30 college freshmen:

19	18	19	22	18	21	20	19	19	18
19	20	19	18	20	15	33	19	19	20
20	19	18	20	19	20	19	26	21	19

Test the hypothesis that the median age for college freshmen is 19.

9.51 A supplier of storm windows claims that the median wind leakage from a 50 mph wind is not more than 12.5%. A sample of nine windows yields the following results:

Window	1	2	3	4	5
Leakage	.13	.17	.13	.18	.14

Window	6	7	8	9
Leakage	.12	.11	.14	.20

Does the sample verify the supplier's claim?

9.52 It is reported that the median income level for social workers with less than 5 years experience is $17,500. A random sample of 25 social workers from North Carolina (with less than 5 years experience) yielded the following incomes:

$15,200	16,500	16,700	17,900	18,000
15,500	12,200	14,700	17,000	16,500
17,700	11,500	16,000	13,500	18,500
17,000	14,500	17,600	18,100	16,400
17,900	15,900	17,600	16,100	17,400

Is there evidence to suggest that the North Carolina social workers are being underpaid?

9.6 Test of a Population Proportion

We now turn our attention from the mean of a population of measurements to a simple proportion in a Bernoulli population. Recall that a Bernoulli population is one in which each outcome is classified as either a success or failure, and we are concerned with the proportion of successes. In this section a conjecture about the proportion of successes will be investigated with a test of hypothesis. For example, is it true that 60% of American adults are against nuclear reactors? Is it true that no more than 40% of school-aged children receive regular dental care? Notice that we are not concerned with any type of measurement, but rather with the proportion of successes in a success/failure setting.

Letting π denote the true proportion of successes in the population, it is clear that the null and alternative hypotheses can take one of three forms:

Left-Tailed—$H_0: \pi \geq \pi_0$ vs. $H_a: \pi < \pi_0$

Right-Tailed—$H_0: \pi \leq \pi_0$ vs. $H_a: \pi > \pi_0$

Two-Tailed—$H_0: \pi = \pi_0$ vs. $H_a: \pi \neq \pi_0$

where π_0 represents the value of π that one wishes to test.

Because we are dealing with the proportion of successes in the population, the test statistic will be p, the proportion of successes in the sample. Assuming that the sample is large, the Central Limit theorem guarantees that the sampling distribution of p is approximately normal with a mean of π and standard error of $\sqrt{\pi(1-\pi)/n}$. Using

$$z = \frac{p - \pi_0}{\sqrt{\pi_0(1 - \pi_0)/n}}$$

as the standardized test statistic, one can find the p-value in the standard normal probability table. Note that the standard error of p is $\sqrt{\pi(1-\pi)/n}$, but under the assumption that the null hypothesis is true, the standard error becomes $\sqrt{\pi_0(1-\pi_0)/n}$, the denominator of z.

The complete test is summarized at the top of page 354.

EXAMPLE 9.11

A news reporter in a major eastern city reported that 80% of all violent crimes in that city involve firearms. A member of the NRA doubted this claim and set out to disprove it. He collected data on all violent crimes in the city for the past 2 years. He found that of 280 violent crimes, 240 involved firearms. Was he justified in doubting the news report?

SOLUTION   The claim made by the news reporter is that the percent is 80; therefore we will set up the hypotheses as a two-tailed test.

$$H_0: \pi = .80 \quad \text{vs.} \quad H_a: \pi \neq .80$$

Because we are dealing with the proportion in a Bernoulli population, the test statistic is p, the sample proportion. The standardized form of the test statistic is

$$z = \frac{p - \pi_0}{\sqrt{\pi_0(1 - \pi_0)/n}}$$

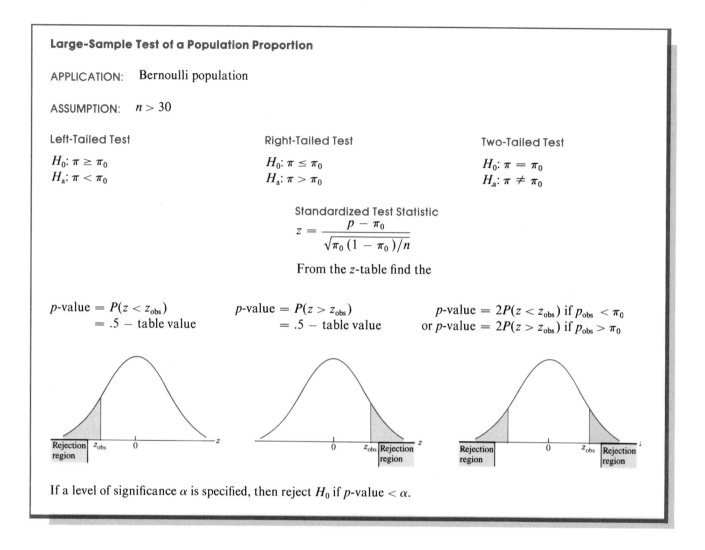

Large-Sample Test of a Population Proportion

APPLICATION: Bernoulli population

ASSUMPTION: $n > 30$

Left-Tailed Test	Right-Tailed Test	Two-Tailed Test
$H_0: \pi \geq \pi_0$	$H_0: \pi \leq \pi_0$	$H_0: \pi = \pi_0$
$H_a: \pi < \pi_0$	$H_a: \pi > \pi_0$	$H_a: \pi \neq \pi_0$

Standardized Test Statistic

$$z = \frac{p - \pi_0}{\sqrt{\pi_0(1 - \pi_0)/n}}$$

From the z-table find the

p-value $= P(z < z_{obs})$	p-value $= P(z > z_{obs})$	p-value $= 2P(z < z_{obs})$ if $p_{obs} < \pi_0$
$= .5 -$ table value	$= .5 -$ table value	or p-value $= 2P(z > z_{obs})$ if $p_{obs} > \pi_0$

If a level of significance α is specified, then reject H_0 if p-value $< \alpha$.

where $p = 240/280 = .857$ and $\pi_0 = .8$. Therefore we have

$$z = \frac{p - .8}{\sqrt{(.8)(.2)/280}} = \frac{.857 - .8}{.0239} = 2.38$$

Because $p > .8$, the p-value is found as it would be in a right-tailed test and then doubled as follows:

$$p\text{-value} = 2P(z > 2.38) = 2(.5 - .4913) = 2(.0087) = .0174$$

This p-value is significant at a level of significance of .02, and thus the NRA proponent was successful in rejecting the news reporter's claim that 80% of all violent crimes involve firearms. After a closer investigation, however, he realized that he actually showed that the percent is greater than 80%, contrary to what he wished to show. This happens sometimes.

EXAMPLE 9.12

A recent report stated that no more than 20% of all college graduates obtain work in their field of study. A random sample of 500 graduates was surveyed to find that 135 obtained work in their field of study. Is there statistical evidence that the reported percentage is too low?

SOLUTION   If we let π denote the percent of college graduates that obtain work in their field of study, then the null and alternative hypotheses will be stated as follows:

$$H_0: \pi \le .20 \quad \text{vs.} \quad H_a: \pi > .20$$

The test statistic is p, the proportion of successes in the sample, which is found to be

$$p = 135/500 = .27$$

The standardized form of the test statistic is

$$z = \frac{p - .20}{\sqrt{(.2)(.8)/500}} = \frac{.27 - .20}{.018} = 3.89$$

Thus we have

$$p\text{-value} = P(z > 3.89) < .5 - .4998 = .0002$$

Clearly this is enough evidence to reject the null hypothesis. On the basis of a random sample of 500 college graduates, there is overwhelming evidence that more than 20% find work in their field of study.

EXERCISES 9.6

9.53 In Exercise 9.11, the psychologist suspected that more than 10% of the adult population is illiterate. She gave the Wechsler's Adult Intelligence Scale to a sample of 1,600 adults. Based upon the results of WAIS, she classified 180 of the 1,600 as illiterate. Is this sufficient evidence to verify her claim?

9.54 In Exercise 9.12, a claim was made that at least 60% of the adult population feels that there is too much violence on TV. A random sample of 200 adults showed that 110 felt that there is too much violence on TV. Is this evidence to reject the previous claim?

9.55 The government believes that no more than 25% of all college students favor reducing the penalties for the use of marijuana. A sample of 2,400 college students revealed that 750 favor reducing the penalties. Do we have statistical evidence to reject the government's claim?

9.56 In Exercise 9.55, using the government's estimate of 25%, how large a sample is needed to estimate the percent to within 1%, using a 95% confidence interval?

9.57 We wish to determine whether a greater proportion of students than local citizens favor the sale of beer in the county.

We know that half of the local citizens oppose the sale of beer. In a random sample of 400 students we find that 230 favor the sale of beer. Is there sufficient evidence to suggest that a greater proportion of students than local citizens favor the sale of beer in the county?

9.58 In polling 1,000 randomly selected voters, 469 were encouraged by the recent military actions in the Mideast. The President stated that he has the support of the majority of the voters. Is his claim valid?

9.59 A manufacturer of autos purchases machine bolts from a supplier who claims that his bolts are no more than 5% defective. From a random sample of 400 bolts it is found that 28 are defective. Is there sufficient evidence to reject the supplier's claim?

9.60 A psychologist has developed a new aptitude test and believes that 80% of the public should score above 50 on the test. From a sample of 200 people, 164 scored above 50. Is there statistical evidence that the claim made by the psychologist is not valid?

9.61 A presidential aide said that over half of the nation agreed with the U.S. invasion of Grenada in 1984. However a Roper poll of 2,000 Americans found that only 34% supported the invasion. Did the 34% simply happen by chance, or is there statistical evidence to imply that the aide was in error?

9.62 One third of all ex-convicts return to jail within 3 years. One state is trying a new rehabilitation system on its prisoners. After releasing 200 prisoners who had participated in the new system, only 48 returned within 3 years. Is there statistical evidence that the new system is working to reduce the proportion of repeat offenders?

9.63 Dr. Quacko claims that at least 80% of his patients never have significant arthritic pain after a 6-month stay in his Caribbean treatment center. Out of a random sample of 200 of his former patients, 65 said that his treatment was ineffective and that they now have arthritic pain. See if there is statistical evidence to refute the doctor's claim.

9.7 Hypothesis Testing versus Confidence Intervals

Both confidence interval estimation and hypothesis testing deal with inferences about unknown population parameters. Deciding which to use in a particular application depends on the intent of the investigation. Do we need to gather information about the parameter, or do we ultimately have to make a decision concerning the parameter? Gathering information about the parameter involves constructing confidence intervals. Determining the truth of a particular conjecture about the parameter involves hypothesis testing. Following are several examples that illustrate the point:

1. A politician wants to know what percent of the voters in his district are in favor of his running for a second term. Here the politician will make a decision on the basis of the results of a poll, but the question at hand is, "What percent are in favor?" Thus a confidence interval is appropriate.

2. The FDA is testing a new dietary supplement to see whether it dissolves cholesterol deposits in arteries. Here they will make a decision that either the supplement is effective or it is not. Therefore a test of hypothesis is appropriate.

3. Government agencies that investigate such quantities as unemployment rates, inflation, gross national product, and so on, simply need estimates via confidence intervals.

4. A claim is made that a greater proportion of women smoke than men. A test of hypothesis would shed light on this question.

5. A textile company wishes to compare a new manufacturing process with the old one. Is the new technique an improvement over the old one? This clearly indicates that hypothesis testing is appropriate.

6. A consumer agency analyzing the rising cost of medical care investigates the mean fee for various operations with confidence intervals.

7. Does Vitamin C help to prevent colds? An experiment could be designed where Vitamin C is compared with a placebo and analyzed using hypothesis testing.

8. A congressional committee investigating fatal traffic accidents caused by drinking most likely will be interested in confidence intervals.

9. In the study of a virus, 60 mice were injected with the virus but not treated. A like group of 60 mice was injected with the virus and treated with an experimental

drug. A comparison study of the effectiveness of the experimental drug could be conducted with hypothesis testing.

10. A business report of the inventory value of a warehouse of household carpet most likely will consist of confidence interval estimates of the mean inventory value of the carpet.

In general, confidence intervals give information and hypothesis testing helps make decisions.

Although confidence intervals and hypothesis testing are different types of inference procedures, they are closely related. They are different ways of expressing the same information contained in a sample. For example, if a 99% confidence interval for the proportion of fatal accidents caused by drinking is given by the limits .45 to .55, then a test of the hypothesis that the proportion is .5 would certainly not be rejected. On the other hand, the hypothesis that it is .6 would be rejected. In fact, it can be said that any claim that the proportion is a specific value between .45 and .55 will be accepted and outside those limits will be rejected at the 1% level of significance. In this sense, confidence interval inference is a more comprehensive inference procedure than is hypothesis testing.

Computer Session (optional)

There are two Minitab commands that relate to the test of hypothesis about the population mean μ. Both tests are based on $\bar{y}$, one being the large-sample test of μ, and the other being the small-sample test of μ when the parent population is normally distributed.

ZTEST

As in Section 9.2, the z-test is appropriate when we have a large sample and the parent distribution does not have tails that are unusually long. Here it is also assumed that the value of σ is known.

The ZTEST command will test the null hypothesis about μ against a two-sided alternative. The form of the command is

```
ZTEST MU = K, SIGMA = K, ON DATA IN C,C,...,C
```

SIGMA is the value of the known population standard deviation. If a one-sided alternative is desired, the subcommand ALTERNATIVE can be used. For a left-tailed test specify ALTERNATIVE $= -1$ and for a right-tailed test specify ALTERNATIVE $= +1$.

EXAMPLE 9.13

The Statistical Insight problem at the beginning of the chapter was concerned with the mileage ratings of new automobiles. The miles per gallon for a Honda Civic was recorded on 35 different occasions with the following results:

48.3	49.8	39.6	43.5	46.8	49.4	52.6
45.3	49.7	45.3	40.7	45.6	48.2	50.3
48.2	40.9	43.5	54.2	50.0	51.2	50.8
45.5	49.2	40.5	45.3	50.2	44.8	47.3
48.7	49.2	47.4	51.4	49.6	50.3	48.5

Use the ZTEST command with $\sigma = 5$ to test the null hypothesis that the mean miles per gallon is at least 49 mpg.

SOLUTION   The data will be stored in the computer with the SET command, after which they will be printed back to check for errors. A stem and leaf plot and box plot will be constructed to detect any peculiarities in the data (see Figure 9.9). Note that the p-value is .036, which is significant. Thus there is statistical evidence that the number of miles per gallon for a Honda Civic is less than 49 mpg.

FIGURE 9.9

Minitab output for Example 9.13

```
MTB > PRINT C1

C1
    48.3    49.8    39.6    43.5    46.8    49.4    52.6
    45.3    49.7    45.3    40.7    45.6    48.2    50.3
    48.2    40.9    43.5    54.2    50.0    51.2    50.8
    45.5    49.2    40.5    45.3    50.2    44.8    47.3
    48.7    49.2    47.4    51.4    49.6    50.3    48.5
```

[handwritten margin notes:] SUBTRACT ONE COLUMN FROM ANOTHER AND STORE INFORMATION IN A NEW COLUMN. DO A "WTEST" ON THIS NEW COLUMN. CAN USE "HELP WTEST". "WTEST" IS WILCOXON TEST

```
MTB > STEM C1

STEM-AND-LEAF OF C1      N = 35
Leaf Unit = 1.0

     1     3 9
     4     4 000
     6     4 33
    12     4 455555
    15     4 677
   (11)    4 88888999999
     9     5 0000011
     2     5 2
     1     5 4

MTB > BOXPLOT C1


                                  -----------------
                 --------------------I         +     I--------------
                                  -----------------
             --+---------+---------+---------+---------+---------+----C1
               39.0      42.0      45.0      48.0      51.0      54.0


MTB > ZTEST MU = 49, SIGMA = 5, C1;
SUBC> ALTERNATIVE = -1.

TEST OF MU = 49.000 VS MU L.T. 49.000&
THE ASSUMED SIGMA = 5.00

              N      MEAN     STDEV    SE MEAN        Z    P VALUE
C1           35    47.480     3.561      0.845    -1.80      0.036
```

TTEST

The *t*-test is for testing μ when we have a small sample and the parent population is normally distributed.

The form of the command is

```
TTEST MU = K, ON DATA IN C,C,...,C
```

EXAMPLE 9.14

The following scores on the Miller Personality test were obtained for a sample of 25 college students.

21	18	20	25	23
19	30	24	29	14
25	22	35	26	23
16	33	25	22	18
34	22	31	27	25

Use the TTEST command to test the hypothesis that the mean score is 25.

SOLUTION   The data will be stored in the computer with the SET command. We will print out the data to check for errors and then construct a stem and leaf plot. Before testing the hypothesis, a box plot will be constructed to check out the lengths of the tails of the distribution to be assured that we do not have a long-tailed or skewed distribution. The stem and leaf plot in Figure 9.10 appears to be long tailed. However, the box plot shows no outliers and appears not to deviate from normality.

FIGURE 9.10

Minitab output for Example 9.14

```
MTB > PRINT C2

C2
  21   18   20   25   23   19   30   24   29   14   25   22   35
  26   23   16   33   25   22   18   34   22   31   27   25


MTB > STEM C2

STEM-AND-LEAF OF C2     N  = 25
Leaf Unit = 1.0

      1      1 4
      2      1 6
      5      1 889
      7      2 01
     12      2 22233
    (5)      2 45555
      8      2 67
      6      2 9
      5      3 01
      3      3 3
      2      3 45

MTB > BOXPLOT C2


                          -----------------
        -----------------I       +       I---------------------
                          -----------------
        --------+---------+---------+---------+---------+-------C2
           16.0      20.0      24.0      28.0      32.0


MTB > TTEST MU = 25, C2

TEST OF MU = 25.00 VS MU N.E. 25.00

              N       MEAN     STDEV     SE MEAN        T     P VALUE
    C2       25      24.28      5.50        1.10     -0.65       0.52
```

Note that the *p*-value = .52, which is insignificant. There is insufficient evidence to say that the mean is different from 25.

EXERCISES 9.8

9.64 Work through Example 9.6 using Minitab and the ZTEST command.

9.65 Work through Example 9.7 using Minitab and the TTEST command.

9.66 A math instructor claims that if a student scores below 80 on his or her first test then, most likely, the student will fail developmental mathematics. In order to determine if this is a valid predictor, she records the following scores on the first test for all students who failed developmental mathematics in the 1986 fall semester:

84	88	96	87	65	98	41	92	78	70
93	77	39	73	62	74	51	88	100	89
100	79	69	74	69	84	49	65	77	48
61	86	68	90	68	76	67	40	84	77

Is there statistical evidence that the mean score of the students who fail developmental mathematics is less than 80? Use Minitab to analyze these data.

Key Concepts

☑ The *null hypothesis* is the hypothesis being tested in a *test of significance*. The *alternative hypothesis* is the hypothesis believed to be true, and is also called the *research hypothesis*.

☑ A *Type I error* is rejecting a true null hypothesis. A *Type II error* is failing to reject a false null hypothesis. The probability of a Type I error is denoted by α and is called the *level of significance*. The probability of a Type II error is denoted by β.

☑ A *test statistic* is the statistic used to test the null hypothesis. The null hypothesis will be rejected if the test statistic falls in the rejection region.

☑ A *right-tailed test* is a test of hypothesis where the rejection region is on the right tail of the sampling distribution of the test statistic. A *left-tailed test* is a test where the rejection region is on the left tail of the sampling distribution of the test statistic. A *two-tailed test* is one where the rejection region is on both tails of the sampling distribution of the test statistic.

☑ The *p-value* is the probability of observing a value of the test statistic at least as extreme as that given by the observed data, under the assumption that the null hypothesis is true. The test statistic will fall in the rejection region if and only if the *p*-value is smaller than the level of significance. In this case, the null hypothesis is rejected.

☑ The steps to follow in testing hypotheses are:

1. Formulate the null and alternative hypotheses.
2. Decide on an appropriate test statistic.
3. Determine the sampling distribution of the test statistic.
4. Collect the data and calculate the *p*-value.
5. Interpret the results.

☑ The test of a population mean can be based on either $\bar{y}$ or $\bar{y}_T$. The test is based on $\bar{y}$ if the tails of the parent distribution are not excessively long. It is based on $\bar{y}_T$ if the parent distribution is symmetric with long tails and the sample size is large. If the parent distribution is skewed, the inference should be concerned with the population median, and the test based on the sample median. For the small-sample test based on $\bar{y}$, it is assumed that the parent population is normally distributed. The test of a population proportion is based on the z statistic and is similar to the test of the population mean based on $\bar{y}$.

Learning Goals

Having completed this chapter you should be able to:

1. Formulate null and alternative hypotheses.   *Section 9.1*
2. Interpret Type I and Type II errors.   *Section 9.1*
3. Determine which test statistic to use in a testing procedure.   *Section 9.2– Section 9.6*

4. Identify a right-tailed, a left-tailed, and a two-tailed hypothesis test. *Section 9.2*

5. Understand the concept of and the calculation of the *p*-value. *Section 9.2–Section 9.6*

6. Test a population mean, with a large sample, based on $\bar{y}$. *Section 9.2*

7. Test a population mean, with a large sample, based on $\bar{y}_T$. *Section 9.3*

8. Test a population mean, with a small sample, based on $\bar{y}$. *Section 9.4*

9. Test a population median with a large or small sample. *Section 9.5*

10. Test a population proportion with a large sample. *Section 9.6*

Review Questions

To test your skills answer the following questions.

9.67 Suppose that a preadmission algebra exam has a standardized mean of 200 and a standard deviation of 50. A sample of 100 students from a large high school takes the college admission exam. The average score turned out to be 215. Is this unusual? Explain with a test of hypothesis procedure.

9.68 To investigate the self-concept of a group of college administrators, a psychologist administers the TSCS exam to 16 administrators and finds that their average self-concept score is 22 and the standard deviation is 3. If the national mean for the self-concept score is 20, can we say that this group of college administrators has an unusually high self-concept? What assumption did you make about the underlying parent distribution?

9.69 It is believed that the percent of convicted felons having a history of juvenile delinquency is 70%. Is there evidence to contradict this claim if out of 200 convicted felons we find that 154 have a history of juvenile delinquency?

9.70 Do the following data on the STEP science test for a class of ability-grouped students provide sufficient evidence to conclude that they fall significantly below the national median of 80?

58 60 82 80 67 70
65 73 75 77 82 68

SUPPLEMENTARY EXERCISES FOR CHAPTER 9

9.71 A construction engineer would like to determine whether a new type of cement has a better bonding quality than the mix he currently uses. Should he use hypothesis testing or confidence interval estimation?

9.72 A health department official would like to determine the severity of the recent flu epidemic. Should she use hypothesis testing or confidence interval estimation?

9.73 A team of Environmental Protection Agency scientists sampled the water in 225 lakes in the Adirondack mountains in New York to assess the extent of acid rain pollution in the lakes (*USA Today,* October 12, 1984). Should they use hypothesis testing or confidence interval estimation?

9.74 Why is the alternative hypothesis also called the research hypothesis?

9.75 A study showed that more than 70% of the women who undergo breast biopsies to remove benign lumps face no unusual risk of later developing breast cancer (*USA Today,* January 17, 1985). Formulate null and alternative hypotheses to test the validity of this claim.

9.76 The pH factor measures the acidity or alkalinity of a substance. A pH of 5.5 is recommended in the soil for raising Christmas trees. Formulate the null and alternative hypotheses in order to test that the soil pH is at the desired level.

9.77 A class of 50 eighth graders has completed a standardized reading test on which their scores had a mean of 107.5 and a standard deviation of 10.5. The national mean score on the test is 100. Set up the null and alternative hypotheses to test to see if this class is "superior in reading ability."

9.78 A sample of 42 observations yielded $\bar{y} = 642$ and $s = 135$. Use this information to test the following hypotheses, with $\bar{y}$ being the test statistic:

$H_0: \mu = 600$
$H_a: \mu \neq 600$

Be sure to calculate the *p*-value and interpret the results.

9.79 The city government claims that the average residence tax is no more than $800 per year. We wish to test this claim, therefore we randomly select 45 property owners and determine the average residence tax to be $848. Can we reject the claim made by city government if it is known that the distribution of residence tax is near normal with a standard deviation of $200?

9.80 A manufacturer claims that the average life of its washing machines is at least 4 years. It is assumed that the distribution of the life span is approximately normal. A random sample of 32 of these machines had an average life span of 3.6 years and a standard deviation of 1.5 years. Is there statistical evidence to dispute the manufacturer's claim?

9.81 A university administrator believes that their applicants for admission are significantly above the national norm of 450 on SATMath. He randomly pulls 100 records from their applicant pool and finds their average SATMath is 463 and the standard deviation is 42.6. Assuming SAT scores are somewhat normal in shape, is this enough evidence to verify his claim?

9.82 In a survey of 100 randomly selected families in California, it was found that the mean medical expense for the year was $1,640 and the standard deviation was $260. An insurance company claims that the mean medical expense for all California families is at least $1,700. Test their claim using the data from the 100 families.

9.83 A psychologist wishes to evaluate a new technique that he has developed to improve rote memorization. The subjects are asked to memorize 50 word phrases using his technique. He feels that his technique is successful if the subjects average more than 40 correct phrases. How should he set up his null and alternative hypotheses?

9.84 Interpret the Type I and Type II errors for Exercise 9.83.

9.85 According to the norms published for a certain intelligence test, the average for college freshmen is expected to be 65 points and the standard deviation is 10 points. A sample of 81 freshmen at State U. averaged 68.2 points on the test. Is there sufficient evidence to indicate that State U. freshmen are more intelligent, as measured by this test, than the average college freshman?

9.86 In order to study the effect of birth control pills on exercise capacity, a physiologist measures the maximum oxygen uptake during a treadmill session. It is known that the average maximal oxygen uptake for females not on the pill is 36 ml/kg of body weight. A random sample of 36 females on the pill had an average maximal oxygen uptake of 33.6 ml/kg of body weight and a standard deviation of 4.8 ml/kg. Is there statistical evidence that women on the pill have a significantly smaller maximal oxygen uptake than women not on the pill?

9.87 A supermarket is trying to decide whether to accept or reject a shipment of tomatoes. It is impossible to check all the tomatoes for size, but they desire an average weight of 8 ounces (they neither want them too large nor too small). A random sample of 400 tomatoes yields an average weight of 7.85 ounces and a standard deviation of 1.15 ounces. Should the supermarket reject the shipment?

9.88 A group of nine slow-ability-grouped children is given a standardized exam that has a national mean score of 50. Their mean was 42 and the standard deviation was 15. Are these children actually slow-ability children as measured by this test? What assumption must be made about the distribution of the scores on the standardized exam?

9.89 A manufacturer of string has established from several years of experience that the breaking strength of the string he manufactures has a mean of 15.9 pounds and a standard deviation of 2.4 pounds. A change is made in the manufacturing process, after which a sample of 64 pieces is taken, whose mean breaking strength is found to be 16.2 pounds. Assuming breaking strength is normally distributed and the new process has the same standard deviation as the old, can we say that the average breaking strength of the string from the new process is greater than the average from the old process?

9.90 The U.S. Postal Service claims that at least 80% of the letters mailed in New York City destined for Los Angeles are delivered within 2 working days. To verify this claim, suppose that 100 letters were mailed from New York to various destinations in the Los Angeles area, and that 76 were delivered within 2 working days. Is there evidence to dispute the U.S. Postal Service's claim?

9.91 The height of adults in a certain town has a mean of 65.42 inches. A sample of 144 adults living in a depressed section of town is found to have a mean height of 64.82 inches and a standard deviation of 2.32 inches. Does this indicate that the adult height of residents in the depressed area is significantly below that of all residents of the town?

9.92 A spokesperson for a statewide organization for raising the legal drinking age claims that at least 70% of the state population feels that the legal drinking age should be raised from 18 to at least 20 years of age. A random sample of 200 people was asked whether they approve of raising the drinking age to at least 20. Of the 200, 132 favored raising the limit. Is there statistical evidence to doubt the spokesperson's claim?

9.93 A utility company claims that its customers' heating bills average no more than $250 during the winter period of December, January, and February. A consumer group sampled 100 accounts for the 3-month period, and found an average cost of

$263 and a standard deviation of $72. Is there statistical evidence to cast doubt on the utility company's claim?

9.94 A grocery store chain is trying to decide how many checkout aisles there should be in each store. From the following checkout times of 40 randomly selected customers, test the hypothesis that the mean checkout time is no more than 7 minutes per customer (data are recorded in minutes).

5.6	3.8	7.9	1.3	4.2	8.8	5.9	9.4
15.6	4.8	1.2	6.5	3.5	7.2	5.3	7.6
6.6	9.2	8.4	14.1	6.3	9.1	6.3	2.5
6.3	7.1	6.7	7.8	5.8	6.5	4.6	6.8
7.9	5.6	6.4	6.2	6.9	5.5	7.3	5.1

9.95 A chemistry student found a means of manufacturing imitation diamonds. From the process, he determines that he can make a profit if the produced stones with a median weight over .5 carat. From his sample he found the following weights:

.46, .61, .52, .48, .57, .54 carats

Should he go into the diamond-making business or stay in school?

9.96 A factory makes a certain computer part for IBM which, according to specifications, must have a mean length of 1.5 centimeters. In a random sample of 16 parts from a shipment, IBM found an average length of 1.56 centimeters and a standard deviation of .09 centimeter. Should IBM reject this shipment?

9.97 Suppose you own a large racing stable and are contemplating the purchase of a 2-year-old stallion. Based on past experience, you think a horse of this age should be able to run the mile in less than 98.5 seconds. You race the horse and obtain the following times in seconds:

104.6	98.8	101.4	98.2	99.7
102.5	103.6	98.7	101.5	

You decide that you will reject the horse if you feel that his lifetime median time exceeds 98.5 seconds on the mile. Should you reject the horse? Three years later you put the horse out to pasture after determining that his lifetime median time is 100 seconds on the mile. Did you make an error? If so, what type?

10

Estimation and Hypothesis Testing of the Difference between Two Parameters

In the last two chapters we studied inference procedures for a single population. Now we will investigate problems concerning two populations. Samples from each will be used to make inferences about two population parameters. The inference from the two samples may take the form of an estimate or a test of hypothesis.

Tests comparing two population means are among the most widely used statistical procedures available. In addition to comparing two population means we will also study the problem of comparing two population proportions. You will see that by a simple extension of the concepts developed for the single population, the realm of statistical analysis can be expanded enormously.

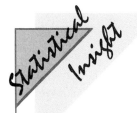

Use of Vitamin C Therapy Loses Status as Cancer Cure

Vitamin C doesn't cure cancer, concludes a new study being published today. The study, conducted at the Mayo Clinic in Rochester, Minn., found that 51 terminally ill cancer patients given massive doses (10 grams) of vitamin C for up to one year fared no better than 49 similar patients who were given useless placebo pills instead of vitamin C. About equal percentages of both groups died within one year—49 percent of the vitamin C group and 47 percent of the placebo group. And 96 of the total 100 patients, all of whom had advanced colorectal cancer, experienced a worsening of the disease during the study, report Dr. Charles Moertel and his associates in the *New England Journal of Medicine*. The study adds evidence that once-touted vitamin C therapy has no benefit. "This fails to show a benefit for high dose vitamin C therapy," the researchers say.

SOURCE: *USA Today*, January 17, 1985. Copyright, 1985 *USA Today*. Reprinted with permission.

As the preceding article suggests, vitamin C is not as effective a cure for cancer as some people believe. In the Mayo Clinic study the vitamin C group did not fare any better than the placebo group.

In this example, the data speak for itself; 49% of the terminally ill cancer patients in the vitamin C group, as compared with 47% in the placebo group, died within 1 year. Clearly there is no statistical evidence to suggest that vitamin C was beneficial in curing cancer of terminally ill patients. But suppose the percentages were different; how do we compare data from two groups? In this chapter we will learn how to compare measurement data from independent samples and from dependent samples. We will also learn how to compare Bernoulli proportions such as those presented in this example.

10.1 Introduction

A very useful tool in statistical analysis is the comparison of two populations. Often we wish to compare a new procedure with an old established procedure, or to compare one product with another, or to compare one treatment with another treatment. The key word is *compare*; we wish to compare the characteristics of one population with those of another population. To compare the two populations, samples from each must be selected and then used to make inferences about the parameters of the two populations.

The observations in the two samples usually come about from an experimental design, as discussed in Chapter 1. Recall from Chapter 1 that in the comparative experiment we usually randomly divide the subjects into two equivalent groups, with one group receiving the treatment and the other group serving as the control. Additionally we observed that two treatments can be compared, each serving as the control for the other.

The technique used to choose and to assign subjects to the treatments is the experimental design. We will consider two types of designs that are appropriate for a comparative experiment. In the first design, the sample assigned to treatment 1 is selected independently of the sample assigned to treatment 2. In the second design, the two samples are dependent, in that the experimental subjects are matched or paired in some sense. The subjects are chosen in pairs or are matched in such a way that, without the treatment, the subjects in each pair will perform almost identically. One member of each pair receives treatment 1 and the other member receives treatment 2. Any observed differences in the responses will be attributed to the treatments.

EXAMPLE 10.1

A study published in the *Journal of the American Medical Association* by Swedish doctors at Sahlgren's Hospital in Göteborg, suggests that nicotine-laced gum helps smokers to abstain from smoking. The study showed that 29 of 106 smokers who chewed nicotine gum while trying to quit remained smoke free for 1 year. In a similar group, 16 of 100 smokers who chewed regular gum remained smoke free for 1 year. Both groups participated in counseling groups to help them quit.

In this example we assume that the two groups of smokers are independent of each other, and we are interested in comparing two Bernoulli proportions. The analysis for this type of problem appears in Section 10.6.

EXAMPLE 10.2

To examine the effects of jogging on the resting pulse rate, a collection of 100 males and females was divided into an experimental and a control group, with each member of one group being paired with a member of the other group according to sex, age, weight, and height. All subjects were asked to continue their everyday activities with the exception that the members of the experimental group were asked to jog 4 days a week. They began jogging a quarter mile each day and built up, over a 6-month period, to 2 miles a day. After 6 months, the resting pulse rate was measured on both members of each pair of subjects.

The pulse rate of a member of the experimental group is related to the pulse rate of the corresponding member of the control group because they were paired according to sex, age, weight, and height. The two samples are dependent. The analysis for dependent samples appears in Section 10.5.

EXAMPLE 10.3

A team of physicians would like to compare the recovery times of two different techniques for a certain type of operation. Subjects are to be randomly and independently assigned to the two techniques, and after the operation their recovery times are to be recorded. The difference in mean recovery times for the two techniques can be evaluated with the procedures in Section 10.2 if the sample sizes are large, or in Section 10.4 if the sample sizes are small.

10.2 Large-Sample Inference on the Difference between Two Population Means Using Independent Samples

The procedures used to estimate and test hypotheses about a single parameter can be modified to apply to two parameters. In this section we develop the large-sample inference procedures for the difference between two population means.

▲ Remember that inferences are in the form of confidence intervals or test of hypotheses.

Let μ_1 and σ_1 denote the mean and standard deviation of population 1 and let μ_2 and σ_2 denote the mean and standard deviation of population 2. Suppose we have two *independent* samples, each of size greater than 30, from the two populations. Let n_1 and n_2 denote the two respective sample sizes. From the two samples we will be able to make inferences about the difference, $\mu_1 - \mu_2$, between the two population means. The first inference problem we will consider is finding a confidence interval for $\mu_1 - \mu_2$.

Large-Sample Confidence Interval for $\mu_1 - \mu_2$ Based on $\bar{y}_1 - \bar{y}_2$

The confidence interval for $\mu_1 - \mu_2$ will be constructed in the same manner that the one-sample confidence interval for μ was constructed. We first need a good point estimator for the parameter $\mu_1 - \mu_2$. We also need to know the sampling distribution of the estimator and the level of confidence desired. As in single-parameter estimation, the choice of an estimator depends on the distributions of the underlying parent populations. First we will consider the case where the two parent populations are symmetric with tails that are not excessively long. Later we will consider the skewed and long-tailed situations.

▲ Recall that if a single population distribution is symmetric with tails that are not excessively long, then $\bar{y}$ is the chosen estimator of μ. Here we are simply extending this to two populations.

Following the single-parameter estimation problem, it might seem that under these conditions, a reasonable estimator of $\mu_1 - \mu_2$ is the difference in the two sample means, $\bar{y}_1 - \bar{y}_2$. In Section 7.5 it was pointed out that for large samples, the sampling distribution of $\bar{y}_1 - \bar{y}_2$ is approximately normal with a mean of $\mu_1 - \mu_2$ and a standard error of

$$\sqrt{\sigma_1^2/n_1 + \sigma_2^2/n_2}$$

Because n_1 and n_2 are both large, the approximation remains valid if σ_1^2 and σ_2^2 are replaced by their estimators s_1^2 and s_2^2, respectively. Then we have that the

$$\text{estimated SE}(\bar{y}_1 - \bar{y}_2) = \sqrt{s_1^2/n_1 + s_2^2/n_2}$$

As before, a confidence interval is constructed in the following manner:

Estimator ± table value × estimated standard error

Because the sampling distribution of our estimator, $\bar{y}_1 - \bar{y}_2$, is approximately normal, the table value will be found in the z (standard normal) table. For a $(1 - \alpha)$ confidence interval the z value is denoted by $z_{\alpha/2}$.

The form of the large-sample confidence interval is summarized as follows:

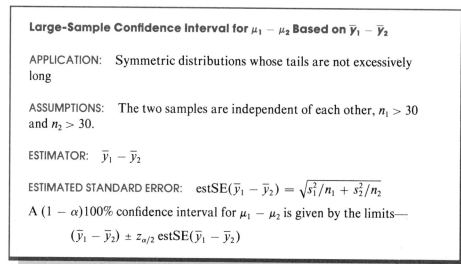

Large-Sample Confidence Interval for $\mu_1 - \mu_2$ Based on $\bar{y}_1 - \bar{y}_2$

APPLICATION: Symmetric distributions whose tails are not excessively long

ASSUMPTIONS: The two samples are independent of each other, $n_1 > 30$ and $n_2 > 30$.

ESTIMATOR: $\bar{y}_1 - \bar{y}_2$

ESTIMATED STANDARD ERROR: $\text{estSE}(\bar{y}_1 - \bar{y}_2) = \sqrt{s_1^2/n_1 + s_2^2/n_2}$

A $(1 - \alpha)100\%$ confidence interval for $\mu_1 - \mu_2$ is given by the limits—

$$(\bar{y}_1 - \bar{y}_2) \pm z_{\alpha/2}\, \text{estSE}(\bar{y}_1 - \bar{y}_2)$$

EXAMPLE 10.4

Suppose two independent random samples are drawn from two separate populations whose distributions are symmetric with tails that are not excessively long. Suppose further that the two samples resulted in the following summarized data:

Sample 1: $n_1 = 32$ $\bar{y}_1 = 105.4$ $s_1 = 15.28$
Sample 2: $n_2 = 40$ $\bar{y}_2 = 112.1$ $s_2 = 28.74$

Construct a 99% confidence interval for the difference between the two population means.

SOLUTION   From the summary data we have

$$\bar{y}_1 - \bar{y}_2 = 105.4 - 112.1 = -6.7$$

and

$$
\begin{aligned}
\text{estimated SE}(\bar{y}_1 - \bar{y}_2) &= \sqrt{(15.28)^2/32 + (28.74)^2/40} \\
&= \sqrt{233.4784/32 + 825.9876/40} \\
&= \sqrt{7.2962 + 20.64969} \\
&\doteq \sqrt{27.94589} \\
&= 5.2864
\end{aligned}
$$

Because a 99% confidence interval is desired, the table value is

$$z_{.005} = 2.58 \qquad (1 - \alpha = .99; \alpha = .01; \alpha/2 = .005)$$

Therefore the 99% confidence interval for $\mu_1 - \mu_2$ is

$$-6.7 \pm (2.58)(5.2864)$$

which gives

$$-6.7 \pm 13.64$$

Thus we are reasonably sure that the difference in the two means is somewhere in the interval

$$(-20.34, 6.94)$$

Large-Sample Test for the Difference between Two Population Means Based on $\bar{y}_1 - \bar{y}_2$

We next develop the large-sample test of hypothesis about the difference between two population means. Following the procedures outlined in the one-sample theory, the null and alternative hypotheses can take one of three forms:

Left-Tailed—$H_0: \mu_1 - \mu_2 \geq 0$ vs. $H_a: \mu_1 - \mu_2 < 0$
Right-Tailed—$H_0: \mu_1 - \mu_2 \leq 0$ vs. $H_a: \mu_1 - \mu_2 > 0$
Two-Tailed—$H_0: \mu_1 - \mu_2 = 0$ vs. $H_a: \mu_1 - \mu_2 \neq 0$

In the left-tailed test, for example, testing

$$H_0: \mu_1 - \mu_2 \geq 0$$

is the same as testing

$$H_0: \mu_1 \geq \mu_2$$

Thus we see that the test of the difference in two population means is appropriate when we wish to determine whether the mean response under condition 1 is greater than (or less than) the mean response under condition 2. Also it is easy to see that the inference is on the difference, $\mu_1 - \mu_2$, of the two population means.

Given that the distributions are symmetric without long tails, the choice of a test statistic is the point estimator, $\bar{y}_1 - \bar{y}_2$. As was stated earlier, its sampling distribution is approximately normally distributed about $\mu_1 - \mu_2$ with an estimated standard error of

$$\sqrt{s_1^2/n_1 + s_2^2/n_2}$$

In testing $H_0: \mu_1 - \mu_2 = 0$, the problem becomes one of determining whether $\bar{y}_1 - \bar{y}_2$ is significantly different from zero. As in single-parameter inference, significant departure of a test statistic from a given value is measured by first computing the "standardized" test statistic and then finding the associated p-value. In this case, the standardized test statistic is

$$z = \frac{(\bar{y}_1 - \bar{y}_2) - (\mu_1 - \mu_2)}{\sqrt{s_1^2/n_1 + s_2^2/n_2}}$$

which has an approximate standard normal sampling distribution. Under the assumption that the null hypothesis is true, we have $\mu_1 - \mu_2 = 0$, so

▲ Testing $H_0: \mu_1 - \mu_2 \geq 0$ is equivalent to testing $H_0: \mu_1 - \mu_2 = 0$.

$$z = \frac{(\bar{y}_1 - \bar{y}_2)}{\sqrt{s_1^2/n_1 + s_2^2/n_2}}$$

The p-value is the probability on the tail of the sampling distribution of the test statistic beyond the observed value of the test statistic. Thus in testing $H_0: \mu_1 - \mu_2 \geq 0$ vs. $H_a: \mu_1 - \mu_2 < 0$ we have the p-value $= P(z < z_{obs})$, where z_{obs} is found by substituting the observed values of $n_1, n_2, \bar{y}_1, \bar{y}_2, s_1^2$, and s_2^2 in the formula for z.

The test is summarized as follows:

Large-Sample Test of $\mu_1 - \mu_2$ Based on $\bar{y}_1 - \bar{y}_2$

APPLICATION: Symmetric distributions whose tails are not excessively long

ASSUMPTIONS: The two samples are independent of each other, $n_1 > 30$ and $n_2 > 30$

Left-Tailed Test	Right-Tailed Test	Two-Tailed Test
$H_0: \mu_1 - \mu_2 \geq 0$	$H_0: \mu_1 - \mu_2 \leq 0$	$H_0: \mu_1 - \mu_2 = 0$
$H_a: \mu_1 - \mu_2 < 0$	$H_a: \mu_1 - \mu_2 > 0$	$H_a: \mu_1 - \mu_2 \neq 0$

Standardized Test Statistic

$$z = \frac{(\bar{y}_1 - \bar{y}_2)}{\sqrt{s_1^2/n_1 + s_2^2/n_2}}$$

From the z-table find

p-value $= P(z < z_{obs})$ p-value $= P(z > z_{obs})$ p-value $= 2P(z < z_{obs})$ if $z_{obs} < 0$

or p-value $= 2P(z > z_{obs})$ if $z_{obs} > 0$

If a level of significance α is specified, then reject H_0 if p-value $< \alpha$.

EXAMPLE 10.5

Wind speed data were gathered for the months of January and July at a site for a proposed wind generator, to determine whether the season affects the production of electricity by the wind generator. Assuming that the distribution of wind speed (in miles per hour) at this location is symmetric without long tails, can we conclude a significant difference in the average wind speed for the two months?

	n	$\bar{y}$	s^2
January	32	23.4	26.42
July	35	16.2	23.875

SOLUTION   Let μ_1 denote the mean wind speed in January and μ_2 denote the mean wind speed in July. The null and alternative hypotheses are stated as follows:

$$H_0: \mu_1 - \mu_2 = 0$$
$$H_a: \mu_1 - \mu_2 \neq 0$$

The standardized test statistic is

$$z = \frac{(\bar{y}_1 - \bar{y}_2)}{\sqrt{s_1^2/n_1 + s_2^2/n_2}}$$

in which case

$$z_{obs} = \frac{23.4 - 16.2}{\sqrt{26.42/32 + 23.875/35}}$$
$$= \frac{7.2}{\sqrt{1.508}}$$
$$= 5.86$$

Because the test is a two-tailed test, the p-value is

$$p\text{-value} = 2P(z > 5.86)$$

however a z-value of 5.86 is beyond the table values. The largest value in the z-table is 5.0, in which case the p-value would have been

$$2(.5 - .4999997) = 2(.0000003) = .0000006$$

Because 5.86 exceeds 5.0, we can say that the p-value is less than .0000006, which indicates that the average wind speeds for the two months are highly significantly different.

When we have large samples, as in this section, it is not necessary to assume that the parent populations are normally distributed, as is the case in Section 10.4, where the sample sizes are small. The large-sample z-test (just presented) is valid over a large spectrum of underlying parent distributions. Care should be exercised in using this test, however. As pointed out previously, the statistical properties of $\bar{y}$ as an estimator of the population mean, μ, tend to deteriorate in the presence of outliers. The result is a loss of precision in estimating μ with the usual confidence interval based on $\bar{y}$. In Chapter 8 it was pointed out that a more precise interval for symmetric long-tailed distributions is obtained if the interval is based on $\bar{y}_T$. The same is true when estimating or testing the difference in population means. If the parent populations are symmetric long-tailed distributions, then the confidence interval and hypothesis testing problems should be based on $\bar{y}_{T_1} - \bar{y}_{T_2}$. This situation is considered in the next section. If the distributions are skewed, we should abandon the mean as the parameter of concern and investigate the population median. The difference in population medians is considered in Section 10.4.

10.1 Construct the specified confidence interval for $\mu_1 - \mu_2$ based on $\bar{y}_1 - \bar{y}_2$ from the following summary data:

(a) 90% interval

$n_1 = 65, \quad \bar{y}_1 = 252.8, \quad s_1 = 48.65$
$n_2 = 62, \quad \bar{y}_2 = 215.4, \quad s_2 = 52.83$

(b) 98% interval

$n_1 = 33, \quad \bar{y}_1 = 15.3, \quad s_1 = 2.55$
$n_2 = 34, \quad \bar{y}_2 = 16.8, \quad s_2 = 3.47$

10.2 With $\bar{y}_1 - \bar{y}_2$ as the test statistic, test the following null hypotheses using the summary data:

(a) $H_0: \mu_1 - \mu_2 \geq 0$

$n_1 = 45, \quad \bar{y}_1 = 115.3, \quad s_1 = 15.92$
$n_2 = 45, \quad \bar{y}_2 = 123.6, \quad s_2 = 17.65$

(b) $H_0: \mu_1 - \mu_2 = 0$

$n_1 = 53, \quad \Sigma x_i = 326, \quad \Sigma x_i^2 = 5528$
$n_2 = 48, \quad \Sigma x_i = 415, \quad \Sigma x_i^2 = 6721$

10.3 An instructor of an introductory psychology course is interested in knowing, in general, if the average grades on the final exam differ from the Fall semester to the Spring semester. From the following data, construct a 90% confidence interval based on $\bar{y}_1 - \bar{y}_2$ for the difference in the two population means.

	Fall class	Spring class
n	150	150
$\bar{y}$	82.4	84.2
s	11.56	11.44

10.4 A national science test for tenth graders has a mean of 75 and a standard deviation of 20. In a certain school district, 123 students were divided randomly into two groups. One group of 61 students received instruction through a traditional lecture class and the other group of 62 students received instruction through an experimental class. The average grade on the science test in the traditional group was 77.2 and the standard deviation was 19.6; the average in the experimental group was 78.6 and the standard deviation was 22.4. Is there statistical evidence of a difference in the two groups?

10.5 Nationally 25% of all college freshmen enroll in some type of remedial math course. To determine whether males and females differ in pre-enrollment ability, a math placement exam (scored from 0 to 40) was given to a sample of 35 incoming freshmen females and a sample of 42 incoming freshmen males. Their scores were as follows:

Females	28	21	4	20	16	19	39
	22	5	18	17	21	19	3
	11	22	19	18	17	35	21
	18	16	5	19	21	20	20
	40	21	19	20	22	38	3

Males	18	22	16	14	2	16	18
	20	22	19	11	35	18	22
	4	21	20	15	19	17	38
	18	16	4	19	18	16	17
	33	16	19	18	21	20	17
	17	13	14	19	15	38	40

Construct a 98% confidence interval for the mean difference in ability for males and females.

10.6 Following is a back-to-back stem and leaf plot of the grades for two introductory statistics classes, one meeting at 9 A.M. and the other at 2 P.M. Is there a statistical difference in the mean grades for the two classes?

9 A.M.		2 P.M.
0	6*	0
8 7 7 6 6 6	6-	
4 2	7*	1 4
8 8 8 7 7	7-	6 6 6 6 6 9 9
4 4 3 3 3 2 1	8*	0 1 2 2 2 2 2 4
9 9 8 7 6 6 5 5 5	8-	5 5 5 6 6 7 8 9 9
4 4 3 2 1	9*	0 1 2 3
8	9-	6

10.7 A study was conducted to determine whether persons in suburban District I have a greater mean annual income than those in suburban District II. Random samples were taken in each district with the following results (measured in $1,000):

District I	District II
$n_1 = 38$	$n_2 = 42$
$\bar{y}_1 = 28.65$	$\bar{y}_2 = 25.94$
$s_1 = 6.38$	$s_2 = 5.47$

Determine with a test of hypothesis whether the mean annual income in District I exceeds the mean annual income in District II.

10.8 The FDA tests tobaccos of two different types of cigars for nicotine content and obtains the following results (in milligrams):

Brand A $n_1 = 33, \quad \bar{y}_1 = 85.3, \quad s_1 = 12.44$
Brand B $n_2 = 35, \quad \bar{y}_2 = 89.8, \quad s_2 = 16.67$

Do these results indicate that there is a difference in the mean nicotine content of the two types of cigars?

10.9 At a large industrial plant, employees were classified according to age and given a leadership exam. Do the following data indicate that the mean leadership exam score for the over 35 age group exceeds the mean score for the under 35 group?

	Under 35 years old								Over 35 years old								
Leadership	25	13	9	46	18	26	11	36	24	31	43	23	26	38	29	9	13
exam	25	30	17	20	30	12	32	54	23	21	42	34	50	41	13	15	14
score	17	20	37	25	24	20	16	8	15	38	30	14	16	68	32	7	
	26	23	20	17	21	37	31	26	45	19	20	27	9	28	30	51	

10.3 A Robust Inference Procedure for the Difference between Two Population Means

Under the condition that the parent distributions are symmetric with long tails, it is suggested that inferences on the difference in means be based on trimmed means. Such inferences are natural extensions of the one-sample inference problems and modifications of the inference procedures based on ordinary means given in the previous section. For large independent samples the inference procedures based on $\bar{y}_1 - \bar{y}_2$ given in Section 10.2 are modified by replacing the summary data: $n_1, \bar{y}_1, s_1$, and $n_2, \bar{y}_2$, and s_2 by the trimmed summary data: $k_1, \bar{y}_{T_1}, s_{T_1}$, and $k_2, \bar{y}_{T_2}, s_{T_2}$, respectively (see Koopmans, 1981, p. 314).

The confidence interval for $\mu_1 - \mu_2$ based on $\bar{y}_{T_1} - \bar{y}_{T_2}$ is summarized as follows:

Large-Sample Confidence Interval for $\mu_1 - \mu_2$ Based on $\bar{y}_{T_1} - \bar{y}_{T_2}$

APPLICATION: Symmetric distributions with long tails

ASSUMPTIONS: The two samples are independent of each other, $n_1 > 30$ and $n_2 > 30$

ESTIMATOR: $\bar{y}_{T_1} - \bar{y}_{T_2}$

ESTIMATED STANDARD ERROR: $\text{estSE}(\bar{y}_{T_1} - \bar{y}_{T_2}) = \sqrt{s_{T_1}^2/k_1 + s_{T_2}^2/k_2}$

A $(1 - \alpha)100\%$ confidence interval for $\mu_1 - \mu_2$ is given by the limits

$$(\bar{y}_{T_1} - \bar{y}_{T_2}) \pm z_{\alpha/2}\text{estSE}(\bar{y}_{T_1} - \bar{y}_{T_2})$$

EXAMPLE 10.6

To compare two methods of producing textiles, a textile mill counted the number of imperfections in the pieces of material produced by the two methods. Random samples of size 31 and 34 were taken from method A and method B, respectively. From the following data construct a 95% confidence interval for the difference in means for the two methods.

Number of imperfections in material																
Method A	7	9	8	10	7	9	8	6	2	11	14	8	9	12	4	6
	9	10	8	11	10	9	11	8	9	8	12	10	7	9	11	
Method B	5	8	7	22	6	7	2	6	6	10	7	6	9	13	4	3	6
	7	7	4	6	8	6	6	8	4	11	9	7	6	17	7	4	6

▲ A back-to-back stem and leaf plot is two stem and leaf plots with a common stem. It is an excellent method for comparing two data sets.

SOLUTION   Following is an ordered back-to-back stem and leaf plot and side-by-side box plots for the two data sets:

Ordered back-to-back stem and leaf plots

Method A		Method B
2	0t	2 3
4	0f	4 4 4 4 5
7 7 7 6 6	0s	6 6 6 6 6 6 6 6 6 6 7 7 7 7 7 7 7 7
9 9 9 9 9 9 9 8 8 8 8 8 8	0–	8 8 8 9 9
1 1 1 1 0 0 0 0	1*	0 1
2 2	1t	3
4	1f	
	1s	7
	HI	22

Quarter summary diagrams

	A				B		
	31				34		
$M(16)$	9			$M(17.5)$	6.5		
$Q(8.5)$	8	10	Q-spread $= 2$	$Q(9)$	6	8	Q-spread $= 2$

Box plot summary diagrams

		3					3		
a	6	12			a	3	11		
f	5	13			f	3	11		
2,4	two	one	14		2	one	one	13	
F	2	16			F	0	14		
	none	none				none	two	17,22	

Side-by-side box plots

Due to the presence of outliers in the data, the confidence interval will be based on 10% trimmed means. (Note: The box plot for sample **B** gives the appearance of being skewed right, which violates the condition that the population distributions are symmetrical; we will continue, however, because the trimmed mean is a robust estimator and yields satisfactory results even when the parent distribution is skewed. If both distributions were skewed in the same direction we would compare the population medians using the procedures in the next section.) From the raw data we calculate the following summary statistics:

	Method A	Method B
n	31	34
$\bar{y}$	8.774	7.353
s	2.390	3.829
k	23	26
$\bar{y}_T$	8.913	6.692
s_T	1.730	2.210

From the 10% trimmed means we have

$$\bar{y}_{T_A} - \bar{y}_{T_B} = 8.913 - 6.692 = 2.221$$

and from the s_T's we have

$$\text{estimated SE}(\bar{y}_{T_A} - \bar{y}_{T_B}) = \sqrt{(1.730)^2/23 + (2.210)^2/26} = .564$$

Because a 95% confidence interval is desired, the table value is

$$z_{.025} = 1.96 \qquad (1 - \alpha = .95; \alpha = .05; \alpha/2 = .025)$$

Therefore the 95% confidence interval for $\mu_1 - \mu_2$ is

$$2.221 \pm (1.96)(.564)$$

which gives

$$2.221 \pm 1.105$$

and therefore the interval is

$$(1.116, 3.326)$$

For comparison purposes let us calculate the confidence interval based on $\bar{y}_1 - \bar{y}_2$.

$$\bar{y}_1 - \bar{y}_2 = 8.774 - 7.353 = 1.421$$

$$\text{estSE}(\bar{y}_1 - \bar{y}_2)$$

$$= \sqrt{(2.390)^2/31 + (3.829)^2/34} = .7845$$

Therefore the 95% confidence interval based on the ordinary means is

$$1.421 \pm (1.96)(.7845)$$

$$1.421 \pm 1.538$$

which gives the interval

$$(-.117, 2.959)$$

FIGURE 10.1

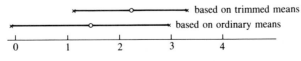

Note in Figure 10.1 that the interval based on the difference in the ordinary means is longer (less precision) and includes 0 inside the interval. The fact that 0 is inside the interval means that in a test of hypothesis of the equality of the two means, the null hypothesis will not be rejected. On the other hand, the interval based on the difference in the trimmed means is more precise and is well above zero. A test based on the trimmed means indicates a significant difference in the two mean responses, which means that method B is a better method (significantly fewer imperfections) of producing textiles. However this is not without a cost. The box plot points out that occasionally method B results in a large number of imperfections. Thus in this example, we see that the trimmed mean is useful in detecting a difference, and the graphical display points out certain peculiarities that might otherwise go undetected.

Large-Sample Test for the Difference between Population Means Based on $\bar{y}_{T_1} - \bar{y}_{T_2}$

Just as in confidence interval estimation, the procedures based on $\bar{y}_1 - \bar{y}_2$, given in Section 10.2, will be modified to construct the test for the difference in population means when the parent distributions are symmetric with long tails.

The test is summarized as follows:

Large-Sample Test of $\mu_1 - \mu_2$ Based on $\bar{y}_{T_1} - \bar{y}_{T_2}$

APPLICATION: Symmetric distributions with long tails

ASSUMPTIONS: The two samples are independent of each other, $n_1 > 30$ and $n_2 > 30$

Left-Tailed Test	Right-Tailed Test	Two-Tailed Test
$H_0: \mu_1 - \mu_2 \geq 0$	$H_0: \mu_1 - \mu_2 \leq 0$	$H_0: \mu_1 - \mu_2 = 0$
$H_a: \mu_1 - \mu_2 < 0$	$H_a: \mu_1 - \mu_2 > 0$	$H_a: \mu_1 - \mu_2 \neq 0$

Standardized Test Statistic

$$z = \frac{(\bar{y}_{T_1} - \bar{y}_{T_2})}{\sqrt{s_{T_1}^2/k_1 + s_{T_2}^2/k_2}}$$

From the z-table find

p-value $= P(z < z_{\text{obs}})$ p-value $= P(z > z_{\text{obs}})$ p-value $= 2P(z < z_{\text{obs}})$ if $z_{\text{obs}} < 0$

or p-value $= 2P(z > z_{\text{obs}})$ if $z_{\text{obs}} > 0$

If a level of significance α is specified, then reject H_0 if p-value $< \alpha$.

EXAMPLE 10.7

Do physically active women have stronger bones than nonactive women as they grow older? Research by University of North Carolina at Chapel Hill doctors indicates that they do, which suggests that active women are less likely to have fractures as they grow older. A study of 300 women, published in the February 1985 issue of the *Journal of Orthopaedic Research,* indicates that older athletic women, age 55 to 75, have arm and spine bone measurements in the same range as younger athletic women.

Suppose that we have a sample of 70 women, 35 of which are physically active and the remaining are considered nonactive. Suppose measurements of bone density yielded the following results:

Active women	213	227	211	208	155	204	216
	219	224	202	207	212	184	214
	245	210	192	218	219	163	230
	214	209	210	226	203	208	207
	217	257	212	203	232	221	215

Nonactive women	201	205	187	208	203	265	201
	210	219	205	173	202	199	216
	192	207	243	194	201	217	270
	209	185	176	202	211	213	208
	236	214	209	162	204	206	213

Is there statistical evidence to support the hypothesis that active women have greater bone density?

SOLUTION   Following is an ordered back-to-back stem and leaf plot and side-by-side box plots for the two data sets:

Ordered back-to-back stem and leaf plots

	Active														Stem	Nonactive

```
                 Active                                         Nonactive

                                    5  | 15 |
                                    3  | 16 | 2
                                       | 17 | 3  6
                                    4  | 18 | 5  7
                                    2  | 19 | 2  4  9
        9 8 8 7 7 4 3 3 2  | 20 | 1 1 1 2 2 3 4 5 5 6 7 8 8 9 9
9 9 8 7 6 5 4 4 3 2 2 1 0 0  | 21 | 0 1 3 3 4 6 7 9
                          7 6 4  | 22 |
                               2 0  | 23 | 6
                                    5  | 24 | 3
                                    7  | 25 |
                                       | 26 | 5
                                       | 27 | 0
```

Quarter summary diagrams

	Active 35					Nonactive 35			
$M(18)$		212			$M(18)$		206		
$Q(9.5)$	207		219	Q-spread $= 12$	$Q(9.5)$	201		213	Q-spread $= 12$

Box plot summary diagrams

		18					18		
	a	192	232			a	185	219	
	f	189	237			f	183	231	
184	one	one		245	173,176	two	two		236,243
	F	171	255			F	165	249	
155,163	two	one		257	162	one	two		265,270

Side-by-side box plots

Both distributions appear symmetric with long tails (an abundance of outliers), suggesting that the inference be based on the difference in trimmed means. We will trim 10%.

The research hypothesis is that the mean bone density for active women is greater than the mean bone density for nonactive women; thus we have

$$H_0: \mu_1 - \mu_2 \le 0 \quad \text{vs.} \quad H_a: \mu_1 - \mu_2 > 0$$

The test statistic is

$$z = \frac{(\bar{y}_{T_1} - \bar{y}_{T_2})}{\sqrt{s_{T_1}^2/k_1 + s_{T_2}^2/k_2}}$$

From the data we have

	Active	Nonactive
$\bar{y}_{T.10}$	212.926	205.778
s_T	9.8198	11.7445

$$\bar{y}_{T_1} - \bar{y}_{T_2} = 212.926 - 205.778$$
$$= 7.148$$

and from the s_T's we have

$$\text{estimated SE}(\bar{y}_{T_1} - \bar{y}_{T_2}) = \sqrt{s_{T_1}^2/k_1 + s_{T_2}^2/k_2}$$
$$= \sqrt{(9.8198)^2/27 + (11.7445)^2/27}$$
$$= 2.95$$

The test statistic becomes

$$z = \frac{7.148}{2.95} = 2.42$$

Being a right-tailed test we have

$$p\text{-value} = P(z > 2.42) = .5 - .4922 = .0078$$

Therefore the null hypothesis is rejected and we have highly significant evidence that the active women have greater bone density than the nonactive women.

10.10 Construct the specified confidence interval for $\mu_1 - \mu_2$ based on 10% trimmed means from the following summary data:

(a) 90% interval

$n_1 = 44, \quad \bar{y}_{T_1} = 36.7, \quad s_{T_1} = 5.24$
$n_2 = 48, \quad \bar{y}_{T_2} = 34.6, \quad s_{T_2} = 6.77$

(b) 98% interval

$n_1 = 50, \quad \bar{y}_{T_1} = 177.5, \quad s_{T_1} = 25.83$
$n_2 = 50, \quad \bar{y}_{T_2} = 185.2, \quad s_{T_2} = 23.68$

10.11 Using 10% trimmed mean, test the specified null hypothesis using the following summary data:

(a) $H_0: \mu_1 - \mu_2 \geq 0$

$n_1 = 35, \quad \bar{y}_{T_1} = 55.8, \quad s_{T_1} = 15.56$
$n_2 = 35, \quad \bar{y}_{T_2} = 62.4, \quad s_{T_2} = 15.38$

(b) $H_0: \mu_1 - \mu_2 = 0$

$n_1 = 120, \quad \bar{y}_{T_1} = 467, \quad s_{T_1} = 79$
$n_2 = 135, \quad \bar{y}_{T_2} = 482, \quad s_{T_2} = 65$

10.12 Two independent random samples were selected from two populations with the following results:

Sample A

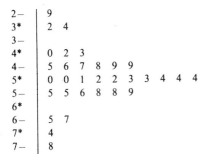

```
2-  | 9
3*  | 2  4
3-  |
4*  | 0  2  3
4-  | 5  6  7  8  9  9
5*  | 0  0  1  2  2  3  3  4  4  4
5-  | 5  5  6  8  8  9
6*  |
6-  | 5  7
7*  | 4
7-  | 8
```

Sample B

```
2-  | 4  9
3*  | 0
3-  | 5
4*  | 0  1  3  3  4  4
4-  | 5  6  6  7  7  8  8  9  9
5*  | 0  1  3  3  4  4
5-  | 5  6  8
6*  | 2
6-  |
7*  | 3  4
7-  | 7
```

Construct a 95% confidence interval for $\mu_1 - \mu_2$ based on 20% trimmed means.

10.13 Use the data in Exercise 10.12 to test the hypothesis of no difference in the population means using the 20% trimmed means.

10.14 Can computers help learning disabled children become better writers? The results of studies by researchers at Claremont Graduate School in Claremont, California, suggest that they can (*USA Today,* October 23, 1984, p. 5D). In a yearlong classroom study, learning disabled children writing on word processors improved their writing skills dramatically as compared with learning disabled children using pen and paper.

Computer group		Pen and paper group	
80	75	75	78
86	66	73	90
92	113	85	70
70	91	76	100
100	83	65	93
88	64	77	61
84	88	55	79
60	93	74	83
85	102	76	74
94	94	105	83
87	87	82	78
89	85	53	79
117	92	77	101
96	89	85	80
82	85	75	60
93	90	50	72

Following are creative writing scores of learning disabled children using computers and those of children using pen and paper: (These are not the data of the Claremont study.)

(a) Organize the data in back-to-back stem and leaf plots and side-by-side box plots.

(b) Comment on the shapes of the two distributions.

(c) Would you base a test of hypothesis concerning the difference in population means on ordinary means or trimmed means?

10.15 Do the data in Exercise 10.14 indicate that the computer tends to improve the creative writing scores of the learning disabled children? Test the hypothesis that the creative writing scores of the computer group exceeded the scores of the noncomputer group. Use the test statistic suggested in part (c) of Exercise 10.14.

10.16 In a study related to the one described in Exercise 10.14, the computer aptitude of 32 learning disabled teens was compared with the test scores of 32 "normal achievers."

(a) Organize the data in back-to-back stem and leaf plots and side-by-side box plots.

(b) Comment on the shapes of the two distributions.

(c) Would you base a test of hypothesis concerning the difference in population means on ordinary means or trimmed means?

10.17 Using the data in Exercise 10.16, test the hypothesis that there is no difference in the computer aptitude scores of the learning disabled and the normal achievers. Use the test statistic suggested in part (c) of Exercise 10.16.

TABLE FOR EXERCISE 10.16

Learning disabled			Normal achievers		
70	73	72	72	69	93
67	76	70	67	60	73
77	70	75	87	71	74
50	79	78	73	56	75
65	73	51	88	79	65
72	73	94	77	78	70
97	74	71	71	71	66
55	66	66	61	69	67
74	74	68	70	74	77
86	69	76	96	76	53
69	79		75	72	

10.4

Small-Sample Inference on the Difference between Two Population Centers Using Independent Samples

In this section, we present small-sample inference procedures for comparing the centers of two population distributions. First we will give the small-sample version of the inference procedures based on $\bar{y}_1 - \bar{y}_2$ given in Section 10.2. Next we will give a nonparametric test for those cases where the test based on $\bar{y}_1 - \bar{y}_2$ is not appropriate.

Small-Sample Inference for $\mu_1 - \mu_2$ Based on $\bar{y}_1 - \bar{y}_2$

When the sample sizes are large, the Central Limit theorem guarantees that the sampling distribution of

$$z = \frac{(\bar{y}_1 - \bar{y}_2) - (\mu_1 - \mu_2)}{\sqrt{s_1^2/n_1 + s_2^2/n_2}}$$

is approximately normally distributed.

If n_1 and n_2 are small (<30), however, the Central Limit theorem no longer applies. Additional assumptions must be made about the parent populations in order to handle the small-sample case.

Assumptions for Small-Sample Inference Based on $\bar{y}_1 - \bar{y}_2$

1. Both parent populations are normally distributed (or very near normal).

2. The population variances, σ_1^2 and σ_2^2, are equal. We say that the variances are homogeneous.

From assumption 2, we can write the standard error of $\bar{y}_1 - \bar{y}_2$ as

$$SE(\bar{y}_1 - \bar{y}_2) = \sqrt{\sigma_1^2/n_1 + \sigma_2^2/n_2}$$
$$= \sigma \sqrt{1/n_1 + 1/n_2}$$

where σ is the common standard deviation.

Because $\sigma_1^2 = \sigma_2^2 = \sigma^2$, we have that both s_1^2 and s_2^2 are unbiased estimators of σ^2. We can combine the two estimators using a weighted or pooled average to obtain a single efficient unbiased estimator of σ^2 as follows:

$$s_p^2 = \frac{(n_1 - 1)\, s_1^2 + (n_2 - 1)\, s_2^2}{n_1 + n_2 - 2}$$

Observe that s_1^2 is weighted by its degrees of freedom, $n_1 - 1$, and s_2^2 is weighted by its degrees of freedom, $n_2 - 1$. The *pooled variance*, s_p^2, then has

$$n_1 - 1 + n_2 - 1 = n_1 + n_2 - 2$$

degrees of freedom, which appears in the denominator. Using $s_p = \sqrt{s_p^2}$ as an estimate of σ, we have

$$estSE(\bar{y}_1 - \bar{y}_2) = s_p \sqrt{1/n_1 + 1/n_2}$$

From the assumptions of normality and common variance, we have that

$$t = \frac{(\bar{y}_1 - \bar{y}_2) - (\mu_1 - \mu_2)}{s_p \sqrt{1/n_1 + 1/n_2}}$$

has a sampling distribution, which is Student's t with $n_1 + n_2 - 2$ degrees of freedom. Consequently the table value for the confidence interval and the p-value for the test of hypothesis will be found in the t-table and not in the z-table.

In light of the previous discussion, we modify the large-sample confidence interval to obtain the small-sample confidence interval for $\mu_1 - \mu_2$ as follows:

▲ Because $\sigma_1^2 = \sigma_2^2$ it can be factored out of the two terms and then taken out from under the square root.

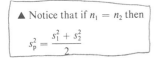

▲ Notice that if $n_1 = n_2$ then

$$s_p^2 = \frac{s_1^2 + s_2^2}{2}$$

Small-Sample Confidence Interval for $\mu_1 - \mu_2$ Based on $\bar{y}_1 - \bar{y}_2$

APPLICATION: Small samples and normally distributed populations

ASSUMPTIONS: Populations are normally distributed with equal variances. Samples are independent.

A $(1 - \alpha)100\%$ confidence interval for $\mu_1 - \mu_2$ is given by the limits

$$(\bar{y}_1 - \bar{y}_2) \pm t_{\alpha/2,\, n_1 + n_2 - 2} s_p \sqrt{1/n_1 + 1/n_2}$$

where

$$s_p = \sqrt{\frac{(n_1 - 1)\, s_1^2 + (n_2 - 1)\, s_2^2}{n_1 + n_2 - 2}}$$

and $t_{\alpha/2,\, n_1 + n_2 - 2}$ is the upper $\alpha/2$ point of the t distribution with $n_1 + n_2 - 2$ degrees of freedom.

EXAMPLE 10.8

Polychlorinated Biphenyls (PCBs) are classified as health hazards and are said to cause serious health problems. To study the concentration of PCBs in a river, a group of environmentalists measured the PCB level (parts per million) in fish at two locations on the river. Their results are summarized as follows:

Location A $n = 8$, $\bar{y} = 25.2$, $s = 3.8$

Location B $n = 8$, $\bar{y} = 23.1$, $s = 4.2$

Assuming that the variability of the concentration is the same in the two areas and that the distributions are close to normal, construct a 95% confidence interval to estimate the difference in the mean concentration levels at the two sites.

SOLUTION   Because $n_1 = n_2 = 8$, the degrees of freedom are $n_1 + n_2 - 2 = 14$, and for a 95% confidence interval we have that $\alpha = .05$, so that $\alpha/2 = .025$. We then find from the t-table that

$$t_{\alpha/2, n_1 + n_2 - 2} = t_{.025, 14} = 2.145$$

The pooled variance is found as follows:

$$s_p^2 = \frac{(n_1 - 1) s_1^2 + (n_2 - 1) s_2^2}{\boxed{n_1 + n_2 - 2}} \quad \text{DEGREES OF FREEDOM}$$

$$= \frac{7(3.8)^2 + 7(4.2)^2}{14}$$

$$= \frac{(3.8)^2 + (4.2)^2}{2}$$

$$= 16.04$$

Therefore the estimated standard error is

$$\text{estSE}(\bar{y}_1 - \bar{y}_2) = s_p\sqrt{1/n_1 + 1/n_2}$$
$$= \sqrt{16.04}\,\sqrt{1/8 + 1/8}$$
$$= (4.005)(.5)$$
$$= 2.0025$$

The confidence interval becomes

$$(25.2 - 23.1) \pm 2.145(2.0025)$$

which reduces to

$$2.1 \pm 4.295$$

which in turn gives the interval

$$(-2.195, 6.395)$$

The hypothesis test for comparing two population means based on small independent samples is constructed in a similar manner. Again the large-sample test is modified to construct the small-sample t-test.

Under the small-sample assumptions and given that the null hypothesis is true ($\mu_1 - \mu_2 = 0$), we have that the test statistic

$$t = \frac{\bar{y}_1 - \bar{y}_2}{s_p \sqrt{1/n_1 + 1/n_2}}$$

has a Student's t distribution with $n_1 + n_2 - 2$ degrees of freedom. Hence the p-value is found in the t-table. This test, commonly called the *two-sample t-test*, is summarized as follows:

Two-Sample *t*-Test Based on $\bar{y}_1 - \bar{y}_2$

APPLICATION: Small samples and normally distributed populations

ASSUMPTIONS: Populations are normally distributed with equal variances. Samples are independent.

Left-Tailed Test	Right-Tailed Test	Two-Tailed Test
$H_0: \mu_1 - \mu_2 \geq 0$	$H_0: \mu_1 - \mu_2 \leq 0$	$H_0: \mu_1 - \mu_2 = 0$
$H_a: \mu_1 - \mu_2 < 0$	$H_a: \mu_1 - \mu_2 > 0$	$H_a: \mu_1 - \mu_2 \neq 0$

Standardized Test Statistic

$$t = \frac{\bar{y}_1 - \bar{y}_2}{s_p \sqrt{1/n_1 + 1/n_2}}$$

From the t-table with $n_1 + n_2 - 2$ degrees of freedom, the p-value is the tail probability most closely associated with t_{obs}. If t_{obs} falls between two table values then give the two associated probabilities as bounds for the p-value.

The p-value for the two-tailed test is calculated as in the single-tailed test and then doubled to account for both tails.

If a level of significance α is specified, then reject H_0 if p-value $< \alpha$.

EXAMPLE 10.9

A random sample of 14 fourth graders took a standardized achievement test immediately after an hour of recess. A second random sample of 16 fourth graders took the same test after a 1-hour rest period. From the following summary data, test the hypothesis of no difference in the mean scores for the two groups.

	n	$\bar{y}$	s
Recess group	14	56.5	6.2
Rest group	16	62.2	9.8

SOLUTION Because the scores on most standardized tests are normally distributed, the test of hypothesis will be based on the difference in the ordinary sample means. The null and alternative hypotheses are stated as follows:

$$H_0: \mu_1 = \mu_2, \quad H_a: \mu_1 \neq \mu_2$$

where μ_1 denotes the mean test score for the recess group and μ_2 denotes the mean test score for the rest group.

The test statistic is

$$t = \frac{\bar{y}_1 - \bar{y}_2}{s_p \sqrt{1/n_1 + 1/n_2}}$$

The sampling distribution is t with degrees of freedom

$$n_1 + n_2 - 2 = 14 + 16 - 2 = 28$$

From the sample data we have the observed value of t to be

$$t_{obs} = \frac{56.5 - 62.2}{\sqrt{[13(6.2)^2 + 15(9.8)^2]/28} \sqrt{1/14 + 1/16}}$$

$$= \frac{-5.7}{(8.3245)(.366)}$$

$$= -1.87$$

From the t-table with 28 degrees of freedom we see that

$$1.701 < 1.87 < 2.048$$

and therefore

▲ The p-value is doubled because this is a two-tailed test.

$$2(.025) < p\text{-value} < 2(.05)$$

which gives

$$.05 < p\text{-value} < .10$$

Thus we can say that the rest group and the recess group scored significantly different on the achievement test.

If the two assumptions (normality and homogeneous variances) are not met, there is a possible alternative to the two-sample t-test. If it can be assumed that the shapes of the parent distributions are similar, then the inference procedure should utilize the nonparametric test called the *Wilcoxon rank sum test*.

The Two-Sample Wilcoxon Rank Sum Test

▲ Remember that the box plot is an excellent tool for detecting distributional shape. Side-by-side box plots are useful in comparing shapes.

For the two-sample *Wilcoxon rank sum test* no distributional assumptions are made except that if a difference exists between the distributions, it is in their locations. Thus if we choose the population median as the measure of location, then, in a test of hypothesis, the null hypothesis will say the medians are equal and the alternative will say that they differ. We will give a modified version of the test (Conover and Inman, 1981) in which the ordinary two-sample t-test is applied to the *rank-transformed* data. The resulting test gives a good approximation to the Wilcoxon test when both sample sizes are at least 10.

To rank-transform a data set means to arrange the data in increasing order and assign ranks of 1, 2, 3, and so on. The test is then conducted on the ranks. If two or more observations have the same value (tied), then average ranks are given. For example, in the following data set

8.6,   11.2,   9.3,   7.4,   11.2,   10.4,   9.1,   12.1

there are two 11.2's, so each is assigned a rank of $(6 + 7)/2 = 6.5$ because they would normally be assigned ranks of 6 and 7. The completed ranking (in the same order as the given data) is

2,   6.5,   4,   1,   6.5,   5,   3,   8

To conduct the two-sample Wilcoxon rank sum test, the two samples are combined as one data set and then ranks are assigned. Then the two-sample t-test is applied to the two sets of ranks.

The Two-Sample Wilcoxon Rank Sum Test

APPLICATION: Comparing locations of general populations.

ASSUMPTIONS: Population distributions are similar except for possibly different locations. Samples are independent. $n_1 \geq 10$ and $n_2 \geq 10$.

The test of hypothesis procedures are the same as the two sample t-test applied to the rank-transformed data.

The following example illustrates the test.

EXAMPLE 10.10

A study was undertaken to determine whether the acquisition of a response is influenced by a particular drug. The dependent measurement is the number of trials required to master a given task. A group of 28 subjects is randomly assigned to the experimental and control groups. Those assigned to the experimental group are given the drug and those assigned to the control group are given a placebo. After a period of time, the following number of trials to complete the task successfully is recorded.

Experimental group	17	15	5	14	18	3	16	
	13	15	16	17	8	19		
Control group	14	10	1	12	11	14	8	10
	2	12	16	12	15	4	12	

Test the hypothesis to see if the drug has an effect on the dependent measurement.

SOLUTION   Following are back-to-back stem and leaf plots and side-by-side box plots:

Experimental   Control

```
                | 0* | 1
              3 | 0t | 2
              5 | 0f | 4
                | 0s |
              8 | 0- | 8
                | 1* | 0  0  1
              3 | 1t | 2  2  2  2
          5 5 4 | 1f | 4  4  5
      7 7 6 6   | 1s | 6
          9 8   | 1- |
```

Quantile summary diagrams

	13		Midsummaries	Spreads
$M(7)$	15		15	
$Q(4)$	13	17	15	4
$E(2.5)$	6.5	17.5	12	11
R	3	19	11	16

	15		Midsummaries	Spreads
$M(8)$	12		12	
$Q(4.5)$	9	13	11	4
$E(2.5)$	3	14.5	8.75	11.5
R	1	16	8.5	15

Box plot summary diagrams

	6	
a	8	19
f	7	23
3,5	two	none
F	1	29
	none	none

	6	
a	4	16
f	3	19
1,2	two	none
F	-3	25
	none	none

Side-by-side box plots

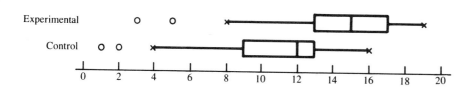

An investigation of the midsummaries and box plots gives an indication that both distributions are skewed left. Because the distributions are not symmetrical (but have the same shape), we will use the two-sample Wilcoxon rank sum test and test the equality of the population medians. From the back-to-back stem and leaf it is easy to rank the combined samples. Following is the same stem and leaf plot with the observations replaced by their ranks:

		0*	1
	3	0t	2
	5	0f	4
		0s	
	6.5	0−	6.5
		1*	8.5, 8.5, 10
	15	1t	12.5, 12.5, 12.5, 12.5
20, 20,	17	1f	17, 17, 20
25.5, 25.5, 23,	23	1s	23
	28, 27	1−	

The averages and standard deviations for the ranks in the two groups are

$$\bar{r}_1 = 18.346 \qquad \bar{r}_2 = 11.167$$

$$s_{r_1} = 8.594 \qquad s_{r_2} = 6.377$$

$$s_p = \sqrt{[12(8.594)^2 + 14(6.377)^2]/26}$$

$$= \sqrt{(886.282 + 569.326)/26}$$

$$= 7.482$$

Given that the drug may have a positive or a negative effect on the number of trials to complete the task, we will conduct a two-tailed test.

Let θ_1 = the median number of trials to complete the task with the drug

θ_2 = median without the drug

The null and alternative hypotheses are stated as follows:

$$H_0: \theta_1 - \theta_2 = 0$$
$$H_a: \theta_1 - \theta_2 \neq 0$$

▲ The null hypothesis for the Wilcoxon rank sum test can be stated more generally as H_0: The population distributions are identical. In this example, the distributions differ at most in their medians, hence the null hypothesis is stated as an equality of population medians.

The test statistic is the usual t-test applied to the rank-transformed data. Thus we have

$$t = \frac{18.346 - 11.167}{7.482\sqrt{1/13 + 1/15}}$$

$$= \frac{7.179}{2.835}$$

$$= 2.53$$

From the t-table with 26 degrees of freedom, we find that $2(.005) < p\text{-value} < 2(.01)$, which gives $.01 < p\text{-value} < .02$. Thus we can declare a significant difference in the two groups. That is, there is sufficient evidence to say that the drug has an effect on the number of trials needed to complete the task.

In Section 10.3 a large-sample robust inference procedure was given for those situations in which the parent distributions are symmetrical with long tails. There the inference was about the difference in population means and was based on trimmed sample means. Recent statistical research suggests that similar techniques can be applied to small samples. Until more concrete evidence is available, however, we shall not attempt to conduct inference procedures based on trimmed means when the sample sizes are small.

As stated earlier, the two-sample Wilcoxon rank sum test is for testing equality of location when the distributions are similar in shape. Thus the Wilcoxon test, as presented here, can be used to test equality of population means when the normality assumption cannot be made and, in particular, when the parent distributions are symmetrical with long tails. The test is illustrated in the following example.

EXAMPLE 10.11

Following are the social adjustment scores for a rural group and a city group of children. Test to see whether there is a significant difference in the mean social adjustment scores for the two groups of children.

Rural	55	57	62	58	34	52	63	84	
	50	56	98	58	54	60	55	51	
City	61	59	64	42	58	65	81	67	69
	23	63	51	65	68	53	61	68	

SOLUTION The null and alternative hypotheses are stated as follows:

$$H_0: \mu_1 = \mu_2, \quad H_a: \mu_1 \neq \mu_2$$

where μ_1 denotes the mean social adjustment score for the rural group and μ_2 denotes the mean social adjustment score for the city group.

Following are back-to-back stem and leaf plots and side-by-side box plots for the two groups:

Social adjustment score

```
            Rural                          City

                          2 | 3
                    4     3 |
                          4 | 2
 8 8 7 6 5 5 4 2 1 0       5 | 1  3  8  9
               3 2 0      6 | 1  1  3  4  5  5  7  8  8  9
                          7 |
                    4     8 | 1
                    8     9 |
```

Quarter summary diagrams

```
                  16                              17
M(8.5)  |       56.5       |          M(9)  |      63       |
Q(4.5)  |  53         61   |  Q-spread = 8   Q(5)  | 58        67  |  Q-spread = 9
```

Box plot summary diagrams

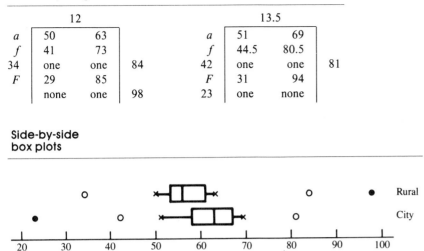

	12				13.5		
a	50	63		a	51	69	
f	41	73		f	44.5	80.5	
34	one	one	84	42	one	one	81
F	29	85		F	31	94	
	none	one	98	23	one	none	

Side-by-side box plots

Because of the long tails, we should not conduct a two-sample t-test. Noting that the two distributions are similar in shape we will compare the population means using the Wilcoxon rank sum test.

Replacing the data in the stem and leaf plot with ranks, we have

Rural		City
	2	1
2	3	
	4	3
15, 15, 13, 12, 10.5, 10.5, 9, 7, 5.5, 4	5	5.5, 8, 15, 17
22.5, 21, 18	6	19.5, 19.5, 22.5, 24, 25.5, 25.5, 27, 28.5, 28.5, 30
	7	
32	8	31
33	9	

From the ranks we have the following averages and standard deviations:

$$\bar{r}_1 = 14.375, \quad \bar{r}_2 = 19.471$$

$$s_{r_1} = 9.099, \quad s_{r_2} = 9.783$$

$$s_p = 9.458$$

The test statistic becomes

$$t = \frac{\bar{r}_1 - \bar{r}_2}{s_p \sqrt{1/n_1 + 1/n_2}}$$

$$= \frac{14.375 - 19.471}{9.458 \sqrt{1/16 + 1/17}}$$

$$= \frac{-5.096}{3.294}$$

$$= -1.55$$

From the t-table and $16 + 17 - 2 = 31$ degrees of freedom, we see that

$$2(.05) < p\text{-value} < 2(.10) \quad \text{or} \quad .10 < p\text{-value} < .20$$

which indicates no significant difference in the rural and city social adjustment scores.

EXERCISES 10.4

10.18 Find a 95% confidence interval for the difference in population means using the following data that have been arranged in a back-to-back stem and leaf plot.

Sample 1   **Sample 2**

	5*	0 1
9 7	5–	5
4	6*	1
8 7 5	6–	5 8
1 0	7*	0 3 4
8 7 6	7–	5 7 8
2 0	8*	0 1
8	8–	
	9*	2
9	9–	7

10.19 Find a 98% confidence interval for the difference in population means using the following data that have been arranged in a back-to-back stem and leaf plot.

Sample 1   **Sample 2**

5	16	1
0	17	3
8 4	18	2 7
8 5 3	19	0 8
2 0	20	1 3 7
9 7	21	2 8
3	22	1

10.20 It is believed that the mean amount of coffee dispensed by vending machine B in the student lounge area is greater than that of machine A in the cafeteria. The following summary statistics were obtained from samples of each machine.

	n	$\bar{y}$	s
Machine A	10	9.8	1.4
Machine B	12	10.1	.8

Is there statistical evidence to support the previous conjecture? Assume the amount dispensed is close to being normally distributed.

10.21 In an experiment to study the effects of a particular drug on the number of errors in maze-learning behavior of rats, the

following results were obtained:

Drug group	Placebo group
$n = 12$	$n = 16$
$\Sigma y_i = 224$	$\Sigma y_i = 256$
$\Sigma y_i^2 = 5{,}516$	$\Sigma y_i^2 = 4{,}352$

Assuming normality, test the hypothesis that the drug has no effect on the errors.

10.22 A store owner wants to know whether an advertising campaign has increased mean daily receipts. The daily receipts for the 2 weeks prior to the campaign were recorded as well as the receipts for a 2-week period after the campaign. From the following data, determine whether the campaign increased mean daily receipts.

Before campaign	After campaign
$n = 12$	$n = 12$
$\bar{y} = \$2{,}277$	$\bar{y} = \$2{,}564$
$s = \$375$	$s = \$438$

Assume that the receipts are close to being normally distributed.

10.23 Strength tests on two types of wool fabric produced the following results:

Type 1	138	127	148	134	135
	136	152	110	137	170

Type 2	134	137	135	140	130
	134	120	150	162	114

Is there a significant difference in the mean strength of the two types of wool?

10.24 Measurements of viscosity for a certain substance were taken with the following results:

First day	35.4	38.3	34.0	36.2	37.2	32.8
	36.0	35.2	36.0	31.3	35.7	

Second day	37.0	38.6	35.1	37.1	36.2	36.8
	37.6	34.8	35.8	38.2	32.1	

Has the population changed from one day to the next?

10.25 A manufacturer feels that the amount of carbon monoxide emitted by their smokestacks is less than their competitor's stacks. The EPA released the following readings for 10 consecutive days:

Manufacturer	3.5	3.1	3.1	3.5	2.5
	3.4	3.4	3.4	2.4	2.5
Competitor	3.7	3.0	3.5	3.8	2.8
	3.5	3.4	3.6	2.7	3.7

Test the manufacturer's claim.

10.5

Inference on the Difference between Two Population Distributions — Matched-Pairs Experiment

Suppose we wish to evaluate a new method of reading instruction for low ability students. An important part of the evaluation is a comparison of reading achievement scores for students using the experimental method with those using the standard method. For the comparison one possibility would be to independently and randomly assign students to the two methods and compare their reading achievement scores at the end of the project. The difference in the mean achievement test scores, $\mu_1 - \mu_2$, can be investigated using the methods of Sections 10.2–10.4.

Although we are dealing with low ability students, they will still have varying levels of ability prior to exposure to the method of instruction. Given that they are independently and randomly assigned to the two groups, it is conceivable that the students assigned to the two groups will not be comparable in reading skills. A difference in reading achievement at the end of the project might only be due to their different levels of skill prior to their instruction. The *matched-pairs experiment* (paired difference) is a method of assignment of subjects to the groups, which prevents this from happening.

Matched-Pairs *t*-Test

Suppose a reading skill test could be given to all students before assigning them to the groups. The students could then be paired according to reading skill and one of each pair assigned to the experimental group and the other assigned to the control group. A comparison of *matched pairs* of achievement test scores would then give a fair evaluation of the new method of instruction.

▲ The matched-pairs test is also referred to as the paired-differences test.

The analysis of the matched-pairs experiment is straightforward. The matched pairs form two *dependent* samples that are reduced to a single sample by taking the difference in the two observations for each pair. These differences form a single sample that can be analyzed using the one-sample techniques of Chapter 9.

Suppose that the matched pairs of observations are

$$(x_1, y_1), (x_2, y_2), \ldots, (x_n, y_n)$$

where x_i is the score for the ith subject in group 1 and y_i is the score for the matching ith subject in group 2. Let $d_i = y_i - x_i$ for $i = 1, 2, \ldots, n$ be the difference in the scores. Then $d_1, d_2, \ldots, d_n$ forms a single sample from which the mean, $\bar{d}$, and the standard deviation, s_d, can be calculated.

If we let $\mu_d = \mu_1 - \mu_2$ be the mean of the population of differences, then the null hypothesis can be stated as

$$H_0: \mu_d = 0$$

If the population of differences is approximately normally distributed, then the test statistic

$$t = \frac{\bar{d}}{s_d/\sqrt{n}}$$

is distributed as Student's t with $n - 1$ degrees of freedom. The remainder of the test proceeds as in the one-sample t-test.

The confidence interval for μ_d and the test of hypothesis are summarized as follows:

Inferences for $\mu_1 - \mu_2$ Based on Dependent Samples

APPLICATION: Dependent (matched) samples

ASSUMPTIONS: The distribution of difference scores is normal.

CONFIDENCE INTERVAL: A $(1 - \alpha)100\%$ confidence interval for $\mu_1 - \mu_2$ based on dependent samples is

$$\bar{d} \pm t_{\alpha/2,\, n-1}\, s_d/\sqrt{n}$$

MATCHED-PAIRS t-TEST:

Left-Tailed Test	Right-Tailed Test	Two-Tailed Test
$H_0: \mu_d \geq 0$	$H_0: \mu_d \leq 0$	$H_0: \mu_d = 0$
$H_a: \mu_d < 0$	$H_a: \mu_d > 0$	$H_a: \mu_d \neq 0$

Standardized Test Statistic

$$t = \frac{\bar{d}}{s_d/\sqrt{n}}$$

From the t-table with $n - 1$ degrees of freedom, find the p-value most closely associated with t_{obs}. If t_{obs} falls between two table values, give the two associated probabilities as bounds for the p-value.

The p-value for the two-tailed test is calculated as in a one-tailed test and then doubled to account for both tails.

If a level of significance α is specified, then reject H_0 if p-value $< \alpha$.

EXAMPLE 10.12

A group of 24 low ability students of the same age was tested for reading skills and then paired according to ability. One member of each of the matched pairs was assigned to an experimental reading group and the other to a control group. The experimental group was taught using a new method of reading instruction, and the control group was taught using the standard method that is normally used to teach low ability students. After the period of instruction all students were given a reading achievement test. The results were as follows:

Reading Achievement Test Scores

Pair	1	2	3	4	5	6	7	8	9	10	11	12
Experimental	82	65	63	71	48	74	61	65	73	92	57	66
Control	78	60	64	66	51	68	61	59	71	88	50	57

Perform a test of hypothesis to determine if the experimental group performed significantly better than the control group.

SOLUTION   Because the data are matched, we will investigate the difference scores. The differences are

Pair	1	2	3	4	5	6	7	8	9	10	11	12
Difference	4	5	−1	5	−3	6	0	6	2	4	7	9

The following dot graph appears to be skewed left; with so few observations, however, it is difficult to discern the shape of the underlying distribution. We will assume normality and conduct the matched-pairs t-test.

Dot graph of the difference scores

To determine whether the experimental group performed better than the control, a test of significance is performed with the hypotheses stated as follows:

$$H_0: \mu_d \leq 0 \quad \text{vs.} \quad H_a: \mu_d > 0$$

where

$$\mu_d = \mu_1 - \mu_2 \text{ and}$$

μ_1 = mean response for the experimental group

μ_2 = mean response for the control group

▲ When calculating $\bar{d}$, be sure that the negative values are added in as negatives.

From the differences we easily find that

$$\bar{d} = 3.667 \quad \text{and} \quad s_d = 3.525$$

and therefore

$$t_{obs} = \frac{3.667}{3.525/\sqrt{12}}$$

$$= 3.604$$

From the t-table with $n - 1 = 11$ degrees of freedom, we find that the

$$p\text{-value} < .005$$

because the *t*-score corresponding to .005 is 3.106, which is smaller than the t_{obs} of 3.604. With such a small *p*-value the evidence is highly significant that the new method of reading instruction for low ability students is better than the old one.

Another application of the matched-pairs design is the *before–after experiment*. The best matching is to have the same subject serve in both groups.

EXAMPLE 10.13

A group of women in a large city was given instructions on self-defense. Prior to the course, however, they were tested to determine their self-confidence. After the course they were given the same test. A high score on the test indicates a high degree of self-confidence. Did the course significantly increase their degree of self-confidence?

Self-confidence score

Woman

	1	2	3	4	5	6	7	8	9
Before course	6	10	8	6	5	4	3	8	5
After course	7	12	7	5	8	6	5	8	6

SOLUTION   The data are matched, so difference scores are found by

$$\text{Difference} = \text{After score} - \text{Before score}$$

The null and alternative hypotheses are stated as

$$H_0: \mu_d \leq 0 \quad \text{vs.} \quad H_a: \mu_d > 0$$

where $\mu_d = \mu_1 - \mu_2$ and

μ_1 = mean self-confidence score after the course

μ_2 = mean self-confidence score before the course

Rejection of the null hypothesis indicates that $\mu_1 > \mu_2$, which says that the self-defense course increased their self-confidence.

To test the hypothesis the differences are found to be

Woman	1	2	3	4	5	6	7	8	9
Difference	1	2	−1	−1	3	2	2	0	1

from which we find

$$\overline{d} = 1.0 \quad \text{and} \quad s_d = 1.414$$

The test statistic becomes

$$t_{obs} = \frac{1.0}{1.414/\sqrt{9}}$$
$$= 2.122$$

From the t-table with 8 degrees of freedom we find that

$$.025 < p\text{-value} < .05$$

because $1.860 < 2.122 < 2.306$. Thus there is evidence to reject H_0, and therefore we can say that the self-defense course was effective in increasing their self-confidence.

It is possible that the distribution of difference scores will deviate from normality. In those cases we have the *Wilcoxon signed rank test.*

Wilcoxon Signed Rank Test for the Matched-Pairs Experiment

▲ A sufficient condition for the difference distribution to be symmetric is for the two distributions to be identical in shape and spread.

The null hypothesis for the Wilcoxon signed rank test states that the population distributions from which the matched-pairs came are identical. If the two distributions differ only in location, then the test is a test of equality of means (as in the matched-pairs t-test) or medians depending on the measure of location that is used. Although this test does not require normality of the difference scores, it does assume that the distribution is symmetric.

As in the Wilcoxon rank sum test, we will give the Conover-Inman (1981) version of the Wilcoxon signed rank test whereby the usual t-test is applied to the ranks. This rank-transform test provides a good approximation to the Wilcoxon signed rank test when the sample size is at least 10.

To conduct a test of hypothesis, the absolute values of the d_i's are ranked and if the original d_i was negative, then the corresponding rank is given a negative sign. The remainder of the test is to apply the ordinary t-test to the signed ranks. The test statistic

$$t = \frac{\bar{r}}{s_r / \sqrt{n}}.$$

where $\bar{r}$ and s_r are the mean and standard deviation of the signed ranks, is closely approximated by Student's t-distribution with $n - 1$ degrees of freedom when $n \geq 10$. The test is illustrated in the following example.

EXAMPLE 10.14

To evaluate a new fabric softener a consumer organization purchased 2 of each type of 10 different garments for children. The 10 pieces of clothing were washed 15 times in the experimental softener. The like 10 pieces of clothing were washed 15 times without a softener. After the washings, the two groups of clothing were measured for softness. Following are the measurements for the two groups of clothing. (A large number indicates that the garment is judged more soft than if a lower number had been assigned.)

Type garment	1	2	3	4	5	6	7	8	9	10
With softener	12	3	12	16	4	24	11	17	19	8
Without softener	8	4	15	14	6	21	10	15	22	7

Test to see whether the softener significantly increases the degree of softness.

SOLUTION   Clearly the data are matched pairs so the differences are found.

Garment	1	2	3	4	5	6	7	8	9	10
Difference	4	-1	-3	2	-2	3	1	2	-3	1

Dot graph of the differences

▲ If the normality assumption is at all suspect, then the Wilcoxon test is recommended over the *t*-test.

The dot graph indicates that the normality assumption might not be met, yet symmetry seems plausible. From these observations it seems reasonable to use the Wilcoxon signed rank test. The null and alternative hypotheses are stated as

$$H_0: \mu_d \leq 0 \quad \text{vs.} \quad H_a: \mu_d > 0$$

where $\mu_d = \mu_1 - \mu_2$ and

$\mu_1 = $ mean softness measurement with the softener

$\mu_2 = $ mean softness measurement without the softener

Table 10.1 is useful in calculating the test statistic.

TABLE 10.1

d_i	$|d_i|$	Rank$|d_i|$	Signed ranks
4	4	10	10
-1	1	2	-2
-3	3	8	-8
2	2	5	5
-2	2	5	-5
3	3	8	8
1	1	2	2
2	2	5	5
-3	3	8	-8
1	1	2	2
			$\bar{r} = .9$
			$s_r = 6.42$

So

$$t_{obs} = \frac{.9}{6.42/\sqrt{10}} = .443$$

From the *t*-table with 9 degrees of freedom, we see that the

p-value $> .10$

which indicates that the softener is not at all effective in making the clothing more soft.

As pointed out in Chapter 1, the matched-pairs experiment is a special case of the randomized block experiment. In the randomized block design, subjects are separated into blocks of subjects based on some extraneous variable that has a confounding effect on the dependent variable. After the blocking is complete, the subjects within each block are randomly assigned to the treatments. In the matched-pairs experiment, each pair of subjects is a block with one of the pair being randomly assigned to group 1 and the other to group 2.

EXERCISES 10.5

10.26 The superintendent of X school district feels that his students, on the whole, have better study habits than the students of Y school district. Eleven students from each district were paired according to IQ and then their study habits were scored by an independent party. The results were as follows:

District X	105	109	115	112	124	107	121
District Y	115	103	110	125	99	121	119
District X	112	104	101	114			
District Y	106	100	97	105			

Test that there is no difference in the study habits of students at the two schools using the matched-pairs t-test.

10.27 To test learning ability, nine randomly selected eighth graders were given a spelling test. After a 2-week course of instruction, they were given a similar test. From the following data and the matched-pairs t-test, determine if there was improvement.

Student	1	2	3	4	5	6	7	8	9
Before	90	72	80	57	64	70	98	76	59
After	95	79	90	60	62	70	99	80	58

10.28 To test the effectiveness of a drug for asthmatic relief, a group of subjects was randomly given a drug and a placebo on two different occasions. After 1 hour an asthmatic relief index was obtained on each subject. Do the following data indicate that the drug significantly reduced the asthmatic relief index? Use the matched-pairs t-test.

Subject	1	2	3	4	5	6	7	8	9
Drug	28	31	17	22	12	32	24	18	25
Placebo	32	33	19	26	17	30	26	19	25

10.29 A study was designed to determine the effect of a certain movie on the moral attitude of young children. The data represent a rating from 0 to 20 on a moral attitude scale recorded before and after viewing the film. A high score is associated with a high morality. Test the hypothesis that the movie had no effect on the moral attitude of the children using the matched-pairs t-test.

Before	14	16	15	18	15	17	19	17	17	16	19	15
After	14	18	16	17	16	19	20	18	19	15	18	16

10.30 In a study using identical twins, one twin was given a drug and then given an intelligence test while under the influence of the drug. The other twin was given the same intelligence test under normal conditions. Following are the results of the test:

Twin A (no drug)	83	74	67	64	70	67	81	64	72
Twin B (drug)	78	74	63	66	68	63	77	65	70

Using the matched-pairs t-test test the hypothesis that the drug had no influence on the test scores.

10.31 A company wished to study the effectiveness of a coffee break on the productivity of its workers. The productivity of each of nine randomly selected workers was measured on a day without a coffee break and later on a day when they were given a 10-minute coffee break. The scores measuring productivity are given as follows:

Worker	1	2	3	4	5	6	7	8	9
Without coffee break	23	35	29	33	43	32	41	38	40
With coffee break	28	38	29	37	42	30	43	37	39

Do the results indicate that a coffee break increases productivity? Use the matched-pairs t-test.

10.32 To evaluate a speed reading course, a group of 15 subjects was asked to read two comparable articles before and after the course. Their scores on a reading comprehension test were:

Before course	After course
57	60
80	90
64	62
72	79
90	95
59	58
76	80
98	99
70	75
57	64
83	94
77	80
46	79
71	73
89	91

Determine whether the course was beneficial using the Wilcoxon signed rank test.

10.33 Twelve sets of identical twins were taught music recognition by two techniques (each twin was taught with a different method). At the end of the course, their improvement scores were recorded.

Twin set	1	2	3	4	5	6	7	8	9	10	11	12
Method 1	7	4	6	1	5	1	6	3	4	6	5	3
Method 2	2	4	3	2	3	4	2	4	4	3	5	2

Is there a statistical difference in the improvement scores for the two methods? Use the Wilcoxon signed rank test.

10.34 A sample of 10 German students was asked to copy a passage written in German. After an experimental course in German, the same 10 students were asked to copy the same passage. From the following data, test the hypothesis that the mean number of errors made before and after the course are not significantly different. Assume the difference scores are normally distributed.

No. of errors before	10	6	8	7	7	12	4	0	7	10
No. of errors after		6	4	5	3	6	8	0	1	8	5

10.35 Two psychiatrists were asked to rate each of 20 prison inmates concerning their rehabilitative potential. Do the following data indicate a significant difference in the mean rating scores given by the two psychiatrists? Perform the analysis using the Wilcoxon signed rank test.

Inmate		1	2	3	4	5	6	7	8	9	10
Psychiatrist 1			7	12	7	5	8	6	5	8	5	9
Psychiatrist 2			5	10	8	6	5	4	5	6	5	8
Inmate	11	12	13	14	15	16	17	18	19	20
Psychiatrist 1	6	7	3	5	7	9	5	11	4	10
Psychiatrist 2	9	7	5	4	8	4	8	9	3	12

10.6 Inference on the Difference between Two Population Proportions

Let π_1 and π_2 denote the proportion of successes in two Bernoulli populations. We will now investigate inference procedures on the difference $\pi_1 - \pi_2$.

Assume that random samples of size n_1 and n_2 are drawn independently from the two populations. Let p_1 and p_2 denote the two sample proportions of success. Clearly the statistic $p_1 - p_2$ is an unbiased estimator of the parameter $\pi_1 - \pi_2$. In Section 7.5 it was pointed out that if n_1 and n_2 are large, then the sampling distribution of $p_1 - p_2$ is approximately normally distributed with a mean of $\pi_1 - \pi_2$ and a standard error of

$$SE(p_1 - p_2) = \sqrt{\pi_1(1 - \pi_1)/n_1 + \pi_2(1 - \pi_2)/n_2}$$

Estimating π_1 and π_2, we obtain an estimated standard error of

$$estSE(p_1 - p_2) = \sqrt{p_1(1 - p_1)/n_1 + p_2(1 - p_2)/n_2}$$

Due to the normality of the sampling distribution, the *p*-value and the table value for a confidence interval will be found in the standard normal probability table.

As in previous sections, we construct the confidence interval for $\pi_1 - \pi_2$ as follows:

Confidence Interval for $\pi_1 - \pi_2$

APPLICATION: Bernoulli populations

ASSUMPTIONS: Independent samples and $n_1 > 30$, $n_2 > 30$

A $(1 - \alpha)100\%$ confidence interval for $\pi_1 - \pi_2$ is given by the limits

$$(p_1 - p_2) \pm z_{\alpha/2} \sqrt{p_1(1 - p_1)/n_1 + p_2(1 - p_2)/n_2}$$

where $z_{\alpha/2}$ is the upper $\alpha/2$ point in the standard normal probability table.

EXAMPLE 10.15

Thirty-four out of 50 third grade students believe in Santa Claus. Twenty-eight out of 50 fourth graders believe in Santa Claus. Find a 98% confidence interval for the difference between the proportion of third graders who believe in Santa Claus and the proportion of fourth graders who believe in Santa Claus.

SOLUTION   From the sample data we have

$$p_1 = 34/50 = .68 \quad \text{and} \quad p_2 = 28/50 = .56$$

from which we obtain

$$\text{estSE}(p_1 - p_2) = \sqrt{(.68)(.32)/50 + (.56)(.44)/50}$$
$$= .0963$$

For a 98% confidence interval we have $\alpha = .02$ so that $\alpha/2 = .01$. From the *z*-table we find

$$z_{\alpha/2} = z_{.01} = 2.33$$

The confidence interval for $\pi_1 - \pi_2$ becomes

$$(.68 - .56) \pm 2.33(.0963)$$

which is

$$.12 \pm .2244$$

so the interval becomes

$$(-.1044, .3444)$$

Note that the confidence interval contains the value 0. This indicates that a test of hypothesis will not find a significant difference in the two proportions.

To test hypotheses, the null (for a two-tailed test) is stated as

$$H_0: \pi_1 - \pi_2 = 0 \qquad (\text{or } \pi_1 = \pi_2)$$

Under the assumption that H_0 is true, that is $\pi_1 = \pi_2$, the standard error of $p_1 - p_2$ becomes

$$SE(p_1 - p_2) = \sqrt{\pi(1-\pi)}\ \sqrt{1/n_1 + 1/n_2}$$

▲ The common value of $\pi(1-\pi)$ is factored out from under the square root.

where π is the common value of π_1 and π_2.

Pooling together the information in the two samples, we obtain p, the pooled estimate of π. We have

$$p = \frac{x_1 + x_2}{n_1 + n_2}$$

where x_1 is the number of successes in sample 1 and x_2 is the number of successes in sample 2. Using this estimate of π, we have

$$\text{estSE}(p_1 - p_2) = \sqrt{p(1-p)}\ \sqrt{1/n_1 + 1/n_2}$$

Because H_0 is assumed to be true (until evidence to the contrary is found), the

---

Large-Sample Test of $\pi_1 - \pi_2$

APPLICATION: Bernoulli populations

ASSUMPTIONS: Independent samples and $n_1 > 30$, $n_2 > 30$

Left-Tailed Test	Right-Tailed Test	Two-Tailed Test
$H_0: \pi_1 - \pi_2 \geq 0$	$H_0: \pi_1 - \pi_2 \leq 0$	$H_0: \pi_1 - \pi_2 = 0$
$H_a: \pi_1 - \pi_2 < 0$	$H_a: \pi_1 - \pi_2 > 0$	$H_a: \pi_1 - \pi_2 \neq 0$

Standardized Test Statistic

$$z = \frac{p_1 - p_2}{\sqrt{p(1-p)}\ \sqrt{1/n_1 + 1/n_2}}$$

Given that

$$p_1 = \frac{x_1}{n_1}, \quad \text{and} \quad p_2 = \frac{x_2}{n_2}, \quad \text{and} \quad p = \frac{x_1 + x_2}{n_1 + n_2}$$

where x_1 is the number of successes in sample 1 and x_2 is the number of successes in sample 2.

From the z-table find

$p\text{-value} = P(z < z_{\text{obs}})$	$p\text{-value} = P(z > z_{\text{obs}})$	$p\text{-value} = 2P(z < z_{\text{obs}})$ if $z_{\text{obs}} < 0$
		or $p\text{-value} = 2P(z > z_{\text{obs}})$ if $z_{\text{obs}} > 0$

If a level of significance α is specified, then reject H_0 if $p\text{-value} < \alpha$.

---

standardized test statistic becomes

$$z = \frac{p_1 - p_2}{\sqrt{p(1-p)}\ \sqrt{1/n_1 + 1/n_2}}$$

which has a standard normal sampling distribution when n_1 and n_2 are sufficiently large. At this point it is easy to see that the remainder of the test is just like the large-sample z-test.

EXAMPLE 10.16

The campaign manager for a presidential candidate wishes to test the claim that the proportion of Ohio voters that favor the manager's candidate is at least as large as the proportion of California voters that favor the candidate. Given the following data, test the manager's claim at the 5% level of significance.

	n	No. who favor the candidate
Ohio	100	36
California	200	84

SOLUTION   Let π_1 denote the proportion of Ohio voters who favor the candidate, and let π_2 denote the proportion of California voters who favor the candidate.

The claim of the campaign manager is that $\pi_1 \geq \pi_2$. To test this claim the null and alternative hypotheses are stated as

$$H_0\colon \pi_1 \geq \pi_2 \quad \text{vs.} \quad H_a\colon \pi_1 < \pi_2$$

Pooling the sample information we find

$$p = \frac{36 + 84}{100 + 200} = 120/300 = .4$$

thus we have

$$\begin{aligned} \text{estSE}(p_1 - p_2) &= \sqrt{(.4)(.6)}\ \sqrt{1/100 + 1/200} \\ &= (.4899)(.12247) \\ &= .06 \end{aligned}$$

The test statistic becomes

$$z_{\text{obs}} = \frac{.36 - .42}{.06} = -1.0$$

From the z-table we find that the

$$p\text{-value} = .5 - .3413 = .1587 > \alpha\,(= .05)$$

Consequently there is insufficient evidence to reject the null hypothesis; therefore there is no statistical information to refute the campaign manager's claim.

10.36 Construct a 95% confidence interval for $\pi_1 - \pi_2$ in each of the following:

(a) $n_1 = 240, \quad x_1 = 160, \quad n_2 = 250, \quad x_2 = 145$
(b) $n_1 = 1200, \quad x_1 = 65, \quad n_2 = 1000, \quad x_2 = 70$
(c) $n_1 = 50, \quad x_1 = 38, \quad n_2 = 50, \quad x_2 = 26$

10.37 Independent random samples of size 200 each were selected from two Bernoulli populations. The number of successes in sample 1 and sample 2 were 65 and 74, respectively. Test the null hypothesis that $\pi_1 = \pi_2$ versus the alternative that $\pi_1 \neq \pi_2$.

10.38 In a survey taken to help understand emotions, 22 of 70 persons under 18 years of age expressed a fear of meeting people, and 23 of 90 persons 18 years old and over expressed the same fear. Test the hypothesis that there is no difference in the proportions of those who fear meeting people in the two age groups.

10.39 Suppose 250 castings produced in mold A contained 19 defective, and 300 castings produced in mold B contained 27 defective. Find a 99% confidence interval for the difference in the proportions of defective produced by the two molds.

10.40 The FDA requires that the mean antibody strength for a measle vaccine exceed 1.6 before it can be placed on the market. In a sample of 55 subjects firm A's vaccine exceeded the requirements 37 times. In a sample of 46 subjects firm B's vaccine exceeded the requirements 33 times. Is there a significant difference in the success rates for the two firms?

10.41 In a random sample of 210 college males it was determined that 65 smoked. In a random sample of 240 college females it was determined that 87 smoked. Is there a significant difference in the percent of college males and females that smoke?

10.42 In order to test the effectiveness of a vaccine against a certain disease, 120 experimental animals were given the vaccine and 180 were not. All 300 animals were then infected with the disease. Among those vaccinated, 15 died as a result of the disease. Among the control group, 36 died. Can we conclude that the vaccine was effective in reducing the mortality rate?

10.43 A sample of 500 adults with ages ranging from 24 through 54 was divided randomly into two groups. Those in group 1 consumed 1000 milligrams of vitamin C each day for 6 months. Those in group 2 consumed a placebo each day for the same 6 months. Of the 238 (12 dropped out of the experiment for various reasons) in group 1, 42 had at least one cold during the period. Of the 241 who completed the study in group 2, 61 had at least one cold during the period. Is there a significant difference in the percent that had at least one cold in each group?

10.44 To be judged proficient in reading at a certain university a student must score 85 or above on the Nelson-Denny Reading test. Of the 1,250 entering freshmen, 190 were randomly selected and given a special 2-week course in reading. After the course, it was determined that 146 of the 190 were judged proficient in reading. Of the remaining 1,060 entering freshmen, 762 were judged proficient in reading. Was the reading course beneficial in increasing reading scores?

In Sections 8.8 and 9.8 we were introduced to the TINTERVAL and TTEST commands. These commands also can be used to analyze the difference of two population means when the samples are dependent as in Section 10.5.

Recall that for the matched-pairs t-test we first compute the difference in the two observations within each pair and then perform the usual one-sample t-test. Minitab can compute the differences and perform the t-test or t-interval.

EXAMPLE 10.17

A psychologist interested in testing the relationship between stress and short-term memory administered a test to 12 subjects prior to their exposure to a stressful situation, and retested them after the stress situation. From the following data can we conclude that the stress situation decreases one's performance on a test measuring short-term memory?

Prestress	13	15	9	13	15	17	13	16	11	13	9	12
Poststress	10	14	7	15	11	14	13	14	9	14	9	10

Store the data in Minitab and compute the difference in scores and perform a one-tailed t-test.

SOLUTION   Let μ_1 denote the mean prestress test score and μ_2 denote the mean poststress test score. Also let $\mu_d = \mu_2 - \mu_1$, then the null and alternative hypotheses can be stated as follows:

$$H_0: \mu_d \geq 0 \quad \text{vs.} \quad H_a: \mu_d < 0$$

Rejection of H_0 establishes the alternative, which says that the mean poststress test score is significantly lower than the mean prestress test score.

In the Minitab program we will store the prestress scores in C1 and the poststress scores in C2. With the LET command we find

$$C3 = C2 - C1$$

and then perform a one-sided t-test on C3. The one-sided test is accomplished by using the subcommand ALTERNATIVE = K. If K is $+1$, then Minitab performs a right-tailed test; if K is -1, then Minitab performs a left-tailed test, which is the case in this example. (See Figure 10.2.)

FIGURE 10.2
Minitab output for Example 10.17

```
MTB > SET PRESTRESS IN C11
DATA> 13 15 9 13 15 17 13 16 11 13 9 12
DATA> END

MTB > SET POSTSTRESS IN C2
DATA> 10 14 7 15 11 14 13 14 9 14 9 10
DATA> END

MTB > LET C3 = C2 - C1
```

```
MTB > PRINT C1 C2 C3

  ROW     C1      C2     C3

    1      13      10     -3
    2      15      14     -1
    3       9       7     -2
    4      13      15      2
    5      15      11     -4
    6      17      14     -3
    7      13      13      0
    8      16      14     -2
    9      11       9     -2
   10      13      14      1
   11       9       9      0
   12      12      10     -2

MTB > TTEST MU = 0 DATA IN C3;
SUBC> ALTERNATIVE = -1.

TEST OF MU = 0.000 VS MU L.T. 0.000

            N      MEAN    STDEV   SE MEAN        T    P VALUE
C3         12    -1.333    1.775     0.512    -2.60      0.012
```

The *p*-value = .012 is significant, and thus we reject the null hypothesis. There is statistical evidence that the mean poststress test score is significantly lower than the mean prestress test score.

When the samples are independent (as in Section 10.4), two additional Minitab commands are available. TWOSAMPLE-T constructs a confidence interval and performs a *t*-test on data that are stored in separate columns. TWOT performs the same analysis yet is for problems with the data stored in one column and the groups specified in another column.

TWOSAMPLE-T

In order for TWOSAMPLE-T to perform the pooled *t*-test that is described in Section 10.4, the subcommand POOLED must be given. Otherwise TWOSAMPLE-T conducts an unpooled *t*-test, which is another alternative when the population variances are not homogeneous. In those cases, however, this text suggests that an analysis of trimmed means be made or that the Wilcoxon rank sum test be conducted. For the Wilcoxon rank sum test the two samples must be combined and ranked before proceeding with a two-sample *t*-test on the ranks. This is best accomplished when the data are in one column and the groups specified in another column. As mentioned earlier, data so arranged are analyzed with the TWOT command.

TWOSAMPLE-T and TWOT both allow for one-sided or two-sided tests. To conduct a one-sided test the subcommand ALTERNATIVE = K must be given. If K = −1, a left-tailed test is performed and if K = +1, a right-tailed test is performed. The following example illustrates TWOSAMPLE-T with the POOLED and ALTERNATIVE subcommands.

EXAMPLE 10.18

A new method of making concrete blocks has been proposed. To test whether or not the new method increases the compressive strength, 10 sample blocks are made by each method. The compressive strengths in pounds per square inch are:

New method	152	147	134	146	138	156	145	137	157	160
Old method	132	146	151	127	137	125	138	141	127	137

Store the data in two columns of Minitab and test the hypothesis that the new method is no better than the old.

SOLUTION If we let μ_1 denote the mean compressive strength with the new method, and μ_2 denote the mean compressive strength with the old method, the null hypothesis is

$$H_0: \mu_1 \le \mu_2$$

and the alternative hypothesis is

$$H_a: \mu_1 > \mu_2$$

Because this is a right-tailed test, K $= +1$ in the ALTERNATIVE subcommand. Note that if the null hypothesis is rejected, the alternative hypothesis is established and says that the mean compressive strength is greater under the new method than under the old method. We now store the data in the computer and perform the t-test as illustrated in Figure 10.3.

FIGURE 10.3

Minitab output for Example 10.18

```
MTB > SET NEW IN C1
DATA> 152 147 134 146 138 156 145 137 157 160
DATA> END

MTB > SET OLD IN C2
DATA> 132 146 151 127 137 125 138 141 127 137
DATA> END

MTB > TWOSAMPLE-T C1 C2;
SUBC> POOLED;
SUBC> ALTERNATIVE = +1.

TWOSAMPLE T FOR C1 VS C2

        N      MEAN     STDEV    SE MEAN
C1     10     147.20     9.00     2.85
C2     10     136.10     8.53     2.70

95 PCT CI FOR MU C1 - MU C2: (2.858, 19.34)

TTEST MU C1 = MU C2 (VS GT): T=2.83 P=0.0055 DF=18.0
```

The p-value of .0055 is highly significant, suggesting that the mean compressive strength is greater under the new method.

TWOT

TWOT performs the same analysis as TWOSAMPLE-T when the data are stored in one column and the groups in a second column. The subcommand POOLED must be given in order to conduct a pooled *t*-test.

EXAMPLE 10.19

A math achievement test is given to a random sample of 25 high school students. The scores and sex (coded as 1 for females and 2 for males) are recorded as follows:

Sex	1	2	2	1	2	1	1	1	1	2	2	1	2
Score	87	68	87	91	67	78	81	72	95	74	81	89	93
Sex		2	2	1	2	1	1	2	1	2	2	1	1
Score		60	78	93	74	83	74	92	75	81	62	85	95

Is there a significant difference in the scores for males and females? Store the data in Minitab and conduct a *t*-test using the TWOT command.

SOLUTION   Let μ_1 denote the mean score for females and μ_2 denote the mean score for males, then the null hypothesis is

$$H_0: \mu_1 = \mu_2$$

and the alternative hypothesis is

$$H_a: \mu_1 \neq \mu_2$$

Rejection of the null hypothesis indicates a difference in the scores for males and females. In Figure 10.4 we store the data in C1 and the sex in C2, and perform the TWOT command with the POOLED subcommand.

FIGURE 10.4

Minitab output for Example 10.19

```
MTB > SET DATA IN C1
DATA> 87 68 87 91 67 78 81 72 95 74 81 89 93 60 78
DATA> 93 74 83 74 92 75 81 62 85 95
DATA> END

MTB > SET SEX IN C2
DATA> 1 2 2 1 2 1 1 1 1 2 2 1 2 2 2 1 2 1 1 2 1 2 2 1 1
DATA> END

MTB > TWOT DATA IN C1 GROUPS IN C2;
SUBC> POOLED.

TWOSAMPLE T FOR C1

C2   N       MEAN      STDEV    SE MEAN
1    13      84.46     8.04      2.23
2    12      76.4      11.0      3.17

95 PCT CI FOR MU 1 - MU 2: (0.1335, 15.96)

TTEST MU 1 = MU 2 (VS NE): T=2.10 P=0.047 DF=23.0
```

Based on a *p*-value = .047, there is evidence of a significant difference in the scores for males and females.

EXERCISES 10.7

10.45 Fourteen employees who have not completed high school were given a reading test. Afterward they were given formal vocabulary training and then a retest of their reading skills was given. From the following test scores determine whether there is any difference in the scores before and after the vocabulary training.

First test	84	55	43	64	72	65	72	52	49
Second test	86	52	50	72	70	67	80	50	62

First test	80	38	93	77	60
Second test	81	56	90	78	64

Store the data in C1 and C2 of Minitab and have the computer take the differences and conduct a *t*-test.

10.46 To test the effect of a physical fitness course on one's physical ability, the number of sit-ups that a person could do in 1 minute, both before and after the course, was recorded. Nine randomly selected participants scored as follows:

Before	28	31	17	22	12	32	24	18	25
After	32	33	19	26	17	30	26	19	25

Using Minitab, test the hypothesis that there was an increase in the number of sit-ups.

10.47 A fourth grade teacher feels that his students are on the whole better spellers than his colleague's students. Ten students were randomly selected from each teacher's class and given a standardized spelling test. From the following results, use Minitab to determine if the fourth grade teacher is justified in his claim.

Fourth grade class	105	109	115	112	124	107
Colleague's class	115	103	110	125	99	121

Fourth grade class	121	112	104	119
Colleague's class	119	106	100	123

10.48 The results of quality control tests on two manufacturing processes are as follows:

Process I	1.5	2.5	3.4	2.3	3.2	2.8	1.9	
Process II	2.5	3.0	2.7	4.0	3.5	2.0	1.8	3.7

Use Minitab to determine whether the mean results of the two processes are equivalent.

10.8 Summary and Review

Key Concepts

☑ Three basic inference problems are addressed in this chapter—the comparison of two population means using independent samples, the comparison of two population means using dependent samples, and the comparison of two population proportions. Both large- and small-sample inference procedures are presented.

☑ The large-sample confidence interval for $\mu_1 - \mu_2$ based on $\bar{y}_1 - \bar{y}_2$ is given by

$$(\bar{y}_1 - \bar{y}_2) \pm z_{\alpha/2}\,\text{estSE}(\bar{y}_1 - \bar{y}_2)$$

where

$$\text{estSE}(\bar{y}_1 - \bar{y}_2) = \sqrt{s_1^2/n_1 + s_2^2/n_2}$$

The test statistic for the large-sample test of $\mu_1 - \mu_2$ is

$$z = \frac{(\bar{y}_1 - \bar{y}_2)}{\text{estSE}(\bar{y}_1 - \bar{y}_2)}$$

which has a standard normal sampling distribution.

☑ Large-sample inference procedures for $\mu_1 - \mu_2$ based on $\bar{y}_{T_1} - \bar{y}_{T_2}$ are the same as those based on $\bar{y}_1 - \bar{y}_2$, except that the ordinary sample means and standard deviations are replaced with the trimmed means and trimmed standard deviations.

☑ The small-sample inference procedures based on $\bar{y}_1 - \bar{y}_2$ are valid only if both parent populations are normally distributed with equal variances. Under those assumptions the two-sample t-test statistic

$$t = \frac{(\bar{y}_1 - \bar{y}_2) - (\mu_1 - \mu_2)}{s_p\sqrt{1/n_1 + 1/n_2}}$$

has a Student's t distribution with $n_1 + n_2 - 2$ degrees of freedom.

☑ The Wilcoxon rank sum test is an alternative to the two-sample t-test when the assumption of normality is not valid. For the Wilcoxon rank sum test the usual t-test is applied to the rank transformed data.

☑ When the samples are dependent, the test of the difference in population means is called the *matched-pairs t-test*. The test statistic is

$$t = \frac{\bar{d}}{s_d/\sqrt{n}}$$

which has a Student's t distribution with $n - 1$ degrees of freedom. Again the assumption of normality is required for this t-test unless the sample size is large. If the assumption is not met, then the *Wilcoxon signed rank test* is recommended. Here the usual t-test is applied to the signed ranks.

☑ Only the large-sample inference procedures for the difference in two population proportions are given. The confidence interval is given by

$$(p_1 - p_2) \pm z_{\alpha/2}\sqrt{p_1(1 - p_1)/n_1 + p_2(1 - p_2)/n_2}$$

and the test statistic for testing the hypothesis that $\pi_1 = \pi_2$ is

$$z = \frac{p_1 - p_2}{\sqrt{p(1 - p)}\ \sqrt{1/n_1 + 1/n_2}}$$

where

$$p = \frac{x_1 + x_2}{n_1 + n_2}$$

is the pooled estimate of π.

Learning Goals

Having completed this chapter you should be able to:

1. Construct a confidence interval for $\mu_1 - \mu_2$ based on $\bar{y}_1 - \bar{y}_2$ using large independent samples. *Section 10.2*

2. Test hypotheses about the difference between two population means based on $\bar{y}_1 - \bar{y}_2$ using large independent samples. *Section 10.2*

3. Perform large-sample inferences about $\mu_1 - \mu_2$ based on $\bar{y}_{T_1} - \bar{y}_{T_2}$. *Section 10.3*

4. Perform small-sample inferences about $\mu_1 - \mu_2$ based on $\bar{y}_1 - \bar{y}_2$. *Section 10.4*

5. Perform a two-sample Wilcoxon rank sum test. *Section 10.4*

6. Perform inferences on the difference between two population means using dependent samples by means of the matched-pairs t-test and the Wilcoxon signed rank test. *Section 10.5*

7. Perform inferences on the difference between two population proportions. *Section 10.6*

Review Questions

To test your skills answer the following questions.

10.49 Construct a 95% confidence interval for $\mu_1 - \mu_2$ based on $\bar{y}_{T_1} - \bar{y}_{T_2}$ using 10% trimming and the following summary data:

$n_1 = 42, \quad \bar{y}_{T_1} = 134.6, \quad s_{T_1} = 26.82$
$n_2 = 44, \quad \bar{y}_{T_2} = 127.8, \quad s_{T_2} = 23.47$

10.50 Assume two normal populations with common variance have means μ_1 and μ_2, respectively. With $\bar{y}_1 - \bar{y}_2$ as the test statistic, test

$H_0: \mu_1 - \mu_2 \le 0$

using the summary data:

$n_1 = 10, \quad \bar{y}_1 = 79.7, \quad s_1 = 11.32$
$n_2 = 10, \quad \bar{y}_2 = 76.4, \quad s_2 = 9.64$

10.51 Test the hypothesis that there is no significant difference in the population medians using the following data that have been arranged in a back-to-back stem and leaf plot. Because both distributions appear to be skewed, use the Wilcoxon rank sum test.

		21	0	4	7
3 6	9	22	5	6	9 9
5	8	23	1	4	8
2 6	9	24	5	8	
0 3 5	8	25	0	7	
3 7	7	26	3		
0	6	27	0	1	
	8	28			
	7	29	2		
	2	30	7		
		31			
	4	32			

10.52 In an experiment to evaluate a new variety of corn, 12 plots of land were divided in half with the new variety being planted on one half and a standard variety being planted on the other half. Following are the yields (in bushels) obtained on the 12 plots of land:

Plot	New variety	Standard
1	110	102
2	103	86
3	95	88
4	94	75
5	87	89
6	119	102
7	102	105
8	93	88
9	87	83
10	98	89
11	105	100
12	117	110

Yield per acre

Assuming normality, test the hypothesis that the new variety has a significantly higher yield per acre than the standard variety.

10.53 In a random sample of 160 females, 94 preferred diet soft drinks to regular soft drinks. In a random sample of 135 males, 71 preferred diet to regular soft drinks. Do the data imply that a significantly greater proportion of females prefer diet soft drinks than males?

10.54 Do students attending private high schools spend more time on homework than students attending public high schools? Following are the number of hours per week spent on homework for a random sample of 15 private high school students and a random sample of 15 public high school students:

Private	21.3	16.8	8.5	12.6	15.8
	19.3	18.5	24.6	18.3	12.9
	15.7	18.4	18.7	22.6	20.5
Public	15.3	17.4	12.3	10.7	16.4
	11.3	17.6	13.9	20.2	16.8
	23.6	14.2	5.7	18.8	9.4

Find a 98% confidence interval for the mean difference in the number of hours per week spent on homework for private and public high school students. Assume that the parent distributions are close to normal.

10.55 Use the data in Exercise 10.54 to test the hypothesis that private high school students spend significantly more time on homework per week than public high school students.

SUPPLEMENTARY EXERCISES FOR CHAPTER 10

10.56 Construct a 90% confidence interval for $\mu_1 - \mu_2$ based on $\bar{y}_1 - \bar{y}_2$ using the following summary data:

$n_1 = 65, \quad \bar{y}_1 = 315, \quad s_1 = 32.7$
$n_2 = 60, \quad \bar{y}_2 = 302, \quad s_2 = 28.4$

10.57 Test the null hypothesis H_0: $\mu_1 - \mu_2 = 0$ using the following summary data:

$n_1 = 134, \quad \Sigma y_i = 785, \quad \Sigma y_i^2 = 10,236$
$n_2 = 136, \quad \Sigma y_i = 793, \quad \Sigma y_i^2 = 12,578$

10.58 Two independent random samples were selected from two normally distributed populations with the following results:

	Sample A							Sample B							
6–	7							4	9						
7*	2							0							
7–								5							
8*	2	3						0	1	4					
8–	6	7						5	6	7	7	8	9	9	
9*	0	1	2	2	3	3	4	0	1	3	4				
9–	5	5	6	8	8	9		5	8						
10*								2							
10–	5	7													
11*								3							
11–	8							8							
12*															
12–	5														

Construct a 95% confidence interval for $\mu_1 - \mu_2$ based on $\bar{y}_1 - \bar{y}_2$.

10.59 In a random sample of 200 youthful drivers, 54 were judged as careless drivers. In a random sample of 200 adult drivers, 38 were judged as careless drivers. Is there a significant difference in the percents of youthful and adult drivers judged careless?

10.60 Sixteen children were selected from a first grade class and paired according to IQ such that one member of each pair had attended kindergarten and the other had not. A reading test was given to all 16 children with the following results:

Pair	1	2	3	4	5	6	7	8
Kindergarten	83	74	67	64	70	67	81	64
No kindergarten	78	74	63	66	68	63	77	65

Assuming normality, is there evidence that kindergarten was beneficial in improving the reading skills?

10.61 A 1984 study by the Centers for Disease Control in Atlanta indicates that giving aspirin to children who have the chicken pox or the flu can cause Reye's syndrome. Dr. Sidney M. Wolfe of the Public Citizen Health Research Group said it is the most convincing study to date to link aspirin and the life-threatening illness. Government health officials say that the study is preliminary, however, and that a more in-depth one has been started.

The study traced 29 cases of Reye's syndrome and 143 control cases of children who did not develop the illness. Twenty-eight of the 29 children who contracted Reye's syndrome took aspirin, whereas only 64 of the control group took aspirin. Construct a 95% confidence interval for the difference in population proportions based on those who took aspirin in the Reye's syndrome group and those who took aspirin in the control group.

10.62 Can the computer help high school students prepare for the SAT exam? An educational consultant, George Hopmeier, in Milton, Florida, studied 90 Florida high school students (*USA Today,* "Putting the Byte on SATs," September 27, 1984). Half of the students, without computer coaching, averaged 370 on the SAT math and the other half that used computer coaching averaged 407. Assuming SAT math has a standard deviation of 100, is there evidence that computer coaching improved SAT math scores?

10.63 A distributor of auto gears wishes to determine with which of two manufacturers of gears he should do business. He obtained, from the two manufacturers, the number of defective gears in the production of 100 gears per day for 20 consecutive days.

Manufacturer A	22	25	15	26	31
	22	17	26	23	20
	28	32	43	18	16
	36	21	16	29	26
Manufacturer B	18	24	28	42	17
	14	26	33	26	24
	32	36	38	26	25
	30	21	16	18	34

Compare the mean number of defectives per day for the two manufacturers by testing the null hypothesis that they are the same.

10.64 Following is the mean arterial blood pressure of 11 subjects before and after receiving oxytocin. Does oxytocin affect arterial blood pressure? Assume the population of difference scores is normally distributed.

Subject	1	2	3	4	5	6	7
Before	95	173	94	97	81	100	97
After	55	90	36	59	46	46	49

Subject	8	9	10	11
Before	104	72	101	83
After	92	23	55	49

10.65 Sixty-four individuals randomly selected from a metropolitan area were asked to indicate their preference for a number of candidates for a political office. A week later, one of the candidates visited each one of the 64 respondents. The week following the visit, the respondents were asked to indicate their preference for the same list of candidates. Thirty-two of the 64 respondents favored the candidate before his visit, whereas 40 of the 64 favored him after his visit. Is there evidence to indicate that personal contact makes a positive impact?

10.66 A developer of housing projects would like to cut costs on kitchen cabinets. He obtained the following estimates (in dollars) from two suppliers in 20 prospective homes:

Home	1	2	3	4	5	6	7	8
Supplier A	380	560	425	389	568	651	595	455
Supplier B	325	470	420	375	574	595	570	475
Home	9	10	11	12	13	14	15	16
Supplier A	540	520	375	468	492	510	379	750
Supplier B	560	500	362	465	445	490	350	780
Home	17	18	19	20				
Supplier A	520	480	394	624				
Supplier B	512	465	382	614				

Do the data indicate a significant difference in the estimates from the two suppliers?

10.67 Are students more interested in business and technology jobs now than they were a decade ago? A study by the American Association for Counseling and Development found that in 1983, 59% of all eleventh graders were interested in business and tech jobs, whereas in 1973 it was only 48%. Suppose we wish to determine whether the percent has changed significantly over the past year. A random sample of 400 eleventh graders last year showed that 226 were interested in business and tech jobs. A random sample of 400 eleventh graders this year found that 251 were interested in similar jobs. With a test of hypothesis, determine whether the percent is significantly greater this year than last.

10.68 Ohio is the pothole capital of the United States, with 6.8 million potholes, according to a survey by the Road Information Program. Potholes are caused by water seeping into cracks, which freezes and swells the pavement. Suppose we wish to compare two types of material for making pavement. A measurement is devised to measure the resistance to water. From the following data determine whether there is a significant difference in the resistance to water for the two types of materials.

	Material A	Material B
n	25	25
$\bar{y}$	63.8	87.2
s	16.55	15.75

10.69 Carolyn S. Hartsough, a University of California–Berkeley, educational psychologist, found that out of 301 hyperactive children, 26% of the mothers had poor health during pregnancy.

This compares with 16% of 191 normal children. Is the percentage for hyperactive children significantly greater than the percentage for normal children? Does this study show that poor health during pregnancy causes hyperactive children?

10.70 High levels of blood fat contribute to atherosclerosis (hardening of the arteries), which is an underlying cause of heart attacks and stroke. Dr. Robert Knopp, of the University of Washington School of Medicine in Seattle, analyzed 381 diabetics and found that women had higher levels of LDL, or "bad cholesterol," than men, and lower levels of HDL, or "good cholesterol."

From the following simulated data, determine whether the level of LDL in women is greater than that in men.

Women	Men
$n_1 = 201$	$n_2 = 180$
$\bar{y}_1 = 52.35$	$\bar{y}_2 = 47.28$
$s_1 = 6.55$	$s_2 = 7.29$

10.71 Dr. Marvin Moser, of Yale University School of Medicine, says that automated blood pressure machines in airports and shopping malls are generally unreliable. He suggests having your blood pressure checked regularly by an expert or the use of an at-home device (*USA Today*, May 1, 1985, p. 2D). Suppose 15 adult males check their blood pressure on an automated machine and then have an expert check it. From the following data determine whether the mean difference in diastolic blood pressure is significant.

Machine	68	82	94	106	92	80	76	74
Expert	72	84	89	100	97	88	84	70
Machine	110	93	86	65	74	84	100	
Expert	103	84	86	63	69	87	93	

Note that this test may not evaluate accuracy. It will tell you whether the machine tends to give high readings or tends to give low readings. The machine can be unreliable, however, in that some readings are unusually high and some are unusually low. The differences may cancel each other out, in which case the test statistic will be insignificant. Can you think of a better way of checking the accuracy of the machines?

11

Regression Analysis

In the previous chapter we were concerned with comparison problems. A comparison problem is one of comparing the distributions of a given variable for two populations. For example, we might compare the grade point averages of fraternity and nonfraternity students. Based on samples from each population we would be able to make a statement concerning the equality of the two distributions.

We will now consider problems of association. An association problem is one of relating two or more variables in a single population. For example, the grade point averages of students in general will most likely be related to their intelligence as measured by an IQ score, their study habits, and their motivation. With the exception of Section 11.7, this chapter is devoted to the most elementary association problem of relating two variables.

As Your Waistline Grows, So Does Heart Disease Risk

By Ellen Hale

MONTEREY, Calif.—Extra weight around your midriff increases your risk of heart disease, according to new research released Thursday.

"Even patients I would not consider to be obese may be at risk of heart disease," warns Swedish researcher Dr. Ulf Smith.

Smith's work, and that of scientists in Wisconsin and Vermont, is showing that your risk of heart attack may be as much as eight times higher if you store your extra weight around your stomach. The findings hold true for both men and women. Smith presented his research this week at the American Heart Association's science writers forum.

For men, the risk increases once their waist measurement equals their hip measurement. For example, if a man has a 32-inch waist, he should probably have about 35-inch hips. Once his waist creeps up to 35 inches, he's putting his health at risk. The risk goes up the more his waistline expands.

For women, the rule of thumb is different. If a woman has 36-inch hips and a 29-inch waist, that's OK. But once her waistline reaches 33 inches—or more than 80 percent of her hip measurement—she's at high risk.

A 13-year study of about 1,500 Swedish men with a high waist-to-hip ratio showed their risk of coronary heart disease, stroke and death was four times that for men of more normal proportions. A similar study of 1,500 women revealed that those with the highest waist-to-hip ratio had eight times the risk of heart attack than slimmer women.

Stored fat is more rapidly released into the bloodstream from fat cells around the stomach than from fat cells in the thighs and hips, Smith says.

In other research presented at the meeting, women involved in some kinds of clerical work—especially those using video display terminals—were more prone to heart problems than other workers because of stress.

A study of 518 women telephone operators and service representatives in North Carolina found that 20 percent of those who spent more than half their day in front of computer terminals suffered angina (chest pain that can be a pre-cursor of heart attack). Of the phone company employees in comparable jobs who didn't use computers, only 10 percent had chest pain.

"These women are asked to sit in front of a terminal for eight hours a day without any interaction from their boss at all," says Suzanne Haynes, chief of medical statistics of the National Center for Health Statistics in Rockville, Md.

"When they make mistakes, it shows up on the computer, and they can be reprimanded. This is the ultimate in high job demand and low control."

These findings support Haynes' earlier study of 1,000 working men and women from Framingham, Mass. The workers were first studied from 1965 to 1967, then followed for 10 years to see if and why they developed heart disease.

Haynes found that clerical workers in high-strain jobs were at highest risk. Only 2.4 percent of female workers in low-demand clerical jobs with regular feedback from their bosses developed heart disease, compared to 31 percent of clerical workers in high-demand and low-control jobs.

SOURCE: *USA Today*, January 18, 1985. Copyright, 1985 *USA Today*. Reprinted with permission.

The Swedish research study mentioned in the preceding article has shown a relationship between the risk of heart disease and the waist-to-hip ratio in males and females. The article suggests that as the waist-to-hip ratio increases so does the risk of heart disease. In this chapter we will learn how to investigate relationships between variables. We will learn how to formulate equations relating the variables, as well as to compute measures of the strength of the relationship.

Regression analysis is the statistical study of the relationships among quantitative variables. In Chapter 2 we saw that the scatterplot is a graphical procedure for exhibiting bivariate data for the purpose of visually detecting relationships. In Chapter 3 we measured the association between the variables with the correlation coefficient. Regression analysis gives the tools to describe the relationship numerically, so that predictions can be made.

College admissions officers attempt to predict the success of their applicants in order to determine who should be admitted to their university. The measure of success used by most is the cumulative grade point average (GPA). Many factors, some of which we can measure and some of which we cannot, affect the GPA of the student. Variables such as high school rank, SAT scores, and IQ scores can be measured for prospective students. However, variables such as motivation and parents' attitudes about college are difficult to measure, yet they will affect their GPA. For those variables that can be measured we wish to determine which ones are related to the GPA and how strong that relationship is.

The example we have given here can be generalized in the following manner. We are concerned with a random variable y that is related to the variables $x_1, x_2, x_3, \ldots, x_k$. The variable y is called the *dependent variable,* because presumably its value depends on the x variables, which are called the *independent variables.* The goal is to find an equation of the form

$$y = \beta_0 + \beta_1 x_1 + \beta_2 x_2 + \cdots + \beta_k x_k$$

that relates y to the independent variables $x_1, x_2, \ldots, x_k$. Typically k, the number of independent variables, will not be more than 4 or 5. In fact, with the exception of Section 11.7 we will consider only the case where $k = 1$.

In this case, the equation

$$y = \beta_0 + \beta_1 x$$

is called a *simple linear regression* equation.

▲ To graph a straight line it is necessary to find only two points that satisfy the equation and then connect the points with a straight line.

The equation is that of a straight line (illustrated in Figure 11.1) with β_0 being the y-intercept and β_1 the slope of the line.

To investigate the regression equation, observations are made by (usually) setting x at values $x_1, x_2, \ldots, x_n$ and observing the corresponding values of y, denoted by $y_1, y_2, \ldots, y_n$. It is not absolutely necessary that x be "set" at values $x_1, x_2, \ldots, x_n$. It is acceptable to observe both x and y. For example, we do not set the high school GPA (x)

FIGURE 11.1

Simple linear regression line

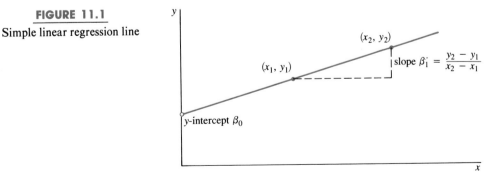

and observe the college GPA (y). We simply observe both the high school and college GPAs of each student in the sample. In any case, the data consist of the n pairs of observations (x_1, y_1), (x_2, y_2), . . . , (x_n, y_n) from which the relationship is evaluated.

As mentioned, the scatterplot is a means by which we can graph the data so that a visual interpretation can be given.

EXAMPLE 11.1

Following are the high school GPA and the college GPA at the end of the freshman year for each of 10 students:

Student	1	2	3	4	5	6	7	8	9	10
High school GPA	2.7	3.1	2.1	3.2	2.4	3.4	2.6	2.0	3.1	2.5
College GPA	2.2	2.8	2.4	3.8	1.9	3.5	3.1	1.4	3.4	2.5

Graph the data in a scatterplot and then comment on the relationship between the two variables.

SOLUTION   Because it is believed that one's college GPA is dependent upon his or her high school GPA, we will assume that the high school GPA is an independent variable x and graph it on the horizontal axis. The college GPA is the dependent variable y and is labeled on the vertical axis. From the data we get the scatterplot shown in Figure 11.2. It appears that the college GPA increases as the high school GPA increases. In fact, the dots can be thought of as clustering around a straight line. We can say that there appears to be a linear relationship between x(high school GPA) and y (college GPA).

FIGURE 11.2

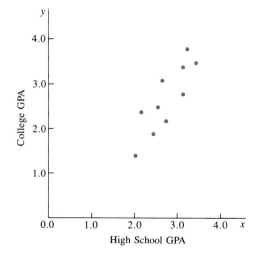

As in the preceding example, there is a linear relationship between x and y when the points of a scatterplot fall in a straight line. However there can be relationships that are not linear as appears in Figure 11.3.

FIGURE 11.3

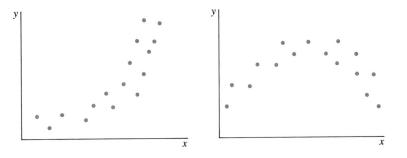

These relationships are said to be *curvilinear*. In some cases it may be possible to make a transformation (such as $\sqrt{y}$) on either x or y or both in such a way that the transformed data are linear. The procedures outlined for the linear case can then be applied to the transformed data. In the next section we will investigate the linear relationship between variables.

EXERCISES 11.1

11.1 Identify the independent and dependent variables in the following situations:

(a) A study of the relationship between air pollution and high blood pressure

(b) A study of the relationship between mental retardation and lead poisoning

(c) A study of the relationship between the percent of drivers who wear seat belts and the death rate for automobile accidents

(d) A study of the relationship between the suicide rate and alcohol consumption among teenagers

(e) A study of the relationship between per capita income and public education expenditures of the states.

11.2 Graph the following straight lines:

(a) $y = 2 + 3x$

(b) $y = -2.6 + 4x$

(c) $y = 7.1 - 8.2x$

11.3 For the following data sets, construct a scatterplot and comment on the relationship between x and y:

(a)
x	2	1	3	2	3	1
y	4	0	6	6	8	2

(b)
x	1	3	4	6	8	9	11	14
y	1	2	4	4	5	7	8	9

(c)
x	1	2	4	5	4	2	6	3
y	5	11	25	39	27	12	50	17

(d)
x	-3	-1	0	1	3
y	12	7	6	4	1

11.4 The amount of energy consumed in a home is related to the size of the home. From the data in the accompanying table, construct a scatterplot and comment on the relationship between the variables.

Size of home (sq ft)	Kilowatt-hours per month
2820	1975
2500	1952
2350	1894
2000	1841
1950	1769
1875	1674
1740	1590
1650	1505
1490	1386
1350	1220
1270	1089
1200	1042

11.5 A study was made to relate aptitude test scores to productivity in a factory after 3 months of personnel training. The following was obtained by testing eight randomly selected applicants and later measuring their productivity.

Applicant	1	2	3	4	5	6	7	8
Aptitude score	9	17	13	19	20	23	12	15
Productivity	23	35	29	33	40	38	25	31

Draw a scatterplot of aptitude score and productivity and comment on the relationship.

11.6 Robbery rate varies in value from precinct to precinct. From the following data construct a scatterplot and comment on the relationship between the robbery rate and the percent of low income in the precinct.

Precinct	1	2	3	4	5	6	7	8
% low income	4.9	7.1	10.1	11.8	13.5	14.8	16.2	11.2
Robbery rate	20	41	165	88	60	120	65	81

11.7 A record of maintenance cost is kept on nine cash registers in a major department store chain.

Age (years)	6	7	1	3	6	2	5	4	3
Cost (dollars)	92	181	23	40	126	35	86	72	51

Construct a scatterplot and comment on the relationship between the variables.

11.8 It is suspected that the life span of a particular electronic component used in a spacecraft is dependent on the heat it might experience. Six such components were tested at various levels of heat with the following results:

Heat (°C)	50	100	150	200	250	300
Life span (hours)	875	884	762	424	365	128

Construct a scatterplot and comment on the relationship between the variables.

11.2 The Method of Least Squares

In Example 11.1 we observed that the data corresponding to high school GPA (x) and college GPA (y) appeared to gather in a straight line. Had the points fallen exactly in a straight line then there would be an exact straight line relationship between the variables of the form

$$y = \beta_0 + \beta_1 x$$

We could say that there is no error in obtaining y from x. However there is no reason to expect the points of the scatterplot to fall in a perfect straight line. There could still be a linear relationship between the variables, and the points could fall off the straight line because of unknown error.

For example, two people with the same high school GPA will be likely to end up with different college GPAs. They differ because of other extraneous variables such as IQ, study habits, motivation, and even parents' attitude about college. In many instances these extraneous variables cannot be measured (or are chosen not to be measured), and thus are lumped together into what we call error. Thus we assume that the random variable y is related to the independent variable x by the *linear regression model*

$$y = \beta_0 + \beta_1 x + \epsilon$$

The ϵ represents the random error, which allows for the possibility that all observations do not fall exactly in a straight line. Some ϵ's are positive, in which case the student has a higher college GPA than expected, and some ϵ's are negative, giving a college GPA lower than expected. Overall they would average out to zero.

Assuming that the errors have expected value (average value) of 0, we have that

$$E(y) = \beta_0 + \beta_1 x$$

Using the sample data we will find estimates of β_0 and β_1 so that for a given x we can compute a *predicted y*.

▲ Recall that an extraneous variable is a variable outside the study that has an effect on the study.

Definition

From the data point (x_i, y_i) the observed value of y is y_i and the **predicted value of** y is obtained by the equation

$$\hat{y}_i = b_0 + b_1 x_i$$

where b_0 and b_1 are estimates of β_0 and β_1, respectively.

The error of the prediction (also called the *residual*) is the difference in the actual y_i and the predicted y_i.

Definition

▲ Note that there is a residual for each data point.

The **residual** associated with the data point (x_i, y_i) is

$$e_i = y_i - \hat{y}_i$$

where $\hat{y}_i = b_0 + b_1 x_i$.

The residuals associated with the data points are depicted as vertical line segments in Figure 11.4. They are, in a sense, estimates of the random error ϵ. Because the expected error is 0, the line is positioned so that some of the sample errors are positive (those above the line) and some are negative (those below the line). To obtain a single value that summarizes the total error, we square the residuals and sum them up over all cases. The resulting value is called the *sum of squares due to error*.

FIGURE 11.4

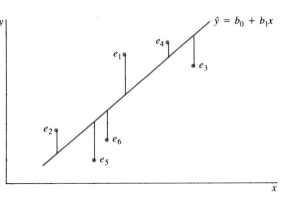

Definition

The **sum of squares due to error** (also called the residual sum of squares) is given by

$$\text{SSE} = \Sigma(y_i - \hat{y}_i)^2$$

Substituting for $\hat{y}_i$, we have

$$\text{SSE} = \Sigma(y_i - b_0 - b_1 x_i)^2$$

▲ To fully understand how the least squares estimates are derived you would need a course in differential calculus. We are not interested in deriving the formulas, however, we just need to know how to use them.

which clearly depends on b_0 and b_1. The method of least squares chooses the estimates b_0 and b_1 so that the value of SSE is as small as possible. (This is desirable because SSE

---

is a combined measure of the discrepancy between the actual y and the predicted y.) The formulas for the values of b_0 and b_1 are given here

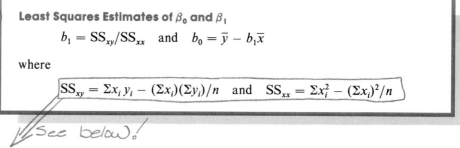

Least Squares Estimates of β_0 and β_1

$$b_1 = SS_{xy}/SS_{xx} \quad \text{and} \quad b_0 = \bar{y} - b_1\bar{x}$$

where

$$SS_{xy} = \Sigma x_i y_i - (\Sigma x_i)(\Sigma y_i)/n \quad \text{and} \quad SS_{xx} = \Sigma x_i^2 - (\Sigma x_i)^2/n$$

See below!

EXAMPLE 11.2

Find the least squares estimates of β_0 and β_1 in the regression line $y = \beta_0 + \beta_1 x$ for the GPA data in Example 11.1. Also find the correlation between the two variables.

SOLUTION The calculations can be simplified by Table 11.1.

TABLE 11.1

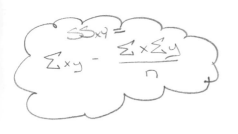

x_i	y_i	x_i^2	y_i^2	$x_i y_i$
2.7	2.2	7.29	4.84	5.94
3.1	2.8	9.61	7.84	8.68
2.1	2.4	4.41	5.76	5.04
3.2	3.8	10.24	14.44	12.16
2.4	1.9	5.76	3.61	4.56
3.4	3.5	11.56	12.25	11.90
2.6	3.1	6.76	9.61	8.06
2.0	1.4	4.00	1.96	2.80
3.1	3.4	9.61	11.56	10.54
2.5	2.5	6.25	6.25	6.25
$\Sigma x_i = 27.1$	$\Sigma y_i = 27.0$	$\Sigma x_i^2 = 75.49$	$\Sigma y_i^2 = 78.12$	$\Sigma x_i y_i = 75.93$

Substituting the totals in the equations, we have

$$SS_{xy} = 75.93 - (27.1)(27.0)/10 = 2.76$$

▲ The value of Σy_i^2 and hence SS_{yy} are not needed to calculate the least squares estimates. It is included here so that the correlation can be calculated.

$$SS_{xx} = 75.49 - (27.1)^2/10 = 2.049$$
$$SS_{yy} = 78.12 - (27.0)^2/10 = 5.22$$
$$\bar{y} = 2.70$$
$$\bar{x} = 2.71$$

$b = \dfrac{SP}{SS_x}$

and the least squares estimates are $a = \bar{y} - b\bar{x}$

$$b_1 = SS_{xy}/SS_{xx} = 2.76/2.049 = 1.347$$
$$b_0 = \bar{y} - b_1\bar{x} = 2.70 - (1.347)(2.71) = -.950$$

The prediction line that relates college GPA to high school GPA is

$$\hat{y} = -.950 + 1.347x$$

FIGURE 11.5

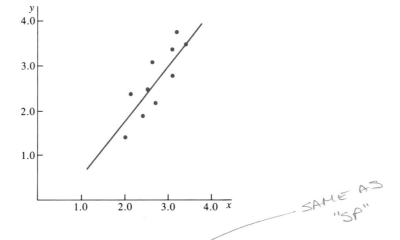

Figure 11.5 is the scatterplot with a graph of the straight line. Recall from Chapter 3 that the correlation between x and y is given by

$$r = \frac{SS_{xy}}{\sqrt{SS_{xx}SS_{yy}}} = \frac{2.76}{\sqrt{(2.049)(5.22)}} = .844$$

which supports the fact that the data points cluster tightly about the regression line.

EXAMPLE 11.3

▲ The solution to this example (and subsequent examples) was obtained using a hand-held calculator with a linear regression mode. The calculator used full-decimal accuracy of b_0 and b_1 to compute $\hat{y}_i$; therefore, the values given may not agree completely with the values that would be found using rounded versions of b_0 and b_1.

Find the residuals for the 10 data points in Example 11.2, and then calculate SSE.

SOLUTION   Substituting the x_i values in the prediction equation, we find the predicted y_i's. They, together with the residuals, are listed in Table 11.2. Squaring the residuals and summing, we find that

$$SSE = (-.487)^2 + (-.425)^2 + \cdots + (.083)^2$$
$$= 1.502$$

We can be assured that for the original data (x_i, y_i) the prediction equation is such that the preceding value of SSE is as small as it possibly can be.

TABLE 11.2

x_i	y_i	$\hat{y}_i$	$e_i = y_i - \hat{y}_i$
2.7	2.2	2.687	−.487
3.1	2.8	3.225	−.425
2.1	2.4	1.878	.522
3.2	3.8	3.360	.440
2.4	1.9	2.282	−.382
3.4	3.5	3.629	−.129
2.6	3.1	2.552	.548
2.0	1.4	1.744	−.344
3.1	3.4	3.225	.175
2.5	2.5	2.417	.083

11.2   THE METHOD OF LEAST SQUARES   **423**

11.9 For the following data sets that appeared in Exercise 11.3, find the least squares estimates of β_0 and β_1 in the model: $y = \beta_0 + \beta_1 x + \epsilon$

(a)

x	2	1	3	2	3	1
y	4	0	6	6	8	2

(b)

x	1	3	4	6	8	9	11	14
y	1	2	4	4	5	7	8	9

(c)

x	1	2	4	5	4	2	6	3
y	5	11	25	39	27	12	50	17

11.10 From the data in Exercise 11.9 find the predicted y for the following values of x.

(a) $x = 0$, $x = 4$
(b) $x = 2$, $x = 5$, $x = 10$, $x = 15$
(c) $x = 0$, $x = 5$, $x = 8$

11.11 Find the residuals associated with the least squares equations found in parts (a), (b), and (c) of Exercise 11.9.

11.12 Find the sum of squares due to error associated with the least squares equations found in parts (a), (b), and (c) of Exercises 11.9 (see Exercise 11.11).

11.13 Consider the following data set:

x	1	3	5	7	9
y	5	7	8	10	13

(a) Find the least squares prediction equation relating x to y.
(b) Use the prediction equation to predict y when $x = 4$.

Consider the following data set:

x	-3	-1	0	1	3
y	12	7	6	4	1

(a) Find the least squares prediction equation relating x to y.
(b) Find the residuals associated with the data.
(c) Find SSE.

11.15 In Exercise 11.5, data relating aptitude test scores to productivity in a factory were given. Use the same data (reproduced here) to obtain the least squares estimates of β_0 and β_1.

Applicant	1	2	3	4	5	6	7	8
Aptitude score	9	17	13	19	20	23	12	15
Productivity	23	35	29	33	40	38	25	31

11.16 In Exercise 11.6, data relating robbery rate and the percent of low income were given. Use the same data (reproduced here) to obtain the least squares estimates of β_0 and β_1.

Precinct	1	2	3	4	5	6	7	8
% low income	4.9	7.1	10.1	11.8	13.5	14.8	16.2	11.2
Robbery rate	20	41	165	88	60	120	65	81

11.17 In Exercise 11.7, data corresponding to maintenance cost on nine cash registers and their ages were recorded as follows:

Age (years)	6	7	1	3	6	2	5	4	3
Cost (dollars)	92	181	23	40	126	35	86	72	51

(a) Find the least squares estimates of β_0 and β_1.
(b) Find the residuals associated with $(1, 23)$ and $(5, 86)$.

11.18 In Exercise 11.8, the data corresponding to the life span of a particular electronic component and the heat it is exposed to are duplicated here:

Heat (°C)	50	100	150	200	250	300
Life span (hours)	875	884	762	424	365	128

(a) Find the least squares estimates of β_0 and β_1.
(b) Find the residuals.
(c) Find SSE.

The line

$$y = b_0 + b_1 x$$

obtained by the method of least squares is an estimate of the linear regression model

$$y = \beta_0 + \beta_1 x + \epsilon$$

that relates y to x. In particular, b_0 is the least squares estimate of β_0, and b_1 is the least squares estimate of the *regression coefficient* β_1. Using b_1 as a test statistic, it is possible to test hypotheses about the parameter β_1. (Although inferences about β_0 are possible, they will not be presented in this book.)

To proceed with the inference about β_1, we must first understand the sampling distribution of its estimate, b_1. To derive the sampling distribution, certain assumptions about the random errors must be made. Those assumptions are

▲ Assumptions about the random errors are necessary because, if you will recall, the formula for b_1 was obtained by minimizing the sum of squares of the random errors.

Assumptions to Make Inferences about the Regression Model

1. y is related to x by the linear regression model

$$y = \beta_0 + \beta_1 x + \epsilon$$

2. The ϵ's are independent.

3. The ϵ's are normally distributed with a mean of 0 and a common standard deviation σ.

Under these assumptions it follows that b_1 is an unbiased estimator of β_1 and the standard error of b_1 is

$$\mathrm{SE}(b_1) = \frac{\sigma}{\sqrt{\mathrm{SS}_{xx}}}$$

Further it can be shown that the statistic

$$t = \frac{b_1 - \beta_1}{s/\sqrt{\mathrm{SS}_{xx}}}$$

is distributed as a t-distribution with $n - 2$ degrees of freedom. The value s, is an estimate of the standard deviation σ and is given by the formula

$$s = \sqrt{\mathrm{SSE}/(n - 2)}$$

In Example 11.3 we saw that the formula $\mathrm{SSE} = \Sigma(y_i - \hat{y}_i)^2$ is somewhat difficult to use. An equivalent equation, which is easier to use, is

$$\mathrm{SSE} = \mathrm{SS}_{yy} - b_1 \mathrm{SS}_{xy}$$

Note that these are the same quantities used in calculating the regression equation and the correlation. Thus at this point in a problem the calculation of SSE is straightforward.

EXAMPLE 11.4

A recent study at the Stanford University School of Medicine suggests that coffee drinking is linked to heart disease. In particular, coffee drinking is strongly connected to elevated levels of apolipoprotein B, a cholesterol-associated protein linked to heart disease. From the data in Table 11.3 on 15 adult males over the age of 35 who drink from one to five cups of coffee per day, calculate the least squares estimates of β_0 and β_1 for the model $y = \beta_0 + \beta_1 x + \epsilon$ and give an estimate of σ.

TABLE 11.3

Subject	No. of cups of coffee per day (x)	Level of apolipoprotein B(y)
1	1	23
2	1	19
3	1	13
4	2	21
5	2	18
6	2	25
7	3	26
8	3	32
9	3	28
10	4	35
11	4	27
12	4	33
13	5	33
14	5	37
15	5	38

▲ It is assumed that the level of apolipoprotein B is the dependent variable because it is presumably dependent on the number of cups of coffee.

SOLUTION   As before, the computations are simplified by arranging the data as in Table 11.4.

TABLE 11.4

▲ If you are using a hand-held calculator with a linear regression mode, then Table 11.4 is not necessary.

x_i	y_i	x_i^2	y_i^2	$x_i y_i$
1	23	1	529	23
1	19	1	361	19
1	13	1	169	13
2	21	4	441	42
2	18	4	324	36
2	25	4	625	50
3	26	9	676	78
3	32	9	1,024	96
3	28	9	784	84
4	35	16	1,225	140
4	27	16	729	108
4	33	16	1,089	132
5	33	25	1,089	165
5	37	25	1,369	185
5	38	25	1,444	190
$\Sigma x_i = 45$	$\Sigma y_i = 408$	$\Sigma x_i^2 = 165$	$\Sigma y_i^2 = 11878$	$\Sigma x_i y_i = 1361$

Therefore we have

$$SS_{xy} = 1361 - (45)(408)/15 = 137$$
$$SS_{xx} = 165 - (45)^2/15 = 30$$
$$SS_{yy} = 11878 - (408)^2/15 = 780.4$$
$$\bar{x} = 3, \quad \bar{y} = 27.2$$

The least squares estimates are

$$b_1 = 137/30 = 4.567$$
$$b_0 = 27.2 - (4.567)(3) = 13.5$$

Furthermore we have

$$SSE = SS_{yy} - b_1 SS_{xy}$$
$$= 780.4 - (4.567)(137)$$
$$= 154.721$$

Then the estimate of σ is

$$s = \sqrt{SSE/(n - 2)} = \sqrt{154.721/13} = 3.45$$

In a regression analysis the aim is to determine whether the value of y depends on the independent variable x. For the linear model $y = \beta_0 + \beta_1 x + \epsilon$ this can be accomplished by testing the null hypothesis

$$H_0: \beta_1 = 0$$

If $\beta_1 = 0$, then it follows that

$$y = \beta_0 + \epsilon$$

That is, the observed value of y is simply the constant value β_0 plus an error ϵ, and the value of y does not depend on x in a linear fashion.

The test of hypothesis about β_1 is summarized on the following page.
We illustrate the test in the following example.

EXAMPLE 11.5

A study to compare the number of years of experience of the New York City street-patrolpersons to the number of tickets they give per week revealed the following data:

▲ The number of tickets is the dependent variable y because presumably it depends on x, the number of years of experience.

Patrolperson	1	2	3	4	5	6	7	8	9	10
No. years experience (x)	3	8	2	15	5	20	1	10	7	12
No. tickets per week (y)	42	30	54	12	32	8	75	28	20	15

Perform a least squares fit for the model

$$y = \beta_0 + \beta_1 x + \epsilon$$

and test the hypothesis that $\beta_1 = 0$.

Test of Hypothesis about β_1

ASSUMPTIONS: The assumptions were listed earlier.

Left-Tailed Test	Right-Tailed Test	Two-Tailed Test
H_0: $\beta_1 \geq 0$	H_0: $\beta_1 \leq 0$	H_0: $\beta_1 = 0$
H_a: $\beta_1 < 0$	H_a: $\beta_1 > 0$	H_a: $\beta_1 \neq 0$

Standardized Test Statistic

$$t = \frac{b_1}{s/\sqrt{SS_{xx}}}$$

From the t-table with $n - 2$ degrees of freedom find the p-value most closely associated with t_{obs}. If t_{obs} falls between two table values, give the two associated probabilities as bounds for the p-value.

The p-value for the two-tailed test is calculated as in the one-tailed test and then doubled to account for both tails.

If a level of significance α is specified, then reject H_0 if p-value $< \alpha$.

SOLUTION   From the data we find that

$$SS_{xy} = 1649 - (83)(316)/10 = -973.8$$
$$SS_{xx} = 1021 - (83)^2/10 = 332.1$$
$$SS_{yy} = 13846 - (316)^2/10 = 3860.4$$
$$\bar{y} = 316/10 = 31.6$$
$$\bar{x} = 83/10 = 8.3$$

from which we find

$$b_1 = -973.8/332.1 = -2.93$$
$$b_0 = 55.94$$

To test the hypothesis

$$H_0\text{: } \beta_1 = 0 \quad \text{vs.} \quad H_a\text{: } \beta_1 \neq 0$$

we need the test statistic

$$t = \frac{b_1}{s/\sqrt{SS_{xx}}}$$

which calls for s and SS_{xx}. SS_{xx} is given above and the formula for s is

$$s = \sqrt{SSE/(n - 2)}$$

Recall that

$$SSE = SS_{yy} - b_1 SS_{xy}$$
$$= 3860.4 - (-2.93)(-973.8)$$
$$= 1007.166$$

Thus

$$s = \sqrt{1007.166/8}$$
$$= 11.22$$

Finally we have

$$t = \frac{-2.93}{11.22/\sqrt{332.1}}$$
$$= -4.76$$

which has 8 degrees of freedom and therefore has a p-value $< 2(.005) = .01$. We reject the null hypothesis and conclude that there is a significant linear relationship between the number of years experience and the number of tickets given per week.

If the null hypothesis

$$H_0: \beta_1 = 0$$

is not rejected, then there is no statistically significant linear relationship between x and y. That is not to say that y does not depend on x at all. It is possible that y could depend on x in a nonlinear way, such as

$$y = \beta_0 + \beta_2 x^2 + \epsilon$$

Using x^2 and y as the data, the significance of β_2 can be analyzed in the same manner as β_1 mentioned earlier. In addition to the test of hypothesis just given, a confidence interval for β_1 can also be constructed.

▲ Note that the form of the confidence interval is the same as that given in previous chapters, namely,

estimate ± (table value)(standard error)

A $(1 - \alpha)100\%$ confidence interval for β_1 is given by the limits

$$b_1 \pm t_{\alpha/2,\, n-2} s/\sqrt{SS_{xx}}$$

EXAMPLE 11.6

From the data in Example 11.4, find a 98% confidence interval for β_1.

SOLUTION   From Example 11.4, we find that

$$b_1 = 4.567$$
$$s = 3.45$$
$$SS_{xx} = 30$$

For a 98% confidence interval we find from the t-table that

$$t_{\alpha/2,\, n-2} = t_{.01,\, 13} = 2.650$$

$n = 15$
$df = 13$

Therefore the confidence interval for β_1 is

$$b_1 \pm t_{\alpha/2,\, n-2} s/\sqrt{SS_{xx}}$$

which gives

$$4.567 \pm (2.650)(3.45)/\sqrt{30}$$

---

or

$$4.567 \pm 1.67$$

Thus the interval is (2.897, 6.237).

EXERCISES 11.3

11.19 From the following data, compute the least squares estimates of β_0 and β_1 in the model $y = \beta_0 + \beta_1 x + \epsilon$.

x	6	9	11	15	11
y	5	10	8	14	12

Test the hypothesis that $\beta_1 = 0$.

11.20 From the following data, test the significance of the slope of the regression line: $y = \beta_0 + \beta_1 x$.

x	11	6	11	15	14	15	8	16
y	8	5	10	12	12	9	5	11

11.21 From the following summary data, find a 95% confidence interval for β_1:

$$n = 20, \quad \Sigma x_i = 56, \quad \Sigma x_i^2 = 524$$

$$\Sigma y_i = 40, \quad \Sigma y_i^2 = 256, \quad \Sigma x_i y_i = 364$$

11.22 A retailer of satellite dishes would like to know the impact that advertising has on his sales. For 6 months he records the number of ads run in the newspaper and the number of sales. The results are as follows:

	March	April	May	June	July	August
No. of ads	0	3	5	8	10	9
No. of sales	5	7	12	15	17	15

Obtain the least squares estimates of β_0 and β_1 for the model $y = \beta_0 + \beta_1 x + \epsilon$, and test the hypothesis that $\beta_1 = 0$.

11.23 In an attempt to cut down on traffic through the city, a free shuttle service from the suburban areas to the downtown area was provided. At the beginning very few people took advantage of the shuttle service, but by the end of 1 year a decision had to be made as to whether or not it would be beneficial to add more shuttles. Following are the numbers of people riding on the shuttle and a count of the number of automobiles in the downtown area on 15 different occasions throughout the year.

No. using shuttle	160	180	240	280	440
No. of autos downtown	2,460	2,730	2,560	2,600	2,290
No. using shuttle	370	490	620	850	840
No. of autos downtown	2,370	2,410	2,040	1,820	1,950
No. using shuttle	970	1,230	1,140	1,290	1,350
No. of autos downtown	1,870	1,460	1,330	1,390	1,250

Obtain the least squares estimates of β_0 and β_1 for the model $y = \beta_0 + \beta_1 x + \epsilon$ and test the hypothesis that $\beta_1 = 0$.

11.24 A cigarette manufacturer recorded the nicotine content and sales figures for eight major brands of cigarettes.

Nicotine (milligrams)	.86	1.38	1.67	.25	.59	1.17	1.00	.66
Sales ($\times$ \$100,000)	24	65	83	34	59	62	85	38

Obtain the least squares estimates of β_0 and β_1 for the model $y = \beta_0 + \beta_1 x + \epsilon$, and test the hypothesis that $\beta_1 = 0$.

11.25 An owner of a chicken farm would like to test a new feed supplement that is supposed to increase egg production. He randomly selects 12 groups of 100 chickens each and feeds them various levels of the feed supplement. Following are the amounts of feed supplement and the number of eggs per day.

Group	1	2	3	4	5	6
Feed supplement	10	10	10	15	15	15
No. of eggs	78	84	81	85	79	95
Group	7	8	9	10	11	12
Feed supplement	20	20	20	25	25	25
No. of eggs	98	96	89	84	93	87

With a test of hypothesis on the regression coefficient, determine whether the feed supplement is linearly related to the production of eggs.

One of the main objectives of regression analysis is to predict y for a given value of x. From the regression model,

$$y = \beta_0 + \beta_1 x + \epsilon$$

the expected y for a given x, say x', is

$$E(y) = \beta_0 + \beta_1 x'$$

This expected y is further estimated by

$$\hat{y} = b_0 + b_1 x'$$

Using $b_0 + b_1 x'$ as an estimate, it is possible to make inferences about $E(y) = \beta_0 + \beta_1 x'$.

In particular, with the basic assumptions of independent, normally distributed errors with common variance, we have

For a given x', a $(1 - \alpha)100\%$ confidence interval for the expected response $E(y)$ is given by the limits

$$b_0 + b_1 x' \pm t_{\alpha/2, n-2} s\sqrt{1/n + (x' - \overline{x})^2/SS_{xx}}.$$

The quantity

$$s\sqrt{1/n + (x' - \overline{x})^2/SS_{xx}}$$

is the standard error of the estimate of $E(y)$.

EXAMPLE 11.7

A manufacturer of thermal pane windows wished to analyze the heat loss through the windows as the outside temperature varied. In a controlled environment the heat loss through three windows was measured at four different outside temperatures. From the following data obtain the least squares estimates of β_0 and β_1 for the model

$$y = \beta_0 + \beta_1 x + \epsilon$$

Outside temp. (°C)	-10	-10	-10	0	0	0	10	10	10	20	20	20
Heat loss	58	62	54	39	41	36	22	26	20	10	13	15

Graph the data in a scatterplot and find a 95% confidence interval for the expected heat loss when the outside temperature is 10 degrees.

SOLUTION   Substituting in the formulas for b_0 and b_1, we find

$$b_1 = -1.52$$
$$b_0 = 40.6$$

so that

$$\hat{y} = b_0 + b_1 x = 40.6 - 1.52x$$

FIGURE 11.6

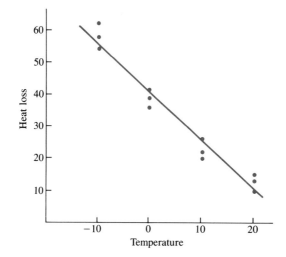

Figure 11.6 is a scatterplot of the data with the prediction line also graphed. When the outside temperature is 10 degrees, we expect the heat loss to be

$$\hat{y} = 40.6 - (1.52)(10) = 25.4$$

For a 95% confidence interval estimate of the heat loss, we add and subtract the margin of error given by

▲ Recall that the margin of error in estimating with a confidence interval is the table value times the standard error.

$$t_{\alpha/2,\,n-2}s\sqrt{1/n + (x' - \overline{x})^2/\text{SS}_{xx}}$$

From the t-table we have

$$t_{.025,\,10} = 2.228$$

Also

$$\overline{x} = 5$$
$$\text{SS}_{xx} = 1500$$
$$s = \sqrt{\text{SSE}/(n-2)} = \sqrt{142.4/10} = 3.774$$

Substituting in all known values, we have that the 95% confidence interval for the expected heat loss when the outside temperature is 10 degrees is

$$25.4 \pm (2.228)(3.774)\sqrt{1/12 + (10 - 5)^2/1500}$$

or

$$25.4 \pm 2.659$$

which gives the interval

$$(22.74, 28.06)$$

EXAMPLE 11.8

Using the same data as in Example 11.7, construct 95% confidence intervals for the expected heat loss when the outside temperature is -10, 0, 10, 20, 30, and 40 degrees. Graph the results together with the least squares regression line.

The solution is tabulated as follows:

x	ŷ	St. error	95% C.I.
−10	55.8	1.82	(51.74, 59.86)
0	40.6	1.19	(37.94, 43.26)
10	25.4	1.19	(22.74, 28.06)
20	10.2	1.82	(6.14, 14.26)
30	−5.0	2.67	(−10.95, .95)
40	−20.2	3.58	(−28.18, −12.22)

▲ Many of the details for this solution are left out. In the exercises, when you are asked to find confidence intervals for several values of x, it is best to use a computer (see Section 11.8) or else have a lot of patience.

Figure 11.7 is the graph of the regression line and the confidence limits for the various values of x. Note that the standard error is smaller when the value of x' is close to $\bar{x}\,(= 5)$. This in turn results in a narrower confidence interval. In general, we can say that predictions are more accurate in the vicinity of $\bar{x}$. The further x' is from $\bar{x}$ the greater the possibility for error when predicting y. Extreme care must be taken when predicting for values of x' beyond the original data (called extrapolation), because the linear relationship may not continue outside the range of available data.

FIGURE 11.7

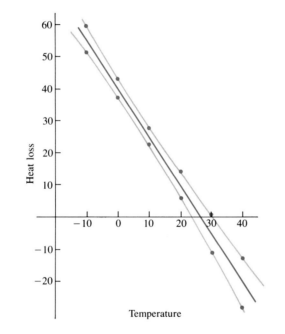

EXERCISES 11.4

11.26 From the following summary data construct the least squares equation relating x and y. Also construct 95% confidence intervals for the expected y for the specified values of x'.

(a) $n = 15$, $SS_{xx} = 265$, $SS_{yy} = 824$, $SS_{xy} = -143$, $\Sigma x_i = 195$, $\Sigma y_i = 330$ confidence intervals for $x' = 10, 12, 14, 15$.

(b) $n = 24$, $SS_{xx} = 1280$, $SS_{yy} = 2460$, $SS_{xy} = 860$, $\Sigma x_i = 480$, $\Sigma y_i = 756$ confidence intervals for $x' = 10, 15, 20, 25, 30$.

11.27 From the data at the top of page 434 construct the least squares estimates of β_0 and β_1 for the model $y = \beta_0 + \beta_1 x + \epsilon$.

x	10	10	10	20	20	20	30	30	30	40	40	40
y	16	20	14	22	26	23	28	30	25	31	35	29

Graph the data in a scatterplot and find 90% confidence intervals for the expected y for $x' = 10$, $x' = 20$, $x' = 30$, $x' = 40$, and $x' = 50$.

11.28 Refer to Exercise 11.9(a) for the least squares estimates of β_0 and β_1 in the model $y = \beta_0 + \beta_1 x + \epsilon$ obtained from the following data:

x	2	1	3	2	3	1
y	4	0	6	6	8	2

Graph the data in a scatterplot and find 95% confidence intervals for the expected y for $x' = 0$, $x' = 1$, $x' = 2$, $x' = 3$, $x' = 4$.

11.29 Refer to Exercise 11.9(b) for the least squares estimates of β_0 and β_1 in the model $y = \beta_0 + \beta_1 x + \epsilon$ obtained from the following data:

x	1	3	4	6	8	9	11	14
y	1	2	4	4	5	7	8	9

Graph the data in a scatterplot and find 95% confidence intervals for the expected y for $x' = 2$, $x' = 4$, $x' = 6$, $x' = 8$, $x' = 10$.

11.30 The following data compare a child's age with the number of gymnastic activities he or she was able to complete successfully.

Age	2	3	4	4	5	6	7	7
No. of activities	5	5	6	3	10	9	11	13

Construct the least squares estimates of β_0 and β_1 for the model $y = \beta_0 + \beta_1 x + \epsilon$. Graph the data in a scatterplot and find 95% confidence intervals for the expected y for $x' = 2$, $x' = 4$, $x' = 6$, $x' = 8$, $x' = 10$.

11.31 An automobile dealer records the number of cars sold and the prime interest rate for six 2-month periods.

Prime rate (%)	15	14	13	12	11	10
Sales (no. autos)	116	132	148	136	155	184

Construct the least squares estimates of β_0 and β_1 for the model $y = \beta_0 + \beta_1 x + \epsilon$. Graph the data in a scatterplot and find 98% confidence intervals for the expected y for $x' = 10$, $x' = 12$, $x' = 14$, $x' = 15$.

11.32 A school psychologist believes that there is a linear relationship between the verbal test scores for eighth graders and the number of library books checked out. Following are data collected on 15 students:

No. of books	12	15	3	7	10	5	22
Verbal scores	77	85	48	59	75	41	94
No. of books	9	13	7	25	17	14	19	20
Verbal scores	72	80	70	98	85	83	96	89

Construct the least squares estimates of β_0 and β_1 for the model $y = \beta_0 + \beta_1 x + \epsilon$. Graph the data in a scatterplot and find 90% confidence intervals for the expected y for $x' = 10$, $x' = 15$, $x' = 20$, $x' = 25$.

11.5 A Robust Alternative to Least Squares — The Resistant Line

In fitting a straight line to a data set, the method of least squares tries to keep the line close to every data point. Thus outliers, if present, tend to have an undue influence on the fit. In an effort to prevent outliers from distorting the analysis, we will construct the *resistant line*. In that the median is resistant to outliers, the construction of the resistant line is based on medians.

For the construction, the data are divided into three groups of approximately the same size. If the sample size n is not divisible by 3 and has a remainder of 1, the extra point is placed in the middle group. If the remainder is 2, then the two points are placed in the first and third groups. If some points have the same x value, then all must go in the same group. Thus it may not be possible to divide the sample into three groups of exactly the same size; however, one must attempt to even them out as much as possible.

After determining the groups, the median of the x data and the median of the y data within each group is found. That is, we find the (x, y) pair of medians in each of

FIGURE 11.8
Location of summary points

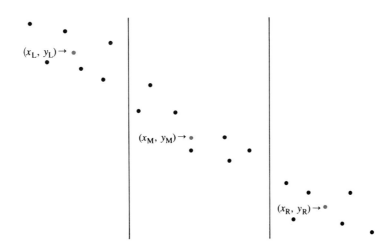

the three groups. They summarize the behavior of the data in their respective groups. We will label these *summary points* as (x_L, y_L), (x_M, y_M), and (x_R, y_R) for the left, middle, and right groups respectively. Figure 11.8 illustrates the dividing lines of the three groups and the location of the summary points.

The equation of the resistant line can be written as

$$\hat{y} = b_0 + b_1 x$$

where b_1 is the slope and b_0 is the y-intercept. Having found the summary points, the line can be constructed with the same slope as a line passing through the left and right median points, by computing the slope by the equation

▲ Often the resistant line can be drawn on a scatterplot by simply eyeballing the location.

$$b_1 = \frac{y_R - y_L}{x_R - x_L}$$

A line can be adjusted up or down by changing the y-intercept. To force the line through the point (x_0, y_0), the y-intercept is computed as

$$b_0 = y_0 - b_1 x_0$$

For the resistant line, however, we will compute the y-intercept as an average of all three summary points by computing

$$b_0 = \tfrac{1}{3}[(y_L + y_M + y_R) - b_1(x_L + x_M + x_R)]$$

EXAMPLE 11.9

Construct the resistant line for the following data set:

x	5.2	7.7	10.8	12.5	13.3	14.1	12.0	16.4	15.3	17.4
y	2.7	18.2	28.9	27.7	24.6	30.9	35.4	28.8	39.1	33.6

x	17.7	18.4	19.6	2.6	11.6	8.3
y	44.3	39.8	36.4	20.4	24.2	24.4

SOLUTION   First we will order the data according to the x variable and then find the medians of the three groups for the x data and for the y data.

x	y
2.6	20.4 ←
5.2	2.7
→ 7.7	18.2
8.3	24.4
10.8	28.9

$(x_L, y_L) = (7.7, 20.4)$

x	y
11.6	24.2
12.0	35.4
→ 12.5	27.7 ←
→ 13.3	24.6
14.1	30.9 ←
15.3	39.1

12.9 29.3 $(x_M, y_M) = (12.9, 29.3)$

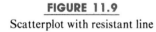
▲ There are six points in the middle group; therefore, the median is the average of the third and fourth smallest observations.

x	y
16.4	28.8
17.4	33.6
→ 17.7	44.3
18.4	39.8
19.6	36.4 ←

$(x_R, y_R) = (17.7, 36.4)$

From the formulas for b_1 and b_0, we have

$$b_1 = \frac{y_R - y_L}{x_R - x_L} = \frac{36.4 - 20.4}{17.7 - 7.7} = 16.0/10 = 1.6$$

and

$$b_0 = \tfrac{1}{3}[(y_L + y_M + y_R) - b_1(x_L + x_M + x_R)]$$
$$= \tfrac{1}{3}[(20.4 + 29.3 + 36.4) - 1.6(7.7 + 12.9 + 17.7)]$$
$$= 8.27$$

FIGURE 11.9
Scatterplot with resistant line

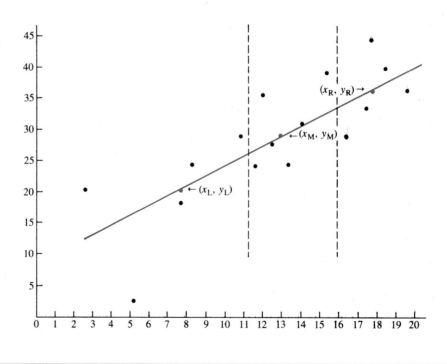

Therefore the resistant line is given by

$$\hat{y} = 8.27 + 1.6x$$

Figure 11.9 gives the scatterplot with the summary points and the resistant line graphed.

In comparison the least squares line for the data in Example 11.9 can be found by the following summary data:

$$\Sigma x = 202.9, \quad \Sigma x^2 = 2930.95, \quad \Sigma xy = 6421.56,$$
$$\Sigma y = 459.4, \quad \Sigma y^2 = 14710.02$$

We find that the equation of the least squares line is

$$\hat{y} = 7.6 + 1.66x$$

(As a point of interest the correlation is

$$r = .8079)$$

We see that the regression line has a larger slope and a lower y-intercept, which is due to the outlier at (5.2, 2.7). The least squares equation is pulled toward the point, whereas the point has no effect on the resistant line.

Residuals

The residuals for the resistant line are computed in the same way as the residuals for the least squares line. That is, the residual for the data point (x_i, y_i) is

$$r_i = y_i - \hat{y}_i$$

where $\hat{y}_i$ is the predicted y from the resistant line.

The purpose of investigating residuals is to detect patterns that were not explained by the regression line. In the case of the resistant line, the residuals can be used to "polish" the line so that it will better fit the data. For a thorough discussion of polishing the fit, see Velleman and Hoaglin (1981).

In the next section we will see how to perform an examination of the residuals. Much of what is presented is directed toward statistical inference and, consequently, applies to the least squares regression line. Presently, there are no inference procedures for the resistant line. If the analysis of residuals suggest that there are problems with the least squares fit, however, it may be that the resistant line will provide a better description of the relationship between the variables.

EXERCISES 11.5

11.33 Construct the resistant line for the following data set:

x	2.4	3.5	4.7	5.3	6.4	7.2	8.1
y	−14.2	−4.3	−2.9	−1.3	2.4	3.5	4.4

x	9.3	10.6	12.1	13.3	14.5
y	5.7	6.5	8.2	9.3	16.3

11.34 Construct the resistant line for the following data set:

x	1.2	1.4	1.6	1.8	2.0	2.2	2.4	2.6	2.8	3.0
y	.44	.65	.63	.71	.64	.83	.81	.88	.85	2.32

11.35 Construct the resistant line for the following data set:

x	120	140	150	160	170	180	190	200	210	220	230
y	32	30	24	21	18	15	33	11	8	6	4

11.36 In Exercise 11.5, data relating aptitude test scores to productivity in a factory were given. Use the same data (reproduced here) to construct the resistant line.

Applicant	1	2	3	4	5	6	7	8
Aptitude score	9	17	13	19	20	23	12	15
Productivity	23	35	29	33	40	38	25	31

Compare your results with the least squares solution found in Exercise 11.15.

11.37 In Exercise 11.6, data relating robbery rate and the percent of low income were given. Use the same data (reproduced here) to construct the resistant line.

Precinct	1	2	3	4	5	6	7	8
% low income	4.9	7.1	10.1	11.8	13.5	14.8	16.2	11.2
Robbery rate	20	41	165	88	60	120	65	81

Compare your results with the least squares solution found in Exercise 11.16.

11.38 In Exercise 11.7, data corresponding to maintenance cost on nine cash registers and their ages were recorded as follows:

Age (years)	6	7	1	3	6	2	5	4	3
Cost (dollars)	92	181	23	40	126	35	86	72	51

Construct the resistant line and compare with the least squares solution found in Exercise 11.17.

11.39 In Exercise 11.8, the data corresponding to the life span of a particular electronic component and the heat it is exposed to are duplicated here:

Heat (°C)	50	100	150	200	250	300
Life span (hours)	875	884	762	424	365	128

Construct the resistant line and compare with the least squares solution found in Exercise 11.18.

11.6　Examination of the Residuals

In the previous sections, we learned how to find the estimates of β_0 and β_1 in the model

$$y = \beta_0 + \beta_1 x + \epsilon$$

Further in the least squares case, we learned how to make inferences about β_1 and the $E(y)$. We should be cautioned, however, that the conclusions drawn by the inference procedures can be seriously misleading if the assumptions made about the model are inadequate. To review, the assumptions are

1. y is related to x by the linear relationship,

$$y = \beta_0 + \beta_1 x + \epsilon$$

2. The ϵ's associated with all elements in the population are independent.

3. The ϵ's are normally distributed with mean 0 and constant standard deviation σ.

A regression study is not complete without an investigation of these assumptions.

As previously noted, the scatterplot can be used to visually check the linearity assumption by observing if the data exhibit a straight line behavior. Pearson's correlation gives a measure of the degree of linearity and r^2, the coefficient of determination, measures the amount of variability in y explained by the independent variable x. The remaining variability that cannot be explained by the model is said to be explained by error and is contained in the residuals. We now present ways to evaluate the residuals.

▲ The coefficient of determination was defined in Section 3.5.

In Section 11.2 the residual associated with data point (x_i, y_i) is defined as

$$e_i = y_i - \hat{y}_i$$

where $\hat{y}_i$ is the predicted y_i. If the assumptions are reasonable, then the e_i's should appear to have come from a normal population with mean 0 and standard deviation σ. It may be convenient to investigate the *standardized residuals*, which are found by e_i/s where $s = \sqrt{\text{SSE}/(n-2)}$ is an estimate of σ. Under the normality assumption the standardized residuals should appear to have come from a standard normal distribution. As such, most of the residuals should be between -2 and $+2$. In most computer programs for linear regression, the standardized residuals are printed out, so it is an easy matter to see if they lie between -2 and $+2$.

▲ Recall that the standard normal distribution has mean 0 and standard deviation 1.

Graphically, the distributional shape of the residuals can be examined in much the same way that raw data were examined in Chapter 6. By using the histogram, stem and leaf plot, box plot, or a simple dot diagram, we can observe whether the data exhibit a normal curve appearance.

As in Chapter 6, the midsummaries can be used to check for symmetry (if the residuals are not symmetrical, then clearly they are not normally distributed) and the pseudostandard deviations can be used to check for normality. Also the box plot can be used to check for outliers and long tails. As might be expected, outliers tend to have an adverse effect on the least squares regression. Not only do they affect the precision of the inference procedure, but they also affect the actual equation of the straight line. Because the least squares procedure minimizes the sum of the squares of the distances of all points to the line, an outlier will tend to draw the line toward it in an unusual fashion. In this situation, a fitting technique such as a *resistant line* is needed. As we saw in Section 11.5, the resistant line tends to reduce, if not eliminate, the effect of outliers.

EXAMPLE 11.10

Table 11.5 lists the research and development (R&D) expenditures and the corresponding sales for a large company.

TABLE 11.5

R&D (in $1 million)	Sales (in $1 million)
1.2	55
2.4	48
3.1	32
4.0	21
4.9	10
6.3	41
1.8	43
2.7	44
3.3	21
3.7	32
4.2	12
5.4	8

Determine whether a linear relationship between the two variables seems plausible.

SOLUTION   The least squares equation relating R&D (x) with sales (y) is

$$\hat{y} = 55.008 - 6.816x$$

The correlation is $r = -.634$, which gives $r^2 = .4025$. Thus approximately 40% of the variability in sales is explained by the R&D expenditures.

Using the least squares equation, we find the predicted y_i's and then the residuals as shown in Table 11.6

TABLE 11.6

x_i	y_i	$\hat{y}_i$	e_i
1.2	55	46.829	8.171
2.4	48	38.649	9.351
3.1	32	33.878	-1.878
4.0	21	27.743	-6.743
4.9	10	21.609	-11.609
6.3	41	12.066	28.934
1.8	43	42.739	.261
2.7	44	36.604	7.396
3.3	21	32.515	-11.515
3.7	32	29.788	2.212
4.2	12	26.380	-14.380
5.4	8	18.201	-10.201

The residuals seem reasonable except for the one large residual corresponding to (6.3, 41). From the scatterplot with the equation graphed (see Figure 11.10) we see that (6.3, 41) is an outlier, which explains the large residual.

▲ Check that the equation for the resistant line is $\hat{y} = 68.95 - 11.48x$

From the scatterplot we can easily fit the resistant line to the median of the three groups, and see that it is a much more reasonable fit than the least squares line.

FIGURE 11.10

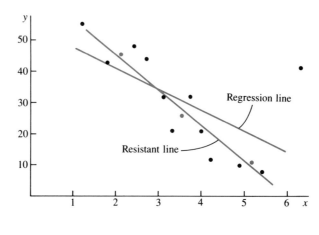

Residual Plot Against Predicted Value

A plot of residuals against the predicted values, $\hat{y}_i$, amounts to an examination of y after removing the linear dependence on x. Figure 11.11 gives three patterns of residual plots that might come up.

FIGURE 11.11
Possible residual plots

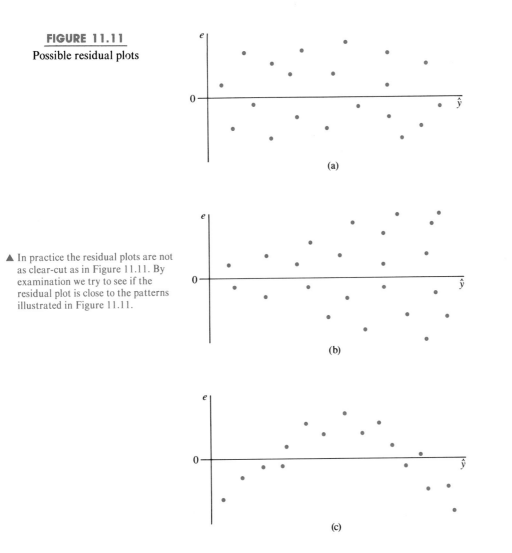

(a)

(b)

(c)

▲ In practice the residual plots are not as clear-cut as in Figure 11.11. By examination we try to see if the residual plot is close to the patterns illustrated in Figure 11.11.

In Figure 11.11(a) the points form a horizontal band about zero as would be expected when there are no abnormalities. In Figure 11.11(b), the width of the band increases as the predicted y increases. This indicates that the variance is not constant. A transformation of data might be in order. In Figure 11.11(c), the residuals exhibit a curvilinear pattern, which suggests that a nonlinear x term should be considered in the model.

In the case of the multiple regression analysis (several independent variables) to be studied in the next section, the residuals can be plotted against each of the independent variables. With a single independent variable, however, the residual plot against the independent variable will give results similar to the plot against the predicted value of y.

EXAMPLE 11.11

In an effort to determine the effective duration of a tranquilizer for animals, the concentration of the substance in blood samples taken at various intervals of time after the injection is measured. From the following data determine the least squares equation relating the concentration to the elapsed time after the injection, and conduct a residual analysis.

Elapsed time (hours)	1	2	3	6	12	18
Concentration (mg/ml)	1.8	1.4	1.2	.9	.5	.1

SOLUTION   From the sample data we find the least squares equation to be

$$y = 1.6052 - .08884x$$

The correlation is $r = -.9625$, which suggests a strong linear relationship. Figure 11.12 is a scatterplot with the graphed equation.

FIGURE 11.12

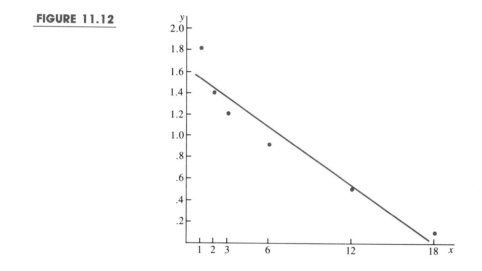

Using the preceding equation, we find the predicted values of y and the residuals as follows:

x_i	y_i	$\hat{y}_i$	e_i
1	1.8	1.5164	.2836
2	1.4	1.4275	-.0275
3	1.2	1.3387	-.1387
6	.9	1.0722	-.1722
12	.5	.5391	-.0391
18	.1	.0061	.0939

Figure 11.13 is a graph of the residuals against the independent variable. The residual plot shows a definite curvilinear pattern suggesting that a higher-ordered x term is needed in the model.

FIGURE 11.13

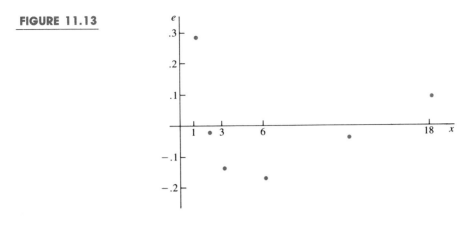

Residual Plot Against Time Often data for a regression study are collected sequentially over a period of time. The observations may be correlated over time, in which case the residuals will be dependent. A plot with time on the horizontal axis and the residuals on the vertical axis will often detect violations of the assumption of independence of the residuals. Figure 11.14 shows a definite relationship between the residuals and time. Clearly time should be incorporated in the model.

FIGURE 11.14
Residual plot against time

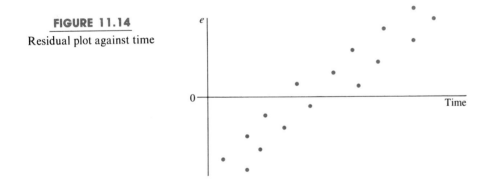

A discussion of *time series models* can be found in Box and Jenkins (1970) or in Box, Hunter, and Hunter (1978).

EXERCISES 11.6

11.40 Following is a stem and leaf plot of residuals:

```
-3 | 05
-2 | 22,  51,  81
-1 | 04,  19,  31,  56,  71
-0 | 02,  15,  24,  48,  61,  75,  93
 0 | 12,  34,  47,  63,  74,  92,  94
 1 | 34,  66,  82
 2 | 21,  47
 3 | 18,  74
```

Do the residuals appear to deviate from the assumptions?

11.41 The figure at the top of page 444 is a residual plot against the independent variable.

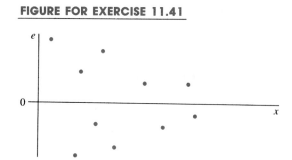

Does it appear that the assumptions have been violated?

11.42 The accompanying figure is a residual plot against time.

What observations can be made about the model?

11.43 From the following data construct the least squares model:

x	1	1	1	2	2	2	3	3	3
y	15	5	3	10	8	3	6	5	4

From a residual plot does it appear that any assumptions have been violated?

11.44 From the following data construct the least squares model:

x	1.2	2.3	3.2	3.9	4.6	6.5	9.8	15.5
y	1.3	2.6	4.8	5.9	7.1	12.1	16.3	24.5

From a residual plot does it appear that any assumptions have been violated?

11.45 Conduct a residual analysis for the least squares solution given in Exercise 11.30. The data, which are repeated here, compare a child's age to the number of gymnastic activities he or she was able to complete successfully.

Age	2	3	4	4	5	6	7	7
No. of activities	5	5	6	3	10	9	11	13

11.46 Conduct a residual analysis for the least squares solution given in Exercise 11.31. The data, which are repeated here, compare an automobile dealer's record of the number of cars sold and the prime interest rate for six 2-month periods.

Prime rate (%)	15	14	13	12	11	10
Sales (no. autos)	116	132	148	136	155	184

11.47 Conduct a residual analysis for the least squares solution given in Exercise 11.32. The data, which are repeated here, give the verbal test scores for eighth graders and the number of library books they checked out.

No. of books	12	15	3	7	10	5	22	9
Verbal scores	77	85	48	59	75	41	94	72
No. of books	13	7	25	17	14	19	20	
Verbal scores	80	70	98	85	83	96	89	

11.7 Multiple Regression

In many regression problems it may be that the dependent variable y is dependent on more than one independent variable. For example, one's blood pressure is dependent on his or her age, weight, and physical condition. To obtain a meaningful prediction model, all variables that might affect the dependent variable should be measured. Then the variables $x_1, x_2, x_3, \ldots, x_k$ that have a significant effect on the dependent variable y can be incorporated into a regression model of the form

$$y = \beta_0 + \beta_1 x_1 + \beta_2 x_2 + \cdots + \beta_k x_k + \epsilon$$

The equation is called a *multiple linear regression model.*

Data are collected on variables $y, x_1, x_2, \ldots, x_k$ and used to find the least squares estimates $b_0, b_1, b_2, \ldots, b_k$ of the parameters $\beta_0, \beta_1, \beta_2, \ldots, \beta_k$, respectively. The prediction equation is

$$y = b_0 + b_1 x_1 + b_2 x_2 + \cdots + b_k x_k$$

There are techniques for confidence interval estimation and hypothesis testing for the multiple regression coefficients, but the formulas are more difficult to work with than those in the simple case of a single independent variable. However a multiple regression analysis is easily performed with the aid of the standard statistical packages such as MINITAB, BMDP, SPSS, and SAS. We will illustrate the multiple regression in the following example.

[handwritten margin notes:]

MULTIPLE REGRESSION:

① DEPENDENT

② CONTINUOUS NOT DISCRETE

③ AT LEAST INTERVAL IN SCALE (OR RATIO)

④ ERROR TERMS ARE NORMALLY DISTRIBUTED WITH A MEAN = 0

⑤ HOMOSCEDASTICITY => COMMON DISPERSION OF Y ABOUT (COMMON VARIANCE) THE PREDICTION LINE ($\hat{y}$).

HETEROSCEDASTICITY => OPPOSITE OF HOMOSCEDASTICITY

⑥ EXPECT EQUAL VARIANCES OF X (NOT COMMON) AT EVERY POINT ON GRAPH

EXAMPLE 11.12

⑦ ERRORS ARE ALL INDEPENDENT OF EACH OTHER

To study the effects of age and weight on the systolic blood pressure of adult males the data in Table 11.7 were recorded from a sample of 15 adult males.

[handwritten margin notes:]

MULTICOLINEARITY = WHEN X_1 IS LINEARLY RELATED TO X_2 AS WELL AS THE DEPENDENT VARIABLE Y. (SEVERAL LINEAR RELATIONSHIPS!)

TABLE 11.7

Age (x_1)	Weight (x_2)	Blood pressure (y)
48	175	143
50	159	131
35	191	135
41	174	131
33	165	121
25	157	115
51	182	133
53	164	126
40	157	120
31	168	128
35	138	120
46	191	160
28	155	124
32	171	143
44	164	128

Perform a multiple regression analysis, using Age and Weight as independent variables and Blood pressure as the dependent variable.

SOLUTION To perform the analysis with Minitab, the data for x_1, x_2, and y must be stored into C1, C2, and C3 with the following command:

[IN THE BI-VARIATE CASE: $r^2 = R^2$.]

R^2 = COEFFICIENT OF MULTIPLE DETERMINATION

IS A DETERMINATION OF GOODNESS OF FIT, THE ADEQUACY OF A SET OF DATA TO FIT A STRAIGHT LINE.

```
READ DATA IN C1-C3

48  175  143
50  159  131
35  191  135
  .
  .
  .
 etc.
  .
  .
  .
```

After the data are stored, the analysis is begun by the command

```
REGRESS Y IN C3 ON 2 PREDICTORS IN C1 AND C2
```

The commands and the output of the program are shown in Figure 11.15. The BRIEF command controls the amount of output from the REGRESS command. BRIEF = 3 gives a complete output.

We see that the fitted regression equation is

$$\hat{y} = 27.5 + .239 \text{ Age} + .559 \text{ Weight}$$

FIGURE 11.15
Minitab output for Example 11.12

```
MTB > BRIEF 3
MTB > REGRESS C3 ON 2 PREDICTORS IN C1 AND C2

The regression equation is
C3 = 27.5 + 0.239 C1 + 0.559 C2

Predictor        Coef        Stdev     t-ratio
Constant        27.45        25.39        1.08
C1             0.2392       0.2502        0.96
C2             0.5594       0.1586        3.53

s = 7.946      R-sq = 58.7%      R-sq(adj) = 51.8%

Analysis of Variance

SOURCE        DF         SS          MS
Regression     2    1077.97      538.99
Error         12     757.76       63.15
Total         14    1835.73

SOURCE        DF     SEQ SS
C1             1     292.49
C2             1     785.48
```

```
Obs.       C1        C3       Fit  Stdev.Fit   Residual    St.Resid
 1        48.0     143.00   136.83     2.94       6.17        0.84
 2        50.0     131.00   128.35     3.89       2.65        0.38
 3        35.0     135.00   142.67     4.70      -7.67       -1.20
 4        41.0     131.00   134.59     2.28      -3.59       -0.47
 5        33.0     121.00   127.64     2.57      -6.64       -0.88
 6        25.0     115.00   121.26     4.04      -6.26       -0.91
 7        51.0     133.00   141.46     3.71      -8.46       -1.20
 8        53.0     126.00   131.87     4.13      -5.87       -0.86
 9        40.0     120.00   124.84     2.66      -4.84       -0.65
10        31.0     128.00   128.84     2.97      -0.84       -0.11
11        35.0     120.00   113.02     4.90       6.98        1.12
12        46.0     160.00   145.30     4.14      14.70        2.17R
13        28.0     124.00   120.85     3.58       3.15        0.44
14        32.0     143.00   130.76     2.95      12.24        1.66
15        44.0     128.00   129.72     2.48      -1.72       -0.23
```

R denotes an obs. with a large st. resid.

▲ Given an estimate and its standard error, you can test hypotheses and find confidence intervals.

The estimated coefficients and their standard errors are

$$b_0 = 27.45 \qquad \text{SE}(b_0) = 25.39$$
$$b_1 = 0.2392 \qquad \text{SE}(b_1) = 0.2502$$
$$b_2 = 0.5594 \qquad \text{SE}(b_2) = 0.1586$$

The value of $s = 7.946$ is an estimate of σ, the standard deviation associated with ϵ. The number of degrees of freedom associated with s is given by

$$\text{df} = n - (\text{no. of independent variables}) - 1$$
$$= n - 2 - 1$$
$$= n - 3$$
$$= 15 - 3$$
$$= 12$$

Assuming that the ϵ's are independent and normally distributed about 0 with a standard deviation of σ, interval estimation and hypothesis testing about the regression coefficients β_0, β_1, and β_2 can be performed. In particular, a 95% confidence interval for β_1 is given by

$$b_1 \pm t_{\alpha/2, n-3}\text{SE}(b_1)$$

From the data in the printout, we have

$$.2392 \pm t_{.025,12}(0.2502)$$
$$.2392 \pm (2.179)(.2502)$$
$$.2392 \pm .5452$$

which gives the interval

$$(-.306, .7844)$$

Note that the interval contains the value of 0. This means that the coefficient is insignificant. The same results can be found by testing the hypothesis that $\beta_1 = 0$. We will illustrate the test on β_2.

To test $H_0: \beta_2 = 0$ vs. $H_a: \beta_2 > 0$ we use the t-ratio

$$t = \frac{b_2}{\text{SE}(b_2)}$$

as the test statistic. Its significance is evaluated by comparing to the tail probabilities associated with a t distribution with $n - 3$ degrees of freedom. From the printout we see that the observed value of the test statistic is

$$t_{\text{obs}} = 3.53$$

From the t distribution table with $n - 3 = 12$ degrees of freedom, we find that $t_{.005,12} = 3.055$. Because

$$t_{\text{obs}} = 3.53 > 3.055$$

we have that p-value $< .005$. Therefore there is significant evidence to reject H_0 and conclude that the blood pressure of adult males is related to their weight. Because the alternative is $\beta_2 > 0$, the blood pressure increases as weight increases.

From the computer printout we see a quantity called R-squared. The R-squared, called the *multiple coefficient of determination,* is analogous to r^2, the coefficient of determination in the simple linear regression case. That is, it is interpreted as being the percent of variability in the dependent variable that is explained by the regression model. In this example,

$$R\text{-squared} = 58.7\%$$

so that 58.7% of the variability in blood pressure is expained by age and weight of the adult male. The remaining 41.3% is explained by other factors.

The Analysis of Variance table gives the total variability associated with the dependent variable, SS_{yy}, and shows how it is divided between the regression model and the residual (error). That is, we have the total variability of y being

$$\text{SS}_{yy} = 1835.73$$

of which

$$\text{SS}_{\text{reg}} = 1077.97$$

is due to the regression model and

$$\text{SS}_{\text{res}} = \text{SSE} = 757.76$$

is due to the residual or error. Thus the percent explained by the regression model is

$$R\text{-squared} = \text{SS}_{\text{reg}}/\text{SS}_{yy}$$
$$= 1077.97/1835.73$$
$$= .5872$$

Note that there is a total of $n - 1 = 14$ degrees of freedom, 2 of which belong to regression and the remaining 12 belong to residual. It is interesting to note that

$$s = 7.946$$

can be found by

$$s = \sqrt{\text{MS for residual}}$$
$$= \sqrt{63.15}$$
$$= 7.9467 \quad \text{(the slight difference is due to round-off error)}$$

which is the estimate of the standard deviation σ.

Polynomial Regression

In Figure 11.11(c) the residual pattern suggests that a nonlinear x term should be considered in the model. This can easily be done through the multiple regression model,

$$y = \beta_0 + \beta_1 x_1 + \beta_2 x_2 + \epsilon$$

We simply associate the variable x with x_1 and its square, x^2, with x_2. So the *second-degree polynomial model*

▲ The degree is the highest exponent in the equation.

$$y = \beta_0 + \beta_1 x + \beta_2 x^2 + \epsilon$$

is a special case of the multiple regression model. To analyze the data with Minitab all that is required is to read the data for y and x in the computer, perform the squared transformation on x and store the results in a third column, and then regress y on the two predictors x and x^2.

EXAMPLE 11.13

In Example 3.20 a very high linear correlation (.9775) between the growth stimulant and weight gain in lab animals was found. The scatterplot also shows the linear relationship. To further study the experiment, additional data were taken. In addition, the weight gains for doses 7, 8, 9, and 10 were measured as follows:

Dosage	0	1	2	3	4	5	6	7	8	9	10
Weight gain	1.0	1.2	2.0	2.4	3.4	4.9	5.1	4.7	3.5	2.5	1.8

With the additional data the correlation between dosage and weight gain is only .410. Construct a scatterplot of the data (see Figure 11.16) and suggest an alternative method of analysis of the data.

FIGURE 11.16

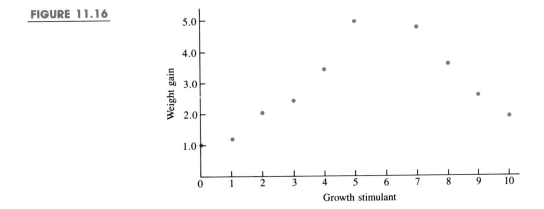

SOLUTION   The scatterplot reveals the reason for the drastic drop in correlation when additional data are measured. Clearly, the data are not exhibiting a linear relationship as they did when only doses 0 through 6 were measured. We see that when the dose increases from 6 to 10, the weight gain begins to fall off. The scatterplot also suggests that a quadratic term (second-degree) should be considered in the model. The Minitab solution shown in Figure 11.17 gives the least squares estimates of β_0, β_1, and β_2 in the second-degree polynomial model

$$y = \beta_0 + \beta_1 x + \beta_2^2 + \epsilon$$

FIGURE 11.17

Minitab output for Example 11.13

```
MTB > SET DOSAGE IN C1
DATA> 0 1 2 3 4 5 6 7 8 9 10
DATA> END

MTB > SET WEIGHT IN C2
DATA> 1.0 1.2 2.0 2.4 3.4 4.9 5.1 4.7 3.5 2.5 1.8
DATA> END

MTB > CORR C1 C2

Correlation of C1 and C2 = 0.410

MTB > MULT C1 C1 C3

MTB > REGRESS C2 2 C1 C3

The regression equation is
C2 = 0.133 + 1.46 C1 - 0.128 C3


Predictor          Coef          Stdev        t-ratio
Constant         0.1329        0.5388           0.25
C1               1.4569        0.2507           5.81
C3              -0.12751       0.02415         -5.28

s = 0.7073      R-sq = 81.5%      R-sq(adj) = 76.8%

Analysis of Variance

SOURCE         DF            SS             MS
Regression      2       17.5855         8.7928
Error           8        4.0018         0.5002
Total          10       21.5873

SOURCE         DF        SEQ SS
C1              1        3.6364
C3              1       13.9491
```

Note that for this model the coefficient of determination is

$$R\text{-squared} = 81.5\%$$

meaning that 81.5% of the variability in weight gain is explained by x and x^2. The coefficient of determination for the simple linear model

$$y = \beta_0 + \beta_1 x + \epsilon$$

is only

$$r^2 = (.410)^2 = .168 = 16.8\%$$

We see that the addition of x^2 to the model gives a reasonable explanation of the weight gain.

If necessary, an x^3 term can be added to the model. Polynomials with any power of x can be fitted in the same way; using more than x, x^2, or x^3, however, runs the risk of fitting a curve to random fluctuations in the scatterplot.

EXERCISES 11.7

11.48 A sample of $n = 25$ data points was used to find the least squares estimates of β_0, β_1, β_2, and β_3 in the multiple regression model:

$$y = \beta_0 + \beta_1 x_1 + \beta_2 x_2 + \beta_3 x_3 + \epsilon$$

From the data it was found that

$$b_0 = 5.6, \quad b_1 = -3.1, \quad b_2 = 4.9, \quad b_3 = -1.1$$

with

$$SE(b_1) = 2.67, \quad SE(b_2) = 1.34, \quad SE(b_3) = .48$$

(a) Use the least squares equation to estimate y when $x_1 = 5$, $x_2 = .35$, and $x_3 = 1.4$.
(b) Find a 95% confidence interval for β_1.
(c) Test the hypothesis that $\beta_3 = 0$ against the alternative $\beta_3 < 0$.

11.49 A sample of $n = 20$ data points was used to find the least squares estimates of β_0, β_1, and β_2 in the second-degree polynomial model

$$y = \beta_0 + \beta_1 x + \beta_2 x^2 + \epsilon$$

The least squares estimates were

$$b_0 = 6.8, \quad b_1 = 2.5, \quad \text{and} \quad b_2 = 3.8$$

with

$$SE(b_1) = .94 \quad \text{and} \quad SE(b_2) = 2.65$$

(a) Determine whether the second-degree term belongs in the model.

(b) Determine whether the first-degree term belongs in the model.

11.50 A study was conducted to evaluate several factors on teachers' attitudes toward extracurricular activities. Two variables considered important are the years of experience and the size of school. Give the form of the linear regression model that relates the teachers' attitudes to years of experience and size of school.

11.51 In an effort to study the effects of different doses of a drug on the pulse rate of human subjects, three dose levels were given to 15 subjects—5 subjects to each dose level. It is believed that a second-degree polynomial model is needed. Give the form of the linear regression model that relates pulse rate to drug dose.

11.52 In a poverty area a study was conducted to investigate the effects of education level and years living in the region on family income. Give the form of the linear regression model that relates family income to education level and years residing in the region.

11.53 A computer printout associated with the problem described in Exercise 11.50 follows.

(a) How many independent variables are there?
(b) What is the regression equation?
(c) What is the value of R-squared?
(d) Give an estimate of σ.
(e) Construct a 98% confidence interval for β_2.
(f) Test the hypothesis that $\beta_1 = 0$.
(g) Make a comment about the results.

Minitab output

```
MTB > READ ATTITUDE, YEARS, SIZE IN C1 C2 C3
DATA> 72  8 73
DATA> 81  7 61
DATA> 65 12 84
DATA> 94  4 38
DATA> 87  3 52
DATA> 89  7 57
DATA> 40  2 103
DATA> 98 10 41
DATA> 86 12 50
DATA> 80  3 65
DATA> 75  7 72
DATA> 79  4 74
DATA> 90  8 46
DATA> 85  6 62
DATA> 60 15 78
DATA> END
      15 ROWS READ


MTB > REGRESS C1 2 C2 C3


The regression equation is
C1 = 132 - 0.232 C2 - 0.810 C3


Predictor       Coef        Stdev      t-ratio
Constant      132.032       5.620       23.49
C2             -0.2322      0.3468       -0.67
C3             -0.81005     0.07494     -10.81


s = 4.906      R-sq = 90.7%     R-sq(adj) = 89.1%

Analysis of Variance

SOURCE        DF          SS           MS
Regression     2        2814.1       1407.0
Error         12         288.8         24.1
Total         14        3102.9


SOURCE        DF        SEQ SS
C2             1           1.5
C3             1        2812.6


Obs.     C2          C1      Fit Stdev.Fit  Residual   St.Resid
  1      8.0       72.00    71.04    1.48      0.96       0.20
  2      7.0       81.00    80.99    1.29      0.01       0.00
  3     12.0       65.00    61.20    2.62      3.80       0.92
  4      4.0       94.00   100.32    2.59     -6.32      -1.52
  5      3.0       87.00    89.21    2.14     -2.21      -0.50
```

6	7.0	89.00	84.23	1.37	4.77	1.01
7	2.0	40.00	48.13	3.62	-8.13	-2.46R
8	10.0	98.00	96.50	2.31	1.50	0.35
9	12.0	86.00	88.74	2.30	-2.74	-0.63
10	3.0	80.00	78.68	1.93	1.32	0.29
11	7.0	75.00	72.08	1.41	2.92	0.62
12	4.0	79.00	71.16	1.83	7.84	1.72
13	8.0	90.00	92.91	1.85	-2.91	-0.64
14	6.0	85.00	80.42	1.34	4.58	0.97
15	15.0	60.00	65.37	3.21	-5.37	-1.45

R denotes an obs. with a large st. resid.

11.54 A computer printout associated with the problem described in Exercise 11.51 follows.

FIGURE FOR EXERCISE 11.54

Minitab output

```
MTB > REGRESS C2 1 C1

The regression equation is
C2 = 58.3 + 5.82 C1

Predictor        Coef        Stdev      t-ratio
Constant       58.250       2.936        19.84
C1              5.825       1.072         5.43

s = 4.795       R-sq = 67.8%      R-sq(adj) = 65.5%

Analysis of Variance

SOURCE         DF          SS           MS
Regression      1        678.61       678.61
Error          14        321.83        22.99
Total          15       1000.44

Obs.     C1          C2       Fit Stdev.Fit   Residual    St.Resid
  1     1.00       60.00     64.07     2.01      -4.07      -0.94
  2     1.00       58.00     64.07     2.01      -6.07      -1.39
  3     1.00       65.00     64.07     2.01       0.93       0.21
  4     1.00       62.00     64.07     2.01      -2.07      -0.48
  5     2.00       72.00     69.90     1.31       2.10       0.46
  6     2.00       71.00     69.90     1.31       1.10       0.24
  7     2.00       69.00     69.90     1.31      -0.90      -0.20
  8     2.00       75.00     69.90     1.31       5.10       1.11
  9     3.00       81.00     75.72     1.31       5.28       1.14
 10     3.00       74.00     75.72     1.31      -1.72      -0.37
 11     3.00       84.00     75.72     1.31       8.28       1.79
```

12	3.00	83.00	75.72	1.31	7.28	1.58
13	4.00	74.00	81.55	2.01	-7.55	-1.73
14	4.00	79.00	81.55	2.01	-2.55	-0.59
15	4.00	78.00	81.55	2.01	-3.55	-0.82
16	4.00	80.00	81.55	2.01	-1.55	-0.36

(a) What is the regression equation?

(b) What is the value of R-squared?

(c) Graph the standardized residuals against x.

(d) Should a higher order of x appear in the model?

11.55 A computer printout associated with Exercise 11.51 and that is a continuation of Exercise 11.54 follows.

(a) What is the regression equation?

(b) What is the value of R-squared?

(c) Construct a 90% confidence interval for β_1.

(d) Test the hypothesis that $\beta_2 = 0$.

(e) After viewing the standardized residuals, does it appear that any assumptions have been violated?

FIGURE FOR EXERCISE 11.55
Minitab output

```
MTB > MULT C1 C1 C3

MTB > REGRESS C2 2 C1 C3

The regression equation is
C2 = 41.7 + 22.4 C1 - 3.31 C3

Predictor        Coef         Stdev      t-ratio
Constant        41.687        4.669        8.93
C1              22.388        4.259        5.26
C3              -3.3125       0.8386      -3.95

s = 3.354       R-sq = 85.4%     R-sq(adj) =  83.1%

Analysis of Variance

SOURCE        DF           SS           MS
Regression     2         854.17       427.09
Error         13         146.26        11.25
Total         15        1000.44

SOURCE        DF         SEQ SS
C1             1         678.61
C3             1         175.56

Obs.      C1          C2        Fit Stdev.Fit   Residual    St.Resid
  1      1.00      60.000      60.763    1.635      -0.763      -0.26
  2      1.00      58.000      60.763    1.635      -2.763      -0.94
```

3	1.00	65.000	60.763	1.635	4.237	1.45
4	1.00	62.000	60.763	1.635	1.237	0.42
5	2.00	72.000	73.213	1.244	-1.213	-0.39
6	2.00	71.000	73.213	1.244	-2.213	-0.71
7	2.00	69.000	73.213	1.244	-4.213	-1.35
8	2.00	75.000	73.213	1.244	1.787	0.57
9	3.00	81.000	79.037	1.244	1.963	0.63
10	3.00	74.000	79.037	1.244	-5.037	-1.62
11	3.00	84.000	79.037	1.244	4.963	1.59
12	3.00	83.000	79.037	1.244	3.963	1.27
13	4.00	74.000	78.238	1.635	-4.238	-1.45
14	4.00	79.000	78.238	1.635	0.762	0.26
15	4.00	78.000	78.238	1.635	-0.238	-0.08
16	4.00	80.000	78.238	1.635	1.762	0.60

11.8   Computer Session (optional)

The Minitab REGRESS command was illustrated in Section 11.7. In addition, the command has many useful subcommands, two of which—PREDICT and RESIDU-ALS—will be discussed in this section.

Two other useful Minitab commands to analyze regression data are the PLOT command (illustrated in Section 2.6) to give a scatterplot and the CORRELATION command (illustrated in Section 3.7) to calculate correlations.

EXAMPLE 11.14

Construct a scatterplot of the following data and calculate the correlation coefficient.

x	1	2	2	3	3	4	5	5	6	6	7	8	8	9
y	10	9	8	8	7	6	6	5	3	4	2	1	2	5

FIGURE 11.18
Minitab output for Example 11.14

SOLUTION   See Figure 11.18.

```
MTB > SET DATA IN C1
DATA> 1 2 2 3 3 4 5 5 6 6 7 8 8 9
DATA> END

MTB > SET DATA IN C2
DATA> 10 9 8 8 7 6 6 5 3 4 2 1 2 5
DATA> END

MTB > PLOT C2 C1

      10.0+      *
          -
C2        -           *
          -
          -           *        *
       7.5+
          -                *
          -
          -                 *          *
          -
       5.0+                       *                        *
          -
          -                          *
          -
          -                          *
       2.5+
          -                              *        *
          -
          -                                  *
          -
          --------+---------+---------+---------+---------+---------+C1
              1.5       3.0       4.5       6.0       7.5       9.0
```

```
MTB > CORR C1 C2

Correlation of C1 and C2 = -0.889
```

Other plot commands are LPLOT to plot with labels and MPLOT for multiple plots. See the *Minitab Handbook* for a description of these commands.

REGRESS

In Section 11.7 the REGRESS command was illustrated in the multiple regression case. The same command can be used to perform a simple regression.

EXAMPLE 11.15

To investigate the relationship between the number of books read during the term by third graders and their final spelling scores, an educator collected the data in Table 11.8 on 17 randomly selected third-grade students.

TABLE 11.8

No. of books read	Final spelling score
27	85
11	81
32	98
5	61
17	92
0	36
8	59
15	84
24	90
6	70
4	72
23	95
41	99
7	78
2	58
19	87
13	80

Use the Minitab REGRESS command to perform a simple linear regression on the data.

SOLUTION   See Figure 11.19.

FIGURE 11.19

Minitab output for Example 11.15

```
MTB > SET C1
DATA> 27 11 32 5 17 0 8 15 24 6 4 23 41 7 2 19 13
DATA> END
```

```
MTB > SET C2
DATA> 85 81 98 61 92 36 59 84 90 70 72 95 99 78 58 87 80
DATA> END


MTB > REGRESS C2 1 C1


The regression equation is
C2 = 59.5 + 1.23 C1


Predictor        Coef         Stdev      t-ratio
Constant       59.519         3.876        15.36
C1             1.2330        0.2081         5.93


s = 9.542       R-sq = 70.1%     R-sq(adj) = 68.1%


Analysis of Variance


SOURCE           DF             SS           MS
Regression        1         3197.1       3197.1
Error            15         1365.8         91.1
Total            16         4562.9


Obs.      C1          C2        Fit Stdev.Fit   Residual    St.Resid
  1     27.0       85.00      92.81      3.41      -7.81       -0.88
  2     11.0       81.00      73.08      2.46       7.92        0.86
  3     32.0       98.00      98.97      4.24      -0.97       -0.11
  4      5.0       61.00      65.68      3.10      -4.68       -0.52
  5     17.0       92.00      80.48      2.35      11.52        1.25
  6      0.0       36.00      59.52      3.88     -23.52       -2.70R
  7      8.0       59.00      69.38      2.73     -10.38       -1.14
  8     15.0       84.00      78.01      2.31       5.99        0.65
  9     24.0       90.00      89.11      2.98       0.89        0.10
 10      6.0       70.00      66.92      2.97       3.08        0.34
 11      4.0       72.00      64.45      3.25       7.55        0.84
 12     23.0       95.00      87.88      2.86       7.12        0.78
 13     41.0       99.00     110.07      5.90     -11.07       -1.48 X
 14      7.0       78.00      68.15      2.84       9.85        1.08
 15      2.0       58.00      61.98      3.55      -3.98       -0.45
 16     19.0       87.00      82.95      2.46       4.05        0.44
 17     13.0       80.00      75.55      2.35       4.45        0.48


R denotes an obs. with a large st. resid.
X denotes an obs. whose X value gives it large influence.
```

The PREDICT subcommand computes the predicted value of y and constructs a 95% confidence interval for $E(y)$ for any given value of x or a given column of values of x.

EXAMPLE 11.16

Use the solution to Example 11.15 to find the predicted spelling score for someone who reads 20 books. Find the predicted spelling scores for a group of students who have read 10, 15, 20, 25, and 30 books.

SOLUTION   Because PREDICT is a subcommand of REGRESS, the REGRESS command will have to be issued first (see Figure 11.20). To predict for the group of scores, we will first store the scores in a column and then give the PREDICT subcommand (Figure 11.21). The residuals of a regression fit can be stored in a column for further study with the RESIDUALS subcommand. The form of the complete command is

```
REGRESS C ON K PREDICTORS IN C, C,...,C;
    RESIDUALS IN C.
```

Once the residuals are stored, other Minitab commands can be used to examine the assumptions made about the residuals.

FIGURE 11.20

Minitab output for Example 11.16

```
MTB > REGRESS C2 1 C1;
SUBC> PREDICT 20.

The regression equation is
C2 = 59.5 + 1.23 C1

Predictor        Coef        Stdev      t-ratio
Constant       59.519        3.876        15.36
C1             1.2330       0.2081         5.93

s = 9.542       R-sq = 70.1%      R-sq(adj) = 68.1%

Analysis of Variance

SOURCE          DF          SS          MS
Regression       1       3197.1      3197.1
Error           15       1365.8        91.1
Total           16       4562.9

    Fit   Stdev.Fit       95% C.I.           95% P.I.
  84.18        2.54   ( 78.76,  89.60)  ( 63.13, 105.23)
```

FIGURE 11.21

Minitab output for Example 11.16

```
MTB > SET C7
DATA> 10 15 20 25 30
DATA> END
```

```
MTB > REGRESS C2 1 C1;
SUBC> PREDICT C7.


The regression equation is
C2 = 59.5 + 1.23 C1


Predictor        Coef        Stdev      t-ratio
Constant       59.519        3.876        15.36
C1             1.2330        0.2081        5.93


s = 9.542       R-sq = 70.1%     R-sq(adj) = 68.1%

Analysis of Variance


SOURCE        DF          SS           MS
Regression     1        3197.1       3197.1
Error         15        1365.8         91.1
Total         16        4562.9


     Fit   Stdev.Fit         95% C.I.           95% P.I.
   71.85      2.53    (  66.45,   77.25)  (  50.80,   92.90)
   78.01      2.31    (  73.08,   82.95)  (  57.08,   98.95)
   84.18      2.54    (  78.76,   89.60)  (  63.13,  105.23)
   90.34      3.12    (  83.69,   97.00)  (  68.94,  111.75)
   96.51      3.90    (  88.20,  104.81)  (  74.53,  118.48)
```

FIGURE 11.22

Minitab output for Example 11.17

```
MTB > REGRESS C3 2 C1 C2;
SUBC> RESIDUALS C5.


The regression equation is
C3 = 27.5 + 0.239 C1 + 0.559 C2


Predictor        Coef        Stdev      t-ratio
Constant        27.45        25.39         1.08
C1             0.2392        0.2502         0.96
C2             0.5594        0.1586         3.53


s = 7.946       R-sq = 58.7%     R-sq(adj) = 51.8%


Analysis of Variance


SOURCE        DF          SS           MS
Regression     2        1077.97       538.99
Error         12         757.76        63.15
Total         14        1835.73
```

```
SOURCE          DF        SEQ SS
C1               1        292.49
C2               1        785.48

Obs.      C1         C3       Fit  Stdev.Fit  Residual   St.Resid
  1     48.0     143.00    136.83     2.94       6.17       0.84
  2     50.0     131.00    128.35     3.89       2.65       0.38
  3     35.0     135.00    142.67     4.70      -7.67      -1.20
  4     41.0     131.00    134.59     2.28      -3.59      -0.47
  5     33.0     121.00    127.64     2.57      -6.64      -0.88
  6     25.0     115.00    121.26     4.04      -6.26      -0.91
  7     51.0     133.00    141.46     3.71      -8.46      -1.20
  8     53.0     126.00    131.87     4.13      -5.87      -0.86
  9     40.0     120.00    124.84     2.66      -4.84      -0.65
 10     31.0     128.00    128.84     2.97      -0.84      -0.11
 11     35.0     120.00    113.02     4.90       6.98       1.12
 12     46.0     160.00    145.30     4.14      14.70       2.17R
 13     28.0     124.00    120.85     3.58       3.15       0.44
 14     32.0     143.00    130.76     2.95      12.24       1.66
 15     44.0     128.00    129.72     2.48      -1.72      -0.23
```

R denotes an obs. with a large st. resid.

```
MTB > PRINT C5
C5
    6.1744     2.6463    -7.6667    -3.5920    -6.6442    -6.2557    -8.4588
   -5.8681    -4.8432    -0.8440     6.9810    14.7025     3.1456    12.2386
   -1.7156
```

```
MTB > PLOT C5 C1

           -                                        *
    14.0+
           -                      *
C5         -
           -
           -
     7.0+                    *                              *
           -
           -          *                                          *
           -
     0.0+
           -             *                           *
           -
           -                          *    *
           -       *                                              *
    -7.0+                    *    *
           -                                              *
           -
           +---------+---------+---------+---------+---------+------C1
         25.0      30.0      35.0      40.0      45.0      50.0
```

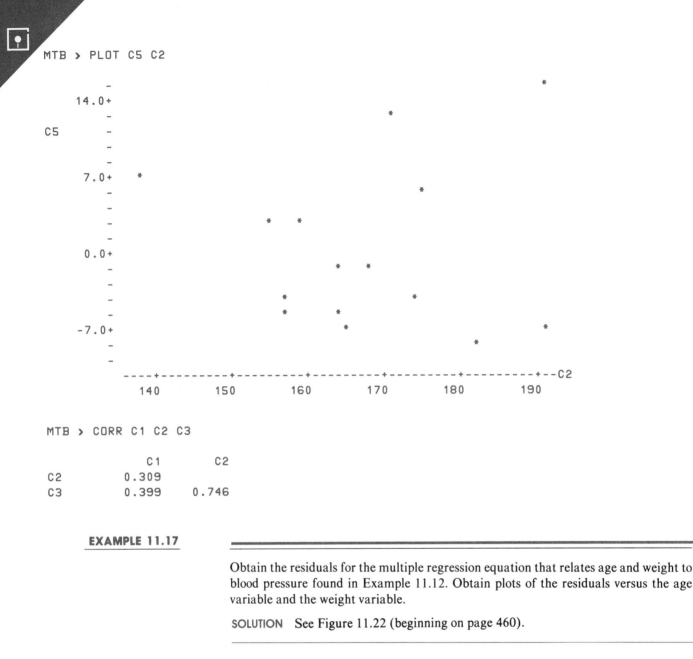

```
MTB > PLOT C5 C2


         -                                                                        *
    14.0+
         -                                                          *
C5       -
         -
         -
     7.0+    *                                                *
         -
         -
         -               *        *
         -
     0.0+
         -                             *        *
         -
         -            *                              *
         -            *         *
    -7.0+                                 *
         -                                        *
         -
         ----+---------+---------+---------+---------+---------+--C2
            140       150       160       170       180       190


MTB > CORR C1 C2 C3

              C1        C2
C2         0.309
C3         0.399     0.746
```

EXAMPLE 11.17

Obtain the residuals for the multiple regression equation that relates age and weight to blood pressure found in Example 11.12. Obtain plots of the residuals versus the age variable and the weight variable.

SOLUTION   See Figure 11.22 (beginning on page 460).

EXERCISES 11.8

11.56 Use Minitab to construct a scatterplot of the following data.

x	20	25	30	35	40	45	50	55	60
y	1.6	2.1	2.7	3.3	4.0	4.7	4.5	4.9	5.1

11.57 Use data in Exercise 11.56 and the REGRESS command

to find the least squares estimates of β_0 and β_1 in the model

$$y = \beta_0 + \beta_1 x + \epsilon$$

11.58 An owner of a large retail store feels that the monthly gross sales figures of her employees are related to the length of employment. The accompanying table shows the July gross sales figures and the length of employment for 10 employees:

No. of months employed	July gross sales
10	3860
22	4230
8	2650
16	5170
31	4970
2	4780
13	3120
36	4690
18	4920
6	2150

Use Minitab to

(a) find the least squares estimates of β_0, β_1

(b) find the predicted gross sales for employees who have been employed for 1 year and for 2 years

(c) find the coefficient of determination

(d) test the hypothesis that $\beta_1 = 0$

11.59 A study was conducted to investigate the relationship of aptitude test scores and years of education on productivity in a factory. The results on 8 employees are as follows:

Education level	8	12	13	11	8	14	12	10
Aptitude score	9	17	20	19	20	23	18	15
Productivity	23	35	29	33	40	32	28	26

Use the Minitab REGRESS command to

(a) find the least squares estimates of β_0, β_1, and β_2.

(b) find the coefficient of determination.

(c) test the hypothesis that $\beta_1 = 0$.

(d) test the hypothesis that $\beta_2 = 0$.

(e) Considering your answers to parts (c) and (d), should the model be modified? Explain.

11.9 Summary and Review

Key Concepts

☑ Regression is the study of the relationship between a *dependent variable y* and the *independent variables* $x_1, x_2, \ldots, x_k$. The regression equation is of the form

$$y = \beta_0 + \beta_1 x_1 + \beta_2 x_2 + \cdots + \beta_k x_k + \epsilon$$

☑ If $k = 1$, the equation is referred to as a *simple linear regression equation*. From the data $(x_1, y_1), (x_2, y_2), \ldots, (x_n, y_n)$ the *least squares estimates* of β_0 and β_1 are found.

☑ From the least squares equation the *predicted value of y* is found by the equation

$$\hat{y}_i = b_0 + b_1 x_i$$

where b_0 and b_1 are the least squares estimates.

☑ The *residuals* associated with the data point (x_i, y_i) is

$$e_i = y_i - \hat{y}_i$$

All the residuals are squared and summed to form the *sum of squares due to error*.

☑ The *linear correlation coefficient* is a numerical measure of the strength of the linear relationship between two variables. The *coefficient of determination* is the square of the correlation coefficient and gives the percent of the variability in the dependent variable that is explained by the independent variable.

☑ Tests of hypothesis about the regression coefficient, β_1, can be performed. The test statistic is given by

$$t = \frac{b_1}{s/\sqrt{SS_{xx}}}$$

and is distributed as a *t*-distribution with $n - 2$ degrees of freedom.

☑ Confidence intervals for β_1 can be constructed with the formula

$$b_1 \pm t_{\alpha/2, n-2} s/\sqrt{SS_{xx}}$$

☑ Confidence intervals for the expected response, $E(y)$ are found with the formula

$$b_0 + b_1 x' \pm t_{\alpha/2, n-2} s\sqrt{1/n + (x' - \overline{x})^2/SS_{xx}}$$

☑ The *resistant line* is a robust alternative to the least squares regression line. The slope of the resistant line is

$$b_1 = \frac{y_R - y_L}{x_R - x_L}$$

The *y*-intercept is $b_0 = \frac{1}{3}[(y_L + y_M + y_R) - b_1(x_L + x_M + x_R)]$.

☑ An *analysis of the residuals* makes it possible to check out the assumptions that are made in a regression analysis. It also can point out possible deficiencies in the regression model and suggest alternatives.

☑ *Multiple regression* is the study of the relationship between the dependent and several independent variables. The computations are best handled with a computer.

Having completed this chapter you should be able to:

1. Identify the independent and dependent variables in a regression problem. *Section 11.1*
2. Construct a scatterplot and comment on the relationship between the variables. *Section 11.1*
3. Formulate the least squares equation relating the independent and dependent variables. *Section 11.2*
4. Calculate the predicted values of *y* associated with the *x* values. *Section 11.2*
5. Calculate the sum of squares due to error. *Section 11.2*
6. Test hypotheses about the regression coefficient. *Section 11.3*
7. Compute a confidence interval for the regression coefficient. *Section 11.3*
8. Compute a confidence interval for the expected response $E(y)$. *Section 11.4*
9. Construct a resistant line. *Section 11.5*
10. Perform an analysis of residuals. *Section 11.6*
11. Analyze the computer output associated with a multiple regression problem. *Section 11.7*

Review Questions

To test your skills answer the following questions.

11.60 Graph the equation $y = -7.8 + 6.3x$.

11.61 Identify the independent variable and dependent variable in a study that involves the time necessary for a subject to react and the number of alternatives given.

11.62 Given the following data:

x	2.5	3.4	4.7	5.2	6.8	7.6
y	10.3	14.2	17.5	22.6	24.8	29.0

Construct a scatterplot and comment on the general relationship between *x* and *y*.

11.63 Given the least squares equation

$\hat{y} = 3.5 - 6.8x$

find the predicted *y* when $x = 3.5$.

11.64 A manufacturer of a new insulation medium examined the heat loss through the insulation as the outside temperature dropped.

Outside temp. (°C)	−10	−10	0	0	10	10	20	20	30	30
Heat loss	96	91	84	82	68	75	49	51	28	24

Construct a scatterplot and find the least squares estimates of β_0 and β_1 in the model $y = \beta_0 + \beta_1 x + \epsilon$.

11.65 In Exercise 11.64, find the predicted heat loss when the outside temperature is 15 degrees.

11.66 Find a 95% confidence interval for β_1 in Exercise 11.64.

11.67 Construct the resistant line from the following data:

IQ (x)	115	132	125	120	119	132	105	114	106	139	127	118
GPA (y)	2.2	3.3	3.0	2.6	2.9	3.5	2.2	2.7	3.7	1.8	3.7	2.4

Would a least squares equation better fit the data?

SUPPLEMENTARY EXERCISES FOR CHAPTER 11

11.68 Identify the independent and dependent variables in the following:

(a) A study of the relationship between robbery rates and population density

(b) A study of the relationship between attitude scores and academic achievement scores

(c) A study of the relationship between growth rate of rainbow trout and number of fish per cubic yard of water

(d) A study of the relationship between expenditures per student and teacher's salary

11.69 Graph the following linear equations:
(a) $y = 3.1 + 4.7x$
(b) $y = -7.3 + 5.5x$
(c) $y = -8.3 - 2.1x$

11.70 In the following identify the slope and y-intercept:
(a) $2x + 3y = 6$
(b) $-3.1y + 7.4x = 12$
(c) $-5.6x - 4.1y = 10$

11.71 Graph the equations in Exercise 11.70.

11.72 Construct a scatterplot for the following data sets and comment on the relationship between x and y.

(a)
x	100	110	120	130	140	150
y	7.8	6.1	5.4	5.2	4.7	3.5

(b)
x	2.3	3.4	1.6	6.4	4.2	3.1	5.6	4.9
y	.65	.82	.47	1.23	.92	.74	1.08	1.01

(c)
x	41	52	37	26	45	32	49	55	22	30
y	1.2	2.8	.6	-1.7	2.9	-1.1	3.1	2.7	-2.4	-1.3

11.73 Find the least squares estimates of β_0 and β_1 in the linear model $y = \beta_0 + \beta_1 x + \epsilon$ for the data in Exercise 11.72.

11.74 For the data in Exercise 11.72 and the solutions found in Exercise 11.73, find the predicted y for the following values of x.
(a) $x = 100$ and $x = 160$
(b) $x = 1.5$ and $x = 5.0$
(c) $x = 30$ and $x = 60$

11.75 Using the solution to Exercise 11.73, find the residuals associated with the following data points:
(a) (120, 5.4) and (150, 3.5)
(b) (1.6, .47) and (3.1, .74)
(c) (52, 2.8) and (32, -1.1)

11.76 Given the following data:

x	1	2	3	4	5
y	8	6	5	3	1

(a) Find the least squares estimates of β_0 and β_1 in the linear model $y = \beta_0 + \beta_1 x + \epsilon$.
(b) Find the residuals.
(c) Find the sum of squares due to error.

11.77 In order to make inference statements about the regression coefficient in the simple linear model

$$y = \beta_0 + \beta_1 x + \epsilon$$

what assumptions must be made about the random errors?

11.78 According to *USA Today* (January 18, 1985) the prices for oranges in 1985 were expected to rise sharply due to low projected harvests and higher advertising costs. Following are the average prices California growers charged for a 75-pound box of navel oranges and the size harvest for the previous 6 years.

Harvest (in millions of boxes)	72	69	58	70	65	54
Price per box	$5.40	6.10	9.30	6.50	7.20	13.40

Construct a scatterplot and find the least squares estimates of β_0 and β_1 in the model $y = \beta_0 + \beta_1 x + \epsilon$.

11.79 In Exercise 11.78, find the predicted price for a box of oranges if the projected harvest is 60 million boxes; 30 million boxes.

11.80 Find the residuals associated with the least squares equation found in Exercise 11.78. Plot the residuals and comment on the results.

11.81 How much confidence do we have that the press reports facts accurately? One study suggests that the confidence level is related to the amount of education one has. Following are the education levels of 20 randomly selected persons and their degree of confidence in the press (a score from 0 to 100, the higher the score the more confidence exhibited).

Education (yr.) x	12	12	14	8	10	12	11	12	16	14
Confidence y	28	36	22	58	41	32	30	62	14	21
Education (yr.) x	8	12	15	12	9	12	10	14	12	16
Confidence y	42	48	25	31	40	42	57	28	16	18

Construct a scatterplot and find the least squares estimates of β_0 and β_1 in the model $y = \beta_0 + \beta_1 x + \epsilon$.

11.82 In Exercise 11.81, find the predicted confidence level for someone whose education is 12 years; 16 years.

11.83 Find the residuals associated with the least squares equation found in Exercise 11.81. Plot the residuals and comment on the results.

11.84 From the following summary data find a 98% confidence interval for β_1 in the regression model $y = \beta_0 + \beta_1 x + \epsilon$:

$n = 30$, $\Sigma x_i = 16.4$, $\Sigma x_i^2 = 22.8$
$\Sigma y_i = 320$, $\Sigma y_i^2 = 4280$, $\Sigma x_i y_i = 88.5$

11.85 The owner of a department store decided to investigate the relationship between the amount lost due to shoplifting and the number of customers in the store. Unable to get a count on the number of shoppers, he measured the sales receipts.

Sales receipts ($1,000) x	8.4	7.1	9.3	12.3	10.8	8.1	6.5	7.8
Shoplifting loss ($100) y	16.2	12.3	19.8	18.4	14.6	15.8	11.4	13.1

Construct a scatterplot and find the least squares estimates of β_0 and β_1 in the model $y = \beta_0 + \beta_1 x + \epsilon$.

11.86 In Exercise 11.85, find the predicted loss due to shoplifting if the sales receipts are $10,000.

11.87 Find the residuals associated with the least squares equation found in Exercise 11.85. Plot the residuals and comment on the results.

11.88 In Exercise 11.85 test the hypothesis that $\beta_1 = 0$ versus $\beta_1 \neq 0$.

11.89 Sutton (1984) reports a negative correlation between stress level and job satisfaction, suggesting a tendency for teachers to report a higher level of job satisfaction when stress levels are low. Construct a scatterplot and find the least squares estimates of β_0 and β_1 in the model $y = \beta_0 + \beta_1 x + \epsilon$ for the following data:

WSPT*	90	78	85	65	94	82	96	79	80
Job satisfaction	3.6	5.3	4.7	8.9	3.2	4.0	3.8	6.2	6.5

*Wilson Stress Profile for Teachers

11.90 Test the hypothesis that $\beta_1 = 0$ in Exercise 11.89.

12

Analysis of Categorical Data

CONTENTS

Most problems we have considered up to this point have been situations where the data were quantitative in nature. Statistical inference procedures for variables such as IQ, reaction time, and earnings per share are presented in Chapters 8 through 11. We now investigate statistical procedures for qualitative variables. The results of surveys and some experiments can only be classified and not quantified. The objective of this chapter is to present methods of analyzing data of this type.

Seat Belt Usage

Many states have instituted mandatory seat belt laws in recent years. Following are the results of a 1984 Gallup poll on the issue of wearing seat belts.

Question: "Thinking about the last time you got into a car, did you use a seat belt, or not?"

May 18–21, 1984

	Yes, did	No, did not	No opinion	Number of interviews
National	25%	74%	1%	(1,516)
Sex				
Male	25	74	1	755
Female	25	74	1	761
Age				
Total under 30	28	71	1	329
18–24 years	25	75	*	173
25–29 years	33	66	1	156
30–49 years	26	74	*	569
Total 50 & older	22	77	1	609
50–64 years	23	76	1	331
65 & older	20	78	2	278
Region				
East	24	75	1	382
Midwest	26	74	*	395
South	18	81	1	416
West	35	64	1	323
Race				
Whites	26	73	1	1,320
Nonwhites	19	80	1	196
Blacks	18	82	*	165
Hispanics	31	69	*	65
Education				
College graduates	39	60	1	310
College incomplete	31	69	*	359
High school grad	21	79	*	529
Less than H.S. grad	16	83	1	316
Politics				
Republicans	32	67	1	410
Democrats	23	77	*	654
Independents	16	83	1	430
Occupation				
Professional & Business	32	67	1	433
Clerical & sales	23	76	1	115
Manual workers	21	78	1	545
Nonlabor force	20	79	1	289

(table continues)

(table continued)

	May 18–21, 1984			
	Yes, did	No, did not	No opinion	Number of interviews
Income				
$40,000 & over	35	64	1	226
30,000–39,999	30	70	*	182
20,000–29,999	24	76	*	299
10,000–19,999	23	76	1	427
Under $10,000	17	83	*	306
Religion				
Protestants	21	78	1	876
Catholics	28	72	*	420
Labor Union				
Union family	26	73	1	333
Nonunion family	25	74	1	1,183
Urbanization				
Center cities	28	72	*	440
Suburbs	29	70	1	514
Rural areas	17	82	1	562

SOURCE: Gallup Report No. 226, July 1984. Used with permission.

*Less than 1%.

From the data, is there statistical evidence to say, for instance, that the frequency of wearing seat belts is the same for all regions of the country? Is one age group more likely to wear seat belts than others? These questions can be addressed with the statistical procedures presented in this chapter.

12.1 Introduction

Thus far, the inference procedures presented have been applied to quantitative variables. Now we consider the analysis of qualitative variables. As described in Chapter 2, a qualitative variable is organized into categories where the data consist of frequency counts for the various categories. For example, of the 703 deaths due to weather conditions reported in 1983, 122 occurred in the Northeast, 379 in the South, 79 in the Midwest, and 123 in the West. These data are organized in the following table. The categories are the various regions of the country and the data consist of the counts associated with the region.

	Region				
	Northeast	South	Midwest	West	Total
No. of deaths due to weather	122	379	79	123	703

For another example, suppose that the Red Cross classifies a group of potential donors according to blood type. The categories are the different blood groups and the data are the number of donors falling in the various blood groups.

Categories can also be defined by ranges of values of a quantitative variable, such as income level being classified as Low, Medium, or High.

When frequency count data are classified according to two or more variables, they are called *cross-tabulated* data, and are displayed in a contingency table. For example, subjects might be classified in a contingency table according to their religious preferences and their attitudes about abortion. Or college students might be classified according to class rank and study habits (Good, Average, Bad). In Example 2.7, workers were classified according to Marital Status and Job Satisfaction.

Categorical data, as described here, are statistically analyzed by means of the *chi-square* (χ^2) *statistic* introduced by Karl Pearson in 1900. The analysis of a single variable is called the chi-square *goodness-of-fit test*. The problem consists of determining whether the frequency counts in the categories of the variable agree with a specified distribution. The data in a contingency table can also be analyzed using the chi-square statistic, as is demonstrated in Sections 12.3 and 12.4.

12.2 The Chi-Square Goodness-of-Fit Test

Consider the problem of determining if a die is fair or not. If it is fair, then the probability of falling on any one side is the same as any other side. That is, all sides have a ⅙ chance of coming up. Suppose we toss the die 60 times and observe the following outcomes:

Side	1	2	3	4	5	6	Total
No. of times observed	8	9	13	7	15	8	60

Ideally, if the die is fair, we expect each side to turn up 10 times. Do the outcomes we observe imply that the die is not fair? To answer this question we need to conduct a statistical test of hypothesis. But first we will describe the *multinomial experiment*, which is an extension of the binomial experiment.

The Multinomial Experiment

1. The experiment consists of n identical, independent trials.

2. The outcome of each trial falls in one of k classes.

3. The probabilities associated with the k classes, denoted by $\pi_1, \pi_2, \ldots, \pi_k$, remain the same from trial to trial. Because there are only k possible classes, we have

$$\pi_1 + \pi_2 + \cdots + \pi_k = 1$$

4. The experimenter is interested in $o_1, o_2, \ldots, o_k$, where o_j ($j = 1, 2, \ldots, k$) is equal to the observed number of trials in which the outcome is in class j. Note that

$$n = o_1 + o_2 + \cdots + o_k$$

The preceding example of tossing a die 60 times fits the description of a multinomial experiment. First the n identical trials are the 60 tosses. Each toss results in a 1, 2, 3, 4, 5, or 6, which are the $k = 6$ classes. The probabilities associated with the 6 sides are denoted as $\pi_1, \pi_2, \pi_3, \pi_4, \pi_5$, and π_6. If the die is fair, then

$$\pi_1 = \pi_2 = \pi_3 = \pi_4 = \pi_5 = \pi_6 = \frac{1}{6}$$

The outcome of any one toss does not depend on any other toss, hence the tosses are independent.

From our example we observe that

$$o_1 = 8, \quad o_2 = 9, \quad o_3 = 13, \quad o_4 = 7, \quad o_5 = 15, \quad o_6 = 8$$

which totals $n = 60$. Thus the toss of a die n times is a multinomial experiment.

It was stated earlier that if the die is fair, then

$$\pi_1 = \pi_2 = \pi_3 = \pi_4 = \pi_5 = \pi_6 = \frac{1}{6}$$

To determine whether it is a fair die, we will state the null hypothesis accordingly; that is,

$$H_0: \pi_1 = \pi_2 = \pi_3 = \pi_4 = \pi_5 = \pi_6 = \frac{1}{6}$$

versus

$$H_a: \text{at least one } \pi_i \neq \frac{1}{6}$$

If H_0 is true, then out of 60 tosses we *expect* that each o_j will be $(\frac{1}{6})(60) = 10$, as our intuition suggested earlier.

Expected Number of Outcomes in a Multinomial Experiment

Out of n trials of a multinomial experiment the **expected number of outcomes** to fall in class j is $e_j = n\pi_j$. Note that the expected cell count is calculated assuming H_0 is true.

Returning to the example, we realize that the die could be fair and not all $o_j = 10$. There could be some discrepancies, but probably not much if the die is indeed fair.

The test statistic should consider the differences in the observed o_j's and those expected under the null hypothesis. The following statistic, proposed by Karl Pearson, measures the amount of disagreement between the observed data and the expected data.

Pearson χ^2 Test Statistic

$$\chi^2 = \Sigma(o_j - e_j)^2/e_j$$

where the sum is over all classes with o_j being the observed frequency count and e_j the expected frequency count in class j.

If the null hypothesis is to be rejected, there will be large discrepancies between o_j and e_j, and thus χ^2 will get large. But how large will it have to get in order that we have significant evidence to reject the null hypothesis? This can be answered only by investigating the sampling distribution of the test statistic.

When n is large and H_0 is true, the sampling distribution of χ^2 is known to be approximately *chi-square with $k - 1$ degrees of freedom*. Figure 12.1 illustrates a typical chi-square density curve. The curve begins at zero and is skewed right. As the degrees of freedom increase the distribution stretches out along the horizontal axis.

▲ Remember that k is the number of classes. There are six sides to a die; therefore, for the die example $k = 6$.

FIGURE 12.1

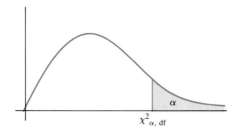

$\chi^2_{\alpha,\ df}$

Table B5 in Appendix B gives the tail probabilities of the χ^2 distribution for various degrees of freedom. For example, when the degrees of freedom are 16, then 10% of the area lies above 23.5, 5% lies above 26.3, and 1% lies above 32.0. The table also gives lower-tail probabilities. For 16 degrees of freedom we see that 95% of the area lies above 7.96; consequently 5% lies below 7.96.

For the χ^2 goodness-of-fit test we are interested only in large values of χ^2 and the upper-tail probabilities.

Returning to the die-toss example, we have

$$\chi^2 = (8 - 10)^2/10 + (9 - 10)^2/10 + (13 - 10)^2/10$$
$$+ (7 - 10)^2/10 + (15 - 10)^2/10 + (8 - 10)^2/10$$

$$= \frac{1}{10}[(-2)^2 + (-1)^2 + (3)^2 + (-3)^2 + (5)^2 + (-2)^2]$$

$$= \frac{1}{10}(4 + 1 + 9 + 9 + 25 + 4)$$

$$= \frac{52}{10} = 5.2$$

From the χ^2 table with $k - 1 = 6 - 1 = 5$ degrees of freedom, we see that 10% of the area lies above 9.24. Thus more than 10% of the area lies above 5.2; that is, the

$$p\text{-value} > .10$$

and therefore there is insufficient evidence to reject H_0. The conjecture made in the null hypothesis is that the die is fair. The observed data seem to fit that conjecture, and hence there is insufficient evidence to conclude that the die is not fair.

In this example we are investigating statistically how well the observed data fit the conjectured probabilities in the null hypothesis. For this reason the test is called the χ^2 goodness-of-fit test. The general form of the test is as follows.

χ^2 Goodness-of-Fit Test

APPLICATION: Multinomial experiments

ASSUMPTIONS:

1. The experiment satisfies the properties of a multinomial experiment.

2. The expected cell counts, e_j, must all be 5 or more. (This is so that the χ^2 approximation will be good.)

H_0: $\pi_1 = p_1, \pi_2 = p_2, \ldots, \pi_k = p_k$, where $p_1, p_2, \ldots, p_k$ are the hypothesized values of the multinomial probabilities.

H_a: At least one of the multinomial probabilities does not equal the hypothesized value.

Test statistic $= \chi^2 = \Sigma(o_j - e_j)^2/e_j$, where $e_j = np_j$

The test is a right-tailed test where the p-value is found in the χ^2 table with $k - 1$ degrees of freedom. Usually the exact value cannot be found, but bounds for it can be located by finding the closest values to the observed value of the χ^2 statistic.

EXAMPLE 12.1

A local grocery store stocks four brands of cola. Suppose that, nationally, brand A commands 40% of the market, brand B has 35%, brand C has 20%, and brand D has 5%. Of 2,000 colas sold 1 week in the store, 615 were brand A, 804 were brand B, 383 were brand C, and 198 were brand D. Do the data collected at the local grocery store fit the national percentage?

SOLUTION   Let

$$\pi_1 = \text{proportion of people who buy brand A at the local grocery store;}$$

$$\pi_2 = \text{proportion of people who buy brand B at the local grocery store;}$$

$$\pi_3 = \text{proportion of people who buy brand C at the local grocery store;}$$

$$\pi_4 = \text{proportion of people who buy brand D at the local grocery store}$$

The null hypothesis, which says that the local percentages are the same as the national percentages, is

$$H_0: \pi_1 = .40, \quad \pi_2 = .35, \quad \pi_3 = .20, \quad \pi_4 = .05$$

and the alternative is

$$H_a: \text{At least one of the proportions differs from its hypothesized value}$$

The test statistic is

$$\chi^2 = \Sigma(o_j - e_j)^2/e_j$$

where

$$e_1 = (2000)(.40) = 800$$
$$e_2 = (2000)(.35) = 700$$
$$e_3 = (2000)(.20) = 400$$
$$e_4 = (2000)(.05) = 100$$

Because all of these values are larger than 5 and the conditions of a multinomial experiment are met, we have a χ^2 goodness-of-fit test.

The observed value of χ^2 is

$$\chi^2_{obs} = (615 - 800)^2/800 + (804 - 700)^2/700$$
$$+ (383 - 400)^2/400 + (198 - 100)^2/100$$
$$= 42.78125 + 15.45143 + .7225 + 96.04$$
$$= 154.99518$$

From the χ^2 table with 3 degrees of freedom we see, as depicted in Figure 12.2, that the p-value $< .005$ because any χ^2 value greater than 12.9 will have a tail probability less than .005.

This is highly significant evidence that H_0 should be rejected. Clearly the shoppers at the local grocery store do not choose their brands of cola the same as the rest of the

FIGURE 12.2

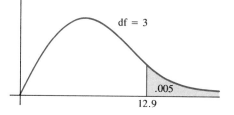

nation. The manager of the grocery store would be well advised not to follow the national market when he stocks his cola.

If the expected cell counts are not 5 or greater, it may be possible to combine cells so that the assumption is satisfied.

EXAMPLE 12.2

Suppose a high school teacher gives a standardized English grammar test to her class of 25 students, with the following results.

56	58	40	77	87
75	61	70	73	71
66	69	67	68	60
72	73	61	64	66
84	72	52	65	67

▲ Percentiles divide data into 100 parts, deciles divide it into 10 parts, and quartiles divide it into 4 parts.

She wishes to compare her class with the national standard. The test manual reports decile scores as follows:

Decile	1	2	3	4	5	6	7	8	9
Score	45.0	56.8	62.5	66.1	68.7	71.3	74.0	78.5	84.2

(that is, 10% scored below 45.0, 10% from 45.0 to 56.8, etc.). Does her class differ significantly from the national standard?

SOLUTION   First we set up the classes as shown in Table 12.1.

TABLE 12.1

Class	Number of scores
below 45.0	1
45.0–56.8	2
56.8–62.5	4
62.5–66.1	4
66.1–68.7	3
68.7–71.3	3
71.3–74.0	4
74.0–78.5	2
78.5–84.2	1
above 84.2	1

Because each class is defined by the decile scores, 10% of the data should be within each class. Therefore the expected number in each class is

$$e_j = 25(.10) = 2.5$$

which is in violation of the basic assumption that each class should have an expected cell count of 5 or more.

A solution to the problem is to combine the 10 classes into five classes, where each class covers 20% of the data. Then the expected number in each class is

$$e_j = 25(.20) = 5$$

Table 12.2 gives the expected number and observed number of scores in each of the combined classes.

TABLE 12.2

Class	No. of scores	Expected no. of scores
below 56.8	3	5
56.8–66.1	8	5
66.1–71.3	6	5
71.3–78.5	6	5
above 78.5	2	5

▲ This example also illustrates how categories can be defined by ranges of values of a quantitative variable so that they can be analyzed with the chi-square test.

The χ^2 statistic can be calculated as follows:

$$\chi^2 = \frac{(3-5)^2 + (8-5)^2 + (6-5)^2 + (6-5)^2 + (2-5)^2}{5}$$

$$= (4 + 9 + 1 + 1 + 9)/5$$

$$= 24/5 = 4.8$$

With 4 degrees of freedom the p-value $> .10$ ($4.8 < 7.78$), and therefore there is insufficient evidence to reject

$$H_0: \pi_1 = \pi_2 = \pi_3 = \pi_4 = \pi_5 = .20$$

Her class does not differ significantly from the national standard.

EXERCISES 12.2

12.1 Find the p-value associated with the following observed values of the chi-square statistic with the specified degrees of freedom.

(a) $\chi^2 = 8.4$, df $= 4$

(b) $\chi^2 = 3.7$, df $= 7$

(c) $\chi^2 = 16.9$, df $= 3$

(d) $\chi^2 = 13.5$, df $= 5$

12.2 Compute the value of the chi-square statistic for each of the following tables of observed and expected values. Does there appear to be a significant difference in the observed and expected frequencies?

(a)

j	1	2	3
o_j	15	24	6
e_j	15	20	10

(b)

j	1	2	3	4
o_j	23	36	51	70
e_j	36	45	63	36

12.3 A multinomial experiment with $k = 5$ and $n = 500$ yielded the following results:

Class	1	2	3	4	5
O_i	92	97	106	85	120

Is there evidence to reject the hypothesis that the classes are equally likely?

12.4 A multinomial experiment with 4 possible outcomes and 100 trials produced the following data:

Class	1	2	3	4
O_i	44	29	21	6

Is there evidence to reject the hypothesis that

$$\pi_1 = 40\% \quad \pi_2 = 30\% \quad \pi_3 = 20\% \quad \pi_4 = 10\%$$

12.5 A local official claims that of all voters in the district, 40% are Democrats, 45% are Republican, 7% are Conservative, 5% are Liberal, and the remaining 3% are classified as Other. The party preferences of a sample of 1,200 voters were found to be as follows:

Party	No. of voters
Democrat	504
Republican	523
Conservative	72
Liberal	70
Other	31

Are the sample results consistent with the claim made by the official?

12.6 In 180 throws of a die, the observed frequencies of the values from 1 to 6 are 34, 27, 41, 25, 18, and 35. Test the hypothesis that the die is fair.

12.7 A large jar has red, black, blue, and white marbles in it. One hundred marbles are drawn from the jar with the following results:

red–28
black–19
blue–22
white–31

Is there statistical evidence that the proportions of marbles of the four colors are the same?

12.8 The number of books borrowed from a public library for a particular week is

Day	M	T	W	T	F
No. of books	125	105	120	114	136

Determine whether the number of books borrowed depends on the day of the week.

12.9 Five different strains of flies were tested for their resistance to a particular chemical agent. A large resort area was sprayed with the chemical, after which 1,000 dead flies were randomly selected and classified according to their strain.

Strain	1	2	3	4	5	Total
No. killed	265	178	301	115	141	1,000

Assuming that each strain is equally prevalent in the resort area, is there evidence that some strains are more resistant to the chemical agent than others?

12.10 According to test theory, the mental ages of females, as measured by a certain test, should be equally distributed in the categories: less than 85, 86 to 95, 96 to 110, 111 to 125, and above 125. A random sample of 35 females recorded the following mental ages:

99	93	87	102	108	93	113
89	115	96	128	105	131	81
109	94	120	136	84	99	126
123	107	95	129	117	84	125
114	136	89	101	140	124	100

Do the data support the theory?

12.3 The Chi-Square Test of Independence

For cross-tabulated data one aim is to determine whether there is any dependency between the classifying variables. For example, suppose that 200 randomly selected people were asked about their views on gun control and what political party they preferred. The chi-square test of independence will determine whether there is any statistical dependence between their views and their political party preference. Table 12.3 is a contingency table giving the distribution of the frequency counts from the sample of 200 people.

TABLE 12.3

Political party	Opinion on gun control			
	Favor	Oppose	No opinion	Total
Democrat	44	48	18	110
Republican	32	48	10	90
Total	76	96	28	200

▲ Note that we have a single sample that is classified two ways. In other words, there are two variables measured on each subject.

The test will be constructed in such a way that the classifying variables are assumed to be independent until the data prove otherwise. The null hypothesis is

H_0: Political party and opinion on gun control are independent

and the alternative hypothesis is

H_a: Political party and opinion on gun control are dependent

To utilize the χ^2 test statistic we must demonstrate that the conditions of a multinomial experiment are satisfied.

Let us denote the probabilities of falling in the six cells (classes) as depicted in Table 12.4.

TABLE 12.4

Political party	Opinion on gun control			
	Favor	Oppose	No opinion	Total
Democrat	π_{D1}	π_{D2}	π_{D3}	π_D
Republican	π_{R1}	π_{R2}	π_{R3}	π_R
Total	π_1	π_2	π_3	1

▲ A cell is the intersection of a specific row and a specific column.

The cell probability π_{D1} denotes the probability that a randomly selected subject will be a Democrat and favor gun control. That is,

$$\pi_{D1} = P(\text{Democrat and Favor})$$

The other five cell probabilities are similarly defined. As far as satisfying the conditions of a multinomial experiment, we have that the 200 people constitute the n independent trials. Each subject will fall in one of the six cells (classes) and the probabilities of falling in the cells are $\pi_{D1}, \pi_{D2}, \pi_{D3}, \pi_{R1}, \pi_{R2},$ and π_{R3} and remain the same for all subjects. Thus we see that the conditions of a multinomial experiment are satisfied.

The row and column probabilities are called *marginal probabilities*. The row marginal probability π_D is the probability that a Democrat is selected regardless of his or her opinion on gun control. We have that

$$\pi_D = P(\text{Democrat}) = \pi_{D1} + \pi_{D2} + \pi_{D3}$$

Similarly we have

$$\pi_R = P(\text{Republican}) = \pi_{R1} + \pi_{R2} + \pi_{R3}$$
$$\pi_1 = P(\text{Favor}) = \pi_{D1} + \pi_{R1}$$
$$\pi_2 = P(\text{Oppose}) = \pi_{D2} + \pi_{R2}$$
$$\pi_3 = P(\text{No opinion}) = \pi_{D3} + \pi_{R3}$$

The existing data suggest that

$$P(\text{Democrat}) = 110/200$$

and

$$P(\text{Republican}) = 90/200$$

Also not accounting for political party, we have

$$P(\text{Favor}) = 76/200$$
$$P(\text{Oppose}) = 96/200$$

and $\qquad P(\text{No opinion}) = 28/200$

▲ Remember the multiplication law that says that if events A and B are independent, then $P(A \cap B) = P(A)P(B)$.

Recall from Chapter 4 that if two events are independent, then the probability of the joint occurrence of the two is the product of their probabilities. Thus if political party and opinion are independent, then the probability of being a Democrat and favoring gun control is

$$\pi_{D1} = P(\text{Democrat and Favor})$$
$$= P(\text{Democrat})P(\text{Favor})$$
$$= (110/200)(76/200)$$

Therefore out of 200 subjects, we would expect

$$200(110/200)(76/200) = (110)(76)/200$$

to fall in the first cell, that is

$$E(\text{Democrat and Favor}) = (110)(76)/200 = 41.8$$

In a similar fashion

$$E(\text{Democrat and Oppose}) = (110)(96)/200 = 52.8$$
$$E(\text{Democrat and No opinion}) = (110)(28)/200 = 15.4$$

Note that an expected cell count under the condition that the two classifying variables are independent is

$$\frac{(\text{row total})(\text{column total})}{\text{grand total}}$$

Thus

$$E(\text{Republican and Favor}) = (90)(76)/200 = 34.2$$
$$E(\text{Republican and Oppose}) = (90)(96)/200 = 43.2$$
$$E(\text{Republican and No opinion}) = (90)(28)/200 = 12.6$$

Table 12.5 gives the observed cell counts and the expected cell counts in parentheses.

TABLE 12.5

Political party	Opinion on gun control			
	Favor	Oppose	No opinion	Total
Democrat	44 (41.8)	48 (52.8)	18 (15.4)	110
Republican	32 (34.2)	48 (43.2)	10 (12.6)	90
Total	76	96	28	200

Because the expected cell counts were calculated under the assumption that H_0 is true (political party and opinion are independent), a large discrepancy between the observed and expected cell counts should lead to rejection of H_0 and thus establish that the two classifications are dependent. To compare the differences between observed and expected, we calculate the χ^2 statistic as in the previous section as follows:

$$\chi^2 = (44 - 41.8)^2/41.8 + (48 - 52.8)^2/52.8 + (18 - 15.4)^2/15.4$$
$$+ (32 - 34.2)^2/34.2 + (48 - 43.2)^2/43.2 + (10 - 12.6)^2/12.6$$
$$= .1158 + .4364 + .4390 + .1415 + .5333 + .5365$$
$$= 2.2025$$

To determine whether the observed value of χ^2 is statistically large, we must determine its p-value from the χ^2 table. But first we must determine the number of degrees of freedom for the test statistic. Note that in Table 12.5 the expected cell counts in the first row sum to 110. Therefore we only had to calculate two of the three expected cell counts in the first row. Also the two expected cell counts in the first column sum to 76, so only one had to be calculated. Without much difficulty we see that of the six expected cell counts, only two had to be calculated, and the remaining four were obtained by subtraction. We say that, of the six pieces of data, only two are free to vary and, therefore, the degrees of freedom are 2. In general, the number of degrees of freedom will be

$$df = (\text{no. of rows} - 1)(\text{no. of columns} - 1)$$

From the χ^2 table with 2 degrees of freedom we see that the observed χ^2 statistic of 2.2025 falls below 4.61, which is the value associated with a tail probability of .10. Thus we have

$$p\text{-value} > .10$$

which says that there is insufficient evidence to reject H_0. We have failed to show a dependence between political party and opinion on gun control.

We now summarize the test procedure.

Chi-Square Test of Independence

APPLICATION: Test the independence of two classifying variables.

ASSUMPTIONS:

1. The experiment satisfies the properties of a multinomial experiment
2. The expected cell counts are greater than or equal to 5.

   H_0: The two classifications are independent.

   H_a: The two classifications are dependent.

TEST STATISTIC: $\chi^2 = \Sigma(o_j - e_j)^2/e_j$ where o_j represents the observed cell frequencies and e_j represents the expected cell frequencies given by

$$e_j = \frac{RC}{n}$$

where

R = row total
C = column total
n = grand total or the total number of subjects

The test is a right-tailed test where the p-value is found in the χ^2 table with $(r - 1)(c - 1)$ degrees of freedom (r denotes the number of rows and c denotes the number of columns).

EXAMPLE 12.3

A random sample of 400 undergraduate college students was classified according to class rank and study habits. From the data in Table 12.6, test to see if the two classifications are independent.

TABLE 12.6
Observed cell count

Class rank	Study habits			
	Good	Average	Bad	Total
Freshmen	20	42	58	120
Sophomore	25	48	32	105
Junior	31	28	35	94
Senior	24	27	30	81
Total	100	145	155	400

SOLUTION The null hypothesis is

H_0: class rank and study habits are independent

---

and the alternative hypothesis is

$$H_a: \text{class rank and study habits are dependent}$$

To find the value of the test statistic we must first find the expected cell counts. The expected cell counts are obtained by the formula

$$e_j = RC/n$$

For the first cell in the first row and first column we have

$$e_1 = (120)(100)/400 = 30$$

The next cell in the first row has expected count

$$e_2 = (120)(145)/400 = 43.5$$

The expected count in the third cell in the first row is

$$e_3 = (120)(155)/400 = 46.5$$

Continuing we have the expected cell counts in the array shown in Table 12.7.

TABLE 12.7

Expected cell count

Class rank	Study habits			
	Good	Average	Bad	Total
Freshmen	30	43.5	46.5	120
Sophomore	26.25	38.06	40.69	105
Junior	23.5	34.08	36.42	94
Senior	20.25	29.36	31.39	81
Total	100	145	155	400

Note that all expected cell frequencies exceed 5, so the condition for the χ^2 approximation is satisfied. Substituting the observed and expected values in the formula for the χ^2 statistic, we find that

$$\chi^2 = (20 - 30)^2/30 + (42 - 43.5)^2/43.5 + \cdots + (30 - 31.39)^2/31.39$$
$$= 3.333 + .052 + 2.844 + .060 + 2.596 + 1.856$$
$$+ 2.394 + 1.085 + .055 + .694 + .190 + .062$$
$$= 15.221$$

The degrees of freedom for the χ^2 statistic are

$$df = (r - 1)(c - 1) = (4 - 1)(3 - 1)$$
$$= (3)(2) = 6$$

From the χ^2 table with 6 degrees of freedom we see that

$$14.5 < 15.221 < 16.8 \quad \text{and thus} \quad .01 < p\text{-value} < .025$$

Consequently there is statistical evidence to reject H_0; therefore the data suggest that the study habits of the students are related to their class rank.

12.11 Test the independence of the two variables displayed in the accompanying contingency table.

	Variable A		
	Level 1	Level 2	Level 3
Variable B Level 1	65	39	16
Variable B Level 2	133	156	61

12.12 Test the independence of the two variables displayed in the accompanying contingency table.

	Variable A		
	Level 1	Level 2	Level 3
Variable B Level 1	572	418	451
Variable B Level 2	352	379	315
Variable B Level 3	256	278	205

12.13 A group of college students was classified as being either left- or right-handed and as being either mathematically inclined or not. From the data in the accompanying table test the hypothesis that the two classifications are independent.

	Predominant hand	
	Left	Right
Mathematically inclined	12	93
Not considered mathematically gifted	7	108

12.14 Suppose a number of patients were treated for cancer with the results as given in the accompanying table:

Toxic reaction	Tumor regression	
	Yes	No
Yes	15	5
No	4	22

Determine whether there is a relationship between the presence of toxicity and tumor regression.

12.15 An experimental psychologist wishes to study the effects that three different drugs have on the ability to learn a list of nonsense syllables. Sixty subjects were categorized as to the type of drug they had been receiving for the past 3 months and their ability to memorize the list of nonsense syllables. Determine whether the type drug is related to the subject's ability to memorize the list of nonsense syllables. Are all of the required assumptions for the χ^2 test met? (See accompanying table.)

12.16 The accompanying table shows the results of an experiment that was designed to study the effects of vaccinating laboratory animals against a particular disease.

TABLE FOR EXERCISE 12.15

Ability to memorize the list of syllables	Drug		
	A	B	C
Low	12	6	6
Medium	8	5	11
High	2	7	3

TABLE FOR EXERCISE 12.16

	Got the disease	Did not get the disease
Vaccinated	12	38
Not vaccinated	21	29

Test the independence of Disease and Vaccination.

12.17 In 1984 history was made when Geraldine Ferraro was nominated as the first woman vice-presidential candidate. For the first time ever, the sex of the candidate was an issue. On August 7 and 9, 1984, *Time* magazine conducted a telephone survey of 1,000 voters. From the following two contingency tables, determine whether the sex of the respondent is independent of their choice.

(a) If the election were tomorrow, for whom would you vote?

	Reagan/ Bush	Mondale/ Ferraro	Undecided
Men	245	140	115
Women	205	160	135

(b) Who would be a better vice president?

	Bush	Ferraro	Undecided
Men	245	155	100
Women	185	235	80

12.18 Does the desire to participate in class projects relate to a child's academic achievement? To study this issue, a sample of 80 third-grade students was asked if they wished to participate in the Science Project program. The students were then classified according to their academic standing. From the data in the accompanying table is there evidence of a relationship between the desire to participate in the Science Project program and their academic standing?

Academic standing	Desire to participate in Science Project	
	Yes	No
Below average	17	9
Average	14	13
Above average	22	5
Total	53	27

12.19 A political pollster would like to determine whether the voters' feelings on a local referendum are related to their views on freedom of the press. A sample of 237 voters was asked how they plan to vote on the referendum and were asked to choose one of the following that most closely represents their view on freedom of the press.

A—The press is at liberty to report anything it sees fit.
B—The press should not report anything that would jeopardize one's life.
C—The press should not report anything that would jeopardize one's life or reputation.

The accompanying contingency table summarizes the number of voters in each category. Do the data show a relationship between opinion on the referendum view on freedom of the press?

Response to question	Opinion on referendum			
	For	Against	Undecided	Total
A	24	29	7	60
B	68	39	12	119
C	47	8	3	58

12.4   The Chi-Square Test of Homogeneity

In the test of independence of the previous section, we attempted to determine whether two characteristics (variables) associated with the subjects in a *single* population were independent. For example, subjects randomly selected from a population were classified according to their views on gun control and their preferred political party. We then determined whether their views on gun control were independent of their preferred political party. The chi-square test of homogeneity, presented in this section, attempts to determine whether *several populations* are similar or *homogeneous* with respect to some variable. With the variable as one classification and the populations as the other, the data form a two-way contingency table. The assumptions and statements of the null and alternative hypotheses are different from those for the test of independence, but the details of the analysis are the same. We will illustrate with an example.

EXAMPLE 12.4

Suppose that over a 2-year period, 120 heart patients were treated with one of two drugs (A or B). After a period of time each patient's condition was rated as either no change, improved, or greatly improved. Table 12.8, a contingency table, gives the distribution of frequency counts. Determine whether the patients' conditions are similar with respect to the two drugs.

SOLUTION   The null and alternative hypotheses are stated as follows:

H_0: The proportions of patients falling in the three categories are the same for drug A and drug B.

TABLE 12.8

	Patient's condition			
	No change	Improved	Greatly improved	Total
Drug A	15	22	33	70
Drug B	20	18	12	50
Total	35	40	45	120

H_a: The proportions of patients falling in the three categories are not the same for drug A and drug B.

If we denote the probabilities of falling in the three categories (cells) for drug A as

$$\pi_{A1}, \pi_{A2}, \text{ and } \pi_{A3}$$

and for drug B as

$$\pi_{B1}, \pi_{B2}, \text{ and } \pi_{B3}$$

then the null hypothesis can be stated as follows:

$$H_0: \pi_{A1} = \pi_{B1}, \pi_{A2} = \pi_{B2}, \text{ and } \pi_{A3} = \pi_{B3}$$

From the two samples, one of size 70 from the population of patients who received drug A and the other of size 50 from the population of patients who received drug B, we wish to determine whether the two populations are homogeneous with respect to the cell probabilities.

Under the assumption that the null hypothesis is true, the expected cell counts are calculated just as they were in the test for independence. Namely, the expected cell count is

$$\frac{(\text{row total})(\text{column total})}{(\text{grand total})}$$

Thus we have

$$E(A \text{ and no change}) = (70)(35)/120 = 20.42$$
$$E(A \text{ and improved}) = (70)(40)/120 = 23.33$$
$$E(A \text{ and greatly improved}) = (70)(45)/120 = 26.25$$
$$E(B \text{ and no change}) = (50)(35)/120 = 14.58$$
$$E(B \text{ and improved}) = (50)(40)/120 = 16.67$$
$$E(B \text{ and greatly improved}) = (50)(45)/120 = 18.75$$

Table 12.9 gives the observed cell counts and the expected cell counts in parentheses. The test statistic is computed exactly as in the previous section. That is,

$$\chi^2 = \Sigma(o_j - e_j)^2/e_j$$

TABLE 12.9

	Patient's condition			
	No change	**Improved**	**Greatly Improved**	**Total**
Drug A	15 (20.42)	22 (23.33)	33 (26.25)	70
Drug B	20 (14.58)	18 (16.67)	12 (18.75)	50
Total	35	40	45	120

From the data we have

$$\chi^2 = (15 - 20.42)^2/20.42 + (22 - 23.33)^2/23.33$$
$$+ (33 - 26.25)^2/26.25 + (20 - 14.58)^2/14.58$$
$$+ (18 - 16.67)^2/16.67 + (12 - 18.75)^2/18.75$$
$$= 1.4386 + .0758 + 1.7357 + 2.0148 + .1061 + 2.743$$
$$= 7.801$$

Because there are two rows and three columns, the degrees of freedom are

$$df = (2 - 1)(3 - 1) = 2$$

From the χ^2 table we see that the observed χ^2 statistic of 7.801 falls between 7.38 and 9.21, the values associated with tail probabilities of .025 and .01, respectively. Thus we have

$$.01 < p\text{-value} < .025$$

which says that there is evidence to reject H_0. We conclude that the patients' conditions depend on the drug received.

Following is a summary of the test procedures.

Chi-Square Test of Homogeneity

APPLICATION: Contingency table with fixed marginal totals.

ASSUMPTIONS:

1. A random sample is selected from each of the row category populations. The sample sizes (row marginal totals) are fixed prior to sampling.

2. The expected cell counts are greater than or equal to 5.

H_0: The populations are homogeneous with respect to the variable of classification.

H_a: The populations are not homogeneous.

TEST STATISTIC: $\chi^2 = \Sigma(o_j - e_j)^2/e_j$

where o_j represents the observed cell frequencies and e_j represents the expected cell frequencies given by

$$e_j = \frac{RC}{n}$$

where

R = row total

C = column total

n = grand total or the total number of subjects

The test is a right-tailed test where the p-value is found in the χ^2 table with $(r - 1)(c - 1)$ degrees of freedom (r denotes the number of rows and c denotes the number of columns).

12.20 From the data in the accompanying table test the hypothesis that the proportions falling in the three categories are the same for the three populations:

	Category			
	1	**2**	**3**	**Total**
Population 1	35	16	29	80
Population 2	29	21	30	80
Population 3	25	27	28	80

12.21 From the data in the accompanying table test the hypothesis that the proportions falling in the four categories are the same for the three populations:

	Category				
	1	**2**	**3**	**4**	**Total**
Population 1	16	38	5	41	100
Population 2	24	41	12	23	100
Population 3	19	36	15	30	100

12.22 A pollster sampled 200 voters, 100 from District 1 and 100 from District 2 to determine their opinion on an upcoming referendum. The results of the survey are given in the accompanying contingency table.

	Opinion on referendum			
	Favor	**Against**	**Undecided**	**Total**
District 1	72	21	7	100
District 2	60	34	6	100

Is there evidence that the two districts will vote differently in the referendum?

12.23 A study was conducted to compare two treatments for smokers who wish to stop smoking. Three hundred smokers were divided between the two methods with the following results. Do the data suggest that the two treatments were equally effective in helping smokers stop smoking?

Treatment	Stopped smoking	Smoke less	Smoke the same
A	44	38	68
B	33	42	75

12.24 The record for single-season average rushing yards per game is held by Marcus Allen, who in 1981 with Southern Cal, rushed for an average of 212.9 yards per game. Generally when a record is broken, the back carrying the ball receives the credit in the press release; many believe, however, that the linemen should receive the credit. Samples of 50 football players and 70 members of the press were asked who should receive the most credit—the back or the linemen. From the accompanying data determine whether their views are similar.

TABLE FOR EXERCISE 12.24

	Who should receive the most credit?			
	Linemen	**Back**	**Both**	**Total**
Football players	22	14	14	50
Press	6	48	16	70

12.25 The average cost of a funeral ranges from $2,085 in the West to $2,705 in the Central states (*1983 Statistical Abstract of Funeral Costs in the United States*). One hundred people in each region who had buried a loved one were surveyed and asked to complete a questionnaire regarding the cost of the funeral. From the accompanying contingency table, determine whether the perceived cost of a funeral is the same across the regions.

Region	Less than expected	About what expected	More than expected
West	15	60	25
Central	20	38	42
South	34	44	22
East	12	40	48

12.26 Recent reports are that a Vietnam veteran is much more likely to commit suicide than a nonveteran. From the list of those eligible for the draft in 1970, a sample of 100 Vietnam veterans and a sample of 100 nonveterans were selected. Each of the 200 was asked if he had ever contemplated suicide. The results are recorded in the following contingency table.

Contemplated suicide	Vietnam veteran	
	Yes	**No**
Yes	32	11
No	68	89

Determine whether there is statistical evidence to suggest that the proportion of veterans who have contemplated suicide is different from nonveterans.

12.27 An experiment is designed to study the side effects of two drugs used as treatments for a certain ailment. A group of 90 subjects is assigned to two drug groups. After being given the specified drug, the side effects are classified as follows:

	Side effects			
	Major	**Minor**	**None**	**Total**
Drug A	13	15	17	45
Drug B	8	21	16	45

Are the side effects distributed the same for the two drugs?

12.28 Do male and female college students differ in their preference of a favorite sport? Random samples of 100 college females and 100 college males were asked what their favorite sport was. The results are recorded in the accompanying contingency table. Are the favorite sports distributed the same for males and females?

TABLE FOR EXERCISE 12.28

	Favorite sport			
	Football	Basketball	Baseball	Tennis
Male	33	38	24	5
Female	38	21	15	26

12.5 Computer Session (optional)

The computations for the χ^2 goodness-of-fit test are not too difficult with a hand-held calculator. In Example 12.3, however, we see that the computations for a contingency table can become tedious when the number of rows and columns becomes large. Minitab makes the task quite easy. All that is required is that the frequency counts be read into the columns of the worksheet with the READ command. The command to perform an analysis of the data in a contingency table is

        CHISQUARE analysis of table in C1,C2,...,Ck.

EXAMPLE 12.5

Using Minitab analyze the data in Example 12.3.

SOLUTION   Aside from round-off error we see that the solution given in Figure 12.3 is the same as that found in Example 12.3.

FIGURE 12.3

Minitab output for Example 12.5

```
MTB > READ THE TABLE INTO C1 C2 C3
DATA> 20 42 58
DATA> 25 48 32
DATA> 31 28 35
DATA> 24 27 30
DATA> END
      4 ROWS READ

MTB > CHISQUARE ANALYSIS OF TABLE IN C1 C2 C3

Expected counts are printed below observed counts

              C1        C2        C3     Total
     1        20        42        58       120
            30.0      43.5      46.5

     2        25        48        32       105
            26.3      38.1      40.7

     3        31        28        35        94
            23.5      34.1      36.4

     4        24        27        30        81
            20.2      29.4      31.4

   Total     100       145       155       400

ChiSq =    3.33 +    0.05 +    2.84 +
           0.06 +    2.59 +    1.85 +
           2.39 +    1.08 +    0.06 +
           0.69 +    0.19 +    0.06 = 15.22
df = 6
```

In the previous example the frequencies for the contingency table were given. In the following example we show how Minitab can construct the table of frequencies from the data.

EXAMPLE 12.6

A survey was conducted in a voting district with (among several other variables) the political party and sex of the respondent recorded and stored in Minitab. The political party and sex of the subject were coded as follows and stored in columns C5 and C8, respectively.

```
    C5                      C8

    1 — democrat            1 — female
    2 — republican          2 — male
    3 — other
```

Organize the data in a contingency table according to the political party and sex variables. Perform a chi-square test of independence on the two variables.

SOLUTION   The data, having been stored in C5 and C8, will be organized with the TABLE command and then analyzed with the CHISQUARE subcommand. The form of the command is

```
    TABLE C5, C8;
    CHISQUARE 2.
```

The 2 says to print observed and expected counts. If it is omitted, only observed counts are printed. If more than 2 columns are specified in the TABLE command, a separate chi-square test is done for each two-way table. Figure 12.4 gives the complete analysis.

FIGURE 12.4

Minitab output for Example 12.6

```
MTB > TABLE C5 C8;
SUBC> CHISQUARE 2.

   ROWS: C5       COLUMNS: C8

                1          2        ALL

      1        42         55         97
             45.01      51.99      97.00

      2        63         73        136
             63.10      72.90     136.00

      3        11          6         17
              7.89       9.11      17.00

    ALL       116        134        250
            116.00     134.00     250.00

CHI-SQUARE =      2.666    WITH D.I

   CELL CONTENTS —
                        COUNT
                        EXP FREQ
```

12.29 A sample of 550 voters was classified by political party and opinion on the administration's foreign policy. Test the hypothesis of independence of political party and opinion on the foreign policy by reading the data in the accompanying table into Minitab and conducting a chi-square analysis.

	Opinion	
	For	Against
Republican	167	103
Democrat	96	145
Other	15	24

12.30 A survey was given to 556 adults 18 years of age and older to determine if premarital sex is permissible. From the results in the accompanying table, determine whether their opinion is independent of their age.

Age	View toward premarital sex		
	OK	Not OK	No opinion
18–21	63	41	2
22–29	58	54	4
30–39	46	82	4
40–54	24	79	6
above 55	11	77	5

Read the data into Minitab and conduct a chi-square analysis.

12.31 Fifty residents were asked if they agree or disagree with the current school board policy on busing. Following are the sex and opinion of each respondent (M—male, F—female; f—for, a—against).

Sex	M	F	F	M	F	M	M	M	F
Opinion	f	a	f	f	f	a	f	a	f
Sex	F	F	M	F	M	F	M	M	F
Opinion	f	a	f	f	f	a	a	f	f
Sex	M	F	F	M	F	F	F	M	M
Opinion	f	a	f	f	a	f	f	a	f
Sex	F	M	F	F	M	F	M	M	M
Opinion	f	f	a	a	f	a	f	f	a
Sex	F	F	M	F	F	M	M	M	F
Opinion	a	f	f	a	f	f	a	f	f
Sex	F	M	F	F	M				
Opinion	a	f	a	a	f				

Store the data in Minitab and construct a two-way table using sex and opinion. Conduct a chi-square test of independence.

12.32 The accompanying contingency table describes the relation between scores above and below the median on an examination and ratings of job performance for 100 employees of a company.

	Rating		
	Below average	Average	Above average
Above median	11	23	36
Below median	14	7	9

Test the hypothesis that job performance is independent of examination results by reading the data into Minitab and conducting a chi-square analysis.

12.6 Summary and Review

Key Concepts

☑ When data consist of frequency counts and satisfy the properties of a multinomial experiment, the statistic given by

$$\chi^2 = \Sigma(o_j - e_j)^2/e_j$$

is used to test hypotheses about the data. In this chapter we discussed three applications of this test—the χ^2 goodness-of-fit test, the χ^2 test of independence, and the χ^2 test of homogeneity. To use the statistic in any of the three cases the expected cell counts must be found. For the goodness-of-fit the expected counts are

$$e_j = n\pi_j$$

where n is the number of subjects and π_j represents the hypothesized probability of falling in class j.

☑ For the test of independence and test of homogeneity the expected cell counts are found by

$$e_j = \frac{RC}{n}$$

where

$$R = \text{row total}$$

$$C = \text{column total}$$

$$n = \text{total number of subjects}$$

The test statistic is approximately distributed as χ^2 with the degrees of freedom being $k - 1$ in the goodness-of-fit test and $(r - 1)(c - 1)$ in the test of independence and test of homogeneity. In all cases, the approximation is valid only if the expected cell counts are 5 or more.

Learning Goals

Having completed this chapter you should be able to:

1. Organize data corresponding to a single qualitative variable into a frequency table. *Sections 12.1 and 12.2*

2. Organize data corresponding to two qualitative variables into a contingency table. *Sections 12.1 and 12.3*

3. Recognize the characteristics of a multinomial experiment. *Section 12.2*

4. Calculate the *p*-value associated with the χ^2 distribution. *Sections 12.2–12.4*

5. Conduct a χ^2 goodness-of-fit test. *Section 12.2*

6. Conduct a χ^2 test of independence. *Section 12.3*

7. Conduct a χ^2 test of homogeneity. *Section 12.4*

To test your skills, answer the following questions.

12.33 A multinomial experiment with three categories resulted in the accompanying table.

Category 1	Category 2	Category 3
48	102	150

Test the hypothesis that $\pi_1 = 1/6, \pi_2 = 2/6, \pi_3 = 3/6$.

12.34 From the accompanying contingency table determine whether the row and column classifications are independent.

	Column 1	Column 2	Column 3	Column 4
Row 1	15	26	18	31
Row 2	19	21	25	32

12.35 From the data in the accompanying table test the hypothesis that the proportions falling in the three categories are the same for the two populations:

	Category 1	Category 2	Category 3	Total
Population 1	22	33	45	100
Population 2	29	28	43	100

12.36 Teenage suicide is a serious problem in the United States. From 38 teenage suicides the family statuses were distributed as in the accompanying table.

Family status			
Upper class	Upper middle class	Lower middle class	Lower class
14	10	8	6

Is there statistical evidence that the proportions in the four classes differ?

12.37 A person is said to have a Type A personality if he or she is outgoing and always on the go. A Type B personality is the opposite. Suppose 50 Type A and 50 Type B persons were classified according to their risk of a heart attack with the results as given in the accompanying table.

	Risk of a heart attack			
	Low	Average	High	Total
Type A	9	27	14	50
Type B	12	31	7	50

Determine whether the risk of a heart attack is distributed the same for Type A and Type B personalities.

12.38 In a local referendum to legalize the sale of beer, a survey was conducted of the voters on their opinion on the referendum and their views on the moral issue of selling alcoholic beverages. From the accompanying contingency table, determine whether their moral values are related to their opinion on the referendum.

	Opinion on referendum		
	For	Against	Undecided
Drinking is			
A—OK	95	83	21
B—tolerated	73	71	18
C—immoral	12	46	8

SUPPLEMENTARY EXERCISES FOR CHAPTER 12

12.39 A multinomial experiment with four categories resulted in the accompanying table:

Category 1	Category 2	Category 3	Category 4	Total
29	74	32	65	200

Test the hypothesis that $\pi_1 = \pi_3 = .20, \pi_2 = \pi_4 = .30$.

12.40 Mutual funds continually revise their holdings, hoping to improve on their portfolio. At one point a certain mutual fund classified its holdings as in the accompanying table.

Income stocks	Growth stocks	High-risk stocks
48	39	43

Is there statistical evidence that its holdings are equally divided among the three types of stocks?

12.41 Physicians recommend one of four types of medication for a certain ailment. To determine whether one medication is preferred over another, a random sample of 60 physicians was polled on what medications they recommended to their patients. The table is a summary of their choices:

	Recommended medication				
	A	B	C	D	Total
No. of physicians recommending	16	12	21	11	60

Do the data suggest that the medications are equally recommended?

12.42 A sample of 120 diamonds from a diamond mine was graded as being low grade, medium grade, and high grade with the results given in the table.

	Low grade	Medium grade	High grade	Total
No. of diamonds	52	46	22	120

It is suspected that the percents of low- and medium-grade diamonds are the same and each is twice that of high grade. Is there evidence to support this conjecture?

12.43 A national standardized test has 350 as the first quartile, 425 as the median, and 560 as the third quartile. A sample of 20 subjects had the following scores:

300	480	402	523	628
324	561	349	476	538
647	572	621	385	491
587	401	346	565	421

Did the sample differ significantly from the national standard?

12.44 Fifty people were asked to choose their favorite California wine from four different brands. Their preferences are listed in the table. Is there statistical evidence that one wine is favored over the others?

Brand	A	B	C	D	Total
No. choosing	18	9	11	12	50

12.45 A company sales manager wishes to know whether all of his salespeople are contributing the same effort to the company. He records the number of sales made by each of the six salespeople over a given period of time. Are the sales equally distributed among the salespeople?

Salesperson	A	B	C	D	E	F	Total
No. of sales	34	21	44	52	37	42	230

12.46 A summary of the birthday selections for the 1970 draft lottery (see Appendix A3) is given in the accompanying table. The data give the number of selections in each month that were for the first half of potential draftees—that is, those with draft numbers less than or equal to 183. If the draft were truly random, then each month should have been selected with a probability proportional to the number of days in the month. Test the hypothesis that the months were selected with probabilities proportional to the number of days in the months.

Month	No. of selections in the first 183
Jan.	13
Feb.	12
March	9
April	11
May	14
June	14
July	14
Aug.	18
Sept.	19
Oct.	13
Nov.	21
Dec.	25

12.47 In a pre-election poll to study the influence that age has on voter preference for two presidential candidates, the following results were obtained:

Age	For Candidate A	For Candidate B	Undecided
20–29	67	117	16
30–49	109	74	17
Over 49	118	64	18

Test the independence of age and choice of candidate.

12.48 On August 3–6 1984, at shopping malls in Manhasset, New York, Houston, Chicago, Puente Hills and Westminster, California, and Kansas City, *People* magazine conducted a poll of 365 men and women of voting age. All subjects were shown photos of famous people such as President Reagan, Geraldine Ferraro, John Chancellor, and Diane Sawyer, and then asked to name them. From the following data determine whether the sex of the respondent is independent of his or her ability to recognize Diane Sawyer.

	Recognized Diane Sawyer	
	Yes	No
Male	20	153
Female	28	164

12.49 Executives of small, medium, and large corporations were polled on the issue of the economy. From the following data determine whether the size of the corporation is related to the outlook on the economy.

	Economic outlook		
	Optimistic	Cautious	Pessimistic
Small corporation	38	22	14
Medium corporation	25	20	11
Large corporation	20	8	12

12.50 A 14-year study of 200 female college graduates divided the participants into two groups according to their career. Those in male-dominated fields such as law, government, business, science, and medicine were classified in the Professional group. Those who never worked or with careers in teaching, nursing, secretarial, and social work were classified in the Traditional group. From the data in the accompanying table determine whether the marital status is independent of career.

	Married	
	Yes	No
Professional	65	35
Traditional	85	15

12.51 A research study was designed to compare three methods of therapy for mental patients. From the following data determine whether the rating is independent of the method of therapy.

	Rating		
	Improved greatly	Improved some	Not Improved
Therapy A	5	16	24
Therapy B	9	21	18
Therapy C	15	24	8

12.52 The following table classifies a group of people according to income bracket and time elapsed since they last consulted with a physician. Is there evidence of an association between the income bracket and the duration of time since the last visit with a physician?

Income bracket	Last consulted with physician		
	Less than 6 mos.	6 mos. to 1 yr.	More than 1 yr.
Less than $10,000	192	35	41
$10,000 to $19,999	124	43	65
$20,000 to $29,999	135	51	64
$30,000 to $39,999	174	62	108
$40,000 or more	121	65	52

12.53 It is believed that in many species of mammals and birds a greater proportion of young males than females die when resources are scarce. For example, due to the greater size, the domestic ram requires 15% more food than ewes. To study this further, a group of scientists used three populations of Red deer in three areas where the food supply was considered Low, Medium, and High, to record the number of deer that die before their first birthday. From the following data determine whether there is any association between the sex of the dead deer and the food supply.

Food supply	Number of deaths of Red deer before their first birthday	
	Males	Females
Low	23	17
Medium	12	10
High	7	8

12.54 Compared with boys, girls begin declining in academic achievement about age 13 and are below boys in math ability by age 17 (based on interviews of 285 educators and students in 10 cities, by the National Coalition of Advocates for Students, reported in *USA Today,* January 29, 1985). Many believe that the decline in math ability of girls is due to their own lack of confidence. Samples of 200 13-year-old girls and 200 13-year-old boys were asked how they perceived their math ability. The results are as follows:

	How do you perceive your math ability?				
	Hopeless	Below average	Average	Above average	Superior
Girls	56	61	54	21	8
Boys	35	43	61	42	19

Is there evidence that girls have less self-confidence than boys as far as math ability is concerned?

13

Analysis of Variance

This chapter extends the methods of Chapters 9 and 10 to the comparison of more than two population distributions. It is one of the most widely used statistical procedures available.

As stated, it is an extension of the one-sample and two-sample problems; the approach is somewhat different, however, because we introduce the idea of analyzing variability. For example, subjects treated for some ailment with different types of drugs have different recovery rates, but is the *variability* in their recovery rates due to the different drugs or is it simply because they are different people? Analysis of variance is a procedure that attempts to determine how much of the variability is due to the treatment and how much is due to other factors, which will be lumped together in what is called error.

This chapter on analysis of variance is an elementary introduction to the vast field of experimental design. A second course in statistical methods begins where this chapter ends.

Lowest in Tar and Nicotine

In almost any magazine one can find advertisements for several different brands of cigarettes, each claiming that theirs is the lowest in tar and nicotine.

To evaluate cigarettes the Federal Trade Commission (FTC) uses "smoking machines," which measure tar, nicotine, and carbon monoxide in each cigarette. Carbon monoxide has been linked to heart disease, tar has been linked to cancer, and nicotine is addictive.

Suppose that tar, measured in milligrams, is recorded on 25 cigarettes randomly selected from each of three brands. Do the accompanying data indicate that one brand is "lowest" in tar?

Because the mean is a typical measurement in a data set, it seems natural to compare the mean tar content for the three brands. In this chapter we will learn how to compare the means of three or more populations at one time.

Brand A	Brand B	Brand C
.61	.43	.52
.48	.71	.48
.44	.52	.67
.53	.65	.49
.71	.63	.38
.40	.55	.57
.53	.64	.70
.32	.41	.79
.61	.58	.47
.52	.63	.55
.40	.53	.49
.51	.57	.56
.53	.68	.51
.69	.44	.60
.49	.32	.41
.48	.53	.65
.58	.52	.57
.46	.43	.39
.59	.57	.54
.50	.48	.61
.47	.55	.53
.51	.68	.72
.71	.44	.50
.56	.53	.58
.61	.77	.49

Introduction

The methods for comparing two population means were presented in Chapter 10. In many applications, as in the Statistical Insight example, we wish to compare more than two means. Some other examples are (1) a college administrator wishes to compare the mean grade point averages of freshmen, sophomores, juniors, and seniors. (2) A building contractor wishes to compare the effectiveness of five different types of insulation. (3) A medical doctor wishes to compare the effectiveness of three different drugs that he or she is using to treat patients. (4) An agricultural experiment station wishes to compare the yields of crops treated with four different types of fertilizer. There is an unending list of experiments in which we wish to compare more than two means. *Analysis of variance* provides us with the tools to compare the means of several populations.

As in previous inference procedures, the analysis performed depends on the shape of the underlying parent distributions. Here, we will consider two alternatives. When it can be assumed that the parent distributions are normally distributed with homogeneous variances, the classical F procedure is performed. This is presented in Section 13.2. If those assumptions cannot be made about the parent distributions, we have the nonparametric Kruskal-Wallis (K-W) test. The K-W test, presented in Section 13.4, is recommended when the distributions are similar in shape but possibly differ in location. For example, the K-W test can be used to compare the medians of skewed populations if they are similar in shape.

▲ Recall that homogeneous variances means that the variances of the different populations are close in value.

13.2

Comparison of Several Means: The Completely Randomized Design

In Sections 10.2 and 10.3 we considered the comparison of two population means using independent samples that were randomly selected from the two populations. The extension of this to more than two populations is the *completely randomized design*.

Definition

The **completely randomized design (CR)** is an experiment in which independent random samples are obtained from each of k populations.

The design is given in Table 13.1.

TABLE 13.1

	Sample 1	Sample 2	...	Sample k	
	y_{11}	y_{12}		y_{1k}	
	y_{21}	y_{22}		y_{2k}	
	y_{31}	y_{32}		y_{3k}	
	.	.		.	
	.	.		.	
	.	.		.	
	$y_{n_1 1}$	$y_{n_2 2}$		$y_{n_k k}$	
Total	T_1	T_2		T_k	Grand total $= T$
Average	$\bar{y}_1$	$\bar{y}_2$		$\bar{y}_k$	Grand mean $= \bar{y}$
Variance	s_1^2	s_2^2		s_k^2	

In the array, y_{ij} denotes the response measurement obtained from experimental unit (subject) i in sample j where $i = 1, 2, \ldots, n_j$ and $j = 1, 2, \ldots, k$. Note that the possibility exists for different sample sizes for the various samples. The grand sample size is $N = n_1 + n_2 + n_3 + \cdots + n_k$. The grand mean is found by dividing the grand total by the grand sample size, that is, $\bar{y} = T/N$.

▲ The grand mean is the average of the individual averages *only* when all sample sizes are equal.

We illustrate the completely randomized design with an example. Suppose we wish to compare the average gas mileage of standard 4-wheel drive pickup trucks manufactured by Chevrolet, Dodge, and Ford. An experiment is designed where five vehicles of each type are randomly and independently selected from the population of 4-wheel drive trucks. (Choosing the samples randomly and independently defines the characteristics of a completely randomized design.) Each vehicle is driven in a stationary position for an equivalent of 500 miles. The gasoline consumed is measured and the miles per gallon are computed. The results appear in Table 13.2.

TABLE 13.2

▲ Because the sample sizes are equal, the grand mean is the average of the three averages.

	Chevy	Dodge	Ford	
	15.2	14.8	15.1	
	15.4	14.4	14.3	
	14.8	14.3	14.6	
	14.4	14.1	13.9	
	14.7	14.4	14.6	
Total	74.5	72.0	72.5	Grand total = 219.0
Average	14.9	14.4	14.5	Grand mean = 14.6
Variance	.16	.065	.195	

We should ask, "Are there any differences in the mean gas mileages of the three types of pickups?" At first glance we would say "yes" because the sample means differ (14.9, 14.4, 14.5). Yet we must remember that they are sample means, and sample means can fluctuate about a population mean. It might be that the population mean miles per gallon for each of the three types of vehicles is 14.6. That is, $\mu_1 = \mu_2 = \mu_3 = 14.6$ where μ_1, μ_2, and μ_3 represent the overall miles per gallon for each of the brands of trucks. Or it may be that there is indeed a difference in the population means, and the sample means are reflecting those differences.

Basically the question is, "Are the differences in the sample means different enough to conclude that the population means are different?" The question can be answered by a test of the null hypothesis

$$H_0: \mu_1 = \mu_2 = \mu_3$$

versus the alternative hypothesis

$$H_a: \text{at least two } \mu\text{'s differ}$$

As in all tests of hypotheses, an appropriate test statistic is needed to give a unique numerical measure of how much the sample means differ.

Consider for the moment that the miles per gallon for the vehicles mentioned were as appears in Table 13.3.

TABLE 13.3

	Chevy	Dodge	Ford	
	14.9	14.4	14.5	
	14.9	14.4	14.5	
	14.9	14.4	14.5	
	14.9	14.4	14.5	
	14.9	14.4	14.5	
Total	74.5	72.0	72.5	Grand total = 219.0
Average	14.9	14.4	14.5	Grand mean = 14.6
Variance	0.0	0.0	0.0	

The data suggest that Chevrolets, in general, get 14.9 miles per gallon, Dodges get 14.4 miles per gallon, and Fords get 14.5 miles per gallon. There is no variability of measurements within any of the samples. Yet there is a clear distinction between the samples and therefore a clear distinction between the population means. Data as in Table 13.3 lead to the rejection of H_0.

On the other hand, suppose now that the data were as in Table 13.4.

TABLE 13.4

	Chevy	Dodge	Ford	
	16.6	13.2	13.5	
	16.8	14.7	16.9	
	13.2	15.7	15.4	
	15.3	12.3	12.8	
	12.6	16.1	13.9	
Total	74.5	72.0	72.5	Grand total = 219.0
Average	14.9	14.4	14.5	Grand mean = 14.6
Variance	3.71	2.63	2.705	

Again the averages are 14.9, 14.4, and 14.5, yet there is no clear distinction between the samples. For example, any measurement in sample 2 could easily have been from sample 1 or 3. It is as if all the data came from a population with a lot of variability and a single mean of 14.6, rather than three samples from distinct populations. There is insufficient information to reject H_0.

The difference in the data sets of Tables 13.3 and 13.4 is the relationship between the variability across the samples and the variability within the samples. In Table 13.3 the variation between (across) the samples is large in comparison with the variation within the samples (there is no variation within the samples). Yet in Table 13.4 the variation between the samples is slight in comparison with the variation within. Figure 13.1 further points out the difference.

We see an obvious difference in the samples in Figure 13.1(a), whereas in 13.1(b) the data from the three samples are intermixed to the point that no distinction can be made between the samples.

Returning to the original data in Table 13.2, it is not clear whether the variation between samples is statistically greater than the variation within the samples. We will investigate with the test of hypothesis. The test statistic used to test the hypothesis

FIGURE 13.1

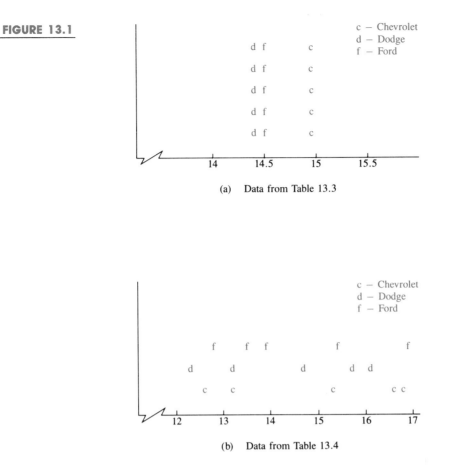

(a)   Data from Table 13.3

(b)   Data from Table 13.4

should be constructed to detect the differences in variation. But first we need to know how to measure the variations *between* and *within* samples.

Within-Sample Variability

If there is only one sample, the sample variance, s^2, is a measure of variability within that sample. In the case of multiple samples, a logical combined measure of variation within the samples is a pool of all the individual sample variances. Indeed, if

$$s_1^2, s_2^2, \ldots, s_k^2$$

represents the sample variances from k samples and $n_1, n_2, \ldots, n_k$ are the associated sample sizes, then

▲ Go back and review the definition of pooled variance in Chapter 10.

$$s_p^2 = \frac{(n_1 - 1)s_1^2 + (n_2 - 1)s_2^2 + \cdots + (n_k - 1)s_k^2}{n_1 + n_2 + \cdots + n_k - k}$$

is the obvious extension of the pooled variance considered in the two-sample t-test of Chapter 10. The numerator of s_p^2 is called the *sum of squares for error* because it is a combined measure of errors within each sample. The denominator is the degrees of freedom associated with s_p^2.

Recall that an s^2 that is obtained from n measurements has $n - 1$ degrees of freedom. Because s_p^2 is obtained from several s^2's, its degrees of freedom will be

$$(n_1 - 1) + (n_2 - 1) + \cdots + (n_k - 1) = n_1 + n_2 + \cdots + n_k - k$$

Dividing the degrees of freedom into the sum of squares gives the variance s_p^2, which is also called a *mean square*.

Definition

The **sum of squares for error** is

$$\text{SSE} = (n_1 - 1)s_1^2 + (n_2 - 1)s_2^2 + \cdots + (n_k - 1)s_k^2$$
$$= \Sigma(y_{i1} - \bar{y}_1)^2 + \Sigma(y_{i2} - \bar{y}_2)^2 + \cdots + \Sigma(y_{ik} - \bar{y}_k)^2$$

and has $n_1 + n_2 + \cdots + n_k - k$ degrees of freedom. Dividing the degrees of freedom into the sum of squares for error yields the **mean square for error**

$$\text{MSE} = s_p^2 = \frac{(n_1 - 1)s_1^2 + (n_2 - 1)s_2^2 + \cdots + (n_k - 1)s_k^2}{n_1 + n_2 + \cdots + n_k - k}$$

EXAMPLE 13.1

Calculate the sum of squares for error for the data in Table 13.2. Also give the degrees of freedom associated with the sum of squares for error and calculate the mean square for error.

SOLUTION   From Table 13.2 we find that

$$s_1^2 = .16, \quad s_2^2 = .065, \quad s_3^2 = .195$$

Moreover each sample size is 5, so that $n_j - 1 = 4$ for $j = 1, 2, 3$.
From the preceding formula

$$\text{SSE} = (4)(.16) + (4)(.065) + (4)(.195) = (4)(.42) = 1.68$$

The degrees of freedom are

$$n_1 + n_2 + \cdots + n_k - k = 5 + 5 + 5 - 3 = 12$$

Dividing SSE by 12

$$\text{MSE} = \text{SSE}/\text{df} = 1.68/12 = .14$$

Between-Sample Variability

To measure the variability between the samples we simply need to calculate the variation across the sample means. If

$$\bar{y}_1, \bar{y}_2, \ldots, \bar{y}_k$$

are the sample means of the k samples, then

$$\frac{\Sigma(\bar{y}_j - \bar{y})^2}{k - 1}$$

is the sample variance where the $\bar{y}_j$'s constitute the sample and $\bar{y}$ $(= T/N)$ is the overall sample mean. Recalling that each of the $\bar{y}_j$'s came from a sample of size n_j, we have

$$s_b^2 = \frac{\Sigma n_j(\bar{y}_j - \bar{y})^2}{k - 1}$$

as the between-sample variation. The numerator

$$\Sigma n_j(\bar{y}_j - \bar{y})^2$$

is called the between-sample sum of squares; or it is more commonly known as the *sum of squares for treatments,* because the various samples arise from the different treatments of the experiment. The denominator, $k - 1$, is the associated degrees of freedom.

Definition

The **sum of squares for treatments** is

$$\text{SST} = \Sigma n_j(\bar{y}_j - \bar{y})^2$$

and has $k - 1$ degrees of freedom. Dividing the degrees of freedom into the sum of squares yields the **mean square for treatments**

$$\text{MST} = s_b^2 = \frac{\Sigma n_j(\bar{y}_j - \bar{y})^2}{k - 1}$$

EXAMPLE 13.2

Calculate the sum of squares for treatments for the data in Table 13.2. Also give the degrees of freedom associated with the sum of squares for treatments and calculate the mean square for treatments.

SOLUTION   From the data we have

$$\bar{y}_1 = 14.9 \quad \bar{y}_2 = 14.4 \quad \bar{y}_3 = 14.5 \quad \text{and} \quad \bar{y} = 14.6$$

Also each $n_i = 5$ so that

$$\begin{aligned}
\text{SST} &= 5(14.9 - 14.6)^2 + 5(14.4 - 14.6)^2 + 5(14.5 - 14.6)^2 \\
&= 5(.09 + .04 + .01) \\
&= .7
\end{aligned}$$

The degrees of freedom are $k - 1 = 3 - 1 = 2$. Dividing into SST, we have

$$\text{MST} = \text{SST}/\text{df} = .7/2 = .35$$

The *F* Distribution

As previously stated, the test statistic used to test equality of population means should compare the relative magnitude of the between-sample variability with that of the within-sample variability. This is accomplished by calculating the ratio of MST and MSE. That is, we calculate the value of

$$F = \frac{\text{MST}}{\text{MSE}}$$

Significantly large values of F indicate that the variability between the samples (MST) is significantly larger than the variability within the samples (MSE), which in turn indicates a difference in population means. Thus large values of F should lead to the rejection of the null hypothesis that the population means are equal.

The next question is, "How large does F have to be in order to reject H_0?" The answer depends on the sampling distribution of F.

The sampling distribution of F is not easily obtainable when sampling from arbitrary populations. If the null hypothesis is true, however, and if the samples are *independent* and from *normal populations* with *equal variances,* then the sampling distribution is called the *F distribution* after the renowned English statistician Sir Ronald Fisher (1890–1962).

Because the F statistic is comprised of the ratio of two mean squares, which have their own associated degrees of freedom, we say that the F distribution has degrees of freedom associated with its numerator and its denominator. As with other sampling distributions, the shape depends on the degrees of freedom. In general, the F distribution begins at zero (a negative F is impossible) and is skewed right. The amount of skewness depends on the number of degrees of freedom in the numerator and the denominator. An F with 2 degrees of freedom in the numerator and 12 in the denominator is given in Figure 13.2.

FIGURE 13.2

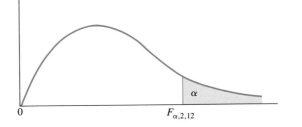

In order to assess the statistical evidence against H_0 we need to determine the tail probability (p-value) of the distribution. The upper-tail F-value, F_{α, df_n, df_d}, for various values of df_n, df_d, and α are given in Table B6 of Appendix B. The value df_n denotes the degrees of freedom for the numerator, and the value df_d denotes the degrees of freedom for the denominator.

EXAMPLE 13.3

Suppose $df_n = 2$ and $df_d = 12$.

1. Find the F-values associated with upper-tail probabilities of .05, .025, .01.
2. Find the approximate p-values associated with an F of 2.5 and 4.6.

SOLUTION

1. From Table B6 we find

$$F_{.05, 2, 12} = 3.89$$
$$F_{.025, 2, 12} = 5.10$$
$$F_{.01, 2, 12} = 6.93$$

2. Also from Table B6 we have $F_{.10, 2, 12} = 2.81$. Because 2.5 is less than 2.81, we have that its p-value is greater than .10. Because 4.6 is between 3.89 and 5.10, we have that its p-value is between .025 and .05.

The *F*-Test

In the CR design we wish to compare several population means. That is, we wish to test

$$H_0: \mu_1 = \mu_2 = \cdots = \mu_k$$

versus

$$H_a: \text{at least two } \mu\text{'s differ}$$

As stated the test statistic is

$$F = \text{MST}/\text{MSE}$$

Certain assumptions must be satisfied, however, in order to conduct an *F*-test. Those assumptions are:

Assumptions for *F* Distribution

1. The samples are randomly and independently selected from their respective populations.

2. The sampled populations are normally distributed.

3. The variances of the sampled populations are equal; that is,

$$\sigma_1^2 = \sigma_2^2 = \cdots = \sigma_k^2$$

When these assumptions are satisfied and when the null hypothesis is true, then

$$F = \text{MST}/\text{MSE}$$

has an *F*-distribution with $k - 1$ degrees of freedom in the numerator and $n_1 + n_2 + \cdots + n_k - k$ degrees of freedom in the denominator.

Because the test is a comparison of two sources of variation, the procedure is called *analysis of variance* (ANOVA).

Analysis of Variance for the Completely Randomized Design

ASSUMPTIONS: As stated above.

$$H_0: \mu_1 = \mu_2 = \cdots = \mu_k$$

$$H_a: \text{at least two } \mu\text{'s differ}$$

TEST STATISTIC: $F = \text{MST}/\text{MSE}$

The test is a right-tailed test where the *p*-value is found in the *F*-table with $k - 1$ and $n_1 + n_2 + \cdots + n_k - k$ degrees of freedom. Usually the exact *p*-value cannot be found, but bounds for it can be found by finding the closest value to the observed value of the *F* statistic.

If a level of significance α is specified, then reject H_0 if *p*-value $< \alpha$.

EXAMPLE 13.4

Returning to the earlier discussion of the gas mileage of pickup trucks, we wish to test the equality of the mean miles per gallon for the three major brands of trucks. From the data in Table 13.2 complete the test.

SOLUTION   Here μ_1, μ_2, and μ_3 denote the mean miles per gallon for Chevrolet, Dodge, and Ford trucks, respectively. The null hypothesis says they are the same. That is,

H_0: $\mu_1 = \mu_2 = \mu_3$

H_a: at least two μ's differ

From Example 13.1 we found that SSE = 1.68 with 12 degrees of freedom, so that MSE = 1.68/12 = .14. From Example 13.2 we found that SST = .7 with 2 degrees of freedom, so that MST = .7/2 = .35.

Now we have

$$F = \text{MST}/\text{MSE} = .35/.14 = 2.5$$

From Example 13.3 we found that the p-value associated with an F of 2.5, with 2, 12 degrees of freedom, to be greater than .10. Because p-value > .10, we fail to reject H_0. Thus there is insufficient evidence to say that the trucks differ in gas mileage.

As we have noted, SST and SSE measure the variability between samples and within samples, respectively. Combining the two we get a measure of the total variability in the data, called the *total sum of squares*.

Definition

The **total sum of squares** (TSS) in a data array consisting of k samples is the total of the squared deviations between each observation and the overall mean. If y_{ij} denotes the ith observation in the jth sample, and $\bar{y}$ denotes the overall mean, then

$$\text{TSS} = \Sigma\Sigma(y_{ij} - \bar{y})^2$$

▲ The double summation denotes a sum over subscripts i and j; that is, a summation over rows and columns of the data array.

The degrees of freedom associated with TSS are the number of observations less one. Thus we have

$$\text{df} = n_1 + n_2 + \cdots + n_k - 1$$

Computational Formulas

(Refer to the notation in Table 13.1.)

Total sum of squares = TSS = $\Sigma\Sigma y_{ij}^2 - T^2/N$

Treatment sum of squares = SST = $\Sigma(T_j^2/n_j) - T^2/N$

Error sum of squares = SSE = TSS − SST

The value $C = T^2/N$ appears frequently and is called the correction term.

An important relationship between TSS, SST, and SSE is

$$TSS = SST + SSE$$

Generally the formula for TSS is easier to use than the formula for SSE, thus this equation provides an alternative way of finding SSE. We have

$$SSE = TSS - SST$$

The calculations for TSS and SST can be simplified further by using formulas similar to the computational formula for a single sample variance s^2.

EXAMPLE 13.5

Use the computational formulas to find TSS, SST, and SSE for the data in Table 13.2.

SOLUTION   To calculate TSS we need to square each observation and then sum. From Table 13.2 we have

$$\begin{aligned}
TSS &= (15.2)^2 + (15.4)^2 + \cdots + (14.6)^2 - (219.0)^2/15 \\
&= 3199.78 - (219.0)^2/15 \\
&= 3199.78 - 3197.4 \\
&= 2.38
\end{aligned}$$

Also, from Table 13.2 we have

$$T_1 = 74.5 \quad T_2 = 72.0 \quad T_3 = 72.5 \quad \text{and} \quad T = 219.0$$

so that

$$\begin{aligned}
SST &= (74.5)^2/5 + (72.0)^2/5 + (72.5)^2/5 - (219.0)^2/15 \\
&= 3198.1 - 3197.4 \\
&= .7
\end{aligned}$$

From TSS and SST we have

$$SSE = 2.38 - .7 = 1.68$$

We see that the values agree with those found earlier.

The results of an analysis of variance are often summarized in an *ANOVA table*. The form of the table for the completely randomized design is presented in Table 13.5.

TABLE 13.5
ANOVA Table

▲ Remember that
$N = n_1 + n_2 + n_3 + \cdots + n_k$.

SV	SS	df	MS	F	p-value
Treatment	SST	$k-1$	MST	MST/MSE	
Error	SSE	$N-k$	MSE		
Total	TSS	$N-1$			

The first column lists the possible *sources of variation* in the data. The second column gives the numerical value of the source (i.e., the *sum of squares*). The third column gives the *degrees of freedom* associated with the sum of squares. The fourth column gives the *mean square* (or the s^2), which is obtained by dividing the sum of squares by the degrees of freedom. The fifth and sixth columns give the *F*-ratio and its *p*-value.

▲ Note that the degrees of freedom for treatment and error add to the total degrees of freedom.

EXAMPLE 13.6

Complete an ANOVA table for the truck mileage data.

SOLUTION Table 13.6 shows the completed analysis of variance table.

TABLE 13.6

SV	SS	df	MS	F	p-value
Treatment	.7	2	.35	2.5	> .10
Error	1.68	12	.14		
Total	2.38	14			

Because the *p*-value > .10, there is insufficient evidence to say that the trucks differ in gas mileage.

After conducting an analysis of variance it is possible to find a confidence interval for any one of the treatment means. The procedure is the same as a one-sample confidence interval based on $\bar{y}$ given in Chapter 8, except that the estimate of σ^2 is obtained from the pooled variance MSE, given in the ANOVA table.

> A $(1 - \alpha)100\%$ confidence interval for the mean of treatment j is given by the limits
>
> $$\bar{y}_j \pm t_{\alpha/2, N-k} \sqrt{MSE/n_j}$$

EXAMPLE 13.7

From the data in Table 13.2, find a 95% confidence interval for the mean miles per gallon for Chevrolet trucks.

SOLUTION From Table 13.2 we have

$$\bar{y}_1 = 14.9 \quad \text{and} \quad n_1 = 5$$

From Example 13.1,

$$MSE = .14$$

From the *t*-table,

$$t_{.025, 12} = 2.179$$

The confidence interval is

$$14.9 \pm 2.179 \sqrt{.14/5}$$

or
$$14.9 \pm .3646$$

which gives the limits
$$(14.5354, 15.2646)$$

EXAMPLE 13.8

An experiment was conducted to study the reaction effects of four drugs on a nervous disorder. Twenty-eight subjects with the nervous disorder were independently and randomly assigned to the four drug groups—seven to each group. Unfortunately two subjects in group 1 and one in group 4 were unable to complete the experiment. The reaction time to an experimental task was recorded for the remaining 25 subjects after having been administered their drug (see Table 13.7).

TABLE 13.7

Drug type			
1	2	3	4
3	5	6	2
5	7	5	4
4	3	7	3
6	4	9	4
4	5	6	2
	3	7	5
	6	8	

Complete an ANOVA table and test the hypothesis of equality of mean reaction time for the four drugs.

SOLUTION   If we let μ_j denote the mean reaction time to drug j for $j = 1, 2, 3$, and 4, the null hypothesis is
$$H_0: \mu_1 = \mu_2 = \mu_3 = \mu_4$$

and the alternative is
$$H_a: \text{at least two } \mu\text{'s differ}$$

The sample sizes are
$$n_1 = 5, \quad n_2 = 7, \quad n_3 = 7, \quad n_4 = 6, \quad \text{so that} \quad N = 25$$

From the data we find
$$\Sigma\Sigma y_{ij}^2 = 685 \quad \text{and} \quad T = 123$$

from which we can compute
$$\begin{aligned} \text{Total sum of squares} = \text{TSS} &= \Sigma\Sigma y_{ij}^2 - T^2/N \\ &= 685 - 123^2/25 \\ &= 79.84 \end{aligned}$$

Further we have

$$T_1 = 22, \quad T_2 = 33, \quad T_3 = 48, \quad T_4 = 20$$

from which we can compute

$$\text{Treatment sum of squares} = \text{SST} = \Sigma(T_j^2/n_j) - T^2/N$$
$$= 22^2/5 + 33^2/7 + 48^2/7 + 20^2/6 - 123^2/25$$
$$= 43.021$$

We also have

$$\text{Error sum of squares} = \text{SSE} = \text{TSS} - \text{SST}$$
$$= 79.84 - 43.021$$
$$= 36.819$$

We summarize the results in the ANOVA table shown in Table 13.8.

TABLE 13.8

SV	SS	df	MS	F	p-value
Treatment	43.021	3	14.34	8.179	< .01
Error	36.819	21	1.7533		
Total	79.84	24			

A p-value $< .01$ means that we should reject H_0 in favor of the alternative. Thus there is significant evidence that the mean reaction effects of the four drugs are different. In the next section we will address the problem of determining where the difference lies.

EXERCISES 13.2

13.1 Find the F-values associated with upper-tail probabilities of .05, .025, and .01 when
(a) $df_n = 4$ and $df_d = 15$
(b) $df_n = 6$ and $df_d = 23$
(c) $df_n = 10$ and $df_d = 30$

13.2 Given that $df_n = 5$ and $df_d = 25$. Find the approximate p-values associated with an F of
(a) 2.4
(b) 3.1
(c) 4.7

13.3 From the following data set calculate SSE, SST, and TSS. Also give the associated degrees of freedom and compute the mean squares.

Group A					
0					
1	0	2	4		
2	3	5	6	7	7
3	1	3	6	7	8
4	0	2	3		
5					
6	4	5			

Group B						
0	8					
1	2					
2	2	4	4	6		
3	0	3	5	5	8	9
4	1	4	5			
5	0	2				

Group C				
0				
1				
2	2	4	6	
3	5	8	9	9
4	0	3	5	6
5	3	5	7	
6	0	2		

13.4 Construct an analysis of variance table for the data in Exercise 13.3. Compute the *p*-value associated with the *F*-value.

13.5 From the following summary data calculate SSE, SST, and TSS. Also give the associated degrees of freedom and compute the mean squares.

Group A	Group B	Group C
$n = 23$	$n = 26$	$n = 25$
$\bar{y} = 67$	$\bar{y} = 76$	$\bar{y} = 72$
$s = 6.4$	$s = 7.9$	$s = 8.3$

13.6 Construct an analysis of variance table for the data in Exercise 13.5. Compute the *p*-value associated with the *F*-value.

13.7 From the data in the following table calculate SSE, SST, and TSS. Also give the associated degrees of freedom and compute the mean squares.

Treatment 1	Treatment 2	Treatment 3
5	5	8
7	1	6
6	3	7
6	5	9
4	3	5
5	6	8
7	4	7
	3	8
	4	

13.8 The accompanying table gives error scores obtained for four groups of experimental animals in running a maze under different experimental conditions. Test the hypothesis that there is no difference in the mean error scores for the four experimental conditions.

Condition A	Condition B	Condition C	Condition D
16	20	9	15
12	18	11	14
15	22	14	18
13	17	15	20
15	21	8	16
14	19	10	17
15	18	11	17
14	18	10	16

13.9 The accompanying table lists the salaries of professional football (in thousands of dollars), baseball, and basketball players selected at random from the list of all professional players. Test the hypothesis that the mean salaries for the three groups are the same.

TABLE FOR EXERCISE 13.9

Football (× $1000)	Baseball (× $1000)	Basketball (× $1000)
70	30	100
120	90	900
85	80	300
200	250	90
60	70	1200
310	55	260
90	180	60

13.10 Three randomly selected groups of chickens are fed three different rations. The weight gained during a specified period of time are as shown in the accompanying table. Test the hypothesis that there is no difference in the average weight gained for the three rations.

Ration 1	Ration 2	Ration 3
4	3	6
4	4	7
7	5	7
3	4	7
2	6	6
5	4	8
4	5	5
5	6	6
2	7	7
3	6	6
6	5	7
4	5	5
5	5	6

13.11 In assessing the impact of the level of impurities in a particular ingredient upon the solubility of a type of aspirin tablet, a statistician wished to test the null hypothesis that the mean dissolving time is the same regardless of the impurity level. In test batches, the dissolving times measured in seconds were recorded in the accompanying table.

	Level of impurity	
1%	5%	10%
2.0	1.9	2.3
1.8	2.3	2.3
1.7	2.2	2.2
1.9	1.9	2.1
2.1	2.2	2.6

Construct an analysis of variance table and test the hypothesis.

13.12 Suppose that four groups of 11 students each used a different method of programmed learning to study statistics. A standardized test was administered to the four groups and was graded on a 15-point scale. Given the results shown in the accompanying table, determine whether there was a significant difference in the results of the four methods.

TABLE FOR EXERCISE 13.12

Method I	Method II	Method III	Method IV
3	5	7	4
5	7	5	6
6	7	6	6
8	7	8	7
4	6	7	6
3	6	6	5
5	8	9	5
6	4	8	5
4	6	7	6
6	7	7	5
3	5	8	4

13.3   Tukey's Multiple Comparison Procedure

The analysis of variance test presented in the previous section was to test the equality of the population means. If the null hypothesis

$$H_0: \mu_1 = \mu_2 = \cdots = \mu_k$$

is accepted, then the problem is essentially finished. If H_0 is rejected, however, the question still remains as to which means differ. For example, if the hypothesis of equality of the gas mileages of the three makes of pickup trucks had been rejected, then we would like to know which truck got the best mileage and which got the worst. Of course, the hypothesis was not rejected, so the question is meaningless because the three makes of trucks get essentially the same gas mileage. On the other hand, when differences exist, we can further analyze the difference with a *multiple comparison test*.

There are several multiple comparison tests, but the one we will consider is Tukey's procedure, which utilizes the *Studentized range,*

$$q = \frac{\bar{y}_{max} - \bar{y}_{min}}{s/\sqrt{n}}$$

▲ The multiple comparison test is needed only after an analysis of variance has led to rejection of the equality of the population means.

We will not study the statistical properties of the Studentized range, but will illustrate its use in Tukey's multiple comparison test.

The upper-tail probability, $q_\alpha(k, \nu)$, of the Studentized range will be needed. It is found in Table B7 in Appendix B by locating the value of k along the top of the table and the value of ν along the left-hand column. The upper-tail probabilities of the Studentized range are given for $\alpha = .05$ and $.01$. For example, if $k = 6$ and $\nu = 20$, then $q_{.05}(6, 20) = 4.45$ and $q_{.01}(6, 20) = 5.51$.

The test is summarized as follows:

Tukey's Multiple Comparison Procedure

For a specified value of α, calculate

$$\omega = q_\alpha(k, \nu) \sqrt{MSE/n}$$

where

n = number of observations in each sample
MSE = the mean square error from the ANOVA table
k = number of different population means
ν = degrees of freedom associated with MSE
$q_\alpha(k, \nu)$ = upper-tail probability of the Studentized range found in Table B7 of Appendix B.

To conduct Tukey's procedure complete the following steps.

1. Rank the sample means from highest to lowest and order the population means in the same order.

2. Compute the difference between the largest and smallest sample means: $\bar{y}_{largest} - \bar{y}_{smallest}$. If the difference exceeds ω, then the corresponding population means are declared significantly different.

   Proceed to compute the difference between the largest and the next smallest sample mean: $\bar{y}_{largest} - \bar{y}_{2nd\,smallest}$. As previously indicated, if the difference exceeds ω, then declare the corresponding population means different.

   Continue to make comparisons with the largest sample mean, $\bar{y}_{largest} - \bar{y}_{3rd\,smallest}$ and so on, until a difference fails to exceed ω. Once a difference between two sample means is less than ω, the difference between the corresponding population means, and all means between, are declared nonsignificant.

3. Next make comparisons with the next largest sample mean, $\bar{y}_{2nd\,largest} - \bar{y}_{smallest}$ and so on, using the same procedures as in step 2. Continue until all possible comparisons are made.

4. Summarize the results by drawing a line under the population means that are declared nonsignificant.

EXAMPLE 13.9

An ecologist wishes to investigate the relative levels of mercury pollution in five major lakes. He catches 10 lake trout from each lake and measures the concentration of mercury present in each fish. A summary of the measurements in parts per million (ppm) is as follows:

	Lake 1	Lake 2	Lake 3	Lake 4	Lake 5
n	10	10	10	10	10
$\bar{y}$	4.1	3.7	2.4	4.6	3.4
s	.82	1.06	.68	1.44	.93

Perform an analysis of variance on the means. If a significant difference exists, analyze the means with Tukey's multiple comparison procedure.

SOLUTION   The null hypothesis is

$$H_0: \mu_1 = \mu_2 = \mu_3 = \mu_4 = \mu_5$$

where μ_j is the mean concentration of mercury present in the fish from lake j and $j = 1$, 2, 3, 4, 5. From the data we have

$$n_1 = n_2 = n_3 = n_4 = n_5 = 10 \quad \text{and} \quad N = 50$$

We also have

$$T_1 = 41, \quad T_2 = 37, \quad T_3 = 24, \quad T_4 = 46, \quad T_5 = 34$$

and

$$T = 182$$

from which we get

$$C = 182^2/50 = 662.48$$

and

$$\begin{aligned} \text{SST} &= (41^2 + 37^2 + 24^2 + 46^2 + 34^2)/10 - C \\ &= 689.8 - 662.48 \\ &= 27.32 \end{aligned}$$

From the values of s we get

$$\begin{aligned} \text{SSE} &= 9(.82)^2 + 9(1.06)^2 + 9(.68)^2 + 9(1.44)^2 + 9(.93)^2 \\ &= 9(5.1969) = 46.7721 \end{aligned}$$

The ANOVA table is as follows:

SV	SS	df	MS	F	p-value
Treatment	27.32	4	6.83	6.6	$< .01$
Error	46.7721	45	1.03938		
Total	74.0921	49			

The p-value indicates that the null hypothesis should be rejected, indicating that the mean concentration of mercury in the various lakes is different. We will now use Tukey's multiple comparison procedure to examine the differences.

We will compute ω using $\alpha = .01$. We have $n = 10$, $k = 5$, $\nu = 45$, and MSE $= 1.03938$. From Table B7 in Appendix B we do not find $q_{.01}(5, 45)$, because the value for 45 degrees of freedom is not present. We will use the closest value of q, which corresponds to 40 degrees of freedom. Thus we have

$$q_{.01}(5, 45) \approx 4.93$$

Note that in going from 40 to 60 degrees of freedom the value of q does not change much. Thus very little is lost by using the value of q for 40 degrees of freedom.

Using the previous information, we have

$$\omega = q_{.01}(5, 45) \sqrt{MSE/n}$$
$$= 4.93 \sqrt{1.03938/10}$$
$$= 1.589$$

Ranking the sample means from highest to lowest, we have:

Population mean	μ_4	μ_1	μ_2	μ_5	μ_3
Sample mean	4.6	4.1	3.7	3.4	2.4

Because $\bar{y}_4$ is the largest, comparisons will be conducted with it first. The analysis is presented in Table 13.9.

TABLE 13.9

Comparison	Difference	ω	Conclusion
$\bar{y}_4 - \bar{y}_3$	2.2	1.589	significant
$\bar{y}_4 - \bar{y}_5$	1.2	1.589	not significant
$\bar{y}_1 - \bar{y}_3$	1.7	1.589	significant
$\bar{y}_1 - \bar{y}_5$	.7	1.589	not significant
$\bar{y}_2 - \bar{y}_3$	1.3	1.589	not significant
$\bar{y}_5 - \bar{y}_3$	1.0	1.589	not significant

We now list the population means in the ranked order and draw a line under the means that we judged not significantly different.

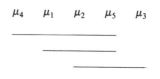

The line from μ_4 to μ_5 signifies that population means μ_4, μ_1, μ_2, and μ_5 do not differ, yet are significantly larger than μ_3.

The line from μ_1 to μ_5 is a subset of the line from μ_4 to μ_5 and, therefore, is not needed. Also the line from μ_5 to μ_3 is a subset of the line from μ_2 to μ_3 and is not needed. The final summary is

$$\mu_4 \quad \mu_1 \quad \mu_2 \quad \mu_5 \quad \mu_3$$

which says that μ_4 and μ_1 are significantly larger than μ_3.

Tukey's procedure can be modified for unequal sample sizes by comparing $\bar{y}_i - \bar{y}_j$ to ω_{ij}, where

$$\omega_{ij} = q_\alpha(k, \nu) \sqrt{MSE/2} \sqrt{1/n_i + 1/n_j}$$

Note that $\omega_{ij} = \omega$ when $n_i = n_j$.

The remainder of the procedure is as before.

EXERCISES 13.3

13.13 Find $q_\alpha(k, \nu)$ for the following values of α, k, and ν.
(a) $\alpha = .01$, $k = 7$, $\nu = 15$
(b) $\alpha = .05$, $k = 3$, $\nu = 14$
(c) $\alpha = .01$, $k = 10$, $\nu = 30$

13.14 The accompanying tables give the summary data that resulted from an experiment involving a treatment with four levels.

Treatment

	A1	A2	A3	A4
n	8	8	8	8
Σy_i	28	22	62	50

ANOVA table

Source	SS	df
Treatment	131.375	3
Error	41.0	28

Analyze the means using Tukey's multiple comparison procedure with $\alpha = .01$.

13.15 Twenty subjects were randomly assigned to four reducing diets (five subjects per diet) for a period of 6 weeks. After the 6-week period the amount of weight lost was recorded and an analysis of variance performed. The results of the test are as follows:

Total weight lost

Diet A	Diet B	Diet C	Diet D
26	42	65	22

ANOVA table

Source	SS	df
Diet	228.55	3
Error	84.1	16

Analyze the means using Tukey's multiple comparison procedure with $\alpha = .05$.

13.16 The following reaction times (measured in tenths of a second) were recorded for a group of subjects after each subject had been given a drug for pain.

Drug A	Drug B	Drug C
4	9	8
7	11	6
6	12	7
3	8	6
4	10	5
3	11	7

Perform an analysis of variance to test the equality of the mean reaction times for the three drugs. If a significant difference is found, analyze the means using Tukey's multiple comparison procedure with $\alpha = .01$.

13.17 Using an α of .05, perform Tukey's multiple comparison procedure on the sample means in Example 13.8.

The completely randomized test presented in Section 13.2 is appropriate when the parent distributions are normal with equal variances. An alternative to the ordinary F-test, when the assumptions are violated, is the nonparametric Kruskal-Wallis test. The only assumption made about the parent distributions is that they are similar in shape. Then the Kruskal-Wallis test can be used to test equality of location. For example, if the parent distributions are similarly skewed, then the Kruskal-Wallis test can be used to test equality of their population medians.

▲ Because the first step is to rank the data, the Kruskal-Wallis test can be applied to ordinal data, whereas the usual F-test can be applied only to interval or ratio data.

The procedure for the Kruskal-Wallis test, which is presented here, is the ordinary F-test applied to the rank summary data, n_j, $\bar{y}_{R_j}$, and s_{R_j}. This is an approximation of the original Kruskal-Wallis test, and therefore it is suggested that there are *five or more observations*. For sample sizes less than 5, exact tables for the Kruskal-Wallis test can be found in Conover (1980). It is also required that the samples be random and independent and can be ranked. As in the two-sample Wilcoxon test of Chapter 10, the ranks are assigned to the combined samples.

EXAMPLE 13.10

A study was conducted to compare the hostility levels between high school students located in rural, suburban, and urban areas. A psychological test, the Hostility Level Test (HLT), was used to measure the degree of hostility. Fifteen students were randomly selected from each type of school and given the HLT. The data are summarized in the following display of side-by-side ordered stem and leaf plots:

```
           Rural                    Suburban                     Urban

  1 | 6
  2 | 1                       2 | 2
  3 | 3                       3 | 7                      3 | 3
  4 |
  5 | 1  3                    5 | 2                      5 | 3  4
  6 | 3  4  4  6  7  8  8     6 | 3  5  7                7 | 2  4  6
  7 | 2  5  7                 7 | 0  2  3  4  6  8  9     8 | 0  2  3  3  4  6  7  8
  8 |                         8 | 2  3                    9 | 2
  9 |
```

From the preceding data, is there evidence to indicate that the hostility levels of students differ in the three school environments?

SOLUTION   First we will look at quantile summary diagrams for each of the three data sets.

Quantile summary diagrams

	Rural				Suburban				Urban		
	15		Midsummaries		15		Midsummaries		15		Midsummaries
$M(8)$	64		64	$M(8)$	72		72	$M(8)$	82		82
$Q(4.5)$	52	68	60	$Q(4.5)$	64	77	70.5	$Q(4.5)$	73	85	79
$E(2.5)$	27	73.5	50.25	$E(2.5)$	44.5	80.5	62.5	$E(2.5)$	53.5	87.5	70.5
R	16	77	46.5	R	22	83	52.5	R	33	92	62.5

▲ The normality assumption has been violated; hence the usual F-test should not be used. The distributions are similarly skewed, so the Kruskal-Wallis test is suggested.

The midsummaries, together with the shapes of the stem and leaf plots, suggest that all distributions are skewed left. Thus to compare the hostility levels of students from the different school environments, it is suggested that a test of equality of the population *medians* be conducted using the Kruskal-Wallis test.

Let θ_1, θ_2, and θ_3 denote the median hostility levels for students of the three school environments. The null hypothesis is

$$H_0: \theta_1 = \theta_2 = \theta_3$$

and the alternative hypothesis is

$$H_a: \text{at least two } \theta_i\text{'s differ}$$

As stated earlier, the form of the Kruskal-Wallis test presented in this text is the usual F-test applied to the ranks. From the side-by-side ordered stem and leaf plots, the ranks can easily be assigned. Remember that the ranks are assigned to the combined samples, and tied values are given the average of the ranks that normally would have been assigned. The rank for each data value is given in italics above the data value.

Rural	Suburban	Urban

```
            Rural                          Suburban                              Urban

         1
   1 |   6
         2                            3                        
   2 |   1                            2                        
         4.5                          6                          4.5
   3 |   3                            7                          3
   4 |                                                         
         7   9.5                      8                          9.5  11
   5 |   1   3                        2                          3    4
        12.5 14.5 14.5  17  18.5 20.5 20.5   12.5 16  18.5       
   6 |   3    4    4    6    7    8    8      3    5   7          
        24  29  32                    22  24  26  27.5 30.5 33 34    24  27.5 30.5
   7 |   2   5   7                    0   2   3   4    6    8  9      2   4    6
        36.5 39                       24   27.5 30.5                 35  36.5 39  39  41  42  43  44
   8 |   2    3                       2    4    6                    0   2    3   3   4   6   7   8
                                                                    45
   9 |                                                              2
```

From the ranks of the data we have the following summary results:

	Rural	Suburban	Urban	
n	15	15	15	
R_j	227.0	336.5	471.5	$T = 1035$
$\bar{y}_R$	15.133	22.433	31.433	
s_R	9.330	11.450	13.436	
s_R^2	87.052	131.102	180.531	

▲ R_j is analogous to T_j in that it denotes the total of all the ranks in group j.

The complete analysis of variance on the ranks is

SV	SS	df	MS	F	p-value
Treatment	1999.90	2	999.950	7.524	$< .01$
Error	5581.60	42	132.895		

The p-value is less than 1% and therefore the null hypothesis is rejected. Thus there is a significant difference in the median hostility scores for the three school environments.

Violations of the assumptions for the ordinary F-test result in a loss of sensitivity of the test. In the case of the long-tailed distribution the loss may be to the point that the test fails to detect differences in populations that actually exist. The Kruskal-Wallis test, which is based on ranks, is resistant to outliers and thus may detect differences that might otherwise go undetected.

EXAMPLE 13.11

Three engineering universities wished to compare the salaries of their graduates 10 years after graduation. Seventeen graduates were randomly and independently selected from each of the three universities. Their salaries (in $1,000) were recorded in the following side-by-side ordered stem and leaf plots:

	University A		University B		University C
2	5		6		5
3	2 6		9		0
4	0 1 1 4 4 5 6 8 8		3 5 7 7		6
5	0 7 9		0 2 2 4 6 8		0 2 4 4 7 7 9
6			0 1 3 5		0 2 2 8
7	9				0 4
8			1		3
9	4				

Is there statistical evidence to indicate that the salaries of the graduates of the three engineering schools differ?

SOLUTION First we will run an ordinary F-test of the equality of population means. Let μ_1, μ_2, and μ_3 denote the mean salaries 10 years after graduation at universities A, B, and C, respectively. The null hypothesis is

$$H_0: \mu_1 = \mu_2 = \mu_3$$

From the data, we have the following summary results:

	University A	University B	University C	
n	17	17	17	
Σy	829	899	963	$T = 2{,}691$
Σy^2	44,835	49,889	57,833	$\Sigma\Sigma y^2 = 152{,}557$
$\bar{y}$	48.765	52.882	56.647	
s	16.6	12.113	14.322	

The completed analysis of variance table is

SV	SS	df	MS	F	p-value
Treatment	528.471	2	264.235	1.263	$> .10$
Error	10,038.71	48	209.140		

A *p*-value greater than 10% indicates that we should not reject the null hypothesis. Therefore the usual *F*-test shows no significant difference in mean salaries.

To further investigate this, we construct quartile summary diagrams and box plot summary diagrams for the three data sets.

Quartile summary diagrams

	University A		University B		University C
	17		17		17
$M(9)$	45	$M(9)$	52	$M(9)$	57
$Q(5)$	41 50	$Q(5)$	47 60	$Q(5)$	52 62

Box plot summary diagrams

	13.5			19.5				15		
a	32	59	a	39	65		a	46	74	
f	27.5	63.5	f	27.5	79.5		f	37	77	
25	one	none	26	one	one	81	25,30	two	one	83
F	14	77	F	8	99		F	22	92	
	none	two	79,94	none	none			none	none	

The following display of side-by-side box plots of the data shows numerous outliers, which is an indication that the assumptions for the usual *F*-test have been violated. The distributions appear to be long-tailed and similar in shape.

Side-by-side box plots

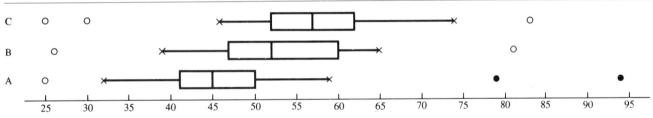

We will now apply the Kruskal-Wallis test to the following rank summary data.

	University A	University B	University C	
n	17	17	17	
Σy	328	457	541	$T = 1{,}326$
Σy^2	9,798.5	15,250	20,464.5	$\Sigma\Sigma y^2 = 45{,}513$
$\bar{y}_R$	19.294	26.882	31.824	
s_R	14.727	13.612	14.248	
s_R^2	216.885	185.287	203.006	

The completed analysis of variance table for the Kruskal-Wallis test is

SV	SS	df	MS	F	*p*-value
Treatment	1,354.235	2	677.12	3.36	$< .05$
Error	9,682.765	48	201.724		

The p-value is less than .05, which indicates that H_0 should be rejected. Thus there is statistical evidence to indicate that the salaries of the three university graduates differ.

As in any analysis of variance, if the null hypothesis is rejected, a multiple comparison test should be run to further analyze the differences. Tukey's multiple comparison test proceeds just as before with the exception that it is applied to the ranks.

EXERCISES 13.4

13.18 Under what conditions would you choose the Kruskal-Wallis test over the ordinary F-test?

13.19 What scale of measurement for the data is necessary for the Kruskal-Wallis test?

13.20 The following summary data were obtained by collectively assigning ranks to three independent samples. Complete an ANOVA table on the ranks for the Kruskal-Wallis test. Include the p-value.

n	10	10	10	
R_j	124.5	147.0	193.5	$T = 465$
$\bar{y}_R$	12.45	14.70	19.35	
s_R	8.67	8.43	8.72	

13.21 Ranks were collectively assigned to three independent samples (see accompanying table). Complete an ANOVA table on the ranks for the Kruskal-Wallis test. Include the p-value.

Sample 1	Sample 2	Sample 3
1	2	5
4	3	9
6	7	12
10	8	18
11	13	20
14	15	22
16	17	23
19	21	24

13.22 Independent random samples were selected from three populations with the following results:

3 –	5		5 6 9	5 6 7 9
4*	2 3 4		0 1 3 4	0 1 2 2 4
4 –	5 6 6 7 8		6 7 8 9	5 6 9
5*	0 1 4		1 3	2
5 –	6			7
6*	4		3	4
6 –	8		5	

(a) Complete side-by-side box plots and comment on the shapes of the distributions.

(b) To test the equality of the centers of the distributions, should an ordinary F-test or the Kruskal-Wallis test be conducted?

13.23 Using the data in Exercise 13.22, test the hypothesis that the population medians are equal using the Kruskal-Wallis test.

13.24 Independent random samples were selected from three populations with the following results:

14		15	7	
15	3	16	5 7	
16	4	17	0 1 2 5 7 9	
17	1 3 6 8	18	0 2 7 9	
18	0 1 1 2 3 5 9	19	4 6	
19	2 4 6			
20	0			

15	3
16	5 7
17	1 5 6 9
18	2 3 4 5 7 9
19	2 6
20	0

(a) Complete side-by-side box plots and comment on the shapes of the distributions.

(b) To test the equality of location of the three distributions, should an ordinary F-test or a Kruskal-Wallis test be conducted?

13.25 Using the data in Exercise 13.24, test the hypothesis that the populations have the same location using the test suggested in part (b) of Exercise 13.24.

13.26 The carbon monoxide level was measured (in parts per million) at three industrial sites at randomly selected times as shown in the accompanying table. Do the data indicate a significant difference in the carbon monoxide levels of the three sites?

Site A	Site B	Site C
.106	.121	.119
.127	.119	.110
.132	.115	.106
.105	.120	.118
.117	.117	.115
.109	.134	.121
.107	.118	.109
.109	.142	.134

TABLE FOR EXERCISE 13.27

	Stimuli		
A	B	C	D
.52	.56	.85	.51
.87	.72	.98	.57
.73	.65	.76	.64
.65	.57	.65	.68
.68	.54	1.09	.59
.77	.68	.89	.94
.52	.83	.79	.68
.88	.63	.94	.66
.72	.76	.80	.73
.94	.74	1.14	.65
.79	.81	.93	.60
.74	.87	.74	.88

13.27 The table at the bottom of the previous column lists the results of an experiment to study the effect of four stimuli on reaction time. The reaction times (in seconds) were as shown. Do the data indicate a significant difference in the reaction times due to the four stimuli?

13.28 The owners of a soon-to-be-built motor lodge wished to evaluate three prospective locations for the business. On randomly selected occasions they measured the amount of traffic passing the prospective sites in 1-hour periods of time (see the accompanying table). Do the data indicate a significant difference in the amount of traffic passing the three points?

Site A	Site B	Site C
162	165	179
154	193	160
174	178	155
148	184	168
150	157	140
148	165	151
185	204	163
157	195	175
164	183	182
172	189	150
159	179	139
193	160	164
160	198	181
159	185	176
203	215	177

13.5 Computer Session (optional)

There are two Minitab commands that will perform an analysis of variance for the completely randomized design. The difference in the two is the form of the input data. AOVONEWAY performs an analysis of variance on data stored in several columns. ONEWAY performs the same analysis; however all data are stored in one column and the levels are stored in a second column.

AOVONEWAY

To use AOVONEWAY the data for the samples are stored in separate columns. The form of the command is

```
AOVONEWAY DATA IN C, C,...,C
```

EXAMPLE 13.12

Perform an analysis of variance on the cigarette data given at the beginning of the chapter.

SOLUTION   First the data must be read into the computer. As always, there are two ways of entering data—the READ command and the SET command; however here the data have already been entered into the computer and saved under the name CIGAR. We will retrieve the data with the RETRIEVE command and then execute the AOVONEWAY command. We see in Figure 13.3 that the F-value is only .38 resulting in a p-value $> .10$. There is insufficient evidence to declare a difference in mean tar content for the three brands of cigarettes.

FIGURE 13.3

Minitab output for Example 13.12

```
MTB > RETRIEVE 'CIGAR'

MTB > PRINT C1 C2 C3

  ROW      C1      C2      C3

    1     0.61    0.43    0.52
    2     0.48    0.71    0.48
    3     0.44    0.52    0.67
    4     0.53    0.65    0.49
    5     0.71    0.63    0.38
    6     0.40    0.55    0.57
    7     0.53    0.64    0.70
    8     0.32    0.41    0.79
    9     0.61    0.58    0.47
   10     0.52    0.63    0.55
   11     0.40    0.53    0.49
   12     0.51    0.57    0.56
```

```
13    0.53    0.68    0.51
14    0.69    0.44    0.60
15    0.49    0.32    0.41
16    0.48    0.53    0.65
17    0.58    0.52    0.57
18    0.46    0.43    0.39
19    0.59    0.57    0.54
20    0.50    0.48    0.61
21    0.47    0.55    0.53
22    0.51    0.68    0.72
23    0.71    0.44    0.50
24    0.56    0.53    0.58
25    0.61    0.77    0.49

MTB > AOVONEWAY C1 C2 C3

ANALYSIS OF VARIANCE
SOURCE      DF         SS          MS          F
FACTOR       2      0.0078      0.0039       0.38
ERROR       72      0.7362      0.0102
TOTAL       74      0.7440
                                    INDIVIDUAL 95 PCT CI'S FOR MEAN
                                    BASED ON POOLED STDEV
LEVEL       N        MEAN       STDEV    -------+---------+---------+---------
C1          25      0.5296      0.0956   (-------------*------------)
C2          25      0.5516      0.1068          (-------------*------------)
C3          25      0.5508      0.1006          (------------*------------)
                                         -------+---------+---------+---------
POOLED STDEV =     0.1011               0.510     0.540      0.570
```

ONEWAY

For this command all the data to be analyzed are stored in one column. Then a second column indicates the sample for which the observation belongs.

EXAMPLE 13.13

An experiment was designed to compare three different cleansing agents. Forty-five subjects with similar skin conditions were randomly assigned to 3 groups of 15 subjects each. A patch of skin on each individual was exposed to a contaminant and then cleansed with one of the cleansing agents. After 8 hours the residual contaminant was measured with the results shown in Table 13.10. Perform an analysis of variance using the ONEWAY command.

We will code agents A, B, and C as 1, 2, and 3, respectively. The data have been stored in C1 and the codes in C2 with a READ command and then a SAVE command. We will retrieve the data named CLEAN and execute the ONEWAY command. The F-statistic ($F = 47.44$) in Figure 13.4 shows clearly that the null hypothesis should be rejected.

TABLE 13.10

Cleansing Agent		
A	B	C
2	6	5
4	7	6
3	9	5
3	8	4
2	6	7
4	6	5
5	8	6
3	6	5
2	7	4
4	8	6
3	5	7
2	6	6
5	7	7
4	8	6
3	8	7

FIGURE 13.4

Minitab output for Example 13.13

```
MTB > RETRIEVE 'CLEAN'

MTB > ONEWAY C1 LEVELS IN C2

ANALYSIS OF VARIANCE ON C1
SOURCE     DF        SS        MS         F
C2          2    108.13     54.07     47.44
ERROR      42     47.87      1.14
TOTAL      44    156.00
                               INDIVIDUAL 95 PCT CI'S FOR MEAN
                               BASED ON POOLED STDEV
LEVEL       N      MEAN     STDEV   -+---------+---------+---------+-----
    1      15     3.267     1.033  (---*---)
    2      15     7.000     1.134                            (---*---)
    3      15     5.733     1.033                  (---*---)
                                   -+---------+---------+---------+-----
POOLED STDEV =     1.068          2.8       4.2       5.6       7.0
```

DOTPLOT

Another Minitab command that is useful when looking at several samples is the DOTPLOT command. As suggested, the command constructs dot plots of the data. The form of the command is

DOTPLOT DATA IN C, C,...,C

If one desires the same scale for the dot plots of several columns, the subcommand

SAME must be given. For example, to construct dot plots of the data in Example 13.12, one would give the following commands:

```
DOTPLOT DATA IN C1, C2, C3;
SAME.
```

EXAMPLE 13.14

Construct dot plots of the data in Example 13.12.

SOLUTION   See Figure 13.5 for the Minitab output.

FIGURE 13.5
Minitab output for Example 13.14

```
MTB > DOTPLOT C1 C2 C3;
SUBC> SAME.


                                   .             .
           .            :    . ..:..:.:  . .. :        . :
        -+---------+---------+---------+---------+---------+-----C1
                             .
           .           . : ::    .  :: : :.      :.. :  .      .
        -+---------+---------+---------+---------+---------+-----C2
                     .
                .. .        ..:..........:. ..   . .   . .     .
        -+---------+---------+---------+---------+---------+-----C3
        0.30      0.40      0.50      0.60      0.70
```

If the data are in a single column with codes in a separate column as in Example 13.13, then the subcommand BY C will have to be given. That is, in Example 13.13 the commands will be

```
DOTPLOT DATA IN C1;
BY C2.
```

EXAMPLE 13.15

Construct dot plots of the data in Example 13.13.

SOLUTION   See Figure 13.6 for the Minitab output.

FIGURE 13.6
Minitab output for Example 13.15

```
MTB > DOTPLOT C1;
SUBC> BY C2.
```

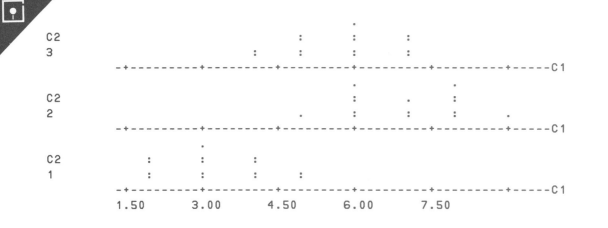

```
C2
3
                                        :       :       :
                        :       :       :       :
   -+---------+---------+---------+---------+---------+-----C1


C2                                      :       .       :
2                               :       :       :       :       .
                        .       :       :       :       .
   -+---------+---------+---------+---------+---------+-----C1


C2                  .
1           :       :       :
            :       :       :       :
   -+---------+---------+---------+---------+---------+-----C1
  1.50      3.00      4.50      6.00      7.50
```

There is a Minitab command for the Kruskal-Wallis test; however the test statistic, although equivalent, is not the same as the one used in this text. If you recall, in this text the test statistic is obtained by applying the usual F-test to the ranks. Thus to perform a Kruskal-Wallis test we must have the data stored in one column and the levels in a second column. The data column is ranked by the Minitab command

```
RANK the data in C put ranks in C
```

and then the analysis of variance is performed by the Minitab command

```
ONEWAY data in C levels in C
```

EXAMPLE 13.16

The dot plots of the data in Example 13.12 (see Example 13.14) suggest that the assumptions for the usual F-test might be violated. Perform a Kruskal-Wallis test on the cigarette data.

SOLUTION First the samples must be combined and then ranked. The samples are combined with the STACK command followed by the subcommand SUBSCRIPTS to identify the samples. See Figure 13.7 for the complete solution.

FIGURE 13.7
Minitab output for Example 13.16

```
MTB > STACK C1 C2 C3 PUT IN C4;
SUBC> SUBSCRIPTS C6.

MTB > RANK C4 PUT IN C5

MTB > ONEWAY C5, LEVELS IN C6
```

```
ANALYSIS OF VARIANCE ON C5
SOURCE       DF        SS        MS         F
C6            2       387       194      0.40
ERROR        72     34695       482
TOTAL        74     35082
                                INDIVIDUAL 95 PCT CI'S FOR MEAN
                                BASED ON POOLED STDEV
LEVEL        N      MEAN     STDEV   ------+---------+---------+---------+
    1       25     34.86     21.49   (----------*-----------)
    2       25     40.16     22.64            (-----------*----------)
    3       25     38.98     21.70         (-----------*----------)
                                    ------+---------+---------+---------+
POOLED STDEV =      21.95          30.0      37.5      45.0      52.5
```

Again the *F* statistic is very small suggesting that the null hypothesis should not be rejected.

EXERCISES 13.5

13.29 Concern over the academic performance of student athletes prompted the following study. The SAT scores (combined Verbal and Math) of athletes at randomly selected major public universities, minor public universities, and private universities were recorded as shown in the accompanying table.

Major university	Minor university	Private university
960	1,040	980
870	790	1,060
1,130	820	920
940	670	1,240
730	840	1,130
640	960	970
820	780	740
650	850	990
920	940	950
840	990	880
940	1,160	1,270
1,050	890	900
750	750	1,050
840	930	840
1,280	870	960
930	1,300	1,370
750	950	1,100
880	800	750
790	840	960
1,030	920	850

(a) Construct stem and leaf plots and box plots of the three samples.

(b) Based on the results of part (a), should the analysis of the data be an ordinary *F*-test or the Kruskal-Wallis test?

13.30 Using the test that you recommended in part (b) of Exercise 13.29, perform an analysis to determine whether a significant difference exists between the SAT scores for student athletes at the different type schools.

13.31 A study was conducted to compare the amount of time it takes to deliver a package for three major overnight delivery companies. For each company, a sample of 20 packages was shipped from the same point to the same destination. The table at the top of page 530 lists the amount of time (in hours) to deliver the packages:

(a) Construct stem and leaf plots and box plots of the three samples.

(b) Based on the results of part (a), should the analysis of the data be an ordinary *F*-test or the Kruskal-Wallis test?

13.32 Using the test that you recommended in part (b) of Exercise 13.31, perform an analysis to determine whether a significant difference exists between the amount of time it takes to deliver packages for the three companies.

Company A	Company B	Company C
26	26	24
24	27	27
34	30	25
25	28	22
23	23	30
20	24	26
25	27	24
26	29	23
30	32	27
22	25	31
27	26	28
23	37	25
25	26	33
22	22	23
28	27	25
21	26	38
24	25	26
31	24	23
24	28	22
25	30	24

Summary and Review

✅ *Analysis of variance* allows one to analyze statistically several population means at one time. The *completely randomized design* is an experiment in which independent random samples are obtained from the several populations. The *total variability* in the data for a completely randomized design is partitioned into two parts, called the *within-sample variability* and *between-sample variability*. The within-sample variability is measured by the *sum of squares for error* and has $n_1 + n_2 + \cdots + n_k - k$ degrees of freedom, where n_j is the sample size for the sample from the jth population. The between-sample variability is measured by the *sum of squares for treatment* and has $k - 1$ degrees of freedom. Dividing the degrees of freedom into the sum of squares yields a *mean square*. Dividing the mean square for treatments by the mean square for error produces the *F statistic,* which is used to test the hypothesis of equality of the population means.

✅ In order to conduct the *F*-test, certain assumptions about the populations should be satisfied. Those assumptions are that the populations should be normally distributed and have homogeneous standard deviations. If the assumptions are not met, then the *Kruskal-Wallis* test should be used.

✅ If an analysis of variance indicates a significant difference in the population means, then *Tukey's multiple comparison procedure* is used to further analyze the means.

Learning Goals

Having completed this chapter you should be able to:

1. Identify a completely randomized design of experiment. *Section 13.2*
2. Work with the data array for the completely randomized design. *Section 13.2*
3. Identify the assumptions that are necessary for an *F*-test. *Section 13.2*
4. Complete an ANOVA table from the data. *Section 13.2*
5. Perform an *F*-test on the data. *Section 13.2*
6. Conduct Tukey's multiple comparison procedure. *Section 13.3*
7. Determine when the Kruskal-Wallis test should be conducted. *Section 13.4*
8. Perform a Kruskal-Wallis test. *Section 13.4*

Review Questions

To test your skills answer the following questions.

13.33 Suppose that an exploratory analysis of data from four populations suggested that the populations are symmetric with tails that are not excessively long. What mode of analysis would you use?

(a) ordinary *F*-test
(b) Kruskal-Wallis test

13.34 Suppose that an exploratory analysis of data from three populations suggested that the populations were similar in shape but their locations were possibly different. What mode of analysis would you use?

(a) ordinary *F*-test
(b) Kruskal-Wallis test

13.35 Find the *F*-values associated with upper-tail probabilities of .05, .025, and .01 when
(a) $df_n = 3$ and $df_d = 24$
(b) $df_n = 5$ and $df_d = 35$
(c) $df_n = 7$ and $df_d = 42$

13.36 Given that $df_n = 4$ and $df_d = 28$. Find the approximate p-values associated with an F of

(a) 2.5   (b) 3.4   (c) 5.6

13.37 Following is a partially completed analysis of variance table:

Source	SS	df	MS	F	p-value
Treatment	428	4			
Total	1460	32			

(a) How many treatment levels are there?

(b) Complete the ANOVA table.

(c) Is there a significant difference in the treatment levels?

13.38 From the following data set, calculate SSE, SST, and TSS. Also give the associated degrees of freedom and compute the mean squares.

Group A		Group B	
10		10	5
11	0	11	4
12	1 4 7	12	2 3 4 7
13	0 3 5 5 8	13	1 2 3 5 8 8
14	1 3 6	14	1 5
15	0 2	15	0
16	4	16	

Group C	
10	6
11	
12	3 4
13	3 5 7 8
14	1 5 8
15	0 3 5
16	0

13.39 Construct an analysis of variance table for the data in Exercise 13.38. Compute the p-value associated with the F-value.

13.40 To compare the efficiency of the pit crews of major NASCAR teams, the duration of pit stops was measured on three of the top crews. The time (in seconds) for 12 randomly selected pit stops yielded the results shown in the accompanying table.

Team A	Team B	Team C
25	25	30
22	30	35
18	24	32
30	26	26
24	22	37
15	15	43
40	32	36
23	46	40
10	20	35
20	28	25
45	35	55
25	25	33

(a) Complete side-by-side box plots and comment on the shapes of the distributions.

(b) To test the equality of the centers of the distributions, should an ordinary F-test or the Kruskal-Wallis test be conducted?

13.41 Using the test that you recommended in part (b) of Exercise 13.40, perform an analysis to determine whether a significant difference exists between the amount of time it takes the three crews to complete a pit stop.

SUPPLEMENTARY EXERCISES FOR CHAPTER 13

13.42 Describe the conditions under which you would use

(a) the ordinary F-test

(b) the Kruskal-Wallis test

13.43 Find $q_\alpha(k, \nu)$ from Table B7 in Appendix B for the following values of α, k, and ν.

(a) $\alpha = .01$, $k = 5$, $\nu = 20$

(b) $\alpha = .05$, $k = 3$, $\nu = 15$

(c) $\alpha = .01$, $k = 8$, $\nu = 40$

13.44 Following are the summary data that resulted from an experiment involving a treatment with four levels.

	Treatment			
	A1	A2	A3	A4
n	12	12	12	12
Σy_i	135	152	147	184

ANOVA table		
Source	SS	df
Treatment	756.5	3
Error	261.3	44

Analyze the means using Tukey's multiple comparison procedure with $\alpha = .01$.

13.45 From the following summary data complete an ANOVA table. Assume that the assumptions for the ordinary F-test are satisfied. Include the p-value.

n	13	13	13	13
$\bar{y}$	16.5	18.3	14.7	19.5
s	6.72	4.94	4.31	5.82

13.46 From the following summary data calculate SSE, SST, and TSS. Also give the associated degrees of freedom and compute the mean squares.

Group A	Group B	Group C
$n = 14$	$n = 16$	$n = 18$
$\bar{y} = 141.8$	$\bar{y} = 136.5$	$\bar{y} = 125.3$
$s = 11.35$	$s = 14.68$	$s = 13.82$

13.47 Construct an analysis of variance table for the data in Exercise 13.46. Compute the p-value associated with the F-value.

13.48 From the following data calculate SSE, SST, and TSS. Also give the associated degrees of freedom and compute the mean squares.

Treatment 1	Treatment 2	Treatment 3
21	41	35
24	44	37
31	38	33
42	37	46
38	42	42
31	48	38
36	39	37
34	32	30

13.49 Construct an analysis of variance table for the data in Exercise 13.48. Compute the p-value associated with the F-value.

13.50 What scale of measurement for the data is necessary for the Kruskal-Wallis test?

13.51 From the following summary data, complete an ANOVA table for an F-test. Include the p-value.

n	14	14	14
$\bar{y}$	81.4	69.6	74.2
s	12.6	14.1	10.4

13.52 Independent random samples were selected from three populations with the following results:

```
3* | 1                       3* | 4
3- | 5 6 6 8 9               3- | 5 6 9
4* | 0 1 1 2 3 4 4           4* | 0 1 1 3 4
4- | 5 6 6 7 8               4- | 5 5 6 7 8 9 9
5* | 0 1                     5* | 1 3
```

```
3* | 1 3
3- | 5 5 6 7 8 8 9
4* | 0 1 2 2 4
4- | 5 6 9
5* | 2
```

(a) Complete side-by-side box plots and comment on the shapes of the distributions.

(b) To compare the distributions should an ordinary F-test or the Kruskal-Wallis test be conducted?

13.53 Using the data in Exercise 13.52, test the hypothesis that the population means are equal using the F-test suggested in part (b) of Exercise 13.52.

13.54 Independent random samples were selected from three populations with the following results:

```
6*  | 2                      5-  | 8
6-  | 8                      6*  | 3
7*  | 0                      6-  | 7
7-  | 6 8 9                  7*  |
8*  | 0 1 3 4 4              7-  | 6 8
8-  | 5 6 7                  8*  | 0 1 3 3 4
9*  | 1 1 3                  8-  | 5 6 7 8 9 9
9-  | 8                      9*  | 1 4
10* | 2                      9-  |
10- | 7                      10* | 2
                             10- | 7
```

```
5-  | 7
6*  | 0
6-  |
7*  | 0 3 4
7-  | 5 6 7 8 8 9
8*  | 0 1 3 4
8-  | 5 8
9*  | 3
9-  | 8
10* | 0
```

(a) Complete side-by-side box plots and comment on the shapes of the distributions.

(b) To compare the distributions, should an ordinary F-test or the Kruskal-Wallis test be conducted?

13.55 Using the data in Exercise 13.54, test the hypothesis that the population means are equal using the test suggested in part (b) of Exercise 13.54.

13.56 The following summary data were obtained by collectively assigning ranks to three independent samples. Complete an ANOVA table on the ranks for the Kruskal-Wallis test. Include the p-value.

n	18	18	18	
R_j	372.5	568.5	544.0	$T = 1485$
$\bar{y}_R$	20.694	31.583	30.222	
s_R	5.336	6.712	8.344	

13.57 Perform Tukey's multiple comparison procedure on the means in Exercise 13.56.

13.58 Independent random samples were selected from three populations with the following results:

0*							0	1		
0–						5 6 7	5	6	8	
1*	1 2 4 4					0 2 4 4	0	1	2	3
1–	5 6 7 8 9					5 6 6 9	5	6	8	
2*	0 2 3					0 3	3			
2–	5					7	5			
3*	0					2	1			
3–	5									

(a) Complete side-by-side box plots and comment on the shapes of the distributions.
(b) To test the equality of the centers of the distributions, should an ordinary F-test or the Kruskal-Wallis test be conducted?

13.59 Using the data in Exercise 13.58, test the hypothesis that the population medians are equal using the test suggested in part (b) of Exercise 13.58.

13.60 The delay time (in minutes) was measured on 20 flights selected randomly from each of four major air carriers (see the accompanying table).

Carrier A	Carrier B	Carrier C	Carrier D
20	15	20	25
14	17	27	17
12	10	22	10
20	36	35	5
17	18	26	22
30	20	24	35
19	5	15	19
7	16	17	24
22	20	10	3
18	13	25	20
10	42	45	15
15	15	20	40
13	8	16	16
5	17	12	10
19	10	5	9
25	4	21	19
45	19	32	45
14	25	23	15
40	12	10	5
10	5	15	10

(a) Construct stem and leaf plots and box plots of the four samples.
(b) Based on the results of part (a), should the analysis of the data be an ordinary F-test or the Kruskal-Wallis test?

13.61 Using the test that you recommended in part (b) of Exercise 13.60, perform an analysis to determine whether a significant difference exists between the delay times for the four air carriers.

13.62 Perform Tukey's multiple comparison procedure on the means found in Exercise 13.61.

13.63 A study was conducted to evaluate three treatments for arthritis. A number of arthritis sufferers were randomly assigned to three groups. After their treatment, the time until they experienced relief was measured (in minutes) with the results listed in the accompanying table.

Treatment A	Treatment B	Treatment C
40	73	50
35	32	75
47	47	34
52	52	47
31	34	87
61	60	45
92	77	38
46	42	25
50	20	86
49	81	39
93	75	42
84	35	30
72	25	75
43	40	36
80	90	90
85	33	32
30	30	89

(a) Construct stem and leaf plots and box plots of the three samples.
(b) Based on the results of part (a), should the analysis of the data be an ordinary F-test or the Kruskal-Wallis test?

13.64 Using the test that you recommended in part (b) of Exercise 13.63, perform an analysis to determine whether a significant difference exists between the relief times for the three treatments for arthritis.

REFERENCES

Bellout, Guy. "Is Vitamin C Really Good for Colds?" *Consumer Reports,* February, 1976, pp. 66–70.

Box, G. E. P., G. Hunter, and J. S. Hunter. *Statistics for Experimentation.* New York: John Wiley, 1978.

Box, G. E. P., and G. M. Jenkins. *Time Series Analysis, Forecasting and Control.* San Francisco: Holden-Day, 1970.

Chambers, J. M., W. S. Cleveland, B. Kleiner, and P. A. Tukey. *Graphical Methods for Data Analysis.* Boston: Duxbury Press, and Belmont, Calif.: Wadsworth International Group, 1983.

Conover, W. J. *Practical Nonparametric Statistics,* 2d ed. New York: John Wiley, 1980.

Conover, W. J., and R. L. Iman. "Rank Transformations as a Bridge Between Parametric and Nonparametric Statistics," *The American Statistician,* 35, no. 3 (1981).

Corley, W. C., director. *Crime in North Carolina—1979—Uniform Crime Report.* Raleigh: State of North Carolina Department of Justice, 1979.

Dixon, W. J., and M. B. Brown, eds. *BMDP Biomedical Computer Programs, P Series.* Berkeley: The University of California Press, 1979.

Feinberg, Stephen E. "Randomization and Social Affairs: The 1970 Draft Lottery." *Science,* Vol. 171, January 22, 1971, pp. 255–261.

Flanagan, T. J., D. J. van Alstyne, and M. R. Gottfredson, eds. *Sourcebook of Criminal Justice Statistics—1981.* Albany, N.Y.: Criminal Justice Research Center, 1982.

Freedman, D., R. Pisani, and R. Purves. *Statistics.* New York: W. W. Norton, 1978.

Gnanadesikan, R., ed. *Statistical Data Analysis—Proceedings of Symposia in Applied Mathematics—Volume 28.* Providence, R.I.: American Mathematical Society, 1983.

Goldman, R. N., and J. S. Weinberg. *Statistics—An Introduction.* Englewood Cliffs, N.J.: Prentice-Hall, 1985.

Gross, A. M. "Confidence Interval Robustness with Long-Tailed Symmetric Distributions," *Journal of the American Statistical Association,* 71, no. 354 (1976).

Haack, D. G. *Statistical Literacy—A Guide to Interpretation.* Boston: Duxbury Press, 1979.

Hartwig, F., and B. E. Dearing. *Exploratory Data Analysis.* Beverly Hills, Calif.: Sage Publications, 1979.

Helwig, J. P., and K. A. Council, eds. *SAS User's Guide.* Cary, N.C.: SAS Institute Inc., P.O. Box 8000, 1979.

Hoaglin, D. C., F. Mosteller, and J. W. Tukey. *Understanding Robust and Exploratory Data Analysis.* New York: John Wiley, 1983.

Hora, S. C., and W. J. Conover. "The F Statistic in the Two-Way Layout with Rank-Score Transformed Data," *Journal of the American Statistical Association,* 79, no. 387 (1984).

Huber, P. J. "Robust Statistics: A Review," *The Annals of Mathematical Statistics,* 43, no. 4 (1972).

Iman, R. L., and W. J. Conover. *A Modern Approach to Statistics.* New York: John Wiley, 1983.

Iman, R. L., S. C. Hora, and W. J. Conover. "Comparison of Asymptotically Distribution-Free Procedures for the Analysis of Complete Blocks," *Journal of the American Statistical Association,* 79, no. 387 (1984).

Johnson, R., and G. Bhattacharyya. *Statistics—Principles and Methods.* New York: John Wiley, 1985.

Kempthorne, O. "Teaching of Statistics: Content Versus Form," *The American Statistician,* 31, no. 1 (1980).

Koopmans, L. H. *An Introduction To Contemporary Statistics.* Boston: Duxbury Press, 1981.

Lane, H. U., M. S. Hoffman, J. Foley, and T. J. McGuire, eds. *The World Almanac & Book of Facts—1985.* New York: Newspaper Enterprise Association, 1984.

Lefkowitz, J. M. *Introduction to Statistical Computer Packages.* Boston: Duxbury Press, 1985.

McClave, J. T., and F. H. Dietrich II. *Statistics,* 3d ed. San Francisco: Dellen Publishing, 1985.

McEntire, A., and A. N. Kitchens. "A New Focus for Educational Improvement Through Cognitive and Other Structuring of Subconscious Personal Axioms," *Education,* 105, no. 2 (1984).

Meier, Paul. "The Biggest Public Health Experiment Ever: The 1954 Field Trial of the Salk Polio Vaccine." *Statistics: A Guide to the Unknown,* 2d edition. Judith Tanur et al (eds.). San Francisco: Holden-Day, 1978.

Mendenhall, W. *Introduction to Probability and Statistics—Sixth Edition.* Boston: Duxbury Press, 1983.

Moore, D. S. *Statistics-Concepts and Controversies.* San Francisco: W. H. Freeman, 1979.

Mosteller, F. "The Teaching of Statistics: Classroom and Platform Performance," *The American Statistician,* 34, no. 1 (1980).

Mosteller, F., and J. W. Tukey. *Data Analysis and Regression, A Second Course in Statistics.* Reading, Mass.: Addison-Wesley, 1977.

Nie, N., C. H. Hull, J. G. Jenkins, K. Steinbrenner, and D. H. Bent. *Statistical Package for the Social Sciences,* 2d ed. New York: McGraw-Hill, 1979.

Noether, G. E. "The Role of Nonparametrics in Introductory Statistics Courses," *The American Statistician,* 34, no. 1 (1980).

Ott, L. *An Introduction to Statistical Methods and Data Analysis,* 2d ed. Boston: Duxbury Press, 1984.

Roll, C. W., Jr., and A. H. Cantril. *Polls: Their Use and Misuse in Politics.* New York: Basic Books, Inc., 1972.

Ryan, B. F., B. L. Joiner, and T. A. Ryan, Jr. *Minitab Handbook Second Edition.* Boston: Duxbury Press, 1985.

Ryan, B. F., B. L. Joiner, and T. A. Ryan, Jr. *Minitab Reference Manual.* University Park, PA.: Minitab Inc., 1981.

Sacks, J., and D. Ylvisaker. "A note on Huber's robust estimation of a location parameter," *The Annals of Mathematical Statistics,* 43, no. 4 (1972).

Stigler, S. M. "Do robust estimators work with 'real' data?" *The Annals of Statistics,* 5, no. 6 (1977).

Sutton, G. W., T. J. Huberty, and R. Price. "An Evaluation of Teacher Stress and Job Satisfaction," *Education,* 105, no. 2 (1984).

Tukey, J. W. *Exploratory Data Analysis.* Reading, Mass.: Addison-Wesley, 1977.

Tukey, J. W., and D. H. McLaughlin. "Less Vulnerable Confidence and Significance Procedures for Location Based on a Single Sample: (Trimming/Winsorization 3)." *Sankhyā Series* A, 25, pp 331–352 (1963).

Velleman, P. F., and D. C. Hoaglin, *Applications, Basics, and Computing of Exploratory Data Analysis.* Duxbury Press, North Scituate, Mass. (1981).

Wade, N. "IQ and Heredity: Suspicion of Fraud Beclouds Classic Experiment," *Science,* 194 (1976), pp. 916–919.

Webster, W. H., director. *Crime in the United States—Uniform Crime Reports for the United States.* Washington, D.C.: Federal Bureau of Investigation, U.S. Department of Justice, 1983.

Wheeler, M. *Lies, Damn Lies, and Statistics.* New York: Liveright, 1976.

Wilner, D. M. et al. "The Housing Environment and Family Life." Baltimore: Johns Hopkins Press, 1962.

Yuen, K., and W. J. Dixon. "The Approximate Behavior and Performance of the Two-Sample Trimmed *t.*" *Biometrika* 60 (2), p. 369 (1973).

A

Data Sets

Executive position	Salary (× $1000)	Corporation	Assets (× $1 million)	Rank	Sales (× $1 million)	Rank	Mkt value (× $1 million)	Rank	Net profit (× $1 million)	Rank
Vice President	2122	Gulf & Western	4425	233	4198	153	2330	145	297.3	103
Cochairman	2110	Phibro-Salomon	42017	16	29757	10	4539	61	470	50
Cochairman	2080	Phibro-Salomon	42017	16	29757	10	4539	61	470	50
Vice President	1950	Phibro-Salomon	42017	16	29757	10	4539	61	470	50
Vice President	1950	Phibro-Salomon	42017	16	29757	10	4539	61	470	50
Vice President	1950	Phibro-Salomon	42017	16	29757	10	4539	61	470	50
President	1717	First Boston	22003	40	1482	422	470		80.2	375
Chairman	1644	Mobil	35072	22	54607	4	11696	12	1503	12
Chairman	1600	First Boston	22003	40	1482	422	470		80.2	375
Chairman	1500	E. F. Hutton	13164	68	2172	303	878	406	110.6	289
Chairman	1490	General Motors	45694	12	74582	2	23419	5	3730.2	4
Chairman	1475	Southern Pacific	11388	80	5976	89	4989	50	333.4	87
Managing Director	1438	First Boston	22003	40	1482	422	470		80.2	375
Chairman	1425	Sears, Roebuck	46176	11	35883	8	13163	7	1300.8	14
Chairman	1421	Ford Motor Co.	23869	33	44455	5	7755	23	1926.9	6
President	1416	Mobil	35072	22	54607	4	11696	12	1503	12
President	1330	General Motors	45694	12	74582	2	23419	5	3730.2	4
Chairman	1262	Shearson/Am Exp	43981	14	9770	45	6961	27	514.7	45
Chairman	1251	Merrill Lynch	26139	27	5687	99	2865	118	313.2	94
Chairman	1235	Chemical Bank	51165	9	4903	118	1325	268	305.6	99
Chairman	1235	Citicorp	134655	2	17037	20	4625	59	860	26
Chairman	1226	AT&T	149530	1	69403	3	59392	2	5746.6	1
Chairman	1210	Exxon	62963	7	88561	1	31623	3	4978	3
Chairman	1191	United Technology	8720	104	14669	25	4334	68	509.2	46
Chairman	1162	Coca-Cola	5228	197	6829	77	7295	25	558.3	42
Chairman	1160	Phillips Petroleum	13094	69	15249	22	5286	46	721	30
Chairman	1158	General Electric	23288	35	26797	13	26653	4	2024	5
President	1147	H. J. Heinz	2421	425	3843	175	2682	121	232.9	137
Chairman	1138	Gannett	1690		1704	363	3149	102	191.7	164
Chairman	1131	Texaco	27199	24	40068	7	9292	18	1233	15
Chairman	1124	City Investing	8361	111	5948	92	1374	256	175	187
Chairman	1115	Philip Morris	9667	94	9466	49	8967	19	903.5	21
President	1113	Ford Motor Co.	23869	33	44455	5	7755	23	1926.9	6
Chairman	1090	McDermott								
Chairman	1080	Burlington Northern	10901	86	4508	137	3678	80	413.2	66
Chairman	1075	NCR	3560	290	3731	180	3383	91	287.7	108
Vice President	1062	E. F. Hutton	13164	68	2172	303	878	406	110.6	289
President	1050	Zayre	908		2614	260	776	459	61.4	440
Vice President	1050	CBS	2990	349	4458	139	1966	163	187.2	169
Chairman	1047	Rockwell Intern.	5339	191	8309	64	5098	47	410.7	67
Chairman	1027	Bristol-Myers	3007	347	3917	168	5760	36	408	69
Chairman	1023	ITT	13967	66	14155	28	6160	33	623.9	37
Chairman	1022	American Brands	4304	242	4436	140	3256	96	390.3	72
Chairman	1020	Fluor	4085	252	5042	113	1357	258	79.3	378
Chairman	1016	Nabisco Brands	3625	282	5985	88	2603	128	322.6	90
Chairman	1014	RCA	7656	127	8977	53	2842	119	240.8	135
Chairman	1010	Dean Witter Reynolds	9874	92	10371	40	6881	28	835	27
President	1001	Merrill Lynch	26139	27	5687	99	2865	118	313.2	94
Chairman	999	Conagra	1290		2485	272	631		53.2	485
Chairman	989	IBM	37243	19	40180	6	74508	1	5485	2
Chairman	989	Anheuser-Busch	4330	241	6034	86	3025	110	348	80

Executive position	Salary (× $1000)	Corporation	Assets (× $1 million)	Rank	Sales (× $1 million)	Rank	Mkt value (× $1 million)	Rank	Net profit (× $1 million)	Rank
President	982	Procter & Gamble	8361	111	12633	32	9469	17	886	24
President	967	Sears, Roebuck	46176	11	35883	8	13163	7	1300.8	14
Chairman	961	GTE	24223	30	12944	31	8341	20	964	20
Chairman	960	Du Pont	24432	29	35173	9	12421	9	1127	16
President	950	Shell Oil	22169	39	19678	16	12364	10	1633	9
Monsanto	950	Monsanto	6427	149	6299	83	4307	69	369	76
Chairman	946	Sears Merchandise	46176	11	35883	8	13163	7	1300.8	14
Chairman	944	Revco Drug Stores	676		2001	326	1112	320	78.4	380
Chairman	942	Allied	7647	1289	10022	44	2980	112	410	68
Chairman	940	Emerson Electric	2519	409	3599	187	4558	60	311	96
President	940	Mobil Oil Co.	35072	22	54607	4	11696	12	1503	12
Chairman	936	Litton Industries	4273	244	5021	114	2956	113	242.7	133
Vice President	925	Ford Motor Comp	23869	33	44455	5	7755	23	1926.9	6
Chairman	923	Standard Oil of California	24010	32	27342	12	11813	11	1590	10
President	905	Union Pacific	10218	90	8353	63	5824	35	414	65
Chairman	904	RJ Reynolds	9874	92	10371	40	6881	28	835	27
Chairman	904	Esmark	3338	314	4587	132	1719	199	128.7	248
Chairman	902	American Express	43981	14	9770	45	6961	27	514.7	45
Chairman	896	J. P. Morgan	58023	8	5764	95	2742	120	460	53
President	885	AT&T	149530	1	69403	3	59392	2	5746	1
President	884	Nabisco Brands	3625	282	5985	88	2603	128	322.6	90
President	883	Exxon	62963	7	88561	1	31623	3	4978	3
Chairman	873	Boeing	7471	129	11129	37	4242	71	355	79
President	871	Household Inter.	8446	110	7912	68	1511	229	206.4	152
Chairman	865	Colt Industries	1177		1576	396	1271	279	99.3	320
Chairman	863	ABC	2090	486	2940	234	1631	205	159.8	212
Chairman	863	American Home Prod.	3086	343	4856	123	7735	24	627.2	35
Chairman	663	IC Industries	3985	256	3724	181	799	448	76.1	387
Chairman	856	CBS	2990	349	4458	139	1966	163	187.2	169
Chairman	851	Procter & Gamble	8361	111	12633	32	9469	17	886	24
Chairman	850	Eaton	2279	449	2674	253	1769	187	109.6	292
Chairman	839	Harris	1583		1474	423	1584	210	56.1	468
Chairman	839	Dart & Kraft	5418	184	9714	47	3653	81	435.1	60
Chairman	835	Johnson & Johnson	4461	228	5973	90	7820	22	547	43
President	833	Searle	1343		946		2165	151	110	291
Vice President	833	Mobil	35072	22	54607	4	11696	12	1503	12
Chairman	831	International Minerals	1980		1493	420	1158	310	84.5	355
President	830	Levi Strauss	1736		2689	250	1744	195	170.3	197
Vice Chairman	829	General Electric	23288	35	26797	13	26653	4	2024	5
Chairman	825	Chase Manhattan	81921	4	8523	57	1581	213	429.6	62
Vice Chairman	825	Santa Fe/Southern	11388	80	5976	89	4989	50	333.4	87
President	825	Hewlett-Packard	4444	232	4933	117	10620	15	442	59
Chairman	820	First Interstate	44423	13	4341	145	1726	196	247.4	126
Chairman	818	Sun	12466	73	14730	24	5081	48	613	38
Chairman	818	Abbott Laboratory	2824	373	2928	236	5480	41	347.6	82
Chairman	517	Colgate-Palmolive	2664	395	4865	122	1763	190	186.2	170
President	811	Coca-Cola	5228	197	6829	77	7295	25	558.3	42
Chairman	811	Tenneco	17994	57	14449	26	5714	37	716	31
President	806	American Express	43981	14	9770	45	6961	27	514.7	45
Average	1090.7		22633.2		17606.2		8127.4		887.44	

SOURCE: *Forbes* magazine, April 30, 1984 and *U.S. News & World Report*, May 2, 1984.

A2 SAT Data

The following data correspond to the grade point averages and SAT scores for 200 high school seniors. Some of the students took a special course to prepare them for the SAT test. They are coded as a 1 for those who took the course and a 0 for those who did not. The students are further classified according to sex (0 = male, 1 = female) and parents' education level (1 = no high school diploma, 2 = high school diploma and no college, 3 = college graduate).

Student ID	Sex	Parents' education	Prep course	GPA	Math	Verbal	Total
1	0	3	0	2.4	496	464	960
2	1	3	1	2.5	377	651	1028
3	1	2	0	3.9	670	651	1321
4	1	2	0	2.7	504	455	959
5	0	2	0	3.4	600	571	1171
6	1	2	0	2.6	533	391	924
7	1	2	0	2.4	649	330	979
8	1	2	0	2.1	382	296	678
9	0	3	0	2.5	357	499	856
10	1	3	1	2.4	487	398	885
11	0	3	0	2.3	581	234	815
12	0	2	1	3.2	489	558	1047
13	1	3	1	2.7	428	471	899
14	0	2	0	2.1	473	331	804
15	0	2	1	2.0	320	429	749
16	0	3	0	3.0	476	472	948
17	0	2	1	2.0	329	608	937
18	1	1	0	3.7	632	627	1259
19	0	3	0	2.9	476	493	969
20	0	2	1	2.1	463	455	918
21	0	3	1	2.2	436	569	1005
22	0	1	0	1.6	407	277	684
23	0	2	1	2.2	540	346	886
24	1	1	0	1.9	439	368	807
25	0	2	0	1.8	344	362	706
26	1	3	1	2.6	391	541	932
27	1	2	1	0.8	504	277	781
28	0	2	0	2.1	420	493	913
29	0	3	1	3.6	549	616	1165
30	1	2	1	2.4	605	433	1038
31	0	3	0	3.2	580	523	1103
32	1	2	1	2.5	454	446	900
33	0	2	0	1.5	415	372	787
34	0	2	0	1.7	264	457	721
35	1	1	1	2.7	521	504	1025
36	0	3	1	2.6	651	437	1088
37	0	2	1	2.3	537	475	1012
38	1	2	1	1.3	493	417	910
39	0	2	0	2.9	576	522	1098
40	0	2	0	2.2	370	560	930
41	1	2	0	2.0	485	333	818
42	0	3	0	3.1	626	375	1001
43	1	3	1	1.3	332	364	696
44	1	1	1	3.2	577	622	1199
45	0	3	1	2.3	508	456	964
46	0	1	1	2.6	525	547	1072
47	0	3	1	2.0	420	418	838
48	0	2	0	3.0	527	415	942
49	0	2	1	3.3	519	469	988
50	1	2	1	2.9	505	555	1060
51	1	3	0	2.1	506	315	821
52	1	2	0	3.2	574	578	1152
53	0	2	1	3.4	564	692	1256
54	1	3	1	2.0	593	647	1240
55	0	3	0	2.0	547	462	1009
56	1	3	1	3.5	566	561	1127
57	0	3	0	2.7	431	601	1032
58	1	2	1	1.9	509	504	1013
59	0	2	0	2.5	291	380	671
60	1	2	0	2.3	350	351	701
61	0	3	1	2.5	362	387	749
62	0	2	0	2.3	460	422	882
63	1	2	1	1.9	395	462	857
64	0	2	0	2.3	521	500	1021
65	1	2	1	1.7	379	499	878
66	0	3	0	2.7	466	454	920
67	1	2	0	1.9	469	365	834
68	0	3	0	3.1	607	420	1027
69	0	3	1	3.2	408	510	918
70	0	2	1	2.6	641	439	1080
71	0	2	0	1.8	383	363	746
72	0	2	0	2.5	437	605	1042
73	0	2	1	3.1	558	564	1122
74	0	2	0	3.0	449	479	928
75	0	2	0	2.1	373	494	867
76	1	3	1	2.9	649	446	1095
77	0	2	1	3.1	507	652	1159
78	1	2	1	3.0	406	448	854
79	0	2	0	2.0	360	381	741
80	0	2	0	2.3	402	366	768
81	1	1	1	1.8	497	554	1051
82	1	3	0	2.3	333	447	780
83	0	2	1	2.7	390	453	843
84	0	1	0	1.9	392	559	951
85	1	2	0	2.9	525	423	948
86	1	3	0	2.9	452	510	962
87	0	3	1	2.4	665	489	1154
88	1	2	1	2.7	423	510	933
89	0	3	0	2.5	622	411	1033
90	1	2	0	0.8	457	599	1056
91	1	2	0	3.0	469	389	858
92	0	1	1	1.5	461	502	963

Student ID	Sex	Parents' education	Prep course	GPA	Math	Verbal	Total
93	0	2	1	2.1	604	382	986
94	0	2	1	2.7	442	633	1075
95	0	2	0	2.3	428	437	865
96	0	2	0	2.6	519	301	820
97	1	2	0	3.0	651	492	1143
98	1	3	0	3.4	634	459	1093
99	1	2	1	2.4	371	491	862
100	0	3	0	3.2	611	421	1032
101	0	2	0	2.9	485	438	923
102	1	2	0	2.7	447	373	820
103	1	1	0	2.1	540	395	935
104	1	1	0	2.9	601	501	1102
105	1	2	0	2.8	438	501	939
106	0	2	1	1.6	426	422	848
107	1	2	0	2.7	468	687	1155
108	0	2	0	3.7	607	562	1169
109	0	2	0	3.6	614	586	1200
110	1	2	1	3.2	490	632	1122
111	0	3	0	3.2	556	468	1024
112	0	2	1	2.0	432	472	904
113	0	2	0	3.7	515	675	1190
114	1	2	1	1.7	388	495	883
115	0	3	0	2.8	598	495	1093
116	0	2	0	2.6	384	467	851
117	0	2	0	3.6	478	571	1049
118	1	2	0	3.3	462	456	918
119	0	3	1	2.1	596	452	1048
120	0	2	0	1.6	574	489	1063
121	0	2	1	3.1	619	539	1158
122	1	2	1	2.0	406	570	976
123	0	2	0	2.7	404	544	948
124	0	2	1	2.8	411	484	895
125	1	3	0	3.0	630	561	1191
126	0	3	1	1.8	410	372	782
127	0	3	0	3.3	473	591	1064
128	1	3	0	3.0	552	637	1189
129	1	2	1	3.0	715	547	1262
130	1	2	0	1.3	450	394	844
131	0	3	0	2.6	436	426	862
132	1	2	1	2.2	480	617	1097
133	1	2	0	2.1	493	357	850
134	1	1	0	0.3	281	168	449
135	0	1	0	2.0	510	613	1123
136	1	3	1	0.6	534	350	884
137	0	2	1	3.3	536	689	1225
138	1	2	1	2.6	409	507	916
139	1	3	1	3.2	504	711	1215
140	1	2	0	2.6	509	315	824
141	1	2	1	2.5	568	484	1052
142	1	1	0	2.6	579	364	943
143	0	2	0	2.3	520	384	904
144	1	2	0	2.5	485	410	895
145	1	2	0	3.0	653	537	1190
146	1	2	0	2.3	396	467	863

Student ID	Sex	Parents' education	Prep course	GPA	Math	Verbal	Total
147	0	3	0	1.6	391	452	843
148	1	2	0	0.9	329	245	574
149	0	2	0	2.5	418	554	972
150	1	2	0	1.7	367	408	775
151	1	2	0	3.5	567	462	1029
152	1	2	1	2.7	363	480	843
153	0	2	1	1.0	391	337	728
154	1	2	1	1.6	575	272	847
155	1	3	1	2.5	463	493	956
156	0	2	0	2.6	602	492	1094
157	0	3	0	3.0	582	492	1074
158	1	2	0	3.2	488	513	1001
159	0	2	1	1.0	306	393	699
160	0	2	0	2.7	587	516	1103
161	1	2	1	2.3	450	597	1047
162	0	2	0	2.0	367	439	806
163	0	3	0	2.6	471	670	1141
164	1	2	1	1.6	435	424	859
165	0	3	1	2.2	492	411	903
166	0	1	0	3.0	599	572	1171
167	0	3	0	2.2	462	396	858
168	0	3	1	2.6	434	532	966
169	0	1	1	2.5	534	666	1200
170	0	3	0	2.3	596	382	978
171	1	2	1	2.1	543	312	855
172	1	3	0	2.5	474	496	970
173	1	3	1	0.9	300	350	650
174	1	2	1	1.4	468	366	834
175	0	2	1	1.8	462	397	859
176	0	2	0	1.9	477	373	850
177	1	1	0	1.8	388	304	692
178	1	3	1	3.1	487	533	1020
179	1	3	0	3.0	443	588	1031
180	1	2	1	1.5	478	492	970
181	1	2	1	2.1	420	505	925
182	1	2	0	2.4	512	326	838
183	1	1	1	1.7	569	471	1040
184	0	2	0	1.4	289	453	742
185	1	2	0	2.4	542	354	896
186	1	2	0	2.5	614	482	1096
187	1	2	0	2.8	502	563	1065
188	0	2	0	3.6	621	741	1362
189	0	1	0	1.2	297	387	684
190	1	3	1	2.6	658	453	1111
191	0	3	0	2.8	559	438	997
192	1	1	0	2.8	436	541	977
193	0	3	0	1.7	269	387	656
194	1	2	0	2.6	415	528	943
195	1	2	0	2.2	475	504	979
196	0	2	0	2.2	508	457	965
197	1	3	1	3.3	506	459	965
198	1	3	0	2.7	497	509	1006
199	0	3	0	3.9	606	739	1345
200	0	1	0	1.6	549	233	782

Day	Draft priority number											
	Jan	Feb	Mar	Apr	May	Jun	Jul	Aug	Sep	Oct	Nov	Dec
1	305	86	108	32	330	249	93	111	225	359	19	129
2	159	144	29	271	298	228	350	45	161	125	34	328
3	251	297	267	83	40	301	115	261	49	244	348	157
4	215	210	275	81	276	20	279	145	232	202	266	165
5	101	214	293	269	364	28	188	54	82	24	310	56
6	224	347	139	253	155	110	327	114	6	87	76	10
7	306	91	122	147	35	85	50	168	8	234	51	12
8	199	181	213	312	321	366	13	48	184	283	97	105
9	194	338	317	219	197	335	277	106	263	342	80	43
10	325	216	323	218	65	206	284	21	71	220	282	41
11	329	150	136	14	37	134	248	324	158	237	46	39
12	221	68	300	346	133	272	15	142	242	72	66	314
13	318	152	259	124	295	69	42	307	175	138	126	163
14	238	4	354	231	178	356	331	198	1	294	127	26
15	17	89	169	273	130	180	322	102	113	171	131	320
16	121	212	166	148	55	274	120	44	207	254	107	96
17	235	189	33	260	112	73	98	154	255	288	143	304
18	140	292	332	90	278	341	190	141	246	5	146	128
19	58	25	200	336	75	104	227	311	177	241	203	240
20	280	302	239	345	183	360	187	344	63	192	185	135
21	186	363	334	62	250	60	27	291	204	243	156	70
22	337	290	265	316	326	247	153	339	160	117	9	53
23	118	57	256	252	319	109	172	116	119	201	182	162
24	59	236	258	2	31	358	23	36	195	196	230	95
25	52	179	343	351	361	137	67	286	149	176	132	84
26	92	365	170	340	357	22	303	245	18	7	309	173
27	355	205	268	74	296	64	289	352	233	264	47	78
28	77	299	223	262	308	222	88	167	257	94	281	123
29	349	285	362	191	226	353	270	61	151	229	99	16
30	164		217	208	103	209	287	333	315	38	174	3
31	211		30		313		193	11		79		100

SOURCE: Stephen E. Feinberg, ``Randomization and Social Affairs: The 1970 Draft Lottery,'' *Science,* vol. 171, January 22, 1971, pp. 255–261.

Day	Draft priority number											
	Jan	Feb	Mar	Apr	May	Jun	Jul	Aug	Sep	Oct	Nov	Dec
1	133	335	14	224	179	65	104	326	283	306	243	347
2	195	354	77	216	96	304	322	102	161	191	205	321
3	336	186	207	297	171	135	30	279	183	134	294	110
4	99	94	117	37	240	42	59	300	231	266	39	305
5	33	97	299	124	301	233	287	64	295	166	286	27
6	285	16	296	312	268	153	164	251	21	78	245	198
7	159	25	141	142	29	169	365	263	265	131	72	162
8	116	127	79	267	105	7	106	49	108	45	119	323
9	53	187	278	223	357	352	1	125	313	302	176	114
10	101	46	150	165	146	76	158	359	130	160	63	204
11	144	227	317	178	293	355	174	230	288	84	123	73
12	152	262	24	89	210	51	257	320	314	70	255	19
13	330	13	241	143	353	342	349	58	238	92	272	151
14	71	260	12	202	40	363	156	103	247	115	11	348
15	75	201	157	182	344	276	273	270	291	310	362	87
16	136	334	258	31	175	229	284	329	139	34	197	41
17	54	345	220	264	212	289	341	343	200	290	6	315
18	185	337	319	138	180	214	90	109	333	340	280	208
19	188	331	189	62	155	163	316	83	228	74	252	249
20	211	20	170	118	242	43	120	69	261	196	98	218
21	129	213	246	8	225	113	356	50	68	5	35	181
22	132	271	269	256	199	307	282	250	88	36	253	194
23	48	351	281	292	222	44	172	10	206	339	193	219
24	177	226	203	244	22	236	360	274	237	149	81	2
25	57	325	298	328	26	327	3	364	107	17	23	361
26	140	86	121	137	148	308	47	91	93	184	52	80
27	173	66	254	235	122	55	85	232	338	318	168	239
28	346	234	95	82	9	215	190	248	309	28	324	128
29	277		147	111	61	154	4	32	303	259	100	145
30	112		56	358	209	217	15	167	18	332	67	192
31	60		38		350		221	275		311		126

SOURCE: Stephen Feinberg, "Randomization and Social Affairs: The 1970 Draft Lottery," *Science,* vol. 171, January 22, 1971, pp. 255–261.

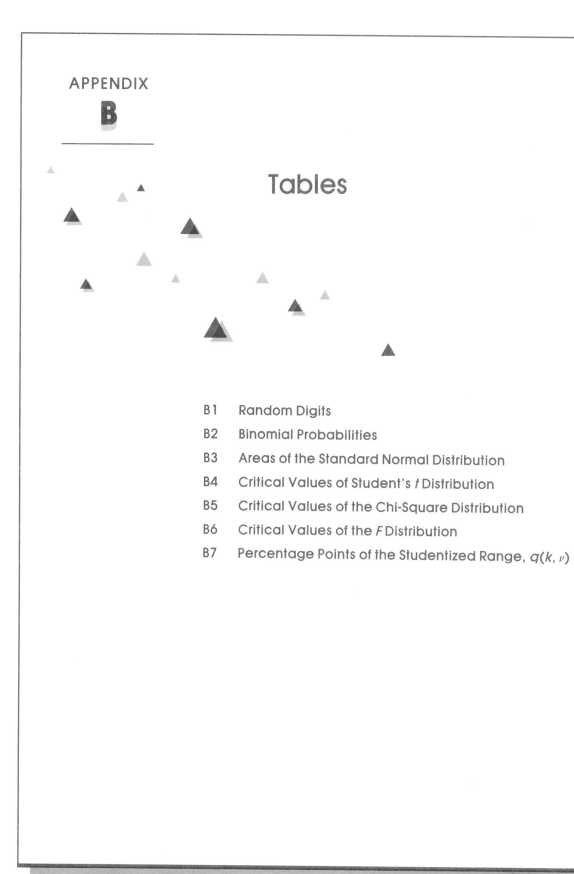

Tables

Row	C1	C2	C3	C4	C5	C6	C7	C8	C9	C10	C11	C12	C13	C14
1	365420	357847	617615	748728	171841	603720	869358	910566	875778	237162	340651	122884	453166	867383
2	807479	590525	37908	993542	431606	51297	874777	873511	775601	577746	64903	363379	620162	525457
3	698421	854995	852681	890200	439806	219253	860013	507636	210787	805258	349531	270813	515447	219540
4	988658	3893	320319	6735	172702	203591	261540	700110	452516	200154	208063	849082	929295	681651
5	598942	163835	349694	757185	222572	922789	988435	550266	704233	975893	586253	565829	164822	119964
6	880316	932343	913504	42349	131161	363401	685239	173127	691767	609496	691473	172069	669154	700246
7	152294	347449	654589	23491	355264	207124	453242	516543	630821	505621	883367	520819	542287	735336
8	281983	908361	249014	768845	730891	262213	408781	854372	610600	121102	271528	533826	474243	993044
9	362141	920311	75490	174415	672344	357025	413645	546516	225311	908317	604994	837242	240428	408946
10	282439	473491	881933	37192	565642	705328	852036	244905	117081	132952	338429	331633	208504	327850
11	655583	627286	933102	655380	912092	796133	778638	259989	684145	881861	216573	186854	961614	304646
12	917495	442367	598727	275128	976076	259791	928675	294640	774286	158996	171438	970092	189198	175380
13	814354	762820	497197	492902	504094	15832	407956	243295	442309	649191	745893	536641	407788	440601
14	833706	320920	381774	403158	336811	614077	67364	805244	690042	404504	820527	402493	820371	952140
15	73895	34208	915871	239555	21291	726479	583352	101489	319395	708958	365221	650262	72649	73989
16	638417	594178	422325	146910	351413	303316	764788	94144	671331	275625	447016	192881	682836	388505
17	612124	587499	463573	884412	785392	341599	692451	488813	967127	129893	250611	84777	864970	924638
18	439176	824029	663015	647964	969181	424802	321693	395317	454922	527258	60484	54282	169765	943851
19	233418	958899	72696	782603	763496	426901	173953	553777	582785	384382	967039	557625	210936	672490
20	823738	762580	791908	935268	501815	807607	645654	186614	595956	843552	83709	15970	971575	680424
21	718508	930802	817782	479101	388896	261581	320656	892958	169208	421586	510729	816251	199597	717023
22	904597	817779	114148	307934	983458	44758	597034	529979	256589	898490	99886	489342	53751	123679
23	841522	734471	827551	699694	601631	774378	995512	708429	376659	669563	432608	339472	683556	389076
24	728982	987530	371449	168377	70179	956579	845438	352075	911638	517616	501707	899824	894024	75145
25	813051	536071	214263	967809	364724	311277	769622	231736	790155	696826	727282	775084	37995	311507
26	619858	786489	902553	315829	837464	645091	733755	285126	408588	909747	303825	220805	156537	708268
27	667615	57182	257227	717820	419998	294881	327325	795612	772480	412289	518150	404203	700251	503653
28	598252	795891	751346	414422	23407	622971	259740	993828	147937	76437	805663	397032	702189	898721
29	708169	8412	994933	572896	327406	760034	882190	428815	132462	627850	674619	127397	186813	669239
30	544309	212770	560912	311231	340567	935093	437448	677033	358035	798509	127711	30987	552878	899095
31	304353	365011	706185	744921	377975	936569	693919	767021	833998	964981	326175	6432	34053	508733
32	495283	656718	175494	264036	535733	858989	549448	318644	265468	571662	602909	188974	632358	968557
33	793651	93392	517536	870723	879711	883373	931635	531012	782058	107090	810899	292647	37901	789385
34	475397	169591	392521	851226	16119	105318	255432	169591	711369	157262	664399	752200	221624	813124
35	758628	336989	290140	874530	231369	271545	768322	208831	609307	371329	839808	81424	792007	695349
36	777525	246180	710168	936857	234600	329210	821399	632144	282369	884545	676092	159200	263476	605831
37	186128	126200	36390	931967	80882	26631	820822	894594	172035	365641	140963	433490	830619	526566
38	298381	929565	823332	790554	746588	609741	117302	544905	708518	71277	560243	942283	124099	241355
39	919012	392259	97731	401007	476540	721293	623885	637539	763206	917419	334340	956040	642390	12680
40	40652	119694	147445	906700	12122	750245	86607	140083	702427	495720	99252	915770	53349	976533
41	397288	974614	507670	771384	295638	278436	574612	887813	817021	461036	387100	631087	221815	721804
42	661110	492815	263699	443886	924501	367893	758235	190232	938494	375112	60465	508473	296146	488215
43	352080	693064	911380	417997	98048	136249	921191	50167	811270	625207	377042	47316	664430	203183
44	6298	610179	142088	425348	396377	809758	217797	856817	824759	930508	882173	360918	581806	241242
45	602294	449340	702347	214809	757494	888740	390217	142403	12985	118757	574631	265483	400418	310933
46	61082	34178	416763	215571	918468	114576	788599	769816	125611	123257	130043	240458	341351	76069
47	730400	880112	373828	226482	905488	874074	706588	38106	503834	589275	898571	47119	509251	478286
48	766496	54240	553904	570745	640474	555699	703296	944211	427555	114795	239577	692869	915891	246856
49	855130	561471	580541	408830	661405	605688	310395	897216	800393	664417	578578	418670	907083	18711
50	794325	413470	976937	482769	149369	556909	252807	533648	901767	370823	646448	898265	417861	924900
51	955394	217932	134065	779616	481469	455712	256290	38812	333203	977034	578068	971487	321686	994900
52	809693	773640	309882	396369	163633	887503	880155	267696	436126	766791	779025	947010	764694	572021

TABLE B1   Random Digits (continued)

Row	C1	C2	C3	C4	C5	C6	C7	C8	C9	C10	C11	C12	C13	C14
53	685414	357749	112873	536284	54266	172952	616470	966122	25697	629885	542144	844753	757538	593670
54	295616	702917	560008	55640	245815	924809	172251	301727	836538	889191	17218	281869	413449	587012
55	900473	341757	400263	429037	360940	704601	143924	547538	131401	330528	910926	295124	735241	260887
56	993694	774150	641770	574125	55729	972187	126415	465964	818800	257351	9749	579520	289460	43941
57	530307	83873	522662	148728	497045	701890	970333	916248	512074	541486	696672	344628	891877	235218
58	195151	957788	649228	748483	636182	164288	727111	587912	901090	679798	411079	404082	205372	33926
59	883497	470043	316860	661173	31418	185760	372546	283224	824256	542769	542621	687883	416504	716405
60	385802	334138	738258	330416	603545	672772	749190	220398	772238	291033	936579	199255	29352	284495
61	481036	730059	60455	382100	668990	969200	555994	302891	228992	104671	743631	424698	844144	906826
62	466325	787873	634856	190992	515626	644960	680671	941619	27606	114091	107779	918015	606178	792961
63	777877	821841	398271	126398	942304	954568	681005	481378	317032	301507	988107	965825	645654	431023
64	19887	223483	629282	49691	36874	255419	362547	422275	121228	800860	591998	497883	829540	778195
65	792481	370821	564689	222091	539512	5422	511662	893169	825951	3210	990729	302155	99442	263528
66	72093	903122	179240	715318	552522	978511	276242	51590	369983	198353	965848	849081	304309	142639
67	694217	304456	993187	840800	991283	467027	380080	443525	621485	611928	669380	381142	556460	37344
68	130970	430766	768476	480860	305194	27590	634441	913788	791205	725395	971291	114031	234716	615551
69	12645	883976	766361	836602	630577	164151	580253	528475	995052	564127	636242	939714	53350	967697
70	572397	515469	876264	161885	74973	714374	424442	60592	204910	635296	995348	4708	912687	961308
71	133043	643948	990994	249599	3112	963609	950378	22134	950005	214527	941782	670682	951210	539611
72	538341	475449	879314	649632	466975	482957	653011	832251	18341	502018	489747	274923	178519	914881
73	71944	634272	114994	548674	441485	804130	630152	305343	725118	71319	425222	203853	892920	405538
74	497288	242459	577815	353200	18020	609720	81616	816046	204986	241241	358223	303187	343292	327506
75	609867	330651	659893	836745	660409	864184	295615	912681	407872	420938	349243	69274	997453	434624
76	185554	649768	635428	598689	857896	699114	575397	853114	867557	698806	989929	184647	71401	356589
77	507114	306129	904903	117695	303840	375877	665261	478851	864623	324325	33189	998877	362123	996839
78	107775	502818	34898	270021	962252	912623	204562	373597	94974	648181	82531	48891	853053	911737
79	137674	501788	298654	635828	14944	337652	105387	762113	410394	316275	564085	297036	185849	762640
80	253485	633586	588855	633788	52346	707592	815124	699274	721276	162435	198385	947033	304403	169974
81	395447	927824	185540	108967	893573	541579	487134	406179	840292	910089	774709	730951	582946	348351
82	814888	355567	592889	929506	864936	203461	629864	487197	483844	789931	274423	483515	699013	939951
83	192166	443871	835274	251738	152664	377290	539809	96725	151807	527090	230465	439432	246668	772360
84	758933	81447	183056	439562	831968	401019	822356	418920	158687	14382	704953	122239	910861	972521
85	408573	382212	499013	700189	739397	587175	359669	788552	773843	80786	974546	761035	409332	5929
86	701254	722349	873247	841183	763359	822715	113369	570215	495473	444614	732837	465575	392764	854588
87	559675	95292	383447	176925	857202	412325	765389	994690	373275	371118	705707	608437	461653	369867
88	119529	417940	311211	401710	305090	551399	451540	611196	911214	19040	579560	445240	89361	328567
89	304740	304770	973847	34564	893200	787082	526571	967188	71654	248945	156234	13814	922979	134469
90	60706	920698	420503	623035	285090	803716	779822	467916	906942	324412	486238	431280	477806	410315
91	661537	813646	618212	774678	651927	611118	563183	84154	497653	210281	58031	685416	492903	684223
92	864871	976628	791876	293491	573353	914926	357158	821787	685658	856693	773843	640913	919893	607974
93	388938	741072	322182	296400	327750	598471	866846	170755	812565	818006	645394	690312	9316	101115
94	890878	120431	880430	890975	704018	106711	813989	666413	607199	759165	394649	897805	168104	985193
95	91152	867117	185178	116975	613257	774150	15003	474852	137327	832501	793078	797859	209231	24262
96	642638	360194	383511	45066	278537	389392	647920	864463	840338	785782	879259	761055	57963	249539
97	755478	472555	156869	474957	576828	886683	537491	700557	561760	749290	267067	375111	496359	253320
98	884158	765488	808915	545033	24186	691263	474471	992813	483996	350594	481754	789951	915310	164648
99	850963	277189	624023	971051	747621	583413	299888	198686	668522	603021	328986	602526	806890	736472
100	850149	158448	348673	288633	230851	638935	90139	540779	108557	146356	117061	827362	52574	365088

SOURCE: Random numbers generated with the Minitab Computer Program.

n	k	.01	.05	.10	.20	.30	.40	π .50	.60	.70	.80	.90	.95	.99
2	0	980	902	810	640	490	360	250	160	090	040	010	002	0+
	1	020	095	180	320	420	480	500	480	420	320	180	095	020
	2	0+	002	010	040	090	160	250	360	490	640	810	902	980
3	0	970	857	729	512	343	216	125	064	027	008	001	0+	0+
	1	029	135	243	384	441	432	375	288	189	096	027	007	0+
	2	0+	007	027	096	189	288	375	432	441	384	243	135	029
	3	0+	0+	001	008	027	064	125	216	343	512	729	857	970
4	0	961	815	656	410	240	130	062	026	008	002	0+	0+	0+
	1	039	171	292	410	412	346	250	154	076	026	004	0+	0+
	2	001	014	049	154	265	346	375	346	265	154	049	014	001
	3	0+	0+	004	026	076	154	250	346	412	410	292	171	039
	4	0+	0+	0+	002	008	026	062	130	240	410	656	815	961
5	0	951	774	590	328	168	078	031	010	002	0+	0+	0+	0+
	1	048	204	328	410	360	259	156	077	028	006	0+	0+	0+
	2	001	021	073	205	309	346	312	230	132	051	008	001	0+
	3	0+	001	008	051	132	230	312	346	309	205	073	021	001
	4	0+	0+	0+	006	028	077	156	259	360	410	328	204	048
	5	0+	0+	0+	0+	002	010	031	078	168	328	590	774	951
6	0	941	735	531	262	118	047	016	004	001	0+	0+	0+	0+
	1	057	232	354	393	303	187	094	037	010	002	0+	0+	0+
	2	001	031	098	246	324	311	234	138	060	015	001	0+	0+
	3	0+	002	015	082	185	276	312	276	185	082	015	002	0+
	4	0+	0+	001	015	060	138	234	311	324	246	098	031	001
	5	0+	0+	0+	002	010	037	094	187	303	393	354	232	057
	6	0+	0+	0+	0+	001	004	016	047	118	262	531	735	941
7	0	932	698	478	210	082	028	008	002	0+	0+	0+	0+	0+
	1	066	257	372	367	247	131	055	017	004	0+	0+	0+	0+
	2	002	041	124	275	318	261	164	077	025	004	0+	0+	0+
	3	0+	004	023	115	227	290	273	194	097	029	003	0+	0+
	4	0+	0+	003	029	097	194	273	290	227	115	023	004	0+
	5	0+	0+	0+	004	025	077	164	261	318	275	124	041	002
	6	0+	0+	0+	0+	004	017	055	131	247	367	372	257	066
	7	0+	0+	0+	0+	0+	002	008	028	082	210	478	698	932
8	0	923	663	430	168	058	017	004	001	0+	0+	0+	0+	0+
	1	075	279	383	336	198	090	031	008	001	0+	0+	0+	0+
	2	003	051	149	294	296	209	109	041	010	001	0+	0+	0+
	3	0+	005	033	147	254	279	219	124	047	009	0+	0+	0+
	4	0+	0+	005	046	136	232	273	232	136	046	005	0+	0+
	5	0+	0+	0+	009	047	124	219	279	254	147	033	005	0+
	6	0+	0+	0+	001	010	041	109	209	296	294	149	051	003
	7	0+	0+	0+	0+	001	008	031	090	198	336	383	279	075
	8	0+	0+	0+	0+	0+	001	004	017	058	168	430	663	923
9	0	914	630	387	134	040	010	002	0+	0+	0+	0+	0+	0+
	1	083	299	387	302	156	060	018	004	0+	0+	0+	0+	0+
	2	003	063	172	302	267	161	070	021	004	0+	0+	0+	0+
	3	0+	008	045	176	267	251	164	074	021	003	0+	0+	0+
	4	0+	001	007	066	172	251	246	167	074	017	001	0+	0+

								π						
n	*k*	.01	.05	.10	.20	.30	.40	.50	.60	.70	.80	.90	.95	.99
	5	0+	0+	001	017	074	167	246	251	172	066	007	001	0+
	6	0+	0+	0+	003	021	074	164	251	267	176	045	008	0+
	7	0+	0+	0+	0+	004	021	070	161	267	302	172	063	003
	8	0+	0+	0+	0+	0+	004	018	060	156	302	387	299	083
	9	0+	0+	0+	0+	0+	0+	002	010	040	134	387	630	914
10	0	904	599	349	107	028	006	001	0+	0+	0+	0+	0+	0+
	1	091	315	387	268	121	040	010	002	0+	0+	0+	0+	0+
	2	004	075	194	302	233	121	044	011	001	0+	0+	0+	0+
	3	0+	010	057	201	267	215	117	042	009	001	0+	0+	0+
	4	0+	001	011	088	200	251	205	111	037	006	0+	0+	0+
	5	0+	0+	001	026	103	201	246	201	103	026	001	0+	0+
	6	0+	0+	0+	006	037	111	205	251	200	088	011	001	0+
	7	0+	0+	0+	001	009	042	117	215	267	201	057	010	0+
	8	0+	0+	0+	0+	001	011	044	121	233	302	194	075	004
	9	0+	0+	0+	0+	0+	002	010	040	121	268	387	315	091
	10	0+	0+	0+	0+	0+	0+	001	006	028	107	349	599	904
11	0	895	569	314	086	020	004	0+	0+	0+	0+	0+	0+	0+
	1	099	329	384	236	093	027	005	001	0+	0+	0+	0+	0+
	2	005	087	213	295	200	089	027	005	001	0+	0+	0+	0+
	3	0+	014	071	221	257	177	081	023	004	0+	0+	0+	0+
	4	0+	001	016	111	220	236	161	070	017	002	0+	0+	0+
	5	0+	0+	002	039	132	221	226	147	057	010	0+	0+	0+
	6	0+	0+	0+	010	057	147	226	221	132	039	002	0+	0+
	7	0+	0+	0+	002	017	070	161	236	220	111	016	001	0+
	8	0+	0+	0+	0+	004	023	081	177	257	221	071	014	0+
	9	0+	0+	0+	0+	001	005	027	089	200	295	213	087	005
	10	0+	0+	0+	0+	0+	001	005	027	093	236	384	329	099
	11	0+	0+	0+	0+	0+	0+	004	020	086	314	569	895
12	0	886	540	282	069	014	002	0+	0+	0+	0+	0+	0+	0+
	1	107	341	377	206	071	017	003	0+	0+	0+	0+	0+	0+
	2	006	099	230	283	168	064	016	002	0+	0+	0+	0+	0+
	3	0+	017	085	236	240	142	054	012	001	0+	0+	0+	0+
	4	0+	002	021	133	231	213	121	042	008	001	0+	0+	0+
	5	0+	0+	004	053	158	227	193	101	029	003	0+	0+	0+
	6	0+	0+	0+	016	079	177	226	177	079	016	0+	0+	0+
	7	0+	0+	0+	003	029	101	193	227	158	053	004	0+	0+
	8	0+	0+	0+	001	008	042	121	213	231	133	021	002	0+
	9	0+	0+	0+	0+	001	012	054	142	240	236	085	017	0+
	10	0+	0+	0+	0+	0+	002	016	064	168	283	230	099	006
	11	0+	0+	0+	0+	0+	0+	003	017	071	206	377	341	107
	12	0+	0+	0+	0+	0+	0+	0+	002	014	069	282	540	886
13	0	878	513	254	055	010	001	0+	0+	0+	0+	0+	0+	0+
	1	115	351	367	179	054	011	002	0+	0+	0+	0+	0+	0+
	2	007	111	245	268	139	045	010	001	0+	0+	0+	0+	0+
	3	0+	021	100	246	218	111	035	006	001	0+	0+	0+	0+
	4	0+	003	028	154	234	184	087	024	003	0+	0+	0+	0+

TABLE B2   Binomial Probabilities (continued)

								π						
n	k	.01	.05	.10	.20	.30	.40	.50	.60	.70	.80	.90	.95	.99
	5	0+	0+	006	069	180	221	157	066	014	001	0+	0+	0+
	6	0+	0+	001	023	103	197	209	131	044	006	0+	0+	0+
	7	0+	0+	0+	006	044	131	209	197	103	023	001	0+	0+
	8	0+	0+	0+	001	014	066	157	221	180	069	006	0+	0+
	9	0+	0+	0+	0+	003	024	087	184	234	154	028	003	0+
	10	0+	0+	0+	0+	001	006	035	111	218	246	100	021	0+
	11	0+	0+	0+	0+	0+	001	010	045	139	268	245	111	007
	12	0+	0+	0+	0+	0+	0+	002	011	054	179	367	351	115
	13	0+	0+	0+	0+	0+	0+	0+	001	010	055	254	513	878
14	0	869	488	229	044	007	001	0+	0+	0+	0+	0+	0+	0+
	1	123	359	356	154	041	007	001	0+	0+	0+	0+	0+	0+
	2	008	123	257	250	113	032	006	001	0+	0+	0+	0+	0+
	3	0+	026	114	250	194	085	022	003	0+	0+	0+	0+	0+
	4	0+	004	035	172	229	155	061	014	001	0+	0+	0+	0+
	5	0+	0+	008	086	196	207	122	041	007	0+	0+	0+	0+
	6	0+	0+	001	032	126	207	183	092	023	002	0+	0+	0+
	7	0+	0+	0+	009	062	157	209	157	062	009	0+	0+	0+
	8	0+	0+	0+	002	023	092	183	207	126	032	001	0+	0+
	9	0+	0+	0+	0+	007	041	122	207	196	086	008	0+	0+
	10	0+	0+	0+	0+	001	014	061	155	229	172	035	004	0+
	11	0+	0+	0+	0+	0+	003	022	085	194	250	114	026	0+
	12	0+	0+	0+	0+	0+	001	006	032	113	250	257	123	008
	13	0+	0+	0+	0+	0+	0+	001	007	041	154	356	359	123
	14	0+	0+	0+	0+	0+	0+	0+	001	007	044	229	488	869
15	0	860	463	206	035	005	0+	0+	0+	0+	0+	0+	0+	0+
	1	130	366	343	132	031	005	0+	0+	0+	0+	0+	0+	0+
	2	009	135	267	231	092	022	003	0+	0+	0+	0+	0+	0+
	3	0+	031	129	250	170	063	014	002	0+	0+	0+	0+·	0+
	4	0+	005	043	188	219	127	042	007	001	0+	0+	0+	0+
	5	0+	001	010	103	206	186	092	024	003	0+	0+	0+	0+
	6	0+	0+	002	043	147	207	153	061	012	001	0+	0+	0+
	7	0+	0+	0+	014	081	177	196	118	035	003	0+	0+	0+
	8	0+	0+	0+	003	035	118	196	177	081	014	0+	0+	0+
	9	0+	0+	0+	001	012	061	153	207	147	043	002	0+	0+
	10	0+	0+	0+	0+	003	024	092	186	206	103	010	001	0+
	11	0+	0+	0+	0+	001	007	042	127	219	188	043	005	0+
	12	0+	0+	0+	0+	0+	002	014	063	170	250	129	031	0+
	13	0+	0+	0+	0+	0+	0+	003	022	092	231	267	135	009
	14	0+	0+	0+	0+	0+	0+	0+	005	031	132	343	366	130
	15	0+	0+	0+	0+	0+	0+	0+	0+	005	035	206	463	860
16	0	851	440	185	028	003	0+	0+	0+	0+	0+	0+	0+	0+
	1	138	371	329	113	023	003	0+	0+	0+	0+	0+	0+	0+
	2	010	146	274	211	073	015	002	0+	0+	0+	0+	0+	0+
	3	0+	036	142	246	146	047	008	001	0+	0+	0+	0+	0+
	4	0+	006	051	200	204	101	028	004	0+	0+	0+	0+	0+
	5	0+	001	014	120	210	162	067	014	001	0+	0+	0+	0+
	6	0+	0+	003	055	165	198	122	039	006	0+	0+	0+	0+
	7	0+	0+	0+	020	101	189	175	084	018	001	0+	0+	0+
	8	0+	0+	0+	006	049	142	196	142	049	006	0+	0+	0+

TABLE B2 Binomial Probabilities (continued)

n	k	.01	.05	.10	.20	.30	.40	.50	.60	.70	.80	.90	.95	.99
	9	0+	0+	0+	001	018	084	175	189	101	020	0+	0+	0+
	10	0+	0+	0+	0+	006	039	122	198	165	055	003	0+	0+
	11	0+	0+	0+	0+	001	014	067	162	210	120	014	001	0+
	12	0+	0+	0+	0+	0+	004	028	101	204	200	051	006	0+
	13	0+	0+	0+	0+	0+	001	008	047	146	246	142	036	0+
	14	0+	0+	0+	0+	0+	0+	002	015	073	211	274	146	010
	15	0+	0+	0+	0+	0+	0+	0+	003	023	113	329	371	138
	16	0+	0+	0+	0+	0+	0+	0+	0+	003	028	185	440	851
17	0	843	418	167	022	002	0+	0+	0+	0+	0+	0+	0+	0+
	1	145	374	315	096	017	002	0+	0+	0+	0+	0+	0+	0+
	2	012	158	280	191	058	010	001	0+	0+	0+	0+	0+	0+
	3	001	042	156	239	124	034	005	0+	0+	0+	0+	0+	0+
	4	0+	008	060	209	187	080	018	002	0+	0+	0+	0+	0+
	5	0+	001	018	136	208	138	047	008	001	0+	0+	0+	0+
	6	0+	0+	004	068	178	184	094	024	003	0+	0+	0+	0+
	7	0+	0+	001	027	120	193	148	057	010	0+	0+	0+	0+
	8	0+	0+	0+	008	064	161	186	107	028	002	0+	0+	0+
	9	0+	0+	0+	002	028	107	186	161	064	008	0+	0+	0+
	10	0+	0+	0+	0+	010	057	148	193	120	027	001	0+	0+
	11	0+	0+	0+	0+	003	024	094	184	178	068	004	0+	0+
	12	0+	0+	0+	0+	001	008	047	138	208	136	018	001	0+
	13	0+	0+	0+	0+	0+	002	018	080	187	209	060	008	0+
	14	0+	0+	0+	0+	0+	0+	005	034	124	239	156	042	001
	15	0+	0+	0+	0+	0+	0+	001	010	058	191	280	158	012
	16	0+	0+	0+	0+	0+	0+	0+	002	017	096	315	374	145
	17	0+	0+	0+	0+	0+	0+	0+	0+	002	022	167	418	843
18	0	835	397	150	018	002	0+	0+	0+	0+	0+	0+	0+	0+
	1	152	376	300	081	013	001	0+	0+	0+	0+	0+	0+	0+
	2	013	168	284	172	046	007	001	0+	0+	0+	0+	0+	0+
	3	001	047	168	230	105	025	003	0+	0+	0+	0+	0+	0+
	4	0+	009	070	215	168	061	012	001	0+	0+	0+	0+	0+
	5	0+	001	002	151	202	115	033	004	0+	0+	0+	0+	0+
	6	0+	0+	005	082	187	166	071	014	001	0+	0+	0+	0+
	7	0+	0+	001	035	138	189	121	037	005	0+	0+	0+	0+
	8	0+	0+	0+	012	081	173	167	077	015	001	0+	0+	0+
	9	0+	0+	0+	003	038	128	186	128	038	003	0+	0+	0+
	10	0+	0+	0+	001	015	077	167	173	081	012	0+	0+	0+
	11	0+	0+	0+	0+	005	037	121	189	138	035	001	0+	0+
	12	0+	0+	0+	0+	001	014	071	166	187	082	005	0+	0+
	13	0+	0+	0+	0+	0+	004	033	115	202	151	022	001	0+
	14	0+	0+	0+	0+	0+	001	012	061	168	215	070	009	0+
	15	0+	0+	0+	0+	0+	0+	003	025	105	230	168	097	001
	16	0+	0+	0+	0+	0+	0+	001	007	046	172	284	168	013
	17	0+	0+	0+	0+	0+	0+	0+	001	013	081	300	376	152
	18	0+	0+	0+	0+	0+	0+	0+	0+	002	018	150	397	835

n	k	.01	.05	.10	.20	.30	.40	.50	.60	.70	.80	.90	.95	.99
19	0	826	377	135	014	001	0+	0+	0+	0+	0+	0+	0+	0+
	1	159	377	285	069	009	001	0+	0+	0+	0+	0+	0+	0+
	2	014	179	285	154	036	005	0+	0+	0+	0+	0+	0+	0+
	3	001	053	180	218	087	018	002	0+	0+	0+	0+	0+	0+
	4	0+	011	080	218	149	047	007	0+	0+	0+	0+	0+	0+
	5	0+	002	027	164	192	093	022	002	0+	0+	0+	0+	0+
	6	0+	0+	007	096	192	145	052	008	0+	0+	0+	0+	0+
	7	0+	0+	001	044	152	180	096	024	002	0+	0+	0+	0+
	8	0+	0+	0+	017	098	180	144	053	008	0+	0+	0+	0+
	9	0+	0+	0+	005	051	146	176	098	022	001	0+	0+	0+
	10	0+	0+	0+	001	022	098	176	146	051	005	0+	0+	0+
	11	0+	0+	0+	0+	008	053	144	180	098	017	0+	0+	0+
	12	0+	0+	0+	0+	002	024	096	180	152	044	001	0+	0+
	13	0+	0+	0+	0+	0+	008	052	145	192	096	007	0+	0+
	14	0+	0+	0+	0+	0+	002	022	093	192	164	027	002	0+
	15	0+	0+	0+	0+	0+	0+	007	047	149	218	080	011	0+
	16	0+	0+	0+	0+	0+	0+	002	018	087	218	180	053	001
	17	0+	0+	0+	0+	0+	0+	0+	005	036	154	285	179	014
	18	0+	0+	0+	0+	0+	0+	0+	001	009	069	285	377	159
	19	0+	0+	0+	0+	0+	0+	0+	0+	001	014	135	377	826
20	0	818	358	122	012	001	0+	0+	0+	0+	0+	0+	0+	0+
	1	165	377	270	058	007	0+	0+	0+	0+	0+	0+	0+	0+
	2	016	189	285	137	028	003	0+	0+	0+	0+	0+	0+	0+
	3	001	060	190	205	072	012	001	0+	0+	0+	0+	0+	0+
	4	0+	013	090	218	130	035	005	0+	0+	0+	0+	0+	0+
	5	0+	002	032	175	179	075	015	001	0+	0+	0+	0+	0+
	6	0+	0+	009	109	192	124	037	005	0+	0+	0+	0+	0+
	7	0+	0+	002	054	164	166	074	015	001	0+	0+	0+	0+
	8	0+	0+	0+	022	114	180	120	036	004	0+	0+	0+	0+
	9	0+	0+	0+	007	065	160	160	071	012	0+	0+	0+	0+
	10	0+	0+	0+	002	031	117	176	117	031	002	0+	0+	0+
	11	0+	0+	0+	0+	012	071	160	160	065	007	0+	0+	0+
	12	0+	0+	0+	0+	004	036	120	180	114	022	0+	0+	0+
	13	0+	0+	0+	0+	001	015	074	166	164	054	002	0+	0+
	14	0+	0+	0+	0+	0+	005	037	124	192	109	009	0+	0+
	15	0+	0+	0+	0+	0+	001	015	075	179	175	032	002	0+
	16	0+	0+	0+	0+	0+	0+	005	035	130	218	090	013	0+
	17	0+	0+	0+	0+	0+	0+	001	012	072	205	190	060	001
	18	0+	0+	0+	0+	0+	0+	0+	003	028	137	285	189	016
	19	0+	0+	0+	0+	0+	0+	0+	0+	007	058	270	377	165
	20	0+	0+	0+	0+	0+	0+	0+	0+	001	012	122	358	818

TABLE B3   Areas of the Standard Normal Distribution

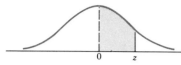

The table value, corresponding to the shaded area under the curve, is the probability that a standard normal random variable assumes a value between 0 and z. Also, the table value is the probability that a nonstandard normal random variable assumes a value from its mean to a point that is z standard deviations above the mean.

Second decimal place in z

z	.00	.01	.02	.03	.04	.05	.06	.07	.08	.09
.0	.0000	.0040	.0080	.0120	.0160	.0199	.0239	.0279	.0319	.0359
.1	.0398	.0438	.0478	.0517	.0557	.0596	.0636	.0675	.0714	.0753
.2	.0793	.0832	.0871	.0910	.0948	.0987	.1026	.1064	.1103	.1141
.3	.1179	.1217	.1255	.1293	.1331	.1368	.1406	.1443	.1480	.1517
.4	.1554	.1591	.1628	.1664	.1700	.1736	.1772	.1808	.1844	.1879
.5	.1915	.1950	.1985	.2019	.2054	.2088	.2123	.2157	.2190	.2224
.6	.2257	.2291	.2324	.2357	.2389	.2422	.2454	.2486	.2517	.2549
.7	.2580	.2611	.2642	.2673	.2704	.2734	.2764	.2794	.2823	.2852
.8	.2881	.2910	.2939	.2967	.2995	.3023	.3051	.3078	.3106	.3133
.9	.3159	.3186	.3212	.3238	.3264	.3289	.3315	.3340	.3365	.3389
1.0	.3413	.3438	.3461	.3485	.3508	.3531	.3554	.3577	.3599	.3621
1.1	.3643	.3665	.3686	.3708	.3729	.3749	.3770	.3790	.3810	.3830
1.2	.3849	.3869	.3888	.3907	.3925	.3944	.3962	.3980	.3997	.4015
1.3	.4032	.4049	.4066	.4082	.4099	.4115	.4131	.4147	.4162	.4177
1.4	.4192	.4207	.4222	.4236	.4251	.4265	.4279	.4292	.4306	.4319
1.5	.4332	.4345	.4357	.4370	.4382	.4394	.4406	.4418	.4429	.4441
1.6	.4452	.4463	.4474	.4484	.4495	.4505	.4515	.4525	.4535	.4545
1.7	.4554	.4564	.4573	.4582	.4591	.4599	.4608	.4616	.4625	.4633
1.8	.4641	.4649	.4656	.4664	.4671	.4678	.4686	.4693	.4699	.4706
1.9	.4713	.4719	.4726	.4732	.4738	.4744	.4750	.4756	.4761	.4767
2.0	.4772	.4778	.4783	.4788	.4793	.4798	.4803	.4808	.4812	.4817
2.1	.4821	.4826	.4830	.4834	.4838	.4842	.4846	.4850	.4854	.4857
2.2	.4861	.4864	.4868	.4871	.4875	.4878	.4881	.4884	.4887	.4890
2.3	.4893	.4896	.4898	.4901	.4904	.4906	.4909	.4911	.4913	.4916
2.4	.4918	.4920	.4922	.4925	.4927	.4929	.4931	.4932	.4934	.4936
2.5	.4938	.4940	.4941	.4943	.4945	.4946	.4948	.4949	.4951	.4952
2.6	.4953	.4955	.4956	.4957	.4959	.4960	.4961	.4962	.4963	.4964
2.7	.4965	.4966	.4967	.4968	.4969	.4970	.4971	.4972	.4973	.4974
2.8	.4974	.4975	.4976	.4977	.4977	.4978	.4979	.4979	.4980	.4981
2.9	.4981	.4982	.4982	.4983	.4984	.4984	.4985	.4985	.4986	.4986
3.0	.4987	.4987	.4987	.4988	.4988	.4989	.4989	.4989	.4990	.4990
3.1	.4990	.4991	.4991	.4991	.4992	.4992	.4992	.4992	.4993	.4993
3.2	.4993	.4993	.4994	.4994	.4994	.4994	.4994	.4995	.4995	.4995
3.3	.4995	.4995	.4995	.4996	.4996	.4996	.4996	.4996	.4996	.4997
3.4	.4997	.4997	.4997	.4997	.4997	.4997	.4997	.4997	.4997	.4998
3.5	.4998									
4.0	.49997									
4.5	.499997									
5.0	.4999997									

SOURCE: Abridged from Table Iii of Fisher and Yates, *Statistical Tables for Biological, Agricultural, and Medical Research*, published by Longman Group UK Ltd., London (1974) 6th ed. (previously published by Oliver and Boyd Ltd., Edinburgh), and by permission of the authors and publishers.

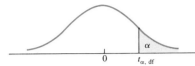

The table values are the critical values for Student's *t* for an area of α in the right-hand tail. Critical values for the left-hand tail are found by symmetry.

	80%	90%	95%	98%	99%
			Amount of α in one-tail		
Degrees of freedom	.1	.05	.025	.01	.005
1	3.078	6.314	12.706	31.821	63.657
2 n−1	1.886	2.920	4.303	6.965	9.925
3	1.638	2.353	3.182	4.541	5.841
4	1.533	2.132	2.776	3.747	4.604
5	1.476	2.015	2.571	3.365	4.032
6	1.440	1.943	2.447	3.143	3.707
7	1.415	1.895	2.365	2.998	3.499
8	1.397	1.860	2.306	2.896	3.355
9	1.383	1.833	2.262	2.821	3.250
10	1.372	1.812	2.228	2.764	3.169
11	1.363	1.796	2.201	2.718	3.106
12	1.356	1.782	2.179	2.681	3.055
13	1.350	1.771	2.160	2.650	3.012
14	1.345	1.761	2.145	2.624	2.977
15	1.341	1.753	2.131	2.602	2.947
16	1.337	1.746	2.120	2.583	2.921
17	1.333	1.740	2.110	2.567	2.898
18	1.330	1.734	2.101	2.552	2.878
19	1.328	1.729	2.093	2.539	2.861
20	1.325	1.725	2.086	2.528	2.845
21	1.323	1.721	2.080	2.518	2.831
22	1.321	1.717	2.074	2.508	2.819
23	1.319	1.714	2.069	2.500	2.807
24	1.318	1.711	2.064	2.492	2.797
25	1.316	1.708	2.060	2.485	2.787
26	1.315	1.706	2.056	2.479	2.779
27	1.314	1.703	2.052	2.473	2.771
28	1.313	1.701	2.048	2.467	2.763
29	1.311	1.699	2.045	2.462	2.756
30	1.310	1.697	2.042	2.457	2.750
40	1.303	1.684	2.021	2.423	2.704
60	1.296	1.671	2.000	2.390	2.660
120	1.289	1.658	1.980	2.358	2.617
∞	1.282	1.645	1.960	2.326	2.576

SOURCE: Abridged from Table III of Fisher and Yates, *Statistical Tables for Biological, Agricultural, and Medical Research*, published by Longman Group UK Ltd., London (1974) 6th ed. (previously published by Oliver and Boyd Ltd., Edinburgh), and by permission of the authors and publishers.

TABLE B5   Critical Values of the Chi-Square Distribution

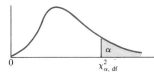

The table values are the critical values for the chi square distribution for which the area to the right under the curve is equal to α.

df	\multicolumn{10}{c}{Amount of α in right-hand tail}
	0.995	0.990	0.975	0.950	0.900	0.100	0.050	0.025	0.010	0.005
1	0.0000393	0.000157	0.000982	0.00393	0.0158	2.71	3.84	5.02	6.63	7.88
2	0.0100	0.0201	0.0506	0.103	0.211	4.61	5.99	7.38	9.21	10.6
3	0.0717	0.115	0.216	0.352	0.584	6.25	7.81	9.35	11.3	12.8
4	0.207	0.297	0.484	0.711	1.0636	7.78	9.49	11.1	13.3	14.9
5	0.412	0.554	0.831	1.15	1.61	9.24	11.1	12.8	15.1	16.7
6	0.676	0.872	1.24	1.64	2.20	10.6	12.6	14.5	16.8	18.5
7	0.989	1.24	1.69	2.17	2.83	12.0	14.1	16.0	18.5	20.3
8	1.34	1.65	2.18	2.73	3.49	13.4	15.5	17.5	20.1	22.0
9	1.73	2.09	2.70	3.33	4.17	14.7	16.9	19.0	21.7	23.6
10	2.16	2.56	3.25	3.94	4.87	16.0	18.3	20.5	23.2	25.2
11	2.60	3.05	3.82	4.58	5.58	17.3	19.7	21.9	24.7	26.8
12	3.07	3.57	4.40	5.23	6.30	18.5	21.0	23.3	26.2	28.3
13	3.57	4.11	5.01	5.90	7.04	19.8	22.4	24.7	27.7	29.8
14	4.07	4.66	5.63	6.57	7.79	21.1	23.7	26.1	29.1	31.3
15	4.60	5.23	6.26	7.26	8.55	22.3	25.0	27.5	30.6	32.8
16	5.14	5.81	6.91	7.96	9.31	23.5	26.3	28.8	32.0	34.3
17	5.70	6.41	7.56	8.67	10.1	24.8	27.6	30.2	33.4	35.7
18	6.26	7.01	8.23	9.39	10.9	26.0	28.9	31.5	34.8	37.2
19	6.84	7.63	8.91	10.1	11.7	27.2	30.1	32.9	36.2	38.6
20	7.43	8.26	9.59	10.9	12.4	28.4	31.4	34.2	37.6	40.0
21	8.03	8.90	10.3	11.6	13.2	29.6	32.7	35.5	38.9	41.4
22	8.64	9.54	11.0	12.3	14.0	30.8	33.9	36.8	40.3	42.8
23	9.26	10.2	11.7	13.1	14.8	32.0	35.2	38.1	41.6	44.2
24	9.89	10.9	12.4	13.8	15.7	33.2	36.4	39.4	43.0	45.6
25	10.5	11.5	13.1	14.6	16.5	34.4	37.7	40.6	44.3	46.9
26	11.2	12.2	13.8	15.4	17.3	35.6	38.9	41.9	45.6	48.3
27	11.8	12.9	14.6	16.2	18.1	36.7	40.1	43.2	47.0	49.6
28	12.5	13.6	15.3	16.9	18.9	37.9	41.3	44.5	48.3	51.0
29	13.1	14.3	16.0	17.7	19.8	39.1	42.6	45.7	49.6	52.3
30	13.8	15.0	16.8	18.5	20.6	40.3	43.8	47.0	50.9	53.7
40	20.7	22.2	24.4	26.5	29.1	51.8	55.8	59.3	63.7	66.8
50	28.0	29.7	32.4	34.8	37.7	63.2	67.5	71.4	76.2	79.5
60	35.5	37.5	40.5	43.2	46.5	74.4	79.1	83.3	88.4	92.0
70	43.3	45.4	48.8	51.7	55.3	85.5	90.5	95.0	100.0	104.2
80	51.2	53.5	57.2	60.4	64.3	96.6	101.9	106.6	112.3	116.3
90	59.2	61.8	65.6	69.1	73.3	107.6	113.1	118.1	124.1	128.3
100	67.3	70.1	74.2	77.9	82.4	118.5	124.3	129.6	135.8	140.2

SOURCE: Adapted from E. S. Pearson and H. O. Hartley, *Biometrika Tables for Statisticians*, vol. I (1970), pp. 136–137. Reprinted by permission of the Biometrika Trustees.

TABLE B6 Critical Values of the *F* Distribution ($\alpha = .10$)

The table values are the critical values of *F* for which the area under the curve to the right is equal to .10.

Degrees of freedom for numerator

df (denom)	1	2	3	4	5	6	7	8	9	10	12	15	20	24	30	40	60	120	∞
1	39.86	49.50	53.59	55.83	57.24	58.20	58.91	59.44	59.86	60.19	60.71	61.22	61.74	62.00	62.26	62.53	62.79	63.06	63.33
2	8.53	9.00	9.16	9.24	9.29	9.33	9.35	9.37	9.38	9.39	9.41	9.42	9.44	9.45	9.46	9.47	9.47	9.48	9.49
3	5.54	5.46	5.39	5.34	5.31	5.28	5.27	5.25	5.24	5.23	5.22	5.20	5.18	5.18	5.17	5.16	5.15	5.14	5.13
4	4.54	4.32	4.19	4.11	4.05	4.01	3.98	3.95	3.94	3.92	3.90	3.87	3.84	3.83	3.82	3.80	3.79	3.78	3.76
5	4.06	3.78	3.62	3.52	3.45	3.40	3.37	3.34	3.32	3.30	3.27	3.24	3.21	3.19	3.17	3.16	3.14	3.12	3.10
6	3.78	3.46	3.29	3.18	3.11	3.05	3.01	2.98	2.96	2.94	2.90	2.87	2.84	2.82	2.80	2.78	2.76	2.74	2.72
7	3.59	3.26	3.07	2.96	2.88	2.83	2.78	2.75	2.72	2.70	2.67	2.63	2.59	2.58	2.56	2.54	2.51	2.49	2.47
8	3.46	3.11	2.92	2.81	2.73	2.67	2.62	2.59	2.56	2.54	2.50	2.46	2.42	2.40	2.38	2.36	2.34	2.32	2.29
9	3.36	3.01	2.81	2.69	2.61	2.55	2.51	2.47	2.44	2.42	2.38	2.34	2.30	2.28	2.25	2.23	2.21	2.18	2.16
10	3.29	2.92	2.73	2.61	2.52	2.46	2.41	2.38	2.35	2.32	2.28	2.24	2.20	2.18	2.16	2.13	2.11	2.08	2.06
11	3.23	2.86	2.66	2.54	2.45	2.39	2.34	2.30	2.27	2.25	2.21	2.17	2.12	2.10	2.08	2.05	2.03	2.00	1.97
12	3.18	2.81	2.61	2.48	2.39	2.33	2.28	2.24	2.21	2.19	2.15	2.10	2.06	2.04	2.01	1.99	1.96	1.93	1.90
13	3.14	2.76	2.56	2.43	2.35	2.28	2.23	2.20	2.16	2.14	2.10	2.05	2.01	1.98	1.96	1.93	1.90	1.88	1.85
14	3.10	2.73	2.52	2.39	2.31	2.24	2.19	2.15	2.12	2.10	2.05	2.01	1.96	1.94	1.91	1.89	1.86	1.83	1.80
15	3.07	2.70	2.49	2.36	2.27	2.21	2.16	2.12	2.09	2.06	2.02	1.97	1.92	1.90	1.87	1.85	1.82	1.79	1.76
16	3.05	2.67	2.46	2.33	2.24	2.18	2.13	2.09	2.06	2.03	1.99	1.94	1.89	1.87	1.84	1.81	1.78	1.75	1.72
17	3.03	2.64	2.44	2.31	2.22	2.15	2.10	2.06	2.03	2.00	1.96	1.91	1.86	1.84	1.81	1.78	1.75	1.72	1.69
18	3.01	2.62	2.42	2.29	2.20	2.13	2.08	2.04	2.00	1.98	1.93	1.89	1.84	1.81	1.78	1.75	1.72	1.69	1.66
19	2.99	2.61	2.40	2.27	2.18	2.11	2.06	2.02	1.98	1.96	1.91	1.86	1.81	1.79	1.76	1.73	1.70	1.67	1.63
20	2.97	2.59	2.38	2.25	2.16	2.09	2.04	2.00	1.96	1.94	1.89	1.84	1.79	1.77	1.74	1.71	1.68	1.64	1.61
21	2.96	2.57	2.36	2.23	2.14	2.08	2.02	1.98	1.95	1.92	1.87	1.83	1.78	1.75	1.72	1.69	1.66	1.62	1.59
22	2.95	2.56	2.35	2.22	2.13	2.06	2.01	1.97	1.93	1.90	1.86	1.81	1.76	1.73	1.70	1.67	1.64	1.60	1.57
23	2.94	2.55	2.34	2.21	2.11	2.05	1.99	1.95	1.92	1.89	1.84	1.80	1.74	1.72	1.69	1.66	1.62	1.59	1.55
24	2.93	2.54	2.33	2.19	2.10	2.04	1.98	1.94	1.91	1.88	1.83	1.78	1.73	1.70	1.67	1.64	1.61	1.57	1.53
25	2.92	2.53	2.32	2.18	2.09	2.02	1.97	1.93	1.89	1.87	1.82	1.77	1.72	1.69	1.66	1.63	1.59	1.56	1.52
26	2.91	2.52	2.31	2.17	2.08	2.01	1.96	1.92	1.88	1.86	1.81	1.76	1.71	1.68	1.65	1.61	1.58	1.54	1.50
27	2.90	2.51	2.30	2.17	2.07	2.00	1.95	1.91	1.87	1.85	1.80	1.75	1.70	1.67	1.64	1.60	1.57	1.53	1.49
28	2.89	2.50	2.29	2.16	2.06	2.00	1.94	1.90	1.87	1.84	1.79	1.74	1.69	1.66	1.63	1.59	1.56	1.52	1.48
29	2.89	2.50	2.28	2.15	2.06	1.99	1.93	1.89	1.86	1.83	1.78	1.73	1.68	1.65	1.62	1.58	1.55	1.51	1.47
30	2.88	2.49	2.28	2.14	2.05	1.98	1.93	1.88	1.85	1.82	1.77	1.72	1.67	1.64	1.61	1.57	1.54	1.50	1.46
40	2.84	2.44	2.23	2.09	2.00	1.93	1.87	1.83	1.79	1.76	1.71	1.66	1.61	1.57	1.54	1.51	1.47	1.42	1.38
60	2.79	2.39	2.18	2.04	1.95	1.87	1.82	1.77	1.74	1.71	1.66	1.60	1.54	1.51	1.48	1.44	1.40	1.35	1.29
120	2.75	2.35	2.13	1.99	1.90	1.82	1.77	1.72	1.68	1.65	1.60	1.55	1.48	1.45	1.41	1.37	1.32	1.26	1.19
∞	2.71	2.30	2.08	1.94	1.85	1.77	1.72	1.67	1.63	1.60	1.55	1.49	1.42	1.38	1.34	1.30	1.24	1.17	1.00

Degrees of freedom for denominator

SOURCE: From E. S. Pearson and H. O. Hartley, *Biometrika Tables for Statisticians*, vol. I (1970), pp. 170–173. Reprinted by permission of the Biometrika Trustees.

TABLE B6   Critical Values of the F Distribution ($\alpha = .05$)

$\alpha = 0.05$

$F_{.05, df_n, df_d}$

The table values are critical values of F for which the area under the curve to the right is equal to .05.

Degrees of freedom for numerator

df (denom)	1	2	3	4	5	6	7	8	9	10	12	15	20	24	30	40	60	120	∞
1	161.4	199.5	215.7	224.6	230.2	234.0	236.8	238.9	240.5	241.9	243.9	245.9	248.0	249.1	250.1	251.1	252.2	253.3	254.3
2	18.51	19.00	19.16	19.25	19.30	19.33	19.35	19.37	19.38	19.40	19.41	19.43	19.45	19.45	19.46	19.47	19.48	19.49	19.50
3	10.13	9.55	9.28	9.12	9.01	8.94	8.89	8.85	8.81	8.79	8.74	8.70	8.66	8.64	8.62	8.59	8.57	8.55	8.53
4	7.71	6.94	6.59	6.39	6.26	6.16	6.09	6.04	6.00	5.96	5.91	5.86	5.80	5.77	5.75	5.72	5.69	5.66	5.63
5	6.61	5.79	5.41	5.19	5.05	4.95	4.88	4.82	4.77	4.74	4.68	4.62	4.56	4.53	4.50	4.46	4.43	4.40	4.36
6	5.99	5.14	4.76	4.53	4.39	4.28	4.21	4.15	4.10	4.06	4.00	3.94	3.87	3.84	3.81	3.77	3.74	3.70	3.67
7	5.59	4.74	4.35	4.12	3.97	3.87	3.79	3.73	3.68	3.64	3.57	3.51	3.44	3.41	3.38	3.34	3.30	3.27	3.23
8	5.32	4.46	4.07	3.84	3.69	3.58	3.50	3.44	3.39	3.35	3.28	3.22	3.15	3.12	3.08	3.04	3.01	2.97	2.93
9	5.12	4.26	3.86	3.63	3.48	3.37	3.29	3.23	3.18	3.14	3.07	3.01	2.94	2.90	2.86	2.83	2.79	2.75	2.71
10	4.96	4.10	3.71	3.48	3.33	3.22	3.14	3.07	3.02	2.98	2.91	2.85	2.77	2.74	2.70	2.66	2.62	2.58	2.54
11	4.84	3.98	3.59	3.36	3.20	3.09	3.01	2.95	2.90	2.85	2.79	2.72	2.65	2.61	2.57	2.53	2.49	2.45	2.40
12	4.75	3.89	3.49	3.26	3.11	3.00	2.91	2.85	2.80	2.75	2.69	2.62	2.54	2.51	2.47	2.43	2.38	2.34	2.30
13	4.67	3.81	3.41	3.18	3.03	2.92	2.83	2.77	2.71	2.67	2.60	2.53	2.46	2.42	2.38	2.34	2.30	2.25	2.21
14	4.60	3.74	3.34	3.11	2.96	2.85	2.76	2.70	2.65	2.60	2.53	2.46	2.39	2.35	2.31	2.27	2.22	2.18	2.13
15	4.54	3.68	3.29	3.06	2.90	2.79	2.71	2.64	2.59	2.54	2.48	2.40	2.33	2.29	2.25	2.20	2.16	2.11	2.07
16	4.49	3.63	3.24	3.01	2.85	2.74	2.66	2.59	2.54	2.49	2.42	2.35	2.28	2.24	2.19	2.15	2.11	2.06	2.01
17	4.45	3.59	3.20	2.96	2.81	2.70	2.61	2.55	2.49	2.45	2.38	2.31	2.23	2.19	2.15	2.10	2.06	2.01	1.96
18	4.41	3.55	3.16	2.93	2.77	2.66	2.58	2.51	2.46	2.41	2.34	2.27	2.19	2.15	2.11	2.06	2.02	1.97	1.92
19	4.38	3.52	3.13	2.90	2.74	2.63	2.54	2.48	2.42	2.38	2.31	2.23	2.16	2.11	2.07	2.03	1.98	1.93	1.88
20	4.35	3.49	3.10	2.87	2.71	2.60	2.51	2.45	2.39	2.35	2.28	2.20	2.12	2.08	2.04	1.99	1.95	1.90	1.84
21	4.32	3.47	3.07	2.84	2.68	2.57	2.49	2.42	2.37	2.32	2.25	2.18	2.10	2.05	2.01	1.96	1.92	1.87	1.81
22	4.30	3.44	3.05	2.82	2.66	2.55	2.46	2.40	2.34	2.30	2.23	2.15	2.07	2.03	1.98	1.94	1.89	1.84	1.78
23	4.28	3.42	3.03	2.80	2.64	2.53	2.44	2.37	2.32	2.27	2.20	2.13	2.05	2.01	1.96	1.91	1.86	1.81	1.76
24	4.26	3.40	3.01	2.78	2.62	2.51	2.42	2.36	2.30	2.25	2.18	2.11	2.03	1.98	1.94	1.89	1.84	1.79	1.73
25	4.24	3.39	2.99	2.76	2.60	2.49	2.40	2.34	2.28	2.24	2.16	2.09	2.01	1.96	1.92	1.87	1.82	1.77	1.71
30	4.17	3.32	2.92	2.69	2.53	2.42	2.33	2.27	2.21	2.16	2.09	2.01	1.93	1.89	1.84	1.79	1.74	1.68	1.62
40	4.08	3.23	2.84	2.61	2.45	2.34	2.25	2.18	2.12	2.08	2.00	1.92	1.84	1.79	1.74	1.69	1.64	1.58	1.51
60	4.00	3.15	2.76	2.53	2.37	2.25	2.17	2.10	2.04	1.99	1.92	1.84	1.75	1.70	1.65	1.59	1.53	1.47	1.39
120	3.92	3.07	2.68	2.45	2.29	2.17	2.09	2.02	1.96	1.91	1.83	1.75	1.66	1.61	1.55	1.50	1.43	1.35	1.25
∞	3.84	3.00	2.60	2.37	2.21	2.10	2.01	1.94	1.88	1.83	1.75	1.67	1.57	1.52	1.46	1.39	1.32	1.22	1.00

Degrees of freedom for denominator

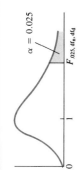

$\alpha = 0.025$

$F_{.025, df_n, df_d}$

The table values are critical values of F for which the area under the curve to the right is equal to .025.

Degrees of freedom for numerator

	1	2	3	4	5	6	7	8	9	10	12	15	20	24	30	40	60	120	∞
1	648	800	864	900	922	937	948	957	963	969	977	985	993	997	1,001	1,006	1,010	1,014	1,018
2	38.51	39.00	39.17	39.25	39.30	39.33	39.36	39.37	39.39	39.40	39.41	39.43	39.45	39.46	39.46	39.47	39.48	39.49	39.50
3	17.44	16.04	15.44	15.10	14.88	14.73	14.62	14.54	14.47	14.42	14.34	14.25	14.17	14.12	14.08	14.04	13.99	13.95	13.90
4	12.22	10.65	9.98	9.60	9.36	9.20	9.07	8.98	8.90	8.84	8.75	8.66	8.56	8.51	8.46	8.41	8.36	8.31	8.26
5	10.01	8.43	7.76	7.39	7.15	6.98	6.85	6.76	6.68	6.62	6.52	6.43	6.33	6.28	6.23	6.18	6.12	6.07	6.02
6	8.81	7.26	6.60	6.23	5.99	5.82	5.70	5.60	5.52	5.46	5.37	5.27	5.17	5.12	5.07	5.01	4.96	4.90	4.85
7	8.07	6.54	5.89	5.52	5.29	5.12	4.99	4.90	4.82	4.76	4.67	4.57	4.47	4.42	4.36	4.31	4.25	4.20	4.14
8	7.57	6.06	5.42	5.05	4.82	4.65	4.53	4.43	4.36	4.30	4.20	4.10	4.00	3.95	3.89	3.84	3.78	3.73	3.67
9	7.21	5.71	5.08	4.72	4.48	4.32	4.20	4.10	4.03	3.96	3.87	3.77	3.67	3.61	3.56	3.51	3.45	3.39	3.33
10	6.94	5.46	4.83	4.47	4.24	4.07	3.95	3.85	3.78	3.72	3.62	3.52	3.42	3.37	3.31	3.26	3.20	3.14	3.08
11	6.72	5.26	4.63	4.28	4.04	3.88	3.76	3.66	3.59	3.53	3.43	3.33	3.23	3.17	3.12	3.06	3.00	2.94	2.88
12	6.55	5.10	4.47	4.12	3.89	3.73	3.61	3.51	3.44	3.37	3.28	3.18	3.07	3.02	2.96	2.91	2.85	2.79	2.72
13	6.41	4.97	4.35	4.00	3.77	3.60	3.48	3.39	3.31	3.25	3.15	3.05	2.95	2.89	2.84	2.78	2.72	2.66	2.60
14	6.30	4.86	4.24	3.89	3.66	3.50	3.38	3.29	3.21	3.15	3.05	2.95	2.84	2.79	2.73	2.67	2.61	2.55	2.49
15	6.20	4.77	4.15	3.80	3.58	3.41	3.29	3.20	3.12	3.06	2.96	2.86	2.76	2.70	2.64	2.59	2.52	2.46	2.40
16	6.12	4.69	4.08	3.73	3.50	3.34	3.22	3.12	3.05	2.99	2.89	2.79	2.68	2.63	2.57	2.51	2.45	2.38	2.32
17	6.04	4.62	4.01	3.66	3.44	3.28	3.16	3.06	2.98	2.92	2.82	2.72	2.62	2.56	2.50	2.44	2.38	2.32	2.25
18	5.98	4.56	3.95	3.61	3.38	3.22	3.10	3.01	2.93	2.87	2.77	2.67	2.56	2.50	2.44	2.38	2.32	2.26	2.19
19	5.92	4.51	3.90	3.56	3.33	3.17	3.05	2.96	2.88	2.82	2.72	2.62	2.51	2.45	2.39	2.33	2.27	2.20	2.13
20	5.87	4.46	3.86	3.51	3.29	3.13	3.01	2.91	2.84	2.77	2.68	2.57	2.46	2.41	2.35	2.29	2.22	2.16	2.09
21	5.83	4.42	3.82	3.48	3.25	3.09	2.97	2.87	2.80	2.73	2.64	2.53	2.42	2.37	2.31	2.25	2.18	2.11	2.04
22	5.79	4.38	3.78	3.44	3.22	3.05	2.93	2.84	2.76	2.70	2.60	2.50	2.39	2.33	2.27	2.21	2.14	2.08	2.00
23	5.75	4.35	3.75	3.41	3.18	3.02	2.90	2.81	2.73	2.67	2.57	2.47	2.36	2.30	2.24	2.18	2.11	2.04	1.97
24	5.72	4.32	3.72	3.38	3.15	2.99	2.87	2.78	2.70	2.64	2.54	2.44	2.33	2.27	2.21	2.15	2.08	2.01	1.94
25	5.69	4.29	3.69	3.35	3.13	2.97	2.85	2.75	2.68	2.61	2.51	2.41	2.30	2.24	2.18	2.12	2.05	1.98	1.91
30	5.57	4.18	3.59	3.25	3.03	2.87	2.75	2.65	2.57	2.51	2.41	2.31	2.20	2.14	2.07	2.01	1.94	1.87	1.79
40	5.42	4.05	3.46	3.13	2.90	2.74	2.62	2.53	2.45	2.39	2.29	2.18	2.07	2.01	1.94	1.88	1.80	1.72	1.64
60	5.29	3.93	3.34	3.01	2.79	2.63	2.51	2.41	2.33	2.27	2.17	2.06	1.94	1.88	1.82	1.74	1.67	1.58	1.48
120	5.15	3.80	3.23	2.89	2.67	2.52	2.39	2.30	2.22	2.16	2.05	1.94	1.82	1.76	1.69	1.61	1.53	1.43	1.31
∞	5.02	3.69	3.12	2.79	2.57	2.41	2.29	2.19	2.11	2.05	1.94	1.83	1.71	1.64	1.57	1.48	1.39	1.27	1.00

Degrees of freedom for denominator

SOURCE: From E. S. Pearson and H. O. Hartley, *Biometrika Tables for Statisticians*, vol. I (1970), pp. 170–173. Reprinted by permission of the Biometrika Trustees.

TABLE B6   Critical Values of the *F* Distribution ($\alpha = .01$)

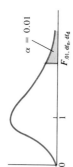

$\alpha = 0.01$

$F_{.01, df_n, df_d}$

The table values are critical values of *F* for which the area under the curve to the right is equal to .01.

								Degrees of freedom for numerator										
	1	**2**	**3**	**4**	**5**	**6**	**7**	**8**	**9**	**10**	**12**	**15**	**20**	**24**	**30**	**40**	**60**	**120**	**∞**
1	4,052	5,000	5,403	5,625	5,764	5,859	5,928	5,981	6,022	6,056	6,106	6,157	6,209	6,235	6,261	6,287	6,313	6,339	6,366
2	98.50	99.00	99.17	99.25	99.30	99.33	99.36	99.37	99.39	99.40	99.42	99.43	99.45	99.46	99.47	99.47	99.48	99.49	99.50
3	34.12	30.82	29.46	28.71	28.24	27.91	27.67	27.49	27.35	27.23	27.05	26.87	26.69	26.60	26.50	26.41	26.32	26.22	26.13
4	21.20	18.00	16.69	15.98	15.52	15.21	14.98	14.80	14.66	14.55	14.37	14.20	14.02	13.93	13.84	13.75	13.65	13.56	13.46
5	16.26	13.27	12.06	11.39	10.97	10.67	10.46	10.29	10.16	10.05	9.89	9.72	9.55	9.47	9.38	9.29	9.20	9.11	9.02
6	13.75	10.92	9.78	9.15	8.75	8.47	8.26	8.10	7.98	7.87	7.72	7.56	7.40	7.31	7.23	7.14	7.06	6.97	6.88
7	12.25	9.55	8.45	7.85	7.46	7.19	6.99	6.84	6.72	6.62	6.47	6.31	6.16	6.07	5.99	5.91	5.82	5.74	5.65
8	11.26	8.65	7.59	7.01	6.63	6.37	6.18	6.03	5.91	5.81	5.67	5.52	5.36	5.28	5.20	5.12	5.03	4.95	4.86
9	10.56	8.02	6.99	6.42	6.06	5.80	5.61	5.47	5.35	5.26	5.11	4.96	4.81	4.73	4.65	4.57	4.48	4.40	4.31
10	10.04	7.56	6.55	5.99	5.64	5.39	5.20	5.06	4.94	4.85	4.71	4.56	4.41	4.33	4.25	4.17	4.08	4.00	3.91
11	9.65	7.21	6.22	5.67	5.32	5.07	4.89	4.74	4.63	4.54	4.40	4.25	4.10	4.02	3.94	3.86	3.78	3.69	3.60
12	9.33	6.93	5.95	5.41	5.06	4.82	4.64	4.50	4.39	4.30	4.16	4.01	3.86	3.78	3.70	3.62	3.54	3.45	3.36
13	9.07	6.70	5.74	5.21	4.86	4.62	4.44	4.30	4.19	4.10	3.96	3.82	3.66	3.59	3.51	3.43	3.34	3.25	3.17
14	8.86	6.51	5.56	5.04	4.69	4.46	4.28	4.14	4.03	3.94	3.80	3.66	3.51	3.43	3.35	3.27	3.18	3.09	3.00
15	8.68	6.36	5.42	4.89	4.56	4.32	4.14	4.00	3.89	3.80	3.67	3.52	3.37	3.29	3.21	3.13	3.05	2.96	2.87
16	8.53	6.23	5.29	4.77	4.44	4.20	4.03	3.89	3.78	3.69	3.55	3.41	3.26	3.18	3.10	3.02	2.93	2.84	2.75
17	8.40	6.11	5.18	4.67	4.34	4.10	3.93	3.79	3.68	3.59	3.46	3.31	3.16	3.08	3.00	2.92	2.83	2.75	2.65
18	8.29	6.01	5.09	4.58	4.25	4.01	3.84	3.71	3.60	3.51	3.37	3.23	3.08	3.00	2.92	2.84	2.75	2.66	2.57
19	8.18	5.93	5.01	4.50	4.17	3.94	3.77	3.63	3.52	3.43	3.30	3.15	3.00	2.92	2.84	2.76	2.67	2.58	2.49
20	8.10	5.85	4.94	4.43	4.10	3.87	3.70	3.56	3.46	3.37	3.23	3.09	2.94	2.86	2.78	2.69	2.61	2.52	2.42
21	8.02	5.78	4.87	4.37	4.04	3.81	3.64	3.51	3.40	3.31	3.17	3.03	2.88	2.80	2.72	2.64	2.55	2.46	2.36
22	7.95	5.72	4.82	4.31	3.99	3.76	3.59	3.45	3.35	3.26	3.12	2.98	2.83	2.75	2.67	2.58	2.50	2.40	2.31
23	7.88	5.66	4.76	4.26	3.94	3.71	3.54	3.41	3.30	3.21	3.07	2.93	2.78	2.70	2.62	2.54	2.45	2.35	2.26
24	7.82	5.61	4.72	4.22	3.90	3.67	3.50	3.36	3.26	3.17	3.03	2.89	2.74	2.66	2.58	2.49	2.40	2.31	2.21
25	7.77	5.57	4.68	4.18	3.85	3.63	3.46	3.32	3.22	3.13	2.99	2.85	2.70	2.62	2.54	2.45	2.36	2.27	2.17
30	7.56	5.39	4.51	4.02	3.70	3.47	3.30	3.17	3.07	2.98	2.84	2.70	2.55	2.47	2.39	2.30	2.21	2.11	2.01
40	7.31	5.18	4.31	3.83	3.51	3.29	3.12	2.99	2.89	2.80	2.66	2.52	2.37	2.29	2.20	2.11	2.02	1.92	1.80
60	7.08	4.98	4.13	3.65	3.34	3.12	2.95	2.82	2.72	2.63	2.50	2.35	2.20	2.12	2.03	1.94	1.84	1.73	1.60
120	6.85	4.79	3.95	3.48	3.17	2.96	2.79	2.66	2.56	2.47	2.34	2.19	2.03	1.95	1.86	1.76	1.66	1.53	1.38
∞	6.63	4.61	3.78	3.32	3.02	2.80	2.64	2.51	2.41	2.32	2.18	2.04	1.88	1.79	1.70	1.59	1.47	1.32	1.00

Degrees of freedom for denominator

SOURCE: From E. S. Pearson and H. O. Hartley, *Biometrika Tables for Statisticians*, vol. I (1970). pp. 170–173. Reprinted by permission of the Biometrika Trustees.

TABLE B7   Percentage Points of the Studentized Range, $q(k, \nu)$, Upper 5%

ν \ k	2	3	4	5	6	7	8	9	10	11	12	13	14	15	16	17	18	19	20
1	17.97	26.98	32.82	37.08	40.41	43.12	45.40	47.36	49.07	50.59	51.96	53.20	54.33	55.36	56.32	57.22	58.04	58.83	59.56
2	6.08	8.33	9.80	10.88	11.74	12.44	13.03	13.54	13.99	14.39	14.75	15.08	15.38	15.65	15.91	16.14	16.37	16.57	16.77
3	4.50	5.91	6.82	7.50	8.04	8.48	8.85	9.18	9.46	9.72	9.95	10.15	10.35	10.52	10.69	10.84	10.98	11.11	11.24
4	3.93	5.04	5.76	6.29	6.71	7.05	7.35	7.60	7.83	8.03	8.21	8.37	8.52	8.66	8.79	8.91	9.03	9.13	9.23
5	3.64	4.60	5.22	5.67	6.03	6.33	6.58	6.80	6.99	7.17	7.32	7.47	7.60	7.72	7.83	7.93	8.03	8.12	8.21
6	3.46	4.34	4.90	5.30	5.63	5.90	6.12	6.32	6.49	6.65	6.79	6.92	7.03	7.14	7.24	7.34	7.43	7.51	7.59
7	3.34	4.16	4.68	5.06	5.36	5.61	5.82	6.00	6.16	6.30	6.43	6.55	6.66	6.76	6.85	6.94	7.02	7.10	7.17
8	3.26	4.04	4.53	4.89	5.17	5.40	5.60	5.77	5.92	6.05	6.18	6.29	6.39	6.48	6.57	6.65	6.73	6.80	6.87
9	3.20	3.95	4.41	4.76	5.02	5.24	5.43	5.59	5.74	5.87	5.98	6.09	6.19	6.28	6.36	6.44	6.51	6.58	6.64
10	3.15	3.88	4.33	4.65	4.91	5.12	5.30	5.46	5.60	5.72	5.83	5.93	6.03	6.11	6.19	6.27	6.34	6.40	6.47
11	3.11	3.82	4.26	4.57	4.82	5.03	5.20	5.35	5.49	5.61	5.71	5.81	5.90	5.98	6.06	6.13	6.20	6.27	6.33
12	3.08	3.77	4.20	4.51	4.75	4.95	5.12	5.27	5.39	5.51	5.61	5.71	5.80	5.88	5.95	6.02	6.09	6.15	6.21
13	3.06	3.73	4.15	4.45	4.69	4.88	5.05	5.19	5.32	5.43	5.53	5.63	5.71	5.79	5.86	5.93	5.99	6.05	6.11
14	3.03	3.70	4.11	4.41	4.64	4.83	4.99	5.13	5.25	5.36	5.46	5.55	5.64	5.71	5.79	5.85	5.91	5.97	6.03
15	3.01	3.67	4.08	4.37	4.59	4.78	4.94	5.08	5.20	5.31	5.40	5.49	5.57	5.65	5.72	5.78	5.85	5.90	5.96
16	3.00	3.65	4.05	4.33	4.56	4.74	4.90	5.03	5.15	5.26	5.35	5.44	5.52	5.59	5.66	5.73	5.79	5.84	5.90
17	2.98	3.63	4.02	4.30	4.52	4.70	4.86	4.99	5.11	5.21	5.31	5.39	5.47	5.54	5.61	5.67	5.73	5.79	5.84
18	2.97	3.61	4.00	4.28	4.49	4.67	4.82	4.96	5.07	5.17	5.27	5.35	5.43	5.50	5.57	5.63	5.69	5.74	5.79
19	2.96	3.59	3.98	4.25	4.47	4.65	4.79	4.92	5.04	5.14	5.23	5.31	5.39	5.46	5.53	5.59	5.65	5.70	5.75
20	2.95	3.58	3.96	4.23	4.45	4.62	4.77	4.90	5.01	5.11	5.20	5.28	5.36	5.43	5.49	5.55	5.61	5.66	5.71
24	2.92	3.53	3.90	4.17	4.37	4.54	4.68	4.81	4.92	5.01	5.10	5.18	5.25	5.32	5.38	5.44	5.49	5.55	5.59
30	2.89	3.49	3.85	4.10	4.30	4.46	4.60	4.72	4.82	4.92	5.00	5.08	5.15	5.21	5.27	5.33	5.38	5.43	5.47
40	2.86	3.44	3.79	4.04	4.23	4.39	4.52	4.63	4.73	4.82	4.90	4.98	5.04	5.11	5.16	5.22	5.27	5.31	5.36
60	2.83	3.40	3.74	3.98	4.16	4.31	4.44	4.55	4.65	4.73	4.81	4.88	4.94	5.00	5.06	5.11	5.15	5.20	5.24
120	2.80	3.36	3.68	3.92	4.10	4.24	4.36	4.47	4.56	4.64	4.71	4.78	4.84	4.90	4.95	5.00	5.04	5.09	5.13
∞	2.77	3.31	3.63	3.86	4.03	4.17	4.29	4.39	4.47	4.55	4.62	4.68	4.74	4.80	4.85	4.89	4.93	4.97	5.01

SOURCE: From E. S. Pearson and H. O. Hartley, (Eds.), *Biometrika Tables for Statisticians*, vol. I, 3rd ed. (Cambridge University Press, 1970), p. 192. Reprinted by permission of the Biometrika Trustees.

TABLE B7   Percentage Points of the Studentized Range, $q(k, v)$, Upper 1%

v \ k	2	3	4	5	6	7	8	9	10	11	12	13	14	15	16	17	18	19	20
1	90.03	135.0	164.3	185.6	202.2	215.8	227.2	237.0	245.6	253.2	260.0	266.2	271.8	277.0	281.8	286.3	290.4	294.3	298.0
2	14.04	19.02	22.29	24.72	26.63	28.20	29.53	30.68	31.69	32.59	33.40	34.13	34.81	35.43	36.00	36.53	37.03	37.50	37.95
3	8.26	10.62	12.17	13.33	14.24	15.00	15.64	16.20	16.69	17.13	17.53	17.89	18.22	18.52	18.81	19.07	19.32	19.55	19.77
4	6.51	8.12	9.17	9.96	10.58	11.10	11.55	11.93	12.27	12.57	12.84	13.09	13.32	13.53	13.73	13.91	14.08	14.24	14.40
5	5.70	6.98	7.80	8.42	8.91	9.32	9.67	9.97	10.24	10.48	10.70	10.89	11.08	11.24	11.40	11.55	11.68	11.81	11.93
6	5.24	6.33	7.03	7.56	7.97	8.32	8.61	8.87	9.10	9.30	9.48	9.65	9.81	9.95	10.08	10.21	10.32	10.43	10.54
7	4.95	5.92	6.54	7.01	7.37	7.68	7.94	8.17	8.37	8.55	8.71	8.86	9.00	9.12	9.24	9.35	9.46	9.55	9.65
8	4.75	5.64	6.20	6.62	6.96	7.24	7.47	7.68	7.86	8.03	8.18	8.31	8.44	8.55	8.66	8.76	8.85	8.94	9.03
9	4.60	5.43	5.96	6.35	6.66	6.91	7.13	7.33	7.49	7.65	7.78	7.91	8.03	8.13	8.23	8.33	8.41	8.49	8.57
10	4.48	5.27	5.77	6.14	6.43	6.67	6.87	7.05	7.21	7.36	7.49	7.60	7.71	7.81	7.91	7.99	8.08	8.15	8.23
11	4.39	5.15	5.62	5.97	6.25	6.48	6.67	6.84	6.99	7.13	7.25	7.36	7.46	7.56	7.65	7.73	7.81	7.88	7.95
12	4.32	5.05	5.50	5.84	6.10	6.32	6.51	6.67	6.81	6.94	7.06	7.17	7.26	7.36	7.44	7.52	7.59	7.66	7.73
13	4.26	4.96	5.40	5.73	5.98	6.19	6.37	6.53	6.67	6.79	6.90	7.01	7.10	7.19	7.27	7.35	7.42	7.48	7.55
14	4.21	4.89	5.32	5.63	5.88	6.08	6.26	6.41	6.54	6.66	6.77	6.87	6.96	7.05	7.13	7.20	7.27	7.33	7.39
15	4.17	4.84	5.25	5.56	5.80	5.99	6.16	6.31	6.44	6.55	6.66	6.76	6.84	6.93	7.00	7.07	7.14	7.20	7.26
16	4.13	4.79	5.19	5.49	5.72	5.92	6.08	6.22	6.35	6.46	6.56	6.66	6.74	6.82	6.90	6.97	7.03	7.09	7.15
17	4.10	4.74	5.14	5.43	5.66	5.85	6.01	6.15	6.27	6.38	6.48	6.57	6.66	6.73	6.81	6.87	6.94	7.00	7.05
18	4.07	4.70	5.09	5.38	5.60	5.79	5.94	6.08	6.20	6.31	6.41	6.50	6.58	6.65	6.72	6.79	6.85	6.91	6.97
19	4.05	4.67	5.05	5.33	5.55	5.73	5.89	6.02	6.14	6.25	6.34	6.43	6.51	6.58	6.65	6.72	6.78	6.84	6.89
20	4.02	4.64	5.02	5.29	5.51	5.69	5.84	5.97	6.09	6.19	6.28	6.37	6.45	6.52	6.59	6.65	6.71	6.77	6.82
24	3.96	4.55	4.91	5.17	5.37	5.54	5.69	5.81	5.92	6.02	6.11	6.19	6.26	6.33	6.39	6.45	6.51	6.56	6.61
30	3.89	4.45	4.80	5.05	5.24	5.40	5.54	5.65	5.76	5.85	5.93	6.01	6.08	6.14	6.20	6.26	6.31	6.36	6.41
40	3.82	4.37	4.70	4.93	5.11	5.26	5.39	5.50	5.60	5.69	5.76	5.83	5.90	5.96	6.02	6.07	6.12	6.16	6.21
60	3.76	4.28	4.59	4.82	4.99	5.13	5.25	5.36	5.45	5.53	5.60	5.67	5.73	5.78	5.84	5.89	5.93	5.97	6.01
120	3.70	4.20	4.50	4.71	4.87	5.01	5.12	5.21	5.30	5.37	5.44	5.50	5.56	5.61	5.66	5.71	5.75	5.79	5.83
∞	3.64	4.12	4.40	4.60	4.76	4.88	4.99	5.08	5.16	5.23	5.29	5.35	5.40	5.45	5.49	5.54	5.57	5.61	5.65

SOURCE: From E. S. Pearson and H. O. Hartley, (Eds.), *Biometrika Tables for Statisticians*, vol. I, 3rd ed. (Cambridge University Press, 1970), p. 193. Reprinted by permission of the Biometrika Trustees.

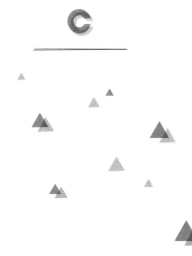

APPENDIX C

Answers to Selected Exercises and to All Review Questions

CHAPTER 1

Section 1.1 (page 7)

1.1   Population—all college students. Sample—500 college students selected from a particular university. Variable—opinion on abortion.

1.2   Population—residents in the neighborhood. Sample—the 1,000 residents of the neighborhood. Variable—opinion on the zoning ordinance.

1.3   Population—accidents investigated by the highway patrol. Sample—the 1,000 accidents investigated. Variable—alcohol related or not.

1.5   Population—freshmen students at this particular university. Sample—the 200 chosen freshmen. Variable—was this university their first choice or not.

1.7   1.1, 1.4, and 1.6.

1.9   No. The rate of fatalities per number of miles traveled.

1.11   Model A: negatively biased, high precision
Model B: unbiased, high precision
Model C: positively biased, low precision

Section 1.2 (page 13)

1.13   Students at this university living on campus.

1.15   Those who live in the Louisiana area and attend football games.

1.16   No, the fact that the neighbor was at home means that they possibly would differ in their opinion on the bond issue. Every effort should be made to obtain the opinion of the person who was not at home. If that person is not contacted, then all those who are not at home at that hour (for whatever reason) would not be represented in the sample.

1.17   (a) Yes
(c) TV log to be filled out by those who watch TV; black-box attached to the TV that electronically monitors what station is being viewed.

1.18   *Strong point:* The pamphlet attempts to educate the public on the uses of nuclear reactors.
Weak point: The pamphlet could possibly be biased in presenting the characteristics of a nuclear reactor. For instance, are the hazards associated with a reactor presented in the pamphlet? Also, the questionnaire is a volunteer questionnaire.

1.20   Not necessarily. Students attending private and public universities differ in income, religion, race, and possibly political party. Thus it is reasonable to suggest that they would have different opinions on drug abuse. The combined results would not represent all college students because there is no indication of where the two universities were selected.

Section 1.3 (page 20)

1.22   (a) It is an experimental drug that might have serious side effects and, therefore, should be tested on only a few subjects. Obviously, the cost involved would prohibit a census.

(b) The cost would prohibit a census.

(c) Time and cost would prohibit a census.

1.23 Target population—all major cities in the United States. Sampled population—the larger cities in the United States.

1.24 Yes. The opinions of residents on corner lots probably differ from the opinions of residents in general because corner homes are usually more expensive.

1.26 (a) Systematic; (b) Physically mixing or lottery; (c) Stratified; (d) Convenience

1.28 Target population—potential shoppers at the new shopping center. Sampled population—people who are listed on the chamber of commerce mailing list.

Section 1.4 (page 28)

1.29 A double-blind experiment is one in which neither the subject nor the one administering the treatment knows who is receiving the actual treatment.

1.31 By conducting a comparative experiment.

1.33 (a) A first grade student; (b) Method of instruction, 2 levels; (c) Score on reading test; (d) Yes; (e) Definitely; (f) No; (g) No. Probably not, because the two classes can serve as control for each other.

1.35 (a) Randomized experiment
(b) If the children are randomly assigned to the two groups so that the groups are equivalent, then most confounding variables should affect the two groups the same and thus nullify their effect.

1.37 (a) Sample survey; (b) Observational study;
(c) Randomized experiment; (d) Observational study;
(e) Randomized experiment

Review Questions Ch. 1 (page 36)

1.42	b	1.43	d	1.44	a
1.45	b	1.46	d	1.47	d
1.48	c	1.49	b	1.50	d

1.51 a

1.52 A placebo

1.53 Double-blind

1.54 Confounded

1.55 (a) Amount of exercise; (b) Risk of heart disease; (c) An individual (police officer or bus driver); (d) Job stress and marital status; (e) Observational study

1.56 (a) T   (b) F   (c) F   (d) F   (e) F

Supplementary Exercises Ch. 1 (page 37)

1.57 (a) Potential customers flying between Charlotte and Dallas; (b) 2,000 past customers; (c) Number of trips per year per person

1.59 No. The number of bankrupt farmers per 1,000 farmers.

1.61 Carefully designed questions. Properly trained interviewers in both telephone and personal interviews.

1.63 (a) Cluster sampling; (b) Stratified sample;
(c) Systematic sample

1.65 (a) Residents of the state; (b) Registered voters in the state; (c) Survey; (d) Possibly not

1.66 Yes, the question starts by leading the subject—"Don't you agree . . ." It should state "Should farmers be allowed . . ."

1.68 (a) No; (b) Black women who read *Essence* magazine

1.69 Yes, it is a volunteer survey.

1.70 Yes. Strike the "Don't you agree . . ." For example, "Should the government limit foreign imports of textile materials?"

1.72 (a) Amount of alcohol; (b) Reaction time; (c) One student; (d) Control group; (e) Yes—weight of the subject, diet prior to test, tolerance level to alcohol

1.73 Randomized experiment

1.75 (a) Observational study; (b) Yes, educational background; (c) Not very reliable

1.76 No, the quality of the pet owners' lives may be better because they have higher incomes.

1.78 Diet

1.80 (a) SAT scores; (b) Success in college; (c) A student; (d) Motivation

1.81 Systematic

CHAPTER 2

Section 2.1 (page 44)

2.1 (a) All welfare recipients in the state
(b) Systematic sample from a list of welfare recipients
(c) Sex, occupation, marital status
(d) Age of head of household—continuous, family income—continuous, number of children—discrete

2.3 (a) All third grade students
(b) Cluster sample of third grade classes
(c) Sex, race, availability of computer at home
(d) Family income—continuous, IQ—continuous, years of parents' education—continuous

2.5 (a) Quantitative—continuous; (b) Qualitative;
(c) Quantitative—continuous; (d) Quantitative—continuous;
(e) Quantitative—continuous; (f) Quantitative—discrete;
(g) Qualitative

2.7 Sex—qualitative
Major—qualitative
Grade point average—quantitative
Number of times using placement service—quantitative
Type of employment—qualitative

Section 2.2 (page 48)

2.10 (a) Qualitative, nominal; (b) Qualitative, nominal; (c) Quantitative, interval/ratio; (d) Qualitative, nominal; (e) Quantitative, interval/ratio; (f) Quantitative, interval/ratio

2.12 (a) Interval/ratio; (b) Nominal; (c) Ordinal; (d) Continuous

2.14 (a) Nominal; (b) Interval/ratio

Section 2.3 (page 55)

2.15

Eye color	Frequency	Relative freq.
Blue	124	.316
Brown	150	.383
Green	15	.038
Hazel	103	.263
Total	392	1.000

2.16 **Eye color of 392 college students**

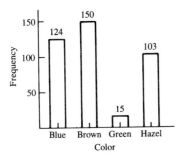

2.18

	Sex		
Occupation	Men	Women	Total
Blue collar	31	9	40
White collar	22	23	45
Farm	15	4	19
Other	4	12	16
Total	72	48	120

(a) $48/120 = .40$; (b) $31/72 = .431$; (c) $9/120 = .075$

2.19 **Occupation group of 120 employed people according to sex**

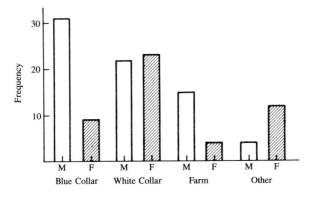

2.20 (a)

Should the possession of a small amount of marijuana be treated as a criminal offense?

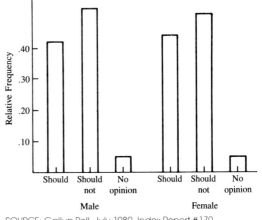

SOURCE: Gallup Poll, July 1980, Index Report #179.

2.21

	Male	**Female**	**Total**
Upper division	30	20	50
Lower division	40	10	50
Total	70	30	100

2.23 (d) Yes, because the frequencies of males and females were the same (1,000 each). If the frequencies had not been the same, we would have had to use the relative frequencies to make comparisons.

2.25

Method	Frequency	Relative frequency
Firearms	422	$422/587 = .719$
Knife	89	$89/587 = .152$
Blunt obj.	30	$30/587 = .051$
Hand, fist	22	$22/587 = .037$
Other	24	$24/587 = .041$
Total	587	$587/587 = 1.000$

(a) Firearm, 71.9%; (b) 28.1%

Section 2.4 (page 62)

2.26

```
1 | 8
2 | 4 8 7 9
3 | 1 6 7 2 7 9 5
4 | 1 2
5 | 4 5 8 5 6
6 | 2
```

2.28

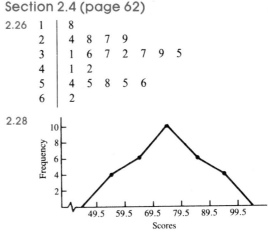

2.29
```
4 | 3  8
5 | 8  5  9
6 | 4  9  3  6  2  8  6
7 | 5  5  9  4  7  3
8 | 1  6  2  3  4
9 | 1
```

2.30
College entrance exam for 24 high school seniors

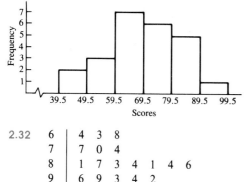

2.32
```
 6 | 4  3  8
 7 | 7  0  4
 8 | 1  7  3  4  1  4  6
 9 | 6  9  3  4  2
10 | 2  9
```

Number of trees on tree farm

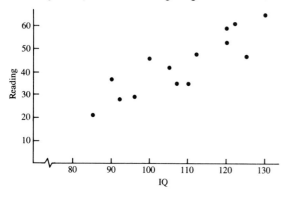

2.34
Reading ability versus IQ for eighth grade students

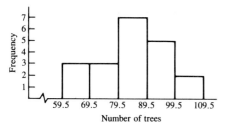

2.37
High, low, and closing price of the Dow Jones Industrial Average

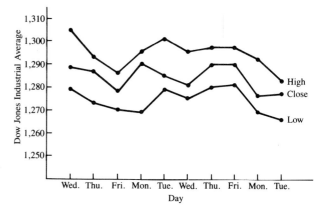

2.39
Median age at their first marriage

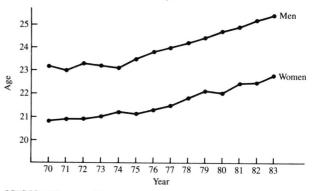

SOURCE: US Bureau of Census

2.41
Percent increase of the GNP from one quarter to the next

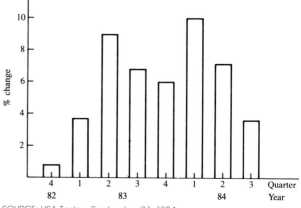

SOURCE: USA Today, September 21, 1984

Review Questions Ch. 2 (page 75)

2.46 A discrete variable is a quantitative variable that can assume a finite or at most a countably infinite number of values. An example of a discrete variable that assumes only a finite number of values could be the number of correct answers on a test containing 20 questions. An example of a discrete variable assuming a countably infinite number of values could be the number of attempts it takes to complete a task. A continuous variable is a quantitative variable that assumes all values in a line interval. By its nature it assumes an infinite number of values. Examples include the time it takes to complete a task and the weight of a boxer at weigh-in time.

2.47 (a) N; (b) N; (c) IR; (d) N; (e) N; (f) O; (g) IR; (h) IR; (i) IR; (j) O

2.48
(a) QL nominal
(b) QN interval
(c) QN interval/ratio
(d) QL nominal
(e) QL nominal
(f) QN interval/ratio

2.49 (a) 70%; (b) 45%; (c) 75.9%; (d) 70.9%

2.50

Type of weapon	No. of robberies
Gun	1,611
Knife	857
Other	356
Unknown	74
Total	2,898

Frequency of use of different weapons in robberies in Atlanta in 1974

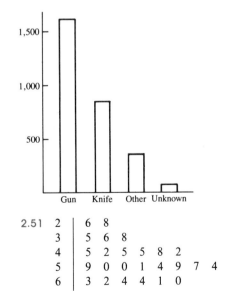

2.51

```
2 | 6  8
3 | 5  6  8
4 | 5  2  5  5  8  2
5 | 9  0  0  1  4  9  7  4
6 | 3  2  4  4  1  0
```

2.52

Class limits	Frequency
20–29	2
30–39	3
40–49	6
50–59	8
60–69	6
Total	25

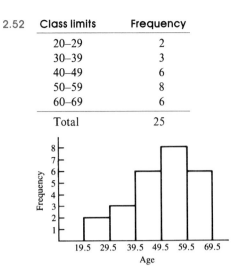

Supplementary Exercises Ch. 2 (page 75)

2.53
(a) QN C IR
(b) QN C IR
(c) QL   N
(d) QN D IR
(e) QL   O
(f) QN D IR

2.55

Major	Frequency
Psychology	10
Sociology	3
Criminal J.	7
Planning	2
Computer S.	3
Other	5
Total	30

Distribution of majors in a class of 30

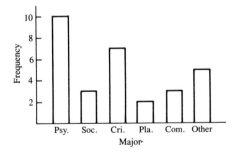

2.57 (a) Quantitative; (b) Qualitative; (c) Quantitative; (d) Qualitative; (e) Quantitative

2.59

0	8
1	
2	9
3	6
4	9 7 5
5	5 7 0 3 0 7 6 2
6	9 1 6 6 7 4 6
7	7 6
8	8 4 0 7 5 2 4
9	9 7 6 1 7 8
10	0 7 5 3
11	3
12	8
13	7
HI	166, 178, 192, 456

2.61 (rotated stem-and-leaf display)

```
      4
      4
      0  4  4
      7  0
      2  2  7
      1
      6  7  8
      4  2  4
      8  7  6  1
      4  2  4
      0  0  1  4
      3
      5  6  6
      0  2  4
      6  5  3
      1  4  6  6
      4  2  4  6  6  6
      0  0  2
      8  8  0  0  7
      0  0  2
      9  9  2  0  0  0  3
      3  4  5  6
```

2.63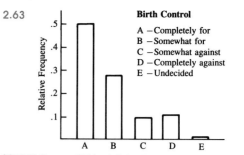

SOURCE: Survey of 516 adults by the American Council of Life Insurance.

2.65 (b) 42%; (c) 59.7%; (d) 33.1%; (e) 28.4%

2.67 (a) Quantitative, discrete; (b) Quantitative, continuous; (c) Qualitative; (d) Quantitative, continuous; (e) Quantitative, continuous

2.69 (a) Nominal; (b) Nominal; (c) Interval/ratio; (d) Interval/ratio

2.71 (a) Sex, age, marital status, type heart received, length of surgery
(b) Qualitative, quantitative, qualitative, qualitative, quantitative
(c) Nominal, interval/ratio, nominal, nominal, interval/ratio

2.73 (a)

Sport	Relative frequency
Baseball	.295
Basketball	.135
Football	.315
Golf	.026
Ice hockey	.062
Soccer	.084
Tennis	.041
Other	.042

CHAPTER 3

Section 3.2 (page 86)

3.1

1–	6 7 8
2*	0 1 2 2 2 3 3
2–	5 5 5 5 6 7 9
3*	0 1 3

3.2

0*	0 0 0 0 0 0 0 0 0 1 1 1 1 1 1 1 1 1 1 1 1 1
0t	2 2 2 2 2 2 2 2 2 2 2 2 3 3 3 3 3 3 3
0f	4 4 4 4 5 5 5
0s	6

3.3

1*	0 0 2 3
1–	6 7 7 8 8 8 9
2*	0 2 3 4
2–	6 7
3*	0 1
3–	6

3.5
```
0s │ 7  6
0- │ 8  9  8
1* │ 1  0  0
1t │ 2  3  2  2
1f │ 4  5  5  4  4  4  5
1s │ 6  7
1- │ 8  9
```

3.7
```
-1* │ 0  2
-0- │ 5  6
-0* │ 2  3  4  4
 0* │ 2  2  3  3  4  4
 0- │ 5  5  5  5  5  6  7  7  8  8  9  9
 1* │ 0  0  1
 1- │ 5
```

3.9
```
1* │ 17
1t │ 29,  32,  36
1f │ 44,  54,  51,  59,  54,  48,  57,  50
1s │ 62,  73,  60,  79,  77,  75,  66,  62
1- │ 97,  83,  90,  84
2* │ 09
```

Weights of 25 soccer players

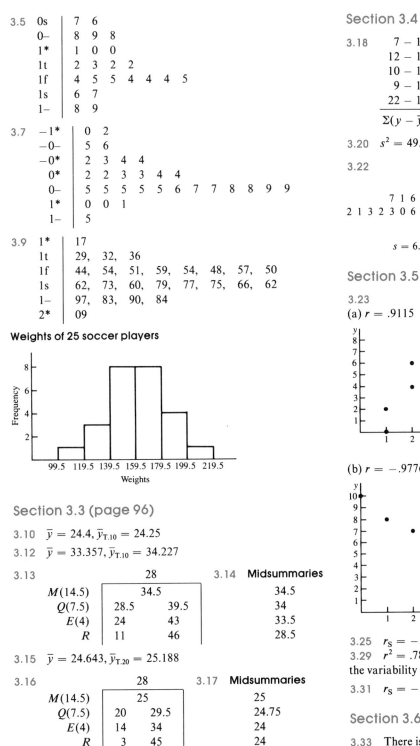

Section 3.3 (page 96)

3.10 $\bar{y} = 24.4$, $\bar{y}_{T.10} = 24.25$

3.12 $\bar{y} = 33.357$, $\bar{y}_{T.10} = 34.227$

3.13

	28	
$M(14.5)$	34.5	
$Q(7.5)$	28.5	39.5
$E(4)$	24	43
R	11	46

3.14 **Midsummaries**

34.5
34
33.5
28.5

3.15 $\bar{y} = 24.643$, $\bar{y}_{T.20} = 25.188$

3.16

	28	
$M(14.5)$	25	
$Q(7.5)$	20	29.5
$E(4)$	14	34
R	3	45

3.17 **Midsummaries**

25
24.75
24
24

Section 3.4 (page 104)

3.18
$$7 - 12 = -5$$
$$12 - 12 = 0$$
$$10 - 12 = -2$$
$$9 - 12 = -3$$
$$22 - 12 = 10$$
$$\overline{\Sigma(y - \bar{y}) = 0}$$

$s = 5.874$

3.20 $s^2 = 49.028$

3.22
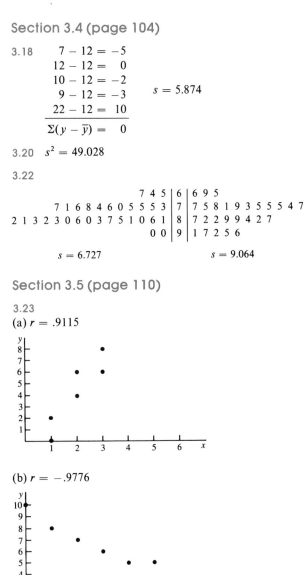

```
                    7 4 5 │ 6 │ 6 9 5
        7 1 6 8 4 6 0 5 5 5 3 │ 7 │ 7 5 8 1 9 3 5 5 5 4 7
2 1 3 2 3 0 6 0 3 7 5 1 0 6 1 │ 8 │ 7 2 2 9 9 4 2 7
                        0 0 │ 9 │ 1 7 2 5 6
```
$s = 6.727$ $s = 9.064$

Section 3.5 (page 110)

3.23
(a) $r = .9115$

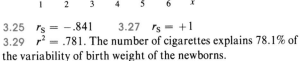

(b) $r = -.9776$

3.25 $r_S = -.841$ 3.27 $r_S = +1$

3.29 $r^2 = .781$. The number of cigarettes explains 78.1% of the variability of birth weight of the newborns.

3.31 $r_S = -.967$

Section 3.6 (page 119)

3.33 There is no way to verify this figure; therefore it cannot be determined whether the figure is reasonable or not.

3.35 People are becoming more aware of sexual abuse of children and are more willing to report individual cases. The number of cases reported does not indicate how many actual cases there were. The number of cases reported in 1982 is approximately five times the number in 1977; yet the child in 1982 is much more than five times larger than the child in 1977.

3.36 No. Not very reliable. No, it applies only to readers of Ann Landers who would respond to a volunteer survey.

Review Questions Ch. 3 (page 127)

3.40 (a) F; (b) F; (c) T; (d) T; (e) F; (f) F

3.41 (a) Sample mean, sample median, sample trimmed mean (b) Sample range, sample standard deviation, sample Q-spread

3.42 (a) $\bar{y} = 10$, $s = 4.093$; (b) $\bar{y}_{T.10} = 10.714$

3.43

0	39,	50,	92
1	12,	21	
2	48,	78	
3	53		
4			
5	49		
6			
7	35		
HI	138.8		

$\times 10^{-1}$

	11	
$M(6)$	24.8	
$Q(3.5)$	10.2	45.1
$E(2)$	5.0	73.5
R	3.9	138.8

3.44 $r = -.959$

Supplementary Exercises Ch. 3 (page 127)

3.45 Parameter   **3.47** Parameter

3.49

0	28,	33,	44						
1	06,	25,	38,	39,	45,	46,	47,	69,	74
2	03,	20,	23,	43,	48,	68,	97		
3	35,	47,	49,	57,	96				
4	33,	33,	37						
5	64								
HI	912								

$\times 10^{-2}$

3.51 (a)

5	3	5	7	
6	4	5	6	8 9
7	0	3	4	7
8	1	2	2	4
9	2	4		
10	1			
11	0			

$\times 10^2$

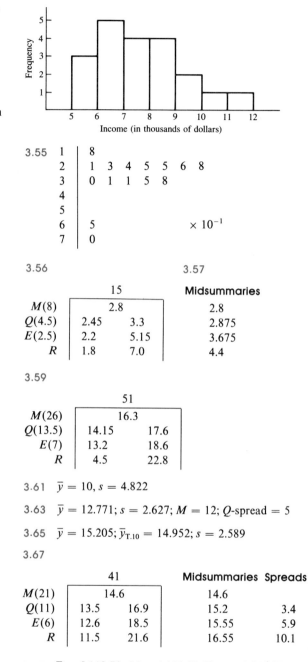

(b)
Per capita incomes for 20 randomly selected North Carolina counties

3.55

1	8							
2	1	3	4	5	5	5	6	8
3	0	1	1	5	8			
4								
5								
6	5							
7	0							

$\times 10^{-1}$

3.56

	15	
$M(8)$	2.8	
$Q(4.5)$	2.45	3.3
$E(2.5)$	2.2	5.15
R	1.8	7.0

3.57

Midsummaries
2.8
2.875
3.675
4.4

3.59

	51	
$M(26)$	16.3	
$Q(13.5)$	14.15	17.6
$E(7)$	13.2	18.6
R	4.5	22.8

3.61 $\bar{y} = 10$, $s = 4.822$

3.63 $\bar{y} = 12.771$; $s = 2.627$; $M = 12$; Q-spread $= 5$

3.65 $\bar{y} = 15.205$; $\bar{y}_{T.10} = 14.952$; $s = 2.589$

3.67

	41		Midsummaries	Spreads
$M(21)$	14.6		14.6	
$Q(11)$	13.5	16.9	15.2	3.4
$E(6)$	12.6	18.5	15.55	5.9
R	11.5	21.6	16.55	10.1

3.69 $\bar{y} = 5{,}118.72$; $M = 4{,}133.50$; $\bar{y}_{T.10} = 4{,}210.07$; $s = 4{,}798.39$

3.71

1	180, 976
2	086, 175, 352, 416, 475, 534, 645, 648, 734, 765, 784, 830, 845, 847, 872
3	127, 160, 170, 249, 320, 420, 640, 679, 743, 820, 869, 970
4	250, 367, 670, 978
5	176, 587, 780
6	870
7	580
8	147

3.73 $\bar{y} = 4{,}049$; $\bar{y}_{T.10} = 4{,}054.38$; $s = 554.48$

3.77 $r = -.868$

CHAPTER 4

Section 4.1 (page 135)

4.1 $2/6 = 1/3$ 4.3 $3/10$

4.5 $25/51$ 4.7 $1/3$

4.9 $1/4$ 4.11 $18/38, 2/38$

Section 4.2 (page 141)

4.13 $A = \{3\}, B = \{1, 2\}, C = \{4, 5, 6\}, D = \{3, 4\}$

4.14 $A = \{(h, h, t), (h, t, h), (t, h, h)\}$
$B = \{(h, t, t), (t, h, t), (t, t, h), (t, t, t)\}$
$C = \{(h, h, h), (h, h, t), (h, t, h), (t, h, h)\}$
$D = \{(t, t, t)\}$

4.15 $\{R, B\}$

4.16 $\{(H_1, M), (H_1, F), (H_2, M), (H_2, F)\}$

4.17 $\{(R, R), (R, B), (B, R), (B, B)\}$

4.19 3

4.21 $\{(B, B), (B, N), (N, B), (N, N)\}$; no

4.23 $S = \{(R, R, R,), (R, R, G), (R, G, R), (G, R, R),$
$(R, G, G), (G, R, G), (G, G, R), (G, G, G)\}$; no

Section 4.3 (page 146)

4.25 (a) $\{3, 4, 5, 6, 7\}$; (b) $\{5\}$; (c) $A = \{3, 5, 7\}$; (d) $\{2, 8\}$;
(e) $\{4, 6\}$; (f) $\{3, 4, 5, 6, 7\}$; (g) $\{5\}$; (h) No, see (b).

4.27 (a) .6; (b) .1; (c) No, A and B have a nonempty intersection.

4.29 Probability of each point is 1/6.
$P(A) = 1/6, P(B) = 2/6, P(C) = 3/6, P(D) = 2/6$

4.31 (a) Yes; (b) .8; (c) .4

Section 4.4 (page 152)

4.33 (a) .7; (b) .6; (c) $.2/.5 = 2/5$; (d) No, $P(A|B) \neq P(A)$.

4.35 $P(NN) = .49, P(NB) = .21, P(BN) = .21,$
$P(BB) = .09$

4.37 $(.6)^3 + 3(.6)^2(.4) = .648$

4.39 $(.7)^3 + 3(.7)^2(.3) = .784$

4.41 No, No

4.43 $.3 + .4 - .12 = .58$

Section 4.5 (page 155)

4.45 .771   4.46 .618

Review Questions Ch. 4 (page 161)

4.51 $1/3, 0$ 4.52 $2/9$

4.53 16, yes, $\{2, 3, 4, 5, 6, 7, 8\}$, no, 6/16

4.54 (a) 3/5; (b) 3/7

4.55 (a) $S = \{(M, M, M), (M, M, F), (M, F, M), (F, M, M),$
$(M, F, F), (F, M, F), (F, F, M), (F, F, F)\}$
(b) .343, .147, .147, .147, .063, .063, .063, .027
(c) $A = \{(M, F, F), (F, M, F), (F, F, M)\}$
(d) .189

Supplementary Exercises Ch. 4 (page 161)

4.56 **1/6**   4.58 **8/36**

4.60 (a) 5/8; (b) 2/7

4.62 (a) $S = \{(W, W, W), (W, W, L), (W, L, W),$
$(L, W, W), (W, L, L), (L, W, L), (L, L, W), (L, L, L)\}$
(b) .064, .096, .096, .096, .144, .144, .144, .216
(c) .352

4.64 (a) Same as Exercise 4.62 with W (win) and L (lose)
replaced by F (female) and M (male)
(b) Same as Exercise 4.62
(c) $A = \{(F, F, M), (F, M, F), (M, F, F)\}$
(d) .288

4.66   (a) Same as Exercise 4.62 with W and L replaced by R (renewed) and C (canceled).
(b) Same as Exercise 4.62.
(c) .352

4.68   .44

4.70   4/7, 1/4

4.72 $S = \{(R, R), (R, G), (R, B), (G, R), (G, G), (G, B),$
$(B, R), (B, G), (B, B)\}$
$A = \{(R, R)\}$
$B = \{(G, G), (G, B), (B, G), (B, B)\}$
$C = \{(R, G), (R, B), (G, R), (B, R)\}$

4.74 $S = \{(N, D), (N, Q), (D, N), (D, Q), (Q, N), (Q, D)\}$
$A = \{(N, Q), (D, Q), (Q, N), (Q, D)\}$
$B = \{(N, D), (D, N)\}$
Yes

4.76   1/4

4.78 $S = \{(A, B, C), (A, C, B), (B, A, C), (B, C, A),$
$(C, A, B), (C, B, A)\}$
If the person guesses, then the simple events will be equally likely. Each will have probability 1/6.

4.80   .64, .04, .32   4.82   1/16

4.84   (a) .29; (b) .444; (c) .0386

CHAPTER 5

Section 5.1 (page 168)

5.1   Range of $x = \{0, 1, 2, 3, 4, 5\}$, discrete

5.3   Range of $w = \{t \mid t$ is a real number greater than 0$\}$, continuous

5.5   Range of $b = \{0, 1, 2, 3, \ldots, 15\}$, discrete

5.7   (a) Continuous; (b) Discrete; (c) Discrete; (d) Continuous; (e) Continuous

Section 5.2 (page 173)

5.8

x	p(x)
0	1/32
1	5/32
2	10/32
3	10/32
4	5/32
5	1/32

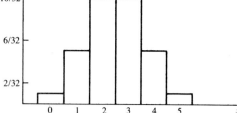

5.9

y	p(x)
2	1/36
3	2/36
4	3/36
5	4/36
6	5/36
7	6/36
8	5/36
9	4/36
10	3/36
11	2/36
12	1/36

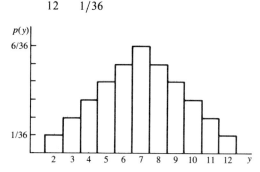

5.11 $P(2) = 1/6, P(3) = 2/6, P(4) = 3/6,$
Total $= 6/6 = 1.0$
Yes it is a probability distribution.

5.13   (a) 4; (b) .556, .444;
(c)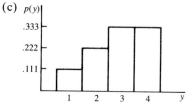

Section 5.3 (page 180)

5.14   95/15, 6.222. The interval (1.344, 11.322) covers at least 75% of the distribution.

5.16   2.9, 1.7. The interval $(-.5, 6.3)$ contains approximately 95% of the distribution.

5.18 $E(y) = .75$, standard deviation $= .75$. The interval $(-.75, 2.25)$ covers at least 75% of the distribution.

Section 5.4 (page 187)

5.20   (a) .201; (b) .147; (c) .312; (d) .069

5.22   (a) 8, 1.2649; (b) 4.5, 1.7748; (c) 3, 1.2247; (d) 3.6, .6

5.23   (a) .849; (b) .901; (c) .549; (d) .943

5.24   Work it out using the binomial probability formula or using a computer.

5.26   (a) .167; (b) 6; (c) 1.549

5.27 .798, 1.6

5.29 $3(1/3)^2(2/3) = 2/9$

5.30 c and e

5.32 P(4 or fewer blacks selected from 12) $= .042 + .012 + .002 = .056$. Because this probability is small, it is doubtful that the jury was selected randomly.

Section 5.5 (page 197)

5.33 (a) .4772; (b) .4953; (c) .4131; (d) .4772; (e) .4222; (f) .4976

5.35 (a) $.3413 + ..2257 = .5670$; (b) $.5 - .4192 = .0808$

5.37 11.84

5.38 (a) Continuous; (b) No; (c) No; (d) 200 and 300; (e) Percent of players that weigh between 200 and 250; (f) Yes

5.39 $.5 - .4987 = .0013$

5.41 (a) .0062; (b) .6915; (c) .0668; (d) .1525

5.42 (a) .0668; (b) .3830; (c) 176.8

5.43 108.2

Review Questions Ch. 5 (page 204)

5.49 (a) Continuous; (b) Discrete; (c) Continuous; (d) Discrete; (e) Continuous

5.50 .4452, .2119

5.52 (a) T; (b) F; (c) F; (d) T; (e) F; (f) F

5.53 .1503, 7

5.54 3.14

Supplementary Exercises Ch. 5 (page 204)

5.55 (a) 5; (b) 6/10; (c) 3/10; (d) 3.6, 2.64

5.57 .656

5.59 (a) Binomial

(b) x	p(x)
0	.168
1	.360
2	.309
3	.132
4	.028
5	.002

(c) 1.5   (d) .837

5.61 .157

5.63 .262

5.65 .952, 14

5.67 $\{1, 2, 3, 4, \ldots\}$, discrete

5.69 .047, 3

5.71 (a) .7888; (b) .2614; (c) .3520; (d) 76.72

5.73 .3085

5.75 (a) .9050; (b) 11.16; (c) 18.84

5.77 (a) .0228; (b) .2743; (c) .3811

5.79 .991

5.81 Mean and standard deviation (or variance)

CHAPTER 6

Section 6.1 (page 214)

6.1

1*	1
1–	6 9
2*	4
2–	7 6 8 9
3*	3 2 4 1 0 2
3–	8 9 5 9 7 6 7
4*	4 1 2 0 3
4–	5 6

The distribution appears to be skewed left.

6.3

LO	144
20	1 9
21	2
22	0
23	7 1 3 3
24	6 8 2
25	2
26	6
27	9
28	3 9 7
29	4 1 2 0
30	4
31	0
HI	340

The distribution appears to be bimodal.

$\times 100$

6.5

0	537
1	596,909,251,200,326,811,718,909
2	977,521,387,804,126,177,560,925,838,372
3	750,222,979,428,552,266,718
4	160,942,764,531,567,772,234,878,020,096
5	768,122,530,531
6	138,409,552,168,868
7	726,167
8	267,071
9	141
HI	19854

$\times 10^{-1}$

The distribution appears to be skewed right.

Section 6.2 (page 218)

6.7

	87		Midsummaries
$M(44)$	23		23
$Q(22.5)$	20	26	23
$E(11.5)$	18	33	25.5
R	14	42	28

From the midsummaries it would appear that the distribution is somewhat skewed to the right.

6.9

	28		Midsummaries
$M(14.5)$	34.5		34.5
$Q(7.5)$	28.5	39.5	34
$E(4)$	24	43	33.5
R	11	46	28.5

The midsummaries decrease, suggesting that the distribution is skewed left.

6.11

	28		Midsummaries
$M(14.5)$	25		25
$Q(7.5)$	20	29.5	24.75
$E(4)$	14	34	24
R	3	45	24

From the midsummaries there is no clear departure from symmetry.

6.13

	25		Midsummaries
$M(13)$	25,200		25200
$Q(7)$	23,300	29,000	26150
$E(4)$	21,200	29,400	25300
R	14,400	34,000	24200

Not symmetrical or skewed.

6.15

	30		Midsummaries
$M(15.5)$	90		90
$Q(8)$	71	98	84.5
$E(4.5)$	62	103	82.5
R	31	128	79.5

Skewed left.

Section 6.3 (page 223)

6.19

	28		
$M(14.5)$	34.5		
$Q(7.5)$	28.5	39.5	Q-spread = 11

**Box plot
summary diagram**

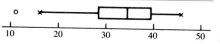

	16.5	
a	16	46
f	12	56
11	one	none
F	−4.5	72.5
	none	none

Box plot

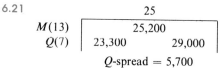

The box plot indicates that the distribution is skewed left.

6.21

	25	
$M(13)$	25,200	
$Q(7)$	23,300	29,000

Q-spread = 5,700

**Box plot
summary diagram**

	8,550	
a	20,100	34,000
f	14,750	37,550
14,400	one	none
F	6,200	46,100
	none	none

Box plot

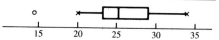

The box plot indicates that the middle 50% of the distribution is slightly skewed right, but then there is a mild outlier on the left tail.

6.23 (a)

0s	6 7
0−	8 8
1*	0 0 0 0 1 1 1
1t	2 2 2 2 2 3 3
1f	4 4 4 4
1s	6 6
1−	8 8
2*	0
2t	
2f	5
2s	
2−	9
3*	
3t	3

(b)

M(15.5)	30		Midsummaries	Spreads
M(15.5)	12		12	
Q(8)	10	16	13	6
E(4.5)	9	19	14	10
R	6	33	19.5	27

(c) From the midsummaries the distribution is skewed right.

(d) **Box plot**
summary diagram

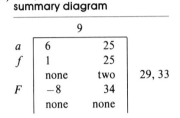

	9		
a	6	25	
f	1	25	
	none	two	29, 33
F	−8	34	
	none	none	

Box plot

The box plot indicates that the distribution is definitely skewed right.

(e) $PSD_q = 6/1.35 = 4.444$, $PSD_e = 10/2.3 = 4.348$, $s = 6.105$

The pseudostandard deviations are very close and smaller than s, which indicates that the distribution has long tails; however, the box plot shows that there is only one long tail.

Section 6.4 (page 227)

6.24 μ 6.25 π 6.26 $\theta_1 - \theta_2$

6.27 θ 6.28 π

6.29 median   6.30 trimmed mean

Section 6.5 (page 234)

6.31

	49		
M(25)	17		
Q(13)	10	23	Q-spread = 13

Box plot
summary diagram

	19.5		
a	5	41	
f	−9.5	42.5	
	none	one	43
F	−29	62	
	none	one	121

Box plot

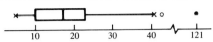

The box plot indicates that the distribution is definitely skewed right, and thus the median is the proper measure of the center of the distribution.

6.33

Quantile summary diagram for 1973–1974 salaries

	50		Midsummaries	Spreads
M(25.5)	9891		9891	
Q(13)	8932	11,121	10026.5	2189
E(7)	8467	11,741	10104	3274
R	7604	15,667	11635.5	8063

Quantile summary diagram for 1983–1984 salaries

	51		Midsummaries	Spreads
M(26)	20657		20657	
Q(13.5)	18572.5	23022	20797.25	4449.5
E(7)	17500	24641	21070.5	7141
R	15895	36564	26229.5	20669

Box plot summary
diagram for 1973–1974

	3283.5		
a	7604	13371	
f	5648.5	14404.5	
	none	one	15667
F	2365	17688	
	none	none	

Box plot summary
diagram for 1983–1984

	6674.25		
a	15895	28877	
f	11898.25	29696.25	
	none	none	
F	5224	36370.5	
	none	one	36564

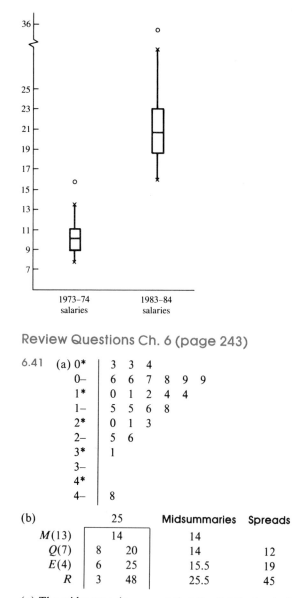

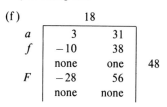

1973–74 salaries   1983–84 salaries

The box plot reveals the skewness to the right.

6.42

	A		B		C
0*	1				
0t			2		
0f					
0s			6 7		
0–	8 8 9		8 8 9 9		8 9 9
1*					
1t	2 2 2		2		2 2 2 2
1f	4 4				4
1s			6		
1–					8

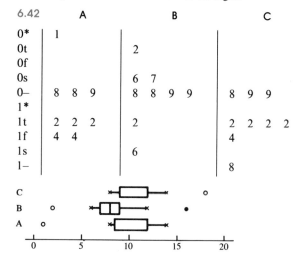

It appears that the educational levels for the employees of Industry B are below those of Industries A and C.

Supplementary Exercises Ch. 6 (page 244)

6.43

11	5 7 7
12	1 3 6 8 8
13	3 4 5 5 6 7 7 8 9 9
14	4 5 6 7 8 9
15	0 7 8 9
16	4 8 9 $\times 10^{-1}$
17	2 3 6
18	0 5
19	0 5 6
20	9
21	6

	41		Midsummaries	Spreads
M(21)		14.6	14.6	
Q(11)	13.5	16.9	15.2	3.4
E(6)	12.6	18.5	15.55	5.9
R	11.5	21.6	16.55	10.1

Skewed right

6.45

1f	5 5 5 5
1s	6 6 6 6 6 7 7 7 7 7
1–	8 8 8 8 8 8 8 8 8 9 9 9 9 9 9 9
2*	0 0 0 0 0 1 1 1 1 1
2t	2 2 2 3
2f	4 4 5 $\times 10^{-1}$
HI	32

Review Questions Ch. 6 (page 243)

6.41 (a)

0*	3 3 4
0–	6 6 7 8 9 9
1*	0 1 2 4 4
1–	5 5 6 8
2*	0 1 3
2–	5 6
3*	1
3–	
4*	
4–	8

(b)

	25		Midsummaries	Spreads
M(13)		14	14	
Q(7)	8	20	14	12
E(4)	6	25	15.5	19
R	3	48	25.5	45

(c) The midsummaries suggest that the distribution is skewed right.
(d) $\bar{y} = 14.96$, $s = 10.224$
(e) $\text{PSD}_q = 8.89$, $\text{PSD}_e = 8.26$
The sample standard deviation is larger than the pseudostandard deviations, which indicates long tails. From the midsummary analysis, however, we see that the distribution has only one long tail.

(f)

	18		
a	3	31	
f	−10	38	
	none	one	48
F	−28	56	
	none	none	

	51		Midsummaries	Spreads
M(26)	1.9		1.9	
Q(13.5)	1.7	2.05	1.875	.35
E(7)	1.6	2.2	1.9	.6
R	1.5	3.2	2.35	1.7

With the exception of the one extreme score of 3.2, the data appear symmetric. The midsummaries indicate that the middle 75% (between E_1 and E_7) is symmetric.

6.47

0	307, 490, 574
1	804
2	228
3	200, 320, 704, 986
4	281, 758, 880
5	247, 279
6	
7	667
8	013
HI	12,000, 20,399

	18		Midsummaries	Spreads
M(9.5)	4133.5		4133.5	
Q(5)	2228	5279	3753.5	3051
E(3)	574	8013	4293.5	7439
R	307	20399	10353	20092

The midsummaries do not reveal any specific shape, although the stem and leaf plot shows a definite skewness to the right.

6.51

2–	800
3*	270, 400
3–	540, 820, 860, 890, 920, 920, 950, 970
4*	180, 250, 260, 370, 420, 470
4–	650, 840
5*	200

	20		Midsummaries	Spreads
M(10.5)	3960		3960	
Q(5.5)	3840	4395	4117.5	555
E(3)	3400	4650	4025	1250
R	2800	5200	4000	2400

The midsummaries indicate that the distribution is symmetric.

CHAPTER 7

Section 7.1 (page 250)

7.1 M—the sample median

7.3 $M_1 - M_2$—the difference in two sample medians

7.5 $\bar{y}$—the sample mean

7.7 (a) 3, 14/3

(b)

$\bar{y}$	$p(\bar{y})$
1	1/27
4/3	3/27
5/3	3/27
2	1/27
8/3	3/27
3	6/27
10/3	3/27
13/3	3/27
14/3	3/27
6	1/27

(c)

M	$p(M)$
1	7/27
2	13/27
6	7/27

(d) The values of $\bar{y}$ are more tightly clustered about the mean of the population ($\mu = 3$) than are the values of M; therefore $\bar{y}$ is the better measure of the center of the distribution.

Section 7.2 (page 258)

7.8 The sampling distribution of $\bar{y}$ is approximately normally distributed with a mean of 500 and a standard error of $100/\sqrt{200}$.

7.10 (a) 25, $8/\sqrt{10}$; (b) 25, $8/\sqrt{16}$; (c) 25, $8/\sqrt{30}$; (d) 25, $8/\sqrt{100}$

7.12 (a) .00003; (b) .0808; (c) .2209

7.14 Yes, because 8,380 is within 2 standard errors of the mean.

7.16 .2483

7.18 An average of $1.21 is a full 3 standard errors above $1.18, suggesting that the $1.18 is low.

7.20 (a) .2347; (b) .8185

Section 7.3 (page 262)

7.22 The sampling distribution of p is approximately normally distributed with a mean of .8 and a standard error of $\sqrt{(.8)(.2)/400} = .02$.

7.24 (a) .1, .0075; (b) .3, .0115; (c) .5, .0125; (d) .7, .0115; (e) .9, .0075; (f) .575, .0124

7.26 (a) No, $z = 2.18$, which is unusual. (b) $.3 + 2\sqrt{(.3)(.7)/100} = .39165$

7.28 .8944

7.30 No, $z = 1.4$, which is not unusual. $P(p > .6) = .0808$

7.32 .00003

7.34 $z = -14.31$. Strong evidence that the aide is in error.

Section 7.4 (page 266)

7.36 (a) .1; (b) .05; (c) .0354; (d) .0224; (e) .0158; (f) .0112

7.38 .05, .0167

7.39 6.25, 2.5

Review Questions Ch. 7 (page 272)

7.42 If $n > 30$ then the sampling distribution of $\bar{y}$ is approximately normally distributed with a mean of μ and a standard error of $\sigma/\sqrt{n}$.

7.43 65, 1.5. Yes, because the sample size is large.

7.44 .0166

7.45 For sufficiently large sample size, the sampling distribution of the sample proportion p will be approximately normally distributed with a mean of π and a standard error of $\sqrt{\pi(1-\pi)/n}$.

7.46 .2451   7.47 .0354   7.48 $\sqrt{30}$

Supplementary Exercises Ch. 7 (page 272)

7.49 Parameter   7.51 Parameter   7.53 Statistic

7.55 $z = 2.11$. Yes, because it is more than 2 standard errors above the mean of 70.

7.57 Approximately normally distributed with a mean of 20 and a standard deviation of .4.

7.59 (66, 78)   7.61 .0089

7.63 $z = P(\bar{y} > 104.6) = .0465$; so it is somewhat unusual.

7.65 .0217   7.67 .0354, .0224   7.69 .98

CHAPTER 8

Section 8.1 (page 280)

8.1 Because the sampling distribution of p is centered at π, it is an unbiased estimator of π.

8.2 The standard error of p is $\sqrt{\pi(1-\pi)/n}$, which approaches 0 as n becomes large.

8.3 (a) yes; (b) yes

8.5 No, $\bar{y}_T$ should be used when the distribution is symmetric with long tails.

8.6 (a) bias-low, efficiency-low; (b) bias-high, efficiency-low; (c) bias-high, efficiency-high; (d) bias-low, efficiency-high

8.7 Mean time for checking reservations. $\bar{y}$ or $\bar{y}_T$. A stem and leaf plot of the data appears to not deviate significantly from normality, hence $\bar{y} = 42.35$ is the preferred statistic.

Section 8.2 (page 288)

8.9 (a) $\bar{y}$; (b) $\bar{y}_T$; (c) $\bar{y}$; (d) M; (e) M

8.11

0–	5	8						
1*	0	0	2	4				
1–	5	5	6	7	8	9	9	
2*	0	0	0	2	3			
2–	5	5	5	8	8	8	9	
3*	0	0	0	0	2	2	3	
3–	5	5	5	6	7	8	8	9	9
4*	0	0	0	1	1	2	3	4
4–	5							

Quantile summary diagram

	50	
$M(25.5)$	29.5	
$Q(13)$	19	38

Box plot summary diagram

	28.5	
a	5	45
f	-9.5	66.5
	none	none
F	-38	95
	none	none

The distribution could be classified as being skewed left; the adjacent values are clearly inside the inner fences, however, and thus we will classify the distribution as short tailed. Consequently $\bar{y} = 27.92$ is a reasonable estimate of the center of the distribution.

8.13

0f	50
0s	70, 75, 75
0–	80, 80, 80, 90, 90
1*	00, 10
1t	20, 25, 25, 35
1f	40, 45, 50, 50, 55
1s	60, 60, 65, 70, 70, 75
1–	80, 80, 80, 80, 90
2*	00, 05, 10, 15
2t	20, 25, 35, 35
2f	45, 50, 50, 50, 55
2s	60, 70, 75
2–	80, 90
HI	400

**Quartile
summary diagram**

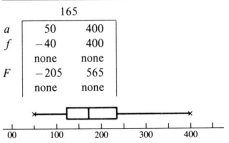

		50		
$M(25.5)$		172.5		
$Q(13)$		125	235	Q-spread $= 110$

**Box plot
summary diagram**

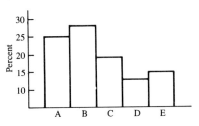

	165		
a	50	400	
f	-40	400	
	none	none	
F	-205	565	
	none	none	

Aside from the one score of 400, the distribution will appear symmetric with short tails. One should check out the score of 400 to see if it is a legitimate score. If it is, the median of 172.5 will be a reasonable estimate of the center of the distribution. The sample mean of 176.5 will also be an acceptable answer.

8.15 No, only for large communities that are similar to the one sampled. Nothing can be said about small or rural communities from a sample taken from a large community.

8.17

The distribution appears to be skewed right; consequently, the median income would be a reasonable measure of the center.

8.18

13	20
14	10
15	48
16	93, 95, 95
17	60, 72, 76, 95, 98
18	10, 13, 30, 60, 64, 96
19	28, 40, 50, 51, 72, 76, 90, 94, 95, 97
20	20, 23, 30, 35, 40, 52, 53, 68, 89
21	25, 32, 38, 46, 50, 76
22	33, 41, 67, 84
23	10, 55
24	20 $\times 10$
25	
26	25

**Quartile
summary diagram**

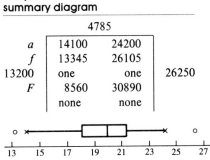

	50		
$M(25.5)$	19945		19945
$Q(13)$	18130	21320	19725
$E(7)$	17600	22410	20005
R	13200	26250	19725

Q-spread $= 3190$

**Box plot
summary diagram**

	4785		
a	14100	24200	
f	13345	26105	
13200	one	one	26250
F	8560	30890	
	none	none	

The distribution is close to normal, and thus $\bar{y} = 19,808$ is the best measure of the center of the distribution.

8.19 The standard deviation, $s = 2427.066$, and the Q-spread $= 3190$. Because the distribution is close to normal (see Exercise 8.18), s is the more reliable estimator of the variability.

Section 8.3 (page 296)

8.20 (a) $94.3 \pm (1.96)(20)/\sqrt{30}$;
 $(87.143, 101.457)$

 (b) $96.4 \pm (1.96)(20)/\sqrt{45}$
 $(90.556, 102.244)$

 (c) $95.6 \pm (1.96)(20)/\sqrt{100}$;
 $(91.68, 99.52)$

 (d) $95.8 \pm (1.96)(20)/\sqrt{200}$
 $(93.028, 98.572)$

8.22 (a) $72.4 \pm (1.96)(11.2)/\sqrt{36}$;
 $(68.74, 76.06)$

 (b) $128.3 \pm (2.33)(32.4)/\sqrt{64}$
 $(118.8635, 137.7365)$

 (c) $465 \pm (2.58)(112)/\sqrt{100}$
 $(436.104, 493.896)$

8.24 $7848 \pm (1.96)(310)/\sqrt{36}$
 $(7,746.733, 7,949.267)$

8.26 $5.2 \pm (2.58)(.24)/\sqrt{40}$
 $(5.102, 5.298)$

8.27 $560 \pm (1.645)(35)/\sqrt{100}$
 $(554.24, 565.76)$

8.29 $(2.33)(2.1)/\sqrt{n} = .3$
$\sqrt{n} = (2.33)(2.1)/.3 = 16.31$
$n = 266.0161 \approx 267$

8.31 $78,460 \pm (1.645)(22,260)/\sqrt{40}$
$(72,670.23,\ 84,249.77)$

Section 8.4 (page 302)

8.32   (a) $(25 - 1.96\sqrt{25})/2 = 7.6 \approx 8$
(b) $(50 - 1.96\sqrt{50})/2 = 18.07 \approx 19$
(c) $(100 - 1.96\sqrt{100})/2 = 40.2 \approx 41$
(d) $(200 - 1.96\sqrt{200})/2 = 86.14 \approx 87$

8.34 $71.22 \pm 2.58\,(15.55)/\sqrt{23}$
$(62.85,\ 79.59)$

8.36 $394.81 \pm 1.645(228.458)/\sqrt{27}$
$(322.48,\ 467.14)$

8.37

3*	1 3 4
3–	5 6 6 6 8 8 8 9 9 9 9
4*	0 1
4–	5 5 6 6 7 7 8 8 8 9 9
5*	0 1 2 2 3 3 4 4
5–	6 7 7 9 9
6*	3 4 4
6–	7 8
7*	1 4
7–	
8*	3 4
8–	8

$\times\ 10^{-1}$

The distribution appears skewed right, and therefore the confidence interval will be an interval for the population median. Location of $C = 19$; $(4.6,\ 5.3)$

8.38   (a) Location of $C = (25 - 1.96\sqrt{25})/2 \approx 8$; $(27, 45)$
(b) Location of $C = (25 - 1.96\sqrt{25})/2 \approx 8$; $(15, 36)$
(c) Location of $C = (24 - 1.96\sqrt{24})/2 \approx 8$; $(12, 36)$

8.40 $406.46 \pm 1.645\,(220.552)/\sqrt{24}$
$(332.40,\ 480.52)$

8.42 $197.30 \pm 1.96\,(57.71)/\sqrt{28}$
$(175.92,\ 218.68)$

8.44   Location of $C = (37 - 2.33\sqrt{37})/2 = 11.4 \approx 12$

4	80, 95
5	24, 42, 76, 95, 96
6	04, 08, 14, 21, 36, 40, 44, 46, 52, 52, 56, 66, 76, 96
7	16, 51, 60, 75, 96
8	34, 44, 50
9	29
10	20, 29, 69
11	53
12	16
13	26, 49

$\times\ 100$

Confidence interval: $(63600,\ 79600)$

Section 8.5 (page 306)

8.45 $p = 180/900 = .2$
$.2 \pm 1.645\sqrt{(.2)(.8)/900}$
$.2 \pm .022 = (.178,\ .222)$
$1.645\sqrt{(.2)(.8)/n} = .03$
$n = 482$

8.47 $p = 722/1183 = .61$
$.61 \pm 2.58\sqrt{(.61)(.39)/1183}$
$.61 \pm .037 = (.573,\ .647)$

8.49 $p = 635/2048 = .31$
$.31 \pm 1.645\sqrt{(.31)(.69)/2048}$
$.31 \pm .0168 = (.293,\ .327)$

8.51 $p = 1440/2000 = .72$
$.72 \pm 2.33\sqrt{(.72)(.28)/2000}$
$.72 \pm .0234 = (.6966,\ .7434)$

8.53 $p = 159/757 = .21$
$.21 \pm 1.645\sqrt{(.21)(.79)/757}$
$.21 \pm .0244 = (.186,\ .234)$

8.55 $n = 1448$

Section 8.6 (page 310)

8.57   (a) 2.160; (b) 2.060; (c) 1.833;
(d) 1.729; (e) 2.831; (f) 2.704

8.59   (a) $27.074 \pm (2.779)(9.0593)/\sqrt{27}$
$27.074 \pm 4.845 = (22.229,\ 31.919)$
(b) $34.607 \pm (2.771)(12.0933)/\sqrt{28}$
$34.607 \pm 6.333 = (28.274,\ 40.94)$
(c) $23.792 \pm (2.807)(7.8959)/\sqrt{24}$
$23.792 \pm 4.524 = (19.268,\ 28.316)$

8.61 $2.99 \pm (1.833)(.2726)/\sqrt{10}$
$2.99 \pm .158 = (2.832,\ 3.148)$

8.63 $250,926.25 \pm (1.796)(34,773.64)/\sqrt{12}$
$250,926.25 \pm 18,028.76 = (232,897.49,\ 268,955.01)$

Section 8.7 (page 313)

8.66   (a) 6;   (b) 5;   (c) 4

8.68   Location of $C = 3$; confidence interval: $(16, 24)$

8.69   Location of $C = 4$; confidence interval: $(20, 45)$

8.71   Location of $C = 2$; confidence interval: $(7.2, 13.8)$

8.72   Location of $C = 4$; confidence interval: $(10.4, 12.6)$

Review Questions Ch. 8 (page 319)

8.78   (a) 1.645; (b) 1.96; (c) 1.753; (d) 2.131

8.79   1068

8.80   (a) (iv); (b) (iii);
(c) Location of $C = (25 - 2.58\sqrt{25})/2 = 6.05 \approx 7$; confidence interval: $(18, 45)$

8.81 (a) (29,698, 34,902); (b) normally distributed; (c) increase the sample size

Supplementary Exercises Ch. 8 (page 320)

8.82 (a) An unbiased estimator is a statistic whose sampling distribution is centered at the parameter being estimated.
(b) Estimator A is more efficient than estimator B when its sampling distribution has a smaller standard error.
(c) Under the assumption of normality, $\bar{y}$ is the best estimator of μ.

8.84 (a) $23 \pm (2.131)(3)/\sqrt{16}$
$(21.4, 24.6)$

8.86 (a) $n = 49$
(b) $128 \pm (2.58)(27)/\sqrt{45}$
$(117.62, 138.38)$

8.88 (a) (.3742, .4258)
(b) $n = 3994$

8.90 (433.42, 538.58)

8.92 44

8.94 (a) 1509; (b) (.147, .229)

8.96 (15.97, 24.03)

8.98 (a) 815; (b) (.282, .358)

8.100 (58.65, 66.15)

8.102 (6.66, 6.74)

8.104 (17, 20)

8.106 Location of $C = (49 - 2.58\sqrt{49})/2 = 15.47 \approx 16$; confidence interval: (11, 22)

8.108 $n = 57$

CHAPTER 9

Section 9.1 (page 329)

9.1 (a) $H_0: \mu = 50, H_a: \mu > 50$
(b) $H_0: \mu = 50, H_a: \mu < 50$
(c) $H_0: \mu = 50, H_a: \mu > 50$
(d) $H_0: \mu = 50, H_a: \mu < 50$
(e) $H_0: \mu = 50, H_a: \mu \neq 50$

9.3 Yes. No, a Type II error is possible only when the null hypothesis is not rejected.

9.5 No, there is the possibility of a Type II error.

9.7 Type I: saying that the patient will not recover when in fact he will.
Type II: saying that the patient will recover when in fact he will not. Clearly, Type I is the more serious.

9.9 $H_0: \pi = .30; H_a: \pi > .30$

9.11 $H_0: \pi = .10; H_a: \pi > .10$

Section 9.2 (page 338)

9.13 (a) Left-tailed test; (b) Two-tailed test; (c) Right-tailed test; (d) Left-tailed test

9.15 (a) $z = -1.64$, p-value $= .0505$
(b) $z = 2.27$, p-value $= .0232$
(c) $z = 7.50$, p-value $= 0^+$

9.17 $z = -.97$, p-value $= .166$. There is insufficient evidence that the average age of food stamp recipients is less than 40 years.

9.19 $z = 1.82$, p-value $= .0344$. There is significant evidence that μ exceeds 25 hours.

9.21 $z = -3.6$, p-value $< .0002$. There is highly significant evidence that this company is paying inferior wages.

9.23 $z = -1.54$, p-value $= .0618$. There is moderately significant evidence that the manufacturer is in error.

Section 9.3 (page 342)

9.25 $z = -1.37$, p-value $= .0853$. There is moderately significant evidence that $\mu < 20$.

9.27 $z = .82$, p-value $= .2061$. There is insufficient evidence that $\mu > 2.5$.

9.29 $z = 1.73$, p-value $= .0418$. There is significant evidence that the average score of 474.6 is high.

9.32 $z = 2.70$, p-value $= 2(.0035) = .0070$

Section 9.4 (page 346)

9.33 $t = 1.38$, $.05 < p$-value $< .10$. There is moderately significant evidence that mean usage exceeds 120 kilowatt hours.

9.35 $t = 1.33$, p-value $> .10$. There is insufficient evidence that the mean carbon dioxide level is more than 4.9 ppm.

9.37 $t = .625$, p-value $> .10$. There is insufficient evidence to claim that the average home is not within 5.5 miles of a fire department. Normally distributed.

9.39 $t = -1.67$, $.05 < p$-value $< .10$. There is mildly significant evidence that the mean life expectancy is less than 10 years. Normally distributed.

9.41 $t = 1.136$, p-value $> .20$. There is no significant evidence.

9.43 $t = -.71$, p-value $> .10$. The data do not provide evidence that the students scored significantly below 72.

Section 9.5 (page 352)

9.44 (a) p-value $= .212$, insufficient evidence that $\theta < 50$
(b) p-value $= .035$, sufficient evidence that $\theta > 3.5$
(c) p-value $= .032$, sufficient evidence that $\theta \neq 100$

9.46 p-value $= .031$, sufficient evidence that $\theta > 50$.

9.48 p-value $= 2(.402) = .804$, insufficient evidence to deny that the median mental age is 100.

9.50 p-value $= 2(.12) = .24$. There is insufficient evidence that the median age is not 19.

9.52 p-value $= .0359$, significant evidence that the North Carolina social workers are underpaid.

Section 9.6 (page 355)

9.53 $z = 1.67$, p-value $= .0475$. There is statistical evidence that her claim is valid.

9.55 $z = 7.07$, p-value $= 0^+$. There is highly significant evidence that the government's claim is invalid.

9.57 $z = 3$, p-value $= .0013$. The evidence is highly significant that a greater proportion of students than local citizens favor the sale of beer in the county.

9.59 $z = 1.84$, p-value $= .0329$. There is significant evidence to reject the supplier's claim.

9.61 $z = -14.31$, p-value $= 0^+$. The evidence is highly significant that the presidential aide was in error.

9.63 $z = -4.42$, p-value $< .00003$. There is highly significant evidence that the doctor's claim is invalid.

Review Questions Ch. 9 (page 363)

9.67 $z = 3$, p-value $= .0026$. The average score of students from this high school is unusually high.

9.68 $t = 2.67$, $.005 < p$-value $< .01$. Significant evidence exists that the administrators have a high self-concept.

9.69 $z = 2.16$, p-value $= 2(.0154) = .0308$. The percent differs significantly from 70%.

9.70 $n = 11$, $T = 2$, p-value $= .032$. There is significant evidence that the group falls below the national median.

Supplementary Exercises Ch. 9 (page 363)

9.71 Hypothesis test

9.73 Confidence interval

9.75 $H_0: \pi \le .70$, $H_a: \pi > .70$

9.77 $H_0: \mu \le 100$, $H_a: \mu > 100$

9.79 $H_0: \mu \le 800$, $H_a: \mu > 800$, $z = 1.61$, p-value $= .0537$. Significant evidence exists to reject the government's claim.

9.81 $z = 3.05$, p-value $= .0011$. The data support the administrator's claim.

9.83 $H_0: \mu \le 40$, $H_a: \mu > 40$.

9.85 $z = 2.88$, p-value $= .002$. There is highly significant evidence that State University freshmen are more intelligent than the average freshman.

9.87 $z = -2.61$, p-value $= 2(.0045) = .0090$. There is highly significant evidence to reject the shipment.

9.89 $z = 1.00$, p-value $= .1587$. Insufficient evidence exists to say that the new string has a greater mean breaking strength than the old.

9.91 $z = -3.10$, p-value $= .001$. There is highly significant evidence that the mean adult height of residents in the depressed area is below that of all residents.

9.93 $z = 1.81$, p-value $= .0351$. There is significant evidence to cast doubt on the utility company's claim.

9.95 $n = 6$, $T = 4$, p-value $= .344$. Insufficient evidence exists that his process produces stones averaging over .5 carat. He should remain in school.

9.97 $n = 9$, $T = 8$, p-value $= .02$. Reject the horse. No.

CHAPTER 10

Section 10.2 (page 374)

10.1 (a) 37.4 ± 14.8 (b) -1.5 ± 1.73
$(22.6, 52.2)$ $(-3.23, .23)$

10.3 1.8 ± 2.18
$(-.38, 3.98)$

10.5 $.414 \pm 4.598$
$(-4.184, 5.012)$

10.7 $z = 2.03$; p-value $= .0212$. Significant evidence exists that the mean income in District I exceeds the mean income in District II.

10.9 $z = 1.15$; p-value $= .1251$. The data fail to show a significant difference in the mean scores for the two groups.

Section 10.3 (page 381)

10.10 (a) 2.1 ± 2.33 (b) -7.7 ± 12.91
$(-.23, 4.43)$ $(-20.61, 5.21)$

10.12 3.17 ± 3.77
$(-.60, 6.94)$

10.14

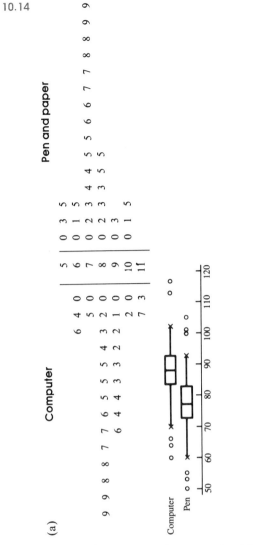

(a)

(b) Both distributions appear to be symmetric with long tails.
(c) Trimmed means.

10.15 $z = 3.98$; p-value $= .00003$. The evidence is highly significant that the creative writing scores of the computer group exceeded the scores of the noncomputer group.

Section 10.4 (page 392)

10.18 1.7125 ± 9.05
   $(-7.3375, 10.7625)$

10.20 $t = .63$; p-value $> .10$. Insufficient evidence exists to support the conjecture that machine B dispenses significantly more than machine A.

10.22 $t = 1.724$; $.025 < p$-value $< .05$. Sufficient evidence exists to claim that the campaign increased mean daily receipts.

10.24   Two-sample Wilcoxon rank sum test. $t = 1.268$; p-value $> .20$. Insufficient evidence exists that the population changed from one day to the next.

Section 10.5 (page 399)

10.26 $t = .647$; p-value $> .20$. Insufficient evidence exists of a difference in the study habits of students of the two school districts.

10.28 $t = -2.75$; $.01 < p$-value $< .025$. There is statistical evidence that the drug significantly reduced the asthmatic relief index.

10.30 $t = -2.4$; $.02 < p$-value $< .05$. There is statistical evidence that the drug affected intelligence test scores.

10.32 $t = 5.0$; p-value $< .005$. The evidence is highly significant that the course was beneficial in raising the reading comprehension scores.

10.34 $t = -3.638$; p-value $< .01$. The difference in the mean number of errors before and after the course is highly significant.

Section 10.6 (page 404)

10.36   (a) $.09 \pm .085$
           $(.005, .175)$
   (b) $-.06 \pm .02$
       $(-.036, .004)$
   (c) $.24 \pm .1822$
       $(.0578, .4222)$

10.38 $z = .81$; p-value $= 2(.209) = .418$. Insufficient evidence exists to declare a difference in the proportion in the two age groups that fear meeting people.

10.40 $z = -.48$; p-value $= .6312$. Insufficient evidence exists to declare a significant difference in the success rates of the two firms.

10.42 $z = -1.69$; p-value $= .0455$. There is significant evidence that the vaccine was effective in reducing the mortality rate.

10.44 $z = 1.40$; p-value $= .0808$. The evidence is moderately significant that the course was beneficial in increasing reading scores.

Review Questions Ch. 10 (page 411)

10.49 $6.8 \pm 12.19 \, (-5.39, 18.99)$

10.50 $t = .702$, p-value $> .10$. Insufficient evidence to reject H_0.

10.51 $t = 1.87$; $.05 < p$-value $< .10$
There is moderately significant difference between the two population medians.

10.52 $t = 3.83$, p-value $< .005$. The yield of the new variety is significantly higher than that of the standard variety.

10.53 $z = 1.06$; p-value $= .1446$. Insufficient evidence exists to claim that females choose diet drinks more frequently than males.

10.54 $(-1.2, 6.65)$

10.55 $t = 1.71$; $.025 < p$-value $< .05$. The data provide significant evidence that private high school students spend more time on homework per week than do public high school students.

Supplementary Exercises Ch. 10 (page 412)

10.56 13 ± 9 $(4, 22)$

10.58 5.652 ± 7.536 $(-1.884, 13.188)$

10.60 $t = 2.116$; $.025 < p$-value $< .05$. The data provide significant evidence that kindergarten was beneficial in improving reading skills.

10.62 $z = 1.76$; p-value $= .0392$. There is significant evidence that computer coaching improved SAT math scores.

10.64 $t = 8.51$; p-value $< .01$. The evidence is highly significant that oxytocin affects arterial blood pressure.

10.66 $t = 2.698$ (Wilcoxon signed rank); $.01 < p$-value $< .02$. The data provide evidence of a significant difference in the estimation from the two suppliers.

10.68 $t = 5.12$; p-value $< .01$. The data indicate that there is a significant difference in the resistance to water for the two pavement materials.

10.70 $z = 7.11$; p-value $= 0^+$. The evidence is highly significant that the level of LDL in women is greater than in men.

CHAPTER 11

Section 11.1 (page 419)

11.1 (a) Independent—air pollution
      Dependent—blood pressure
(b) Independent—lead poisoning
    Dependent—mental retardation
(c) Independent—seat belts
    Dependent—death rate
(d) Independent—alcohol consumption
    Dependent—suicide rate
(e) Independent—per capita income
    Dependent—public education expenditures

11.3 (a) Linear; (b) Linear; (c) Curvilinear; (d) Linear

11.5 Linear

11.7 Linear to curvilinear

Section 11.2 (page 424)

11.9 (a) $b_0 = -1.67$, $b_1 = 3.0$
(b) $b_0 = .5454$, $b_1 = .6363$
(c) $b_0 = -6.849$, $b_1 = 8.918$

11.10 (a) -1.67, 10.33; (b) 1.8182, 3.727, 6.91, 10.091; (c) -6.849, 37.742, 64.497

11.13 (a) $\hat{y} = 3.85 + .95x$; (b) 7.65

11.15 Productivity $= 12.4433 + 1.21$(Aptitude score)

11.17 Cost $= -16.125 + 23.0034$(Age)

Section 11.3 (page 430)

11.19 $\hat{y} = .074 + .935x$
$t = 3.207$; $.02 < p$-value $< .05$. β_1 is significantly different from zero.

11.21 $.6863 \pm .0452$ $(.6411, .7315)$

11.23 $\hat{y} = 2837.23 - 1.151x$
$t = -16.74$; p-value $< .005$.
The linear relationship between the number using the shuttle and the number of autos is highly significant.

11.25 $t = 1.82$; $.05 < p$-value $< .10$. The linear relationship between feed supplement and the number of eggs is mildly significant.

Section 11.4 (page 433)

11.26 (a) $\hat{y} = 29.02 - .54x$
    $x' = 10$: $(18.421, 28.809)$
    $x' = 12$: $(18.19, 26.88)$
    $x' = 14$: $(17.11, 25.80)$
    $x' = 15$: $(16.233, 25.597)$

11.28 $\hat{y} = -1.67 + 3.00x$
    $x' = 0$: $(-5.728, 2.395)$
    $x' = 1$: $(-1.094, 3.761)$
    $x' = 2$: $(2.798, 5.869)$
    $x' = 3$: $(4.906, 9.761)$
    $x' = 4$: $(6.272, 14.395)$

11.30 $\hat{y} = -.03 + 1.64x$
    $x' = 2$: $(.082, 6.408)$
    $x' = 4$: $(4.695, 8.347)$
    $x' = 6$: $(7.727, 11.869)$
    $x' = 8$: $(9.488, 16.661)$
    $x' = 10$: $(10.960, 21.742)$

11.32 $\hat{y} = 45.3 + 2.38x$
    $x' = 10$: $(65.73, 72.61)$
    $x' = 15$: $(77.90, 84.28)$
    $x' = 20$: $(88.48, 97.52)$
    $x' = 25$: $(98.40, 111.44)$

Section 11.5 (page 437)

11.33 $\hat{y} = -8.67 + 1.436x$

11.35 $\hat{y} = 68.76 - .2857x$

11.37 $\hat{y} = 28.7974 + 3.117x$

11.39 $\hat{y} = 1126.875 - 3.165x$

Section 11.6 (page 443)

11.40 No

11.42 There appears to be a relationship between the residuals and time. Time should be incorporated in the model.

11.44 $\hat{y} = -.423 + 1.66x$. There is one unusual residual corresponding to (6.5, 12.1). However $r^2 = 98.9\%$, indicating that most of the variability in y is explained by x. We conclude that this is a reasonable model.

11.46 $\hat{y} = 287 - 11.3x$. A residual plot against the predicted value gives an appearance similar to Figure 11.11 (b), which indicates that the variance is not constant.

Section 11.7 (page 451)

11.48 (a) -9.725; (b) $(-8.6536, 2.4536)$
(c) $t = -2.292$; $.01 < p\text{-value} < .025$

11.50 $y = \beta_0 + \beta_1 x_1 + \beta_2 x_2 + \epsilon$
where y = teacher's attitude
x_1 = years of experience
x_2 = size of school

11.52 $y = \beta_0 + \beta_1 x_1 + \beta_2 x_2 + \epsilon$
where y = family income
x_1 = education level
x_2 = years residing in region

11.55 (a) $\hat{y} = 41.7 + 22.4x - 3.31x^2$
(b) 85.4%
(c) 22.388 ± 1.771 (4.259)
   (14.845, 29.931)
(d) $t = -3.95$, $p\text{-value} < .005$
(e) No

Review Questions Ch. 11 (page 465)

11.60 $y = -7.8 + 6.3x$

11.61 Independent: number of alternatives
Dependent: time to react

11.62 Linear

11.63 -20.3

11.64 $\hat{y} = 81.6 - 1.68x$

11.65 56.4

11.66 $(-1.97, -1.39)$

11.67 $\hat{y} = -2.34 + .043x$

Supplementary Exercises Ch. 11 (page 465)

11.68 (a) Independent: population density
      Dependent: robbery rate
(b) Independent: attitude score
      Dependent: achievement score
(c) Independent: number of fish
      Dependent: growth rate
(d) Independent: expenditures per student
      Dependent: teacher's salary

11.70 (a) Slope: $-\frac{2}{3}$; intercept: 2
(b) Slope: 2.3871; intercept: -3.871
(c) Slope: -1.366; intercept: -2.44

11.73 (a) $\hat{y} = 14.7 - .074x$
(b) $\hat{y} = .2805 + .1485x$
(c) $\hat{y} = -6.45 + .1834x$

11.74 (a) $x = 100$: $\hat{y} = 7.3$
       $x = 160$: $\hat{y} = 2.86$
(b) $x = 1.5$: $\hat{y} = .503$
       $x = 5.0$: $\hat{y} = 1.023$
(c) $x = 30$: $\hat{y} = -.9523$
       $x = 60$: $\hat{y} = 4.55$

11.76 (a) $\hat{y} = 9.7 - 1.7x$
(b) $0, -.3, .4, .1, -.2$
(c) .30

11.78 $\hat{y} = 33.376 - .393x$ where y = harvest and x = price per box

11.80 $.296, -.182, -1.301, .611, -.652, 1.228$

11.81 $y = 85.53 - 4.231x$

11.82 $x = 12: \hat{y} = 34.762$
$x = 16: \hat{y} = 17.839$

11.84 -6.25 ± 2.265 $(-8.515, -3.985)$

11.85 $\hat{y} = 6.105 + 1.035x$

11.86 $1645.49

11.88 $t = 2.29; .05 < p\text{-value} < .10$. The regression coefficient β_1 is significantly different from zero.

11.89 $y = 19.8 - .176x$

11.90 $t = -5.93; p\text{-value} < .01$. The regression coefficient β_1 is significantly different from zero.

CHAPTER 12

Section 12.2 (page 477)

12.1 (a) $.05 < p\text{-value} < .10$; (b) $p\text{-value} > .10$; (c) $p\text{-value} < .005$; (d) $.01 < p\text{-value} < .025$

12.3 $\chi^2 = 7.34; p\text{-value} > .10$. Insufficient evidence exists to reject the hypothesis that the classes are equally likely.

12.5 $\chi^2 = 5.81; p\text{-value} > .10$. Insufficient evidence exists to claim that the percentages are different from those specified by the official.

12.7 $\chi^2 = 3.6; p\text{-value} > .10$. The data support the conjecture that the proportions are the same.

12.9 $\chi^2 = 128.08; p\text{-value} < .005$. The evidence is highly significant that some strains are more resistant to the chemical than others.

Section 12.3 (page 484)

12.11 $\chi^2 = 9.60; .005 < p\text{-value} < .01$. There is statistical evidence that the two variables are dependent.

12.13 $\chi^2 = 1.985; p\text{-value} > .10$. Insufficient evidence exists to reject the hypothesis that the two classifications are independent.

12.15 $\chi^2 = 8.504; .05 < p\text{-value} < .10$. There is mildly significant statistical evidence that the type drug is related to the subjects' ability to memorize.

12.17 (a) $\chi^2 = 6.489; .025 < p\text{-value} < .05$. There is statistical evidence that choice and sex of the respondent are dependent. (b) $\chi^2 = 27.005; p\text{-value} < .005$. There is strong statistical evidence that the sex of the respondent and their choice for Vice President are dependent.

12.19 $\chi^2 = 21.069; p\text{-value} < .005$. There is strong statistical evidence that the opinion of the voter is dependent upon his or her view on freedom of the press.

Section 12.4 (page 488)

12.20 $\chi^2 = 4.621; p\text{-value} > .10$. Insufficient evidence exists to claim that the proportions falling in the three categories are different.

12.22 $\chi^2 = 4.24; p\text{-value} > .10$. Insufficient evidence exists to claim that the proportions in the two districts are different.

12.24 $\chi^2 = 25.274; p\text{-value} < .005$. Highly significant evidence exists that the football players and the press differ in opinion.

12.26 $\chi^2 = 13.065; p\text{-value} < .005$. There is highly significant evidence that the proportion of veterans who have contemplated suicide is different from nonveterans.

12.28 $\chi^2 = 21.553; p\text{-value} < .005$. Highly significant evidence exists that the favorite sports are not distributed the same for males and females.

Review Questions Ch. 12 (page 494)

12.33 $\chi^2 = .12; p\text{-value} > .10$. Insufficient evidence exists to reject the hypothesis that $\pi_1 = 1/6, \pi_2 = 2/6$, and $\pi_3 = 3/6$

12.34 $\chi^2 = 1.9; p\text{-value} > .10$. Insufficient evidence exists to reject the hypothesis of independence.

12.35 $\chi^2 = 6.235; .025 < p\text{-value} < .05$. There is significant evidence that the proportions in the three categories are different.

12.36 $\chi^2 = 3.68; p\text{-value} > .10$. There is insufficient evidence to say that the proportions in the four classes differ.

12.37 $\chi^2 = 3.04; p\text{-value} > .10$. The risk of heart attack is distributed the same for Type A and Type B personalities.

12.38 $\chi^2 = 19.7; p\text{-value} < .005$. The moral values are related to the opinion on the referendum.

Supplementary Exercises Ch. 12 (page 494)

12.39 $\chi^2 = 8.31; .025 < p\text{-value} < .05$. There is sufficient evidence to reject the hypothesis that $\pi_1 = \pi_3 = .20, \pi_2 = \pi_4 = .30$.

12.41 $\chi^2 = 4.13; p\text{-value} > .10$. The data suggest that the medications are equally recommended.

12.43 $\chi^2 = 1.2; p\text{-value} > .10$. No, the sample did not differ significantly from the national standard.

12.45 $\chi^2 = 14.435; .01 < p\text{-value} < .025$. There is statistical evidence that the sales are not equally distributed among the salespeople.

12.47 $\chi^2 = 33.90; p\text{-value} < .005$. The age and choice of candidate are dependent.

12.49 $\chi^2 = 3.91$; p-value $> .10$. Size of corporation and outlook on the economy are independent.

12.51 $\chi^2 = 14.76$; $.005 < p$-value $< .01$. There is evidence that rating and method of therapy are dependent.

12.53 $\chi^2 = .52$; p-value $> .10$. No association exists between sex of the deer and the food supply.

CHAPTER 13

Section 13.2 (page 511)

13.1 (a) 3.06, 3.80, 4.89
(b) 2.53, 3.02, 3.71
(c) 2.16, 2.51, 2.98

13.3

Source	df	SS	MS
Treatment	2	1,069	534
Error	48	8,662	180
Total	50	9,731	

13.5

Source	df	SS	MS
Treatment	2	989.365	494.6825
Error	71	4,114.73	57.954
Total	73	5,104.095	

13.7

Source	df	SS	MS
Treatment	2	51.52	25.76
Error	21	36.48	1.74
Total	23	88.00	

13.9 $F = 2.82$; $.05 < p$-value $< .10$. There is significant evidence that the mean salaries for the three groups differ.

13.11

Source	df	SS	MS	F
Treatment	2	.4000	.2000	6.32
Error	12	.3800	.0317	
Total	14	.7800		

$.01 < p$-value $< .025$

The data suggest that the level of impurity has a significant effect on the solubility of the aspirin tablet.

Section 13.3 (page 517)

13.13 (a) 5.99; (b) 3.70; (c) 5.76

13.15 $\omega = 4.1525$

μ_C μ_B μ_A μ_D

The three diet plans, A, B, and D, do not differ significantly. Diet plan C is significantly better than the other three diet plans.

13.17 μ_3 μ_2 μ_1 μ_4

The mean reaction times due to drugs 2, 1, and 4 do not differ significantly. The mean reaction time due to drug 3 is significantly above the mean reaction times due to the other three drugs.

Section 13.4 (page 522)

13.18 When the normality assumption is not met and the distributions are similar in shape.

13.20

SV	SS	df	MS	F
Treatment	247.65	2	123.825	1.67
Error	2000.5	27	74.1	p-value $> .10$

13.22 (a) The distributions appear similar in shape and are skewed right.
(b) Kruskal-Wallis

13.23 Analysis of variance on ranks

Source	SS	df	MS	F
Treatment	540	2	270	1.61
Error	7033	42	167	p-value $> .10$
Total	7573	44		

There does not exist a significant difference in the population medians.

13.24 (a) The distributions appear similar in shape and are skewed left.
(b) Kruskal-Wallis

13.26

SV	SS	df	MS	F
Treatment	228	2	114	2.61
Error	916.5	21	43.643	
Total	1,144.5	23		

$.05 < p$-value $< .10$

The data suggest that there is a moderately significant difference in the carbon monoxide levels of the three sites.

Review Questions Ch. 13 (page 531)

13.33 (a) ordinary F-test

13.34 (b) Kruskal-Wallis test

13.35 (a) 3.01, 3.72, 4.72
(b) 2.53, 3.03, 3.70
(c) 2.25, 2.62, 3.12

13.36 (a) $.05 < p$-value $< .10$
(b) $.01 < p$-value $< .025$
(c) p-value $< .01$

13.37 (a) 5

(b)

SV	SS	df	MS	F
Treatment	428	4	107	2.903
Error	1032	28	36.857	
Total	1460	32		

(c) Yes, $.025 < p\text{-value} < .05$

13.38

SV	SS	df	MS
Treatment	573	2	286
Error	7291	41	178
Total	7864	43	

13.39 $F = 1.61$; $p\text{-value} > .10$

13.40 (a) Similar in shape and skewed right
(b) Kruskal-Wallis

13.41 Analysis of variance on ranks (Kruskal-Wallis)

Source	SS	df	MS	F
Treatment	1144.8	2	572.4	6.94
Error	2722.7	33	82.5	$p\text{-value} < .01$
Total	3867.5	35		

Supplementary Exercises Ch. 13 (page 532)

13.44 $\omega = 3.306$; $\underline{\mu_4 \quad \mu_2 \quad \mu_3} \quad \mu_1$

μ_4 is significantly greater than μ_1.

13.47

SV	SS	df	MS	F
Treatment	2313.04	2	1156.52	6.38
Error	8154.10	45	181.2	$p\text{-value} < .01$
Total	10467.14	47		

13.49

Source	SS	df	MS	F
Treatment	262.7	2	131.4	4.06
Error	681.2	21	32.4	$.025 < p\text{-value} < .05$
Total	944.0	23		

13.51

SV	SS	df	MS	F
Treatment	990.45	2	495.227	3.19
Error	6054.49	39	155.24	$.05 < p\text{-value} < .10$

13.52 (a)

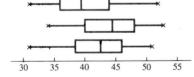

The distributions appear to not deviate significantly from normality.

(b) Ordinary F-test

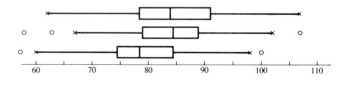

13.54 (a) The distributions appear symmetric with long tails.
(b) Kruskal-Wallis test

13.56

SV	SS	df	MS	F
Treatment	1267.194	2	633.597	13.279
Error	2433.485	51	47.7154	$p\text{-value} < .01$

13.58 (a) The distributions appear similar in shape and are skewed right.
(b) Kruskal-Wallis

13.59 Analysis of variance on ranks

Source	SS	df	MS	F
Treatment	898	2	449	2.83
Error	6672	42	159	$.05 < p\text{-value} < .10$
Total	7570	44		

13.60 (a) The distributions are similar in shape and skewed right.
(b) Kruskal-Wallis

13.61 Analysis of variance on ranks

Source	SS	df	MS	F
Treatment	2019	3	673	1.26
Error	40472	76	533	
Total	42490	79		

$p\text{-value} > .10$. There does not exist a significant difference in the delay times of the four air carriers.

13.64 Analysis of variance on ranks

Source	SS	df	MS	F
Treatment	405	2	202	.91
Error	10635	48	222	
Total	11040	50		

$p\text{-value} > .10$. There does not exist a significant difference between the relief times for the three treatments.

Index

Parameter, 81, 185, 224
for describing a distribution, 224–227
Parent distribution, 210
normal, 282
Pascal, Blaise, 133
Pearson χ^2 statistic, 473
Pearson correlation coefficient, 105
Percentage points of the Studentized range, 561–562
Physical mixing (lottery sampling), 16
Pie chart, 53
Placebo effect, 22
Point estimate, 276
Point estimator, 276–281
Polygon, frequency, 59–60
Polynomial regression, 449
Pooled variance, 383
Population, 4, 10, 14
Bernoulli, 181–182
parent, 210
Population distributions, empirical rule applied to, 179
Population mean(s), 87, 281–282
estimator of, 285
large-sample inference on difference between, using independent samples, 369–375
large-sample test of, 330–339
robust large-sample test of, 339–343, 375–382
small-sample test of, 343–347
trimmed, 89
Population median, 90, 285–286
test of, 347–352
Population parameters, estimators of, 281–289
Population proportion(s), 286–287
confidence interval for, 303–307
inference on the difference between two, 400–404
test of, 353–356
Population variance, 98, 287–288
Precision, lack of, 5
Predicted value, 421
Probability, 131–163
birthday problem, 153–155
computer session, 156–159
conditional probability and independence, 147–153
of an event, 138, 139
event relations and two laws of, 142–146
experiment, sample space, and, 136–141
marginal, 479
multiplication law of, 148, 150

summary and review, 160–161
unconditional, 147
Probability density function, 188
Probability distribution of a discrete random variable, 168–173
Proportion, population, 286–287, 353–356, 400–404
Pseudostandard deviation analysis, 216–217
P-value, 331–332

Q-spread, 103
Qualitative variable, 43–44
scales of measurement for, 45–48
Quantile summary diagram, 94
Quantitative variable, 42–43
scales of measurement for, 45–48
Quartiles, 92–93, 94

Random digits, 547–548
Randomized block design, 23–24
Random sample
simple, 16, 18
stratified, 18–19
Random variable, 166–168
binomial, 182–183
continuous, 188
discrete
expected value of, 174–181
probability distribution of, 168–173
Range, 97–98
Rank-transformed data, 386
Ratio scale, 46–47
Regression analysis, 415–467
computer session, 456–463
examination of the residuals, 438–444
inferences about β_1, 425–430
inferences about $E(y)$, 431–434
method of least squares, 420–424
multiple regression, 444–455
robust alternative to least squares—the resistant line, 434–438
summary and review, 464–465
Rejection region, 329
Relative frequencies, 50
histogram for, 58–59
Research hypothesis, 325
Residual, 421
Resistant, 103
line as alternative to least squares method, 434–438
Response rate, 11
Robust
alternative to least squares, 434–438
confidence interval for the center of a distribution, 297–303